P. MOLIAN

N. P. Mahalik (Ed.)

Micromanufacturing and Nanotechnology

N. P. Mahalik (Ed.)

Micromanufacturing and Nanotechnology

With 300 Figures

Dr. Nitaigour Premchand Mahalik
Universtiy College of Engineering, Burla
(Biju Patnaik University of Technology, Rourkela)
Sambalpur, Orisa, Pin: 768 018
India

Presently at:
Gwangju Institute of Science and Technology
Gwangju, 500 712
Republic of South Korea
nmahalik@gistac.kr

Library of Congress Control Number: 2005929276

ISBN-10 3-540-25377-7 Springer Berlin Heidelberg New York
ISBN-13 978-3-540-25377-8 Springer Berlin Heidelberg New York

Springer is a part of Springer Science+Business Media
springeronline.com

Printed in Germany

Typesetting: Digital data supplied by authors
Final processing by PTP-Berlin Protago-TEX-Production GmbH, Germany
Cover-Design: medionet AG, Berlin
Printed on acid-free paper 62/3141/Yu – 5 4 3 2 1 0

Dedicated to all my

TEACHERS

Especially to

Sri Abhimanyu Mahalik
Smt Jayanti Mahalik
Er Babuchand Mahalik
Sri Prabhakar Barik
Professor P R Moore
Professor S K Lee

and last but not the least

Sri Tathagata Satapathy
(MP – Dhenkanal; Editor – The Dharitri)

Preface

Sophisticated miniaturised components and systems may indeed change all kinds of products and equipment in the most dramatic way. Methodology and design of miniaturisation represent a broad research topic with applications in fundamental physics, chemistry, martial science, computing methods, ultra-precision engineering, fabrication technology, micromachining, and many others based on the principles, characterization, modeling, simulation, sophistication, flexibility of state-of-the-art technology. Micromanufacturing and Nanotechnology (MaN), an advanced product and equipment design concept has emerged and caters to the need for miniaturisation. Technological research in the field of MaN is now expanding; its design phases appear to be highly complex and involve multi-physics and interdisciplinary approaches. The main objective of this book is to provide information on concepts, principles, characteristics, applications, latest technological developments and comparisons with regard to micro/nanosystems and technology. It incorporates research, development, tutorials and case studies. Academic and industrial research and developments in microengineering, micromanufacturing, micromechanics and nanotechnology are being carried out at many different institutions around the world. The technological trend in this domain (e.g., design and development methodology, current application scenarios, pros and cons, etc.) needs to be disseminated extensively so that the MaN revolution can spread to serve society. In particular, the book is intended to focus on describing the implicit concept of micromanufacturing and nanotechnology, multi-physical principle of microelectromechanical systems (MEMS) and micro-opto-electro-mechanical systems (MOEMS), design tips and hints, as much as the techniques and methodology.

Micromanufacturing and nanotechnology are two sides of a coin. There has been confusion and arguments on the terms microengineering, microsystems, ultra-precision engineering, micromachining, nanofinishing, micromechanics, microstructures and microsystems. The authors of the Chapters have attempted to clarify this confusion. The book will undoubtedly enable the readers to understand the underlying technology, philosophy, concepts, ideas, and principles, with regard to broader areas of micromanufacturing and nanotechnology such as application of laser technology, lithography, bulk and surface micromachining, nanofinishing, error compensation, MEMS, MOEMS, carbon nanotubes, micro energy chemical system, fuel cell, microstructure for space propulsion, biosensor, etc. Aspects of microsystems in terms of design process, practice, techniques, platforms, and experimental results have been presented in proper order. The chapters include topical and general description as far as current research and technological developments are concerned. Fundamental methods, initiatives, significant research results as well as references for further study have been presented. Relative merits and demerits are described at the appropriate places so that novices as well as advanced practitioners can use the evaluation to guide their choices. All the contributions have been reviewed, edited, processed and placed appropriately in order to maintain consistency so that irrespective of whether the reader is an advanced practitioner or a new comer he or she can get most out of it. Since this book covers many aspects of interdisciplinary subjects, the importance of the book within the micro and nano domain is considered significant. The roadmap of the book is as follows.

Chapter 1 is a general introduction. Chapter 2 presents the principles of MEMS and MOEMS. Very precise definitions of different physical phenomena and their utilisation with respect to MEMS and MOEMS have been presented. Chapter 3 presents fundamental principles of application of laser technology in micromanufacturing. Geometrical compensation of high precision machine system is of paramount importance. Chapter 4 highlights the basic principle and experimental architecture of computer assisted laser interferometer based method for achieving compensation of errors. In this chapter the model of error and procedural method of its compensation is described. Chapter 5 discusses about the bulk micromachining processes, a fundamental process requirement for microsystems and equipment manufacturing. A step forward to this process is considered as surface micromachining. Chapter 6 adheres the principle of surface micromanufacturing. Chapter 7 discusses latest developments on microsystem conformant OVD (Optically Variable Device), a very demanding device that has long been used for document security applications. Various types of nanofinishing techniques, the important method in manipulating and describing not only micro- but the macrosystems are described in the Chapter 8. The role of micro- and nanotechnology in space applications is presented in Chapter 9. Carbon nanotubes and nanostructures are introduced in Chapter 10 and 11. It has been variously that the future of the computing world will be based on molecular computing. A comprehensive description of molecular logic gates based on fluorescence, absorption and electronic conductance is presented in Chapter 12. Chapter 13 provides some research outcome with regard to the design of microscale cantilever devices that can act as biological sensors. Transportation of micro-energy through microdevices is a challenging breakthrough and is referred to as MECS (Micro Energy and Chemical Systems). Potential applications are microelectronic cooling systems, chemical reactors, fuel processing and heat pumps as outlined in Chapter 14. Next, a detailed description of sculptured thin films is provided. The following two chapters discuss e-beam and optical nanolithography techniques, respectively. Chapter 18 provides some phenomenological description of nanotechnology vis-à-vis fuel cell applications. Derivatisation of carbon nanotubes with amine and chemical crosslinking in C_{60} thin films are presented in Chapter 19 and 20, respectively.

The success story of this book 'Micromanufacturing and Nanotechnology' is in fact due to the direct and indirect involvement of many researchers, advisors, technocrats, academicians, developers, integrators, designers, and last but not the least the well-wishers. Therefore, the editor and hence the publisher acknowledge the potential authors and companies whose papers, reports, articles, notes, study materials, etc. have been referred in this book. Further, many of the authors of the respective chapters gracefully acknowledge their funding agencies, without which their research could not have been completed. In particular persons such as M. Adrian Michalicek, Wassanai Wattanutchariya, Kannachai Kanlayasiri, Joseph Thomas, Hadi Hasan, Nitin Sharma, Patrick Kwon, Sharee McNab, David Melville, Conrad Wolf, Andrew Thompson, Alan Wright, Helen Devereux, Gary Turner and Mike Flaws and the following agencies, institutes, companies and journals are acknowledged.

Ministry of Science & Technology (Project code MS-01-133-01); Marsden Fund of the Royal Society of New Zealand (UOC-604 and UOC-312); EPSRC (UK) (Grant No. HPRN-CT-2000-00028, and GR/N12657/01); NEDO (Japan) through JRCAT; National Council of Science and Technology of Mexico (CONACYT-36317-E & 40399-Y); European Space Agency; National Autonomous University of Mexico (grants DGAPA-IN100402-3 and -IN100303); Ministry of Industry, Government of Orissa; University College of Engineering, Burla; Texas Instruments; University of Colorado at Boulder; Journal of Micromechanics and Microengineering; IEEE Journal of Microelectromechanical Systems; Nature Materials; The FORESIGHT Institute.

Nitaigour Premchand Mahalik

Contents

Authors

B. H. Ahn
Department of Mechatronics, Gwangju Institute of Science and Technology, 1 Oryong-dong Buk-gu, Gwangju 500-712, Republic of South Korea

Brian K. Paul
Oregon State University, Industrial and Manufacturing Engineering, 188 Covell Hall, Corvallis, OR 97331-2407, USA

Chandana Karnati
Chemistry and Institute for Micromanufacturing, Louisiana Tech University, Ruston, LA 71272, USA

Daniela S. Mainardi
Institute for Micromanufacturing, Louisiana Tech University Ruston, 71272, USA

Dae Sung Yoon
Microsystem Research Center, Korean Institute of Science and Technology, Seoul 136-791, Republic of South Korea

Edgar Alvarez-Zauco
Centro de Ciencias Aplicadas y Desarrollo Tecnológico, Universidad Nacional Autónoma de México (UNAM), Circuito Exterior C.U., 04510, Mexico

Elena V. Basiuk
Centro de Ciencias Aplicadas y Desarrollo Tecnológico, UNAM, Circuito Exterior C.U., 04510, Mexico

Giulio Manzoni
Microspace, Via Filzi 6, 34100 Trieste, Italy

Hai-Feng Ji
Chemistry and Institute for Micromanufacturing, Louisiana Tech University, Ruston, LA 71272, USA

Jeong Hoon Lee
Microsystem Research Center, Korean Institute of Science and Technology, Seoul 136-791, Republic of South Korea

John A. Polo
Department of Physics and Technology, Edinboro University of PA, 235 Scotland Road, Edinboro, PA 16444, PA, USA

Jon A Preece
School of Chemistry, University of Birmingham, Edgbaston, Birmingham, B15 2TT, U.K.

K. K. Tan
Department of Electrical and Computer Engineering
National University of Singapore, 4 Engineering Drive 3, 117576, Singapore

M. M. Alkaisi
The MacDiarmid Institute of Advanced Materials and Nanotechnology, Department of Electrical and Computer Engineering, University of Canterbury, Christchurch, New Zealand

N. P. Mahalik
Department of Mechatronics, Gwangju Institute of Science and Technology
1 Oryong-dong Buk-gu, Gwangju 500-712, Republic of South Korea

Paula M Mendes
School of Chemistry, University of Birmingham, Edgbaston, Birmingham B15 2TT, U.K.

Qianwang Chen
Hefei National Laboratory for Physical Sciences at Microscale, University of Science and Technology of China, Hefei 230026, Peoples Republic of China

R. J. Blaikie
The MacDiarmid Institute of Advanced Materials and Nanotechnology, Department of Electrical and Computer Engineering, University of Canterbury, Christchurch, New Zealand

Robert A. Lee
CSIRO Manufacturing and Infrastructure Technology, Normanby Road, Clayton, Victoria 3168, Australia

S. Kumar Bag
Department of Electronics and Communication Engineering, Vellore Institute of Technology (Deemed University), Vellore, India

S. N. Huang
Department of Electrical and Computer Engineering, National University of Singapore, 4 Engineering Drive 3, 117576, Singapore

Sungho Jeong
Department of Mechatronics, Gwangju Institute of Science and Technology 1 Oryong-dong Buk-gu, Gwangju 500-712, Republic of South Korea

Sunil Jha
Department of Mechanical Engineering, Indian Institute of Technology Kanpur, Pin: 208016, India

Sunny E. Iyuke
School of Process and Materials Engineering, Faculty of Engineering and the Built Environment, University of the Witwatersrand, Private Bag 3, Wits 2050, Johannesburg, South Africa

Tae Song Kim
Microsystem Research Center, Korean Institute of Science and Technology, Seoul 136-791, Republic of South Korea

V. K. Jain
Department of Mechanical Engineering, Indian Institute of Technology Kanpur - 208016, India

Vladimir A. Basiuk
Instituto de Ciencias Nucleares, Universidad Nacional Autónoma de México (UNAM), Circuito Exterior C.U., 04510, Mexico

Zhao Huang
Hefei National Laboratory for Physical Sciences at Microscale, University of Science & Technology of China, Hefei 230026, Peoples Republic of China

1 Introduction

N. P. Mahalik

Department of Mechatronics, Gwangju Institute of Science and Technology, Republic of South Korea

1.1 Background

In the year 1959, Feynman proposed the possibility of manufacturing ultraminiaturised systems for a variety of applications, to a level that may engross multiscale formulation methods involving the manipulation of molecules and atoms. Fundamental technological research and development in this domain has been underway since then. Arguably, with regards to past developments, it can be stated that the research and developmental activities broadly fall under the four major categories: precision engineering, ultra-precision engineering, micromanufacturing, and nanotechnology. The scope of research in the first three categories is vast and considered to be discipline dependent. For instance, while the methodological principles of precision engineering and ultra-precision engineering are more closely related to mechanical machining and processes, micromanufacturing in a generalised sense, is applied for producing VLSI-related products and systems. Though recently its scope has been broadened, apparently advocating the interdisciplinary scenario because of the dramatic emergence of MEMS (Microelectromechanical Systems) and MOEMS (Micro-opto-electromechanical Systems) technology. On the other hand, nanotechnology is viewed as a truly interdisciplinary domain accommodating several disciplines including general science. It is not our concern to describe the topical subjects of all these major domains, however at this point it suffices to say that one of the objectives of all these technological advancements is being focused on manufacture and design of miniaturised systems or products that can solve the humane expectations. Many industries in conjunction with academic institutions and R&D sectors are active in this multiphysics-multiengineering sector so that the manufacturing of smaller parts, components, products and systems at the level of micro/nanoscale with more functionalities and capabilities can easily be realised. To some extent this chapter attempts to introduce a road map to the concept of micromanufacturing and nanotechnology, which is being reflected in the following chapters in more detail. Definitions of commonly used terminology will be provided, where necessary.

1.2 Introduction

Manufacturing is the cornerstone of many industrial activities and significantly contributes toward the economic growth of a nation. Generally, the higher the level of manufacturing activity in a country, the better is the standard of living of its citizens. Manufacturing is the process of making large quantities of products by effectively utilizing the raw materials. It is a multidisciplinary design activity simply involving the synergistic integration of production and mechatronics engineering. The products vary greatly from application to application and are prepared through various processes. It encompasses the design and production of goods and systems, using various production principles, methodologies and techniques. The concept is hierarchical in nature in the sense that it inherits a cascade behaviour in which the manufactured product itself can be used to make other products or items. The manufacturing process may produce discrete or continuous products. In general, discrete products mean individual parts or pieces such as nails, gears, steel balls, beverage cans, and engine blocks, for example. Conversely, examples of continuous products are spools of wires, hoses, metal sheets, plastic sheets, tubes, and pipes. Continuous products may be cut into individual pieces and become discrete parts. The scope of manufacturing technology includes the following broad topics:

- Precision engineering and ultra-precision engineering
- Micromanufacturing (Microelectronics and MEMS)
- Nanotechnology

1.2.1 Precision Engineering

The technical field of precision engineering has expanded over the past 25 years. In 1933, the Precision Engineering Society was established in Japan and soon thereafter the activities were accelerated due to new impetus from Europe. The first issue of the journal Precision Engineering appeared in 1979 and the first academic program began in 1982 (Source: American Society of Precision Engineering (ASPE)). According to ASPE, "…precision engineering is dedicated to the continual pursuit of the next decimal place." Precision engineering includes design methodology, uncertainty analysis, metrology, calibration, error compensation, controls, actuators and sensors design. A more complete list is given below (www.aspe.net).

- Controls
- Dimensional metrology and surface metrology

- Instrument/machine design
- Interferometry
- Materials and materials processing
- Precision optics
- Scanning microscopes
- Semiconductor processing
- Standards

Frequently used terms within the domain of precision and ultra-precision engineering are precision processes, scaling, accuracy, resolution and repeatability. The precision process is a concept of design, fabrication, and testing where variations in product parameters are caused by logical scientific occurrences. Identification of these logical phenomena and strategically controlling them is very fundamental to precision manufacturing. Scaling is a parameter that defines the ratio attributes with respect to the prototype model. It is also considered as a fundamental attribute for predicting the behaviour of structures and systems for analysis and synthesis of miniaturised systems. Accuracy defines the quality of nearness to the true value. In the context of machine or production systems, accuracy is the ability to move to a desired position. As an example, if the actual value is 1.123 units and it is recorded as 1.1 units, we are precise to the first decimal place but inaccurate by 0.023 units. Resolution is the fineness of position precision that is attainable by a motion system. The smallest increment that is produced by a servo system is the resolution. There are two types of resolutions, electrical and mechanical. With regard to mechanical resolution, it is defined as the smallest increment that can be controlled by a motion system, i.e., the minimum actual mechanical increment. One can note that mechanical resolution is significantly coarser than that due to the involvement of friction, stiction, deflections, and so on. Repeatability is the variation in measurements obtained when one person takes multiple measurements using the same instruments and techniques. Repeatability is typically specified as the expected deviation, i.e., a repeatability of 1 part in 10,000 or 1:10,000, for example.

1.2.2 Micromilling and Microdrilling

Micromilling and microdrilling are two important processes of precision engineering. The micromilling process is considered versatile and facilitates creating three-dimensional miniaturised structures. The process is characterised by milling tools that are usually in the order of hundreds of micrometers in diameter. These tools are designed by the use of focused-ion beam machining process and are used in a specially designed, high-precision milling machine. The focused-ion beam machining process uses a sharp tungsten needle wetted with gallium metal. The tip of the needle is subjected to a 5-10 kV (sometimes higher) so as to enable the field ionization effect on the gallium. The gallium ions are then accelerated by the use of another energy source and focused into a spot of sub-micrometer order.

The kinetic energy acquired by the ions makes it possible to eject the atoms from the workpiece. This is referred to as a sputtering process. The sputtering yield varies inversely with the strength of the chemical bond in the materials. Either the movement of ions or the workpiece, depending upon the environmental conditions, can be controlled to obtain a wide variety of three-dimensional shapes and structures. It should be pointed out that the machining forces present in micromilling with tools of the order of micrometer diameters are dominated by contact pressure and friction between the tool cutting edges and the workpiece. As a rough calculation, one can note that in the focused-ion beam machining process, for a spot size of 0.45 μm with 2.5 nA of current, the required current density would be approximately 1.65 A/cm^2. The micromilling process is applied for making micromolds and masks to aid in the development of microcomponents. Typically, a high milling rate of 0.65 $\mu m^3/nAs$, corresponding to an average yield of 6.5 atoms/ion, can be obtained at 45 keV, 30° incidence, and 45 scans.

Microdrilling is characterised by the drilling of ultrafine holes. Drilling in the micro ranges, using the special microdrills, requires a precision microdrilling instrument. The end of the microdrill is called the chisel edge, which is indeed removed material cutting at a negative rake angle. Microdrills are made of either micrograin tungsten carbide or cobalt steel. Some coarse microdrilling machines are available that drill holes from the size of 0.03 mm in diameter to 0.50 mm in diameter, with increments of 0.01 mm. However, the present demand is for drills capable of drilling in the order of micrometers. An example of this is a sub-microdrilling technique utilising the phenomenon of ultrafast pulse laser interference. In this regard, for microdrilling and other delicate laser processing applications, Holo-Or Ltd. has released an optical element that creates an output spot in the form of a top hat circle with a diameter of 350 μm. The element accepts a collimated Gaussian incident beam with a diameter of 12 mm from a 10.6-μm CO_2 laser. Smooth 300 nm holes were successfully drilled on a 1000-Å-thick gold film using the interfered laser beam, as compared to micrometer holes ablated using the conventional non-interfered laser beam. The most important parameters considered in microdrilling are: accuracy, sensitivity, quality and affordability. Some of the applications of microdrilling are given below:

- Air bearings and bushings
- EDM tooling
- Electronic components
- Gas and liquid flow
- Microwave components
- Nozzles
- Optical components

The major problem of conventional laser microdrilling is that the process has a short focal depth. It is known that this method typically achieves aspect ratios up to 100 in thick material, such as for a 15-μm hole in 1.5-mm-thick foil, for instance. This problem can be overcome by utilizing a Bessel beam. Deep high-aspect ratio drilling is achieved due to the reason that the Bessel beam is non-diffracting and in practice they do not spread out. In the case of deep high-aspect ratio laser drilling, a pseudo-Bessel beam is generated using a pulsed laser. Some of the examples of microdrilling applications using a laser system developed by ATLASER di Andrea Tappi are presented in Table 1.1. The application of lasers to micromanufacturing has several advantages: noncontact processing, the capability of remote processing, automation, no tool wear and the possibility of machining hard and brittle materials.

Table 1.1. ATLASER di Andrea Tappi microdrilling system performance parameters (Courtesy: ATLASER di Andrea Tappi)

Si Wafer Thickness: 0.54 mm Hole diameter: 25 μm Hole pitch: 50 μm Process time: 0.65 s	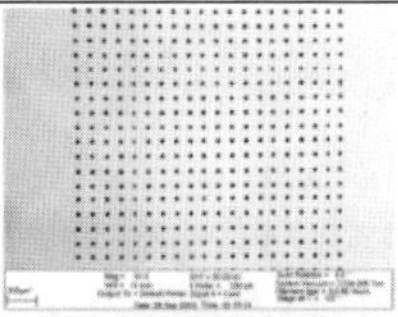
Silicon Carbide Wafer Thickness: 0.64 mm Through Hole: 130x500 μm In width: 130 μm Out width: 110 μm	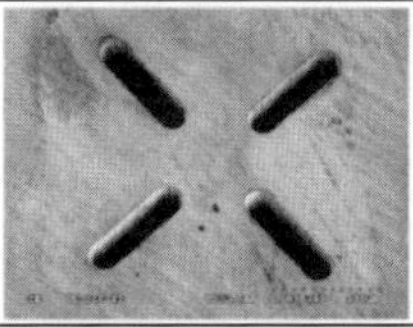
Aluminum Nitride Thickness: 425 μm In side width: 300 μm Out side width: 290 μm Drilling time: 33 s	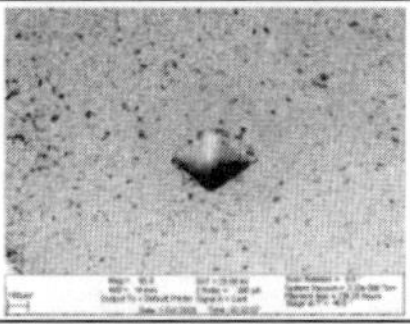
Cu-FR4 sandwich Thickness: 0.5mm Hole dimension: 200 μm Process time: 3.3 s	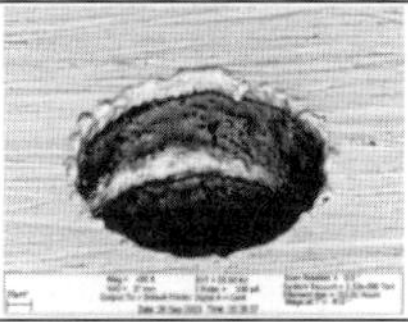
Stainless Steel Sheet Thickness: 120 μm Hole diameter: 9 μm Hole pitch: 50 μm Matrix Process time: 0.15 s	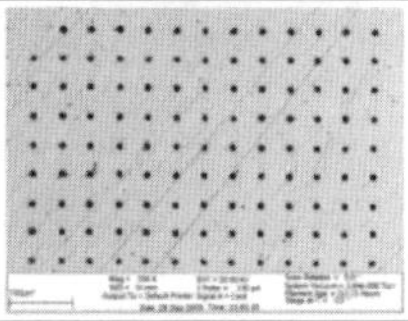

1.3 Microelectromechanical Systems (MEMS)

Microelectromechanical systems (MEMS) have already found significant applications in sectors that include, but are not limited to: automotive industry, aircraft industry, chemical industry, pharmaceuticals, manufacturing, defence, and environmental monitoring. The relative merit for MEM systems lies in the fact that these components are fabricated by batch manufacturing methods similar to microelectronics techniques, which fulfills the added advantage of miniaturization, performance and integrability. The topical areas under MEMS are micromachining methods, microsensors and actuators, magnetic MEMS, RF MEMS, microfluidics, BioMEMS and MOEMS. The progress in microfabrication technologies is transforming the field of solid-state into MEMS. Micromachining is a process for the fabrication of MEMS devices and systems. Various energy transduction principles include thermal, magnetic, optical, electrical and mechanical. These are employed in designing the microsensors and actuators. Radio Frequency (RF) MEMS devices are mostly used in the field of wireless communication. Microfluidic MEMS devices handle and control small volumes of fluids in the order of nano and pico liter volumes. One popular application is a micronozzle for use in printing applications. MEMS technology has applications in the chemical industry, which gives rise to BioMEMS products. Surgical instruments, artificial organs, genomics, and drug discovery systems are based on BioMEMS products.

The miniaturised systems require less reagent, resulting in faster and more accurate systems. MEMS devices have better response times, faster analysis and diagnosis capabilities, better statistical results, and improved automation possibilities with a decreased risk and cost. Since most of the physical phenomenon and activations are to be measured and controlled precisely in a timely predictive manner so as to overcome realtime limitations, miniaturised components will have added advantages because of the inherent temporal behavior they possess. Moreover, prognostic measures in terms of sensor and actuator validation can be achieved through a system-on-a-chip (SOC) design approach. MEMS devices are useful for controlling micro mechanisms such as micromanipulators, micro-handling equipment, microgrippers, microrobot, and others, which are primarily used for clinical, industrial and space applications.

1.3.1 An Example: Microphenomenon in Electrophotography

Electrophotography can by used for a broad range of applications. These include:

- Manufacturing of PCB
- Creating images on a wide variety of receivers
- Thin coating on pharmaceutical tablets and capsules

Electrophotography produces documents and images. The ability to do so requires that micrometer-size marking, toner, and particles be precisely placed on a receiver by adopting micro level technology. In order to produce high quality electrophotographic images, it is necessary to carefully control the forces acting on the toner particles. These forces are predominantly either electrostatic or electrodynamic in nature. The precision in color electrophotography process is extraordinary. Color electrophotography is an advanced technique. It requires that the separations comprising the subtractive primary images be precisely superimposed in order to create a sharp image with excellent color balance.

1.4 Microelectronics Fabrication Methods

One of the major inventions in the last century is microelectronics, called micro devices. Micro devices can be integrated circuits, which are fabricated in sub-micron dimensions and form the basis of all electronic products. Microelectronics design entails the accommodation of essential attributes of modern manufacturing. Fabrication technology, starting from computer assisted off-line design to real fabrication, deals with the processes for producing electronic circuits, solid structures, printing circuits as well as various electronic components, sub-systems and systems of subminiature size.

The design of an IC with millions of transistors and even more interconnections is not a trivial task. Before the real design is manufactured, the circuit is prepared and tested by using EDA (electronic design automation) tools. These tools help in synthesizing and simulating the behavior of the desired circuit by arranging the placement of transistors and interconnections within the chip area. These computer-assisted tools can also verify and validate all defects and conditions, respectively. The technology has been driven by the demands of the computer industry, space technology, the car industry and telecommunications.

The first step in fabrication is always the preparation of a set of photographic masks. The mask represents the features of the various elements and layers of the chip to be manufactured. This procedure is repeated several times to replicate the circuit. The mask appears on the surface of a thin silicon crystal wafer. A single wafer can accommodate several identical chips. Hence, the IC fabrication process is a batch-processing scheme. The preparation of masks can be carried out by the use of a computer-controlled electron beam to expose the photographic mask material in accordance with the desired configuration. The information is supplied to the computer in terms of a design data file.

Then three important fabrication sequences are followed on the wafer surface. These are photography followed by chemical, and thermal operations. This phase is called masking. The mask features are transferred to the wafer by exposing a light-sensitive photoresist coating through the transparent areas of the mask. The material areas of the wafer unprotected by the hardened photoresist are then removed by etching.

Etching techniques are characterised by their selectivity and degree of anisotropy. Etching can be either physical or chemical, or a combination of both. In order to develop active circuit elements such as transistors, n-type and p-type impurities are doped. Two commonly used doping methods are diffusion and ion implantation. Then a thin aluminum layer is deposited on the uppermost layers of the chip in order to allow metal to contact the device elements. The aluminum deposition is often achieved by using the chemical vapor deposition (CVD) method.

1.4.1 Bulk Micromachining

The term micromachining refers to the fabrication of micromechanical structures with the aid of etching techniques to remove part of the substrate or a thin film (Petersen, 1982). Silicon has excellent mechanical properties, making it an ideal material for machining. The fabrication processes fall into the two general categories of bulk micromachining and surface micromachining.

In general, the process uses the bulk material or substrate to form microstructures by etching directly into the bulk material. Bulk micromachining refers to etching through the wafer from the backside in order to form the desired structures. The structures are formed by wet chemical etching or by reactive ion etching (RIE). Usually, suspended microstructures are fabricated using wet chemical micromachining. The advantage of bulk micromachining and chemical etching is that substrate materials such as quartz and single crystal silicon are readily available and reasonably high aspect-ratio structures can be fabricated. It is also compatible to IC technologies, so electronics can be easily integrated. Disadvantages of bulk micromachining include pattern and structure sensitivity and pattern distortion due to different selective etch rates on different crystallographic planes (Bean 1978; Kern 1990; Danel and Delapierre 1991). Further, since both the front side and backside are used for processing, severe limits and constraints are encountered on the minimum feature size and minimum feature spacing.

1.4.2 Surface Micromachining

Surface micromachining is a process of fabrication of MEMS structures out of deposited thin films. The process is also employed for IC fabrication. It involves the formation of mechanical structures in thin films formed on the surface of the wafer. The thin film is primarily composed of three layers: low-pressure chemical-vapor-deposition polycrystalline silicon, silicon nitride, and silicon dioxides. They are deposited in sequence and subsequently selectively removed to build up a three-dimensional mechanical structure integrated with the electronics. The structure is essentially freed from the planar substrate. The process is very complex in the sense that it requires serious attention during the process, as the property of material significantly varies at the microstructure level. In particular, the following issues are dealt with cautiously (James 1998):

- Basic understanding and control of the material properties of thin films
- Fabrication features for hinged structures and high-aspect ratio devices
- Releasing method for the microstructure
- Packaging methods

Micromachining, using doped or undoped polysilicon as the structural material and silicon dioxide as the sacrificial material, is the most frequently used. Silicon nitride is used as an insulator. Hydrofluoric acid is used to dissolve the sacrificial oxide during release. Primarily, the fabrication process involves the following steps:

- Substrate passivation and interconnection
- Sacrificial layer deposition and patterning
- Structural polysilicon deposition and doping
- Microstructure release, rinse, and dry

Some of the examples of polysilicon micromechanical devices are flexible suspensions, gear trains, turbines, cranks, tweezers, and linkages, which have already been fabricated on silicon. Although bulk micromachining is considered to be the older technology, the two developments (bulk and surface) ran parallel (French 1998).

1.5 Microinstrumentation

In order to achieve sophistication and improve availability, miniaturization technology based instrument design concepts are going to be adopted as the advanced instrumentation platform for most of scientific, industrial and academic studies. The topical research fields are material science for microstructures, microinstrumentation devices, transduction principles for microstructures, interfacing; integration, modeling, and performance issues. Microinstrumentation equipment is essentially useful where a higher QoS (Quality of Service) such as sensitivity, resolution, selectivity, fidelity, and repeatability is desired. Miniaturization improves portability, speed and spatial requirements. Miniaturization can help the engineer to measure and analyse the physical, chemical and biological parameters of an application where space and weight are limiting factors. A typical microspectrometer can take less space while satisfying the required capabilities to measure, analyse and provide precise signals for further analysis and processing. Microinstruments can be applied in nuclear reactors, space shuttles, research laboratories, and numerous other places. Other application fields include spectroscopy, surface analysis, tribology studies, topography, microfluidics, microtomography and imaging. Recently developed microinstrumentation equipment includes microoscilloscopes, a microvoltmeter and microradar. The principle of operation of such tools and equipment however, requiring the application of fundamental science and coherent and synergistic technological integration, is of paramount importance.

1.6 Micromechatronics

Mechatronics is a relatively new discipline, but has been firmly established. Technical areas such as motion control, robotics, automotive, intelligent control, actuators and sensors, modeling and design, system integration, vibrations and noise control are studied under this topical subject. This macroscale interdisciplinary research area is considered as a synergistic integration of mechanical engineering with electronics and intelligent control algorithms in the design and manufacture of products process. Micromechatronics deals with microscale machines. Microscale machines in turn incorporate MEMS architecture and methodology along with controls. The synergistic integration aspects with respect to microscale machines are not identical with regards to macroscale machines and systems.

1.7 Nanofinishing

In recent years, the structural modification and finishing methodology has attracted many researchers, system developers and integrators due to the fact that many mechanical and optical parts in the high-precision domain require precise structural perfection and manipulation on the nanometer scale. Nanomachining is a process that can provide advanced material shaping in this realm. Nanofinishing is a subset of the nanomachining process. Final finishing operations in the manufacturing of ultra-precision parts and components are always of concern owing to their physical, critical and controllable nature. In the domain of sophisticated miniature technology, deterministic high precision finishing methods are of utmost importance and are essential to the present micromanufacturing scenario. Proven finishing operations are:

- Grinding, honing and lapping
- Abrasive Flow Machining (AFM) with SiC abrasives
- Magnetic Abrasive finishing (MAF)
- Magnetic Float Polishing (MFP) with CeO_2
- Magnetorheological Finishing (MRF) with CeO_2
- Elastic Emission Machining (EEM) with ZrO_2 abrasives
- Ion Beam Machining (IBM)

It has been predicted that machining accuracies in conventional processes would reach 1 μm, while accuracies in precision and ultra-precision machining would reach 0.01 μm (10 nm) and 0.001 μm (1 nm), respectively.

1.8 Optically Variable Device

There has been a great deal of recent interest in optical security technology. Optically variable devices (OVD) are based on materials and technology that change appearance when viewed from different angles. OVDs are capable of producing sophisticated optical images with an exceptionally high level of security features. The technique allows for the microprecision placement of pixels, which are forensically identifiable for maximum image clarity and security. Holographic technology, foil embossing equipment, application processes and information associated with security devices are the prime concerns in the adoption of these new security measures.

1.9 MECS

For many heat transfer applications, microchannel-based heat exchangers are being developed. The exchangers can be used for thermal management and waste heat recovery. The exchanger system should be compact and performance driven. Microscale combustion systems have been developed which take advantage of microchannel heat transfer to produce extremely compact, reliable, robust, portable and efficient high flux combustion systems. Recent developments have demonstrated heat fluxes several times greater than for conventional systems. This technology can be used within compact heating and cooling schemes or for compact power generation or propulsion. Micro Energy and Chemical System (MECS) is a recently developed concept and methodology, which can offer the required portability, mobility, compactness and flexibility. MECSs are microfluidic microdevices, capable of transferring enhanced rates of heat and mass in the embedded high surface-to-volume ratio microchannels of micro-scale geometries.

1.10 Space Micropropulsion

The application of micro and nanotechnology to the aerospace field requires a very interdisciplinary knowledge and engineering approach. Since the first concept of a nanosatellite to be manufactured on silicon wafers was proposed by The Aerospace Corp. in 1995, many research groups are on the way to produce and demonstrate such an invaluable idea. The building of nanosatellites is related to cost reduction. However, in practice, the space community is still waiting for the boom of the nanosatellite constellations, mainly because the enabling technologies are still not fully available or possibly because a different approach is necessary. Systems for aerospace applications require a sophisticated, one-off technology with more capabilities and functionalities as compared to similar terrestrial devices. This complicates the design or the selection of materials and requires a dedicated manufacturing process.

1.11 e-Beam Nanolithography

Nanoscale dimensions have long been used in semiconductor devices. The thickness of many of the thin films used is often below 1000 angstroms. There are many techniques for depositing thin films. Using electrons to create small structures and devices has long been a productive method of lithography. Electron beams can be used as a tool to pattern surfaces that essentially react in response to electrons. The technique of nanolithography is the patterning of a thin film where the line resolution is at the microscopic level, especially below 100 nm. There are many important techniques that are used for patterning in the nanoscale realm. They are:

- Dip-Pen nanolithography
- Photon nanolithography
- Soft nanolithography
- E-beam nanolithography

While the semiconductor industry struggles with the issues created by the end of optical lithography and the lack of a clear alternative, progress is continuing in critical areas of deep submicron patterning. In this regard, e-beam nanolithography allows for a significantly higher resolution than optical lithography. In many cases the low energy electrons are typically well suited to modify self-assembled monolayers, and e-beam lithography allows for the fabrication of monolayer nanostructures with lateral dimensions down to ~10 nm. E-beam nanolithography also emphasises processes and techniques relevant to the integration of advanced semiconductor nanofabrication technologies with chemical and biological nanosystems. Typical specifications of an e-beam lithography system, Raith 150 for example, is given by:

- Schottky thermal-field emission filament, resolution: 2 nm @ 20 kV
- Acceleration voltage: 200 V – 30 kV
- Beam current: 4 pA – 350 pA
- Working distance: 2 – 12 mm
- 16 bits / 10 MHz pattern generator
- Samples: 1 mm – 6 inches
- Interferometer stage: 2 nm positioning accuracy
- Overlay and stitching accuracy: 50 nm

1.12 Nanotechnology

Nanoscale devices and equipment provide benefits in terms of an improved greener environment, miniaturization, efficiency and resource consciousness. Nanotechnology has accelerated research and development in many disciplines.

However, a key obstacle to its development remains in the need for cost-effective large-scale production methods. Nanotechnology has applications in many fields including automotive, aerospace, household appliances, sporting goods, telecommunication equipment and medical supplies.

1.13 Carbon Nanotubes and Structures

Besides the conventional forms of carbon, graphite and diamond, new forms of carbon such as fullerenes, carbon nanotubes and carbon onions have been discovered. A majority of current research focuses on the potential applications of carbon nanotubes (CNT). Carbon nanotubes are considered as ultra-fine unique devices, which can offer significant advantages over many existing materials due to their remarkable mechanical, electronic and chemical properties. With strong covalent bonding they possess unique one-dimensional structures. Nanotubes can be utilised as electronics devices, super-capacitors, lithium ion batteries, field emission displays, fuel cells, actuators, chemical and biological sensors and electron sources. Some of the R&D areas include: nanoscale phenomena, atomic and mesoscale modeling, carbon nano structures and devices, nano composites, biomaterials and systems, bio nano computational methods, fluidics and nano medicine. CNTs are typically longer in length, usually measuring from about a few tens of nanometers to several micrometers, with a diameter up to 30 nanometers and as small as 2.5 nanometers. Apparently, they are hollow cylinders extremely thin with a diameter about 10,000 times smaller than a human hair. Each nanotube is a single molecule made up of a hexagonal network of covalently bonded carbon atoms. Freestanding carbon nanotubes can be grown by chemical vapor deposition (CVD) across the predefined trenches. The trenches can be fabricated lithographically in SiO_2 and then by depositing Pt over the sample to serve as the conducting substrate. Explicitly, they adhere to metallic semi-conducting properties along with good thermal conductivity. Other essential properties they possess are:

- High tensile strength and high resilience
- High current densities

The ongoing research that is being carried out all over the world is based on the study of conductive and high-strength composites, energy storage and energy conversion devices, sensors for field emission displays and radiation sources, hydrogen storage media and nanometer-sized semiconductor devices such as probes and interfacings. Nanotube-based design scenarios anticipate the design of gears and bearings and hence the development of machines at the molecular level.

The team at the Paul Pascal Research Center has been working on overcoming the obstacle of the formation the carbon nanotubes and has developed a process for aligning them in the form of fibers and strips. Variorum structural constructions can be formed through appropriate methods. Continuous sputtering of carbon atoms from the nanotubes lead to a dimensional change, which facilitates surface reconstruction with annealing. An X-like junction with diverse angles between the branches can be formed. Under careful irradiation one of the branches of the X-junction can be removed, thereby creating Y- and T-like junctions. This new class of carbon junctions exhibits an intrinsic non-linear transport behavior, depending mainly on its pure geometrical configuration and on the kind of topological defects (Terrones, 2000; Andriotis, 2002; Bir´o, 2002). Calculation and measurement of characteristic curves, like current versus voltage of different sets of Y junctions, show robust rectification properties, giving rise to the possibility of using these junctions as nanoscale three-point transistors (Latg´e et al. 2004). Advanced computational techniques, including large-scale parallelisable molecular dynamic simulations of the growth mechanism and first-principles calculations of the electronic structure, are being applied to model the self-assembly and the electronic properties of nanostructures (Lambin 1995). Based on computational results, Han et al. at NASA Ames Research Center have suggested that nanotube-based gear (Fig.1.1(b)) can be made and operated and the gears can work well if the temperature is lower than 600-1000 K and the rotational energy is less than the teeth tilting energy at 20°.

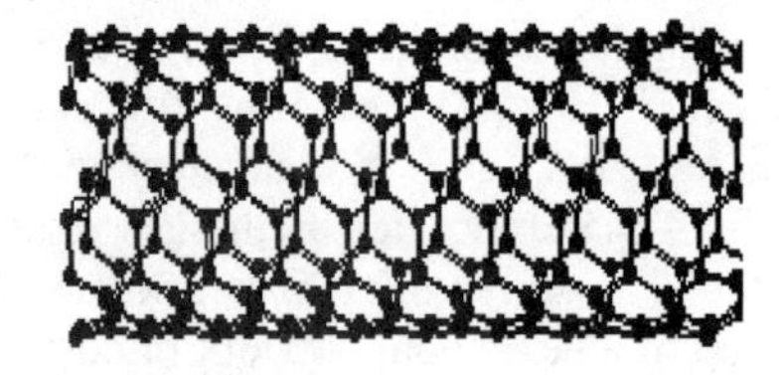

Fig. 1.1. (a) A carbon nanotube (b) A carbon nanotube-based gear (Han 2004)

1.14 Molecular Logic Gates

Logic gates are the fundamental components of digital circuits, which process binary data encoded, in electrical signals. The increasing need for the miniaturisation of logical circuits is reaching the physical limit of the metal oxide semiconductor field effect transistors (FET). There has been tremendous advancement as far as a search for alternatives is concerted. For example, a molecule-based scheme for logical operations, with different in its features, has been proposed. Various other approaches are (Prasanna and Nathan 2004):

- Chemically-controlled fluorescent and transmittance-based switches
- DNA oligonucleotides with fluorescence readout
- Oligonucleotide reactions with DNA-based catalysts
- Chemically-gated photochromics and reversibly denaturable proteins
- Molecular machines with optical and electronic signals
- Two-photon fluorophores

Because of their discrete orbital levels, molecules at room temperature possess large energy level separations. At the nanometer size, the levels make themselves independent of broadband properties. Electronically transduced photochemical switching is possible in organic monolayers and thin films, enzyme monolayers, redox-enzymes tethered with photoisomerisable groups, enzymes reconstituted onto photoisomerisable FAD-cofactors among others. It is found that the photoinduced electron transfer process (PET) occurs in many organic compounds, leading to an on/off mechanism. For instance, pyrazole derivative emits electrons only when the concentration of H^+ is low. Photo-induced electron transfer from the central pyrazoline unit to the pendant benzoic acid quenches the fluorescence of the protonated form. Therefore, the emission intensity switches from a high to low value when the corresponding concentration of H^+ changes from a low to high value. A detailed description is found in this book.

1.15 Microdevices as Nanolevel Biosensors

Innovative enhanced tools for genetic analysis for biomedical and industrial needs are still under development. One development currently being studied is the eSensor™ DNA Detection System by Motorola Inc., by which reliable DNA testing can be achieved. The living being uses DNA to store its genetic code that describes everything about the organism. Specific fractions of the human DNA reveal valuable data. These data can be utilised for health care and medication. Some of the biosensor applications include:

- Forensics and genetic screening
- Stress response analysis
- Antibody gene expression in transgenic cells
- Identification of patients with high tumor risk
- Biowarfare agents
- Pathogens (e.g. throat bacteria)
- Drug discovery

The development of DNA sensors (a DNA sensor is a biosensor) is considered to be the most innovative molecular biology technology due to the fact that these micro or nano systems allow for an easy, accurate, fast and reliable way to analyse many DNA samples simultaneously. The basic element of the DNA sensor is the DNA probe(s). The technology is based on the recognition of oligonucleotides (An oligonucleotide corresponds to a single very short DNA strand) by their complementary genomic sequences (The relative order of base pairs, whether in a DNA fragment, gene, chromosome, or an entire genome). The process of recognition is called hybridisation (Fig.1.2(a)). Hybridisation can be achieved by the optical method, the surface stress method or by electrical detection. The optical method is a traditional method in which the color of the solution changes. The surface stress based hybridization process uses a probe of micro cantilever structures, coated with a detector film that reacts with the specific biomolecules being tested. A biochemical reaction on the cantilever surface changes in the surface stress. The surface stress causes a bending of the cantilever. A piezoresistor integrated with the cantilever reflects the bending variable, and hence detects the presence of DNA. The stress-based detection method is ideal where limited sample preparation is expected as the electrical characteristic based process uses microelectrodes. In this case, the DNA probe is attached to electrodes. There are two opposite electrodes and the electrical resistance between the electrodes is infinite. Similar to the cantilever based method, each probe must be prepared beforehand, i.e., the DNA has to be coated with the probe. The probe has a specific molecular sequence. The active element made from pairs of gold electrodes is sometimes placed in a circular compartment onto which a DNA strand with known sequences is attached. A solution of unknown sequence (target) DNA strands is then brought. Compatible DNA strands bind to the attached probes and the unreacted strands do not. Any DNA that hybridises with the probe is the DNA to be detected. Upon hybridisation, a bridge-like structure is formed on the electrodes. The formation of the bridge the causes electrical resistance between the electrodes to form, which then drops the resistance from infinite to few thousand ohms. This change in resistance is the measure of the presence of DNA. Appropriate instrumentation can be integrated to observe the effect. Such a type of detection method is called a single molecule based detection scheme and is very useful for the detection of specific DNA molecules. Fig.1.2(b) shows how a typical bridge has been formed by the use of an electrical characteristic based hybridisation process. The electrical characteristic based detection of a specific single DNA molecule fragment is very efficient as compared to the surface stress based method. Note that the sensor is a microstructure, but the hybridization phenomenon is at the nanoscale level.

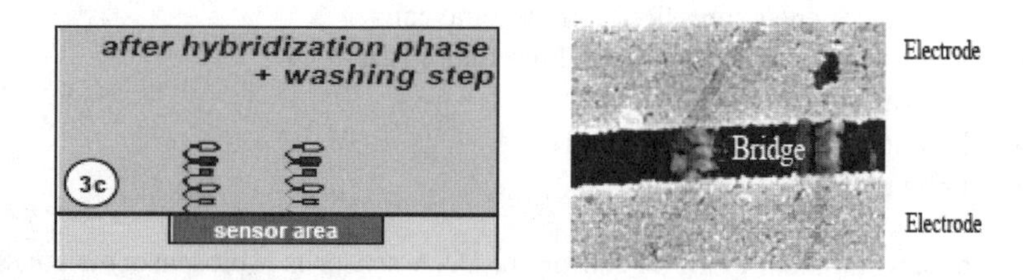

Fig. 1.2. DNA hybridisation (Source: Fuller, Rochester institute of Technology)

1.16 Crosslinking in C_{60} and Derivatisation

The treatment of fullerene films on silicon substrates with 1,8-octanediamine vapor at ca. 120°C produces insoluble polymeric films. They are similar in stability to photopolymerised C_{60} films for potential photovoltaic and Nanocell applications, but the undesirable formation of other carbonaceous phases can be avoided. The solvent-free (gas-phase) treatment with amines under temperatures of ca. 150°C is a simple one-step method of derivatisation of oxidised and pristine carbon nanotubes. It can be useful for the solubilisation/dispersion of CNTs, the synthesis of chemical linkers for other compound immobilisation on the nanotubes, the fabrication of AFM cantilevers and other nano applications.

1.17 Fuel Cell

A fuel cell is primarily an electrochemical system that generates electricity by converting chemical energy into electrical energy. The fuel cell can be thought of as a highly efficient next generation micro power system for a variety of applications. Several challenges to the widespread implementation of fuel cell technology are encountered, although novel inexpensive and long-lasting electrocatalyst materials are becoming major factors in design and development. Recent advances in the application of nanostructured carbon-based materials have suggested the possibility of using carbon nanotubes as novel electrocatalyst supports.

1.18 References

Aronson RB (2004) Micromanufacturing is growing, Manufacturing Engineering 132:4

Bean KE (1978) Anisotrophic etching of silicon, IEEE Trans. Electron Devices 25:1185-1192

Chyr I, Steckl AJ (1999) Focused ion beam micromilling of GaN and related substrate materials (Sapphire, SiC, and Si). American Vacuum Society

Crocombe RA (2004) MEMS technology moves process spectroscopy into a new dimension, Spectroscopy Europe. pp 16-19

Danel JS, Delapierre G (1991) Quartz: A material for microdevices, J. Micromechanics and Microengineering 1:187-198

deSilva AP, Nathan DM (2004) Molecular scale logic gates, Chemistry, A European Journal 10:574-586

Dresselhaus M, Dresselhaus G (1998) Carbon nanotubes. Physics web/world

Friedrich CR, Vasile MJ (1996) Development of the micromilling process for high-aspect-ratio microstructures, Journal of Microelectromechanical Systems 5:33-38

French PJ and Sarro PM (1998) Surface versus bulk micromachining: The contest for suitable applications, J. Micromechanics and Microengineering 8:45–53

Gregory TA, Maluf NI, Peterson KE (1998) Micromachining of silicon. IEEE Proceedings, 86

Han J, Jaffe R, Globus A, Deardorff G (2004) Molecular dynamics simulations of carbon nanotube based gears, NASA Ames Research Center. available at www.nas.nasa.gov

Harasima F, Tomizuka M (1996) Mechatronics; What is it, why and how, IEEE Transaction on Mechatronics 1:1-7

http://www.aspe.net/about/index.html

http://www.microe.rit.edu/research/nanotechnology/dnasensor.htm

James M, Bustilo M, Roger TH, Muller RS (1998) Surface micromachining for microelectromechanical systems. IEEE Proceedings, 86, p 8

Keppeler CR (2003) Micromilling for mold fabrication. LMA Technical Reports, University of California at Berkeley, pp 99-101

Kern W (1967) Chemical etching of silicon, Germanium and Gallium Phosphate, RCA Review 39:278-307

Lambin PH, Fonseca A, Vigneron JP, Nagy JB, Lucas AA (1995) Structural and electronic properties of bent carbon nanotubes, Chem. Phys. Lett. p 245

Latg´e A, Muniz RB and Grimm D (2004) Carbon nanotube structures: Y-junctions and nanorings, Brazilian Journal of Physics 34:2B

Mahalik NP (ed) (2003) Fieldbus Technology. Springer Verlag, ISSN 3540401830, Germany

Mahalik NP (2004) Mechatronics: Principle, concepts and application, McGraw-Hill International Edition (Higher Education), ISSN 007-048374-4

Petersen KE (1982) Silicon as a mechanical material, IEEE Proceedings, 70, pp 420–57

Ray H, Anvar AZ, Walt AH, (2002) Carbonnanotubes; The route toward applications, Science 297

Regan MT, Rimai DS (2004) Electrophotography: A micromanufacturing technology, NexPress Solutions LLC, available at www.nano.neu.edu

Research Trend (2002) Fully electronic DNA sensor arrays, Corporate Research, USA

Steger RB, Neurath A, Messner S, Sandmaier H, Zengerle R, Koltay P (2004) A highly parallel nanoliter dispensing system fabricated by high-speed micromilling of polymers. Technical Report, University of Freiburg and HSG-IMIT, Germany

Tönshoff HK, Momma C, Ostendorf A, Nolte A, Kamlage G (2000) Microdrilling of metals with ultra-short laser pulses, Journal of Laser Applications 12:23-27

2 Principles of MEMS and MOEMS

N. P. Mahalik[1], S. E. Iyuke[2], and B. H. Ahn[1]

[1]Department of Mechatronics, Gwangju Institute of Science and Technology, Republic of South Korea
[2]School of Process and Materials Engineering, University of Witwatersrand, South Africa

2.1 Introduction

A considerable amount of research is being carried out concerning the design and development of existing systems that reach down into micro- and nanometer scale levels. A technology that considers microscale sensors, actuators, valves, gears, and mirrors embedded in semiconductor chips is referred to as microelectromechanical systems, MEMS in short. In essence, MEMS are small, integrated devices that combine electronics, electrical as well as mechanical elements (Fig. 2.1). The size is in the order of a micrometer level. MEMS design technology is an extended form of traditional fabrication techniques used for IC (Integrated Circuit) manufacturing. MEMS add passive elements such as capacitors and inductors including mechanical elements such as springs, gears, beams, flexures, diaphragms, etc. MEMS are thus the integration of these elements on a single substrate (wafer) developed through more advanced microfabrication and micromachining technology. While the ICs are fabricated by the use of IC process, the mechanical micro components are fabricated using micromachining processes. This process helps in etching away the parts of the selected portions of the wafer. The process can also add new structural layers to form mechanical as well as electromechanical components. Thus, MEMS technology promises to revolutionise many products by combing microfabrication-based microelectronics with micromachining process sequences on silicon, making it possible for the realisation of a complete *systems-on-a-chip* (SoC). The technology allows for the development of smart systems and products inheriting increased computational capability, perception and control attributes. Smart systems can lead to expand the scope of possible solutions to diagnostics for target applications. It has been mentioned that microelectronic integrated circuits can be thought of as the brains of a system while MEMS augments the decision-making capability with eyes and arms, to allow microsystems to sense and control the environment (http://www.memsnet.org/ mems/what-is.html). MEMS devices are manufactured by the use of batch fabrication techniques similar to those used for IC. Therefore, unparalleled levels of superiority, sophistication, functionality, reliability and availability can be achieved on a small silicon chip at a relatively low cost. Two important microsystems are microsensors and microactuators.

Sensors gather information from the environment. The commonly used transduction principles are chemical, thermal, biological, optical, magnetic and mechanical phenomena. Accordingly, there are various types of microsensors. The integrated electronics process the information derived from the sensors. In many cases the decision-making logics are integrated into the devices. The decision is mostly transmitted to the actuator in order to achieve moving, positioning, regulating, pumping or filtering actions. In this way, the environment can be controlled depending on the desired purpose. The study of MEMS accommodates the topics listed below. These principles are presented in this chapter.

- Fabrication processes
- Mechanical sensors and actuators
- Thermal MEMS
- Magnetic MEMS
- Micro-opto-electromechanical systems (MOEMS)

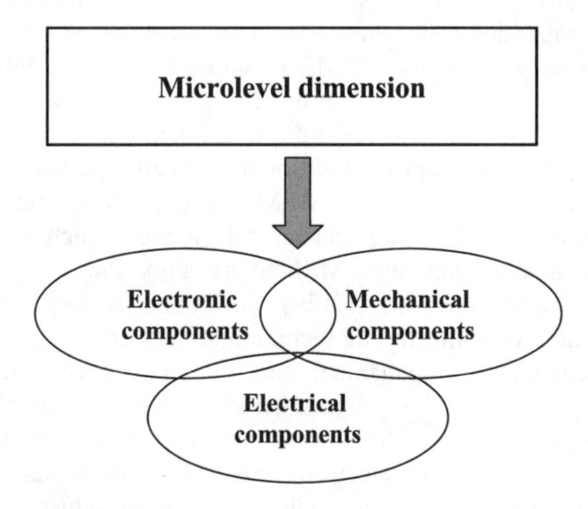

Fig. 2.1. Microelectromechanical systems

2.2 Driving Principles for Actuation

The driving principles used to drive various types of micromachined MEMS systems are primarily four types, namely,

- Electrostatic drive
- Magnetic drive
- Piezoelectric drive
- Electrothermal drive

Each driving principle has specific advantages and disadvantages with respect to deflection range, required force, environmental durability, and most importantly the response time. Furthermore, the required power supply mainly depends on the driving principles involved (Table 2.1).

Table 2.1 typical power requirements for respective driving principles

	Voltage	Current
Electrostatic Drive	tens of volts ~ hundreds of volts	nA ~ μA
Piezoelectric Drive	tens of volts ~ hundreds of volts	nA ~ μA
Electromagnetic Drive	about 1 V	hundreds of mAs
Electrothermal Drive	a few volts ~ tens of volts	mA ~ tens of mAs

Electrostatic drive is based on electrostatic forces between the microelectrodes. When an external voltage is applied between the electrodes, a potential energy is stored which enables the actuation. The electrostatic forces act perpendicular to the parallel electrode. Electromagnetic actuation is primarily a current controlled process. The driving mechanism again requires currents of the order of several hundreds of milliamps, and voltages in the range of less than one volt. Magnetic drive is an attractive driving principle and very suitable for applications like dust-filled environments and in environments where low driving voltages are acceptable or desired. Piezoelectric driving is based on the material properties of crystals, ceramics, polymers, and liquid crystals. In a piezoelectric material, the internal dielectric displacement is developed via an applied electric field and mechanical stress. Electrothermal devices use electrically generated heat as an energy source for actuation. The electrothermal effects can be divided into three different categories: shape memory alloys, electrothermal bimorphs, and thermo-pneumatic actuators. Many electrostatic MEMS actuators that are investigated include micromotors, comb drive actuators and microvalves. Accelerometers, ink jet printer heads, color projection displays, scanning probe microscopes, pressure, temperature, chemical and vibration sensors, light reflectors, switches, vehicle control, pacemakers and data storage devices are examples of high-end applications of MEMS sensors and actuators.

2.3 Fabrication Process

MEMS support the principle of miniaturisation, multiplicity, and microelectronics. Miniaturisation is preferred in order to achieve faster response times and less space, whereas multiplicity is essential in order to advocate batch fabrication. Batch processing is a processing technique that allows for thousands of components to be simultaneously manufactured in order to significantly reduce the cost of the device. Microelectronics embeds the real part of the MEMS device. These can be sensors, actuators, signal conditioning, signal processing or even logic circuits.

The design process involved for IC manufacturing is called microfabrication. The sequences of microfabrication include film growth, doping, lithography, etching, dicing and packaging. A polished silicon wafer is mainly used as the substrate. A thin film is grown on the substrate. Then the properties of the layer are modulated by appropriately introducing doped material in a controllable manner. The doping can be achieved by thermal diffusion. The subsequent process is called lithography, which refers to the creation of a masking pattern. The pattern on a mask is transferred to the film by means of a photoresist. A mask usually consists of a glass plate, which is coated with a patterned layer, which is usually chromium film. The subsequent process is called etching. Etching is a process of removing the portions of material from an insulating base by chemical or electrolytic means. The two types of etching are wet and dry etching. In wet etching the material is dissolved by immersing it in a chemical solution. On the other hand, in dry etching, the material is dissolved by using reactive ions or a vapor phase etchant (Vittorio 2001). The finished wafer has to be segmented into a small dice-like structure. Finally, the individual sections are packaged. Packaging is a complex process that involves physically locating, connecting, and protecting a component or whole device. MEMS design also considers all the process sequences employed for microfabrication, however in this case it is considered as an extension of the IC fabrication process. The following methods are common as far as manufacturing of MEMS designs are concerned:

- Bulk micromachining
- Surface micromachining
- Micromolding

Bulk micromachining makes micromechanical devices by etching deeply into the silicon wafer. It is a subtractive process that involves the selective removal of the wafer materials to form the microstructure, which may include cantilevers, holes, grooves, and membranes.

The majority of currently used MEMS processes involve bulk etching. In light of newly introduced dry etching methods, which are compatible with complementary metaloxidesemiconductors, it is unlikely that bulk micromachining will decrease in popularity in the near future (Kovacs 1998). The available etching methods fall into three categories in terms of the state of the etchant: wet, vapor, and plasma. The etching reactions rely on the oxidation of silicon to form compounds that can be physically removed from the substrate (Kovacs, 1998). Conversely, surface micromachining technology makes thin micromechanical devices on the surface of a silicon wafer.

The surface micromachining sequences are as follows (Madou www. mmadou.eng.uci.edu/Classes/MSE621/MSE62101(12).pdf),

- Wafer cleaning
- Blanket n+ diffusion of Si substrate
- Passivation layer formation
- Opening up of the passivation layer for contacts
- Stripping of resist in piranha
- Removal of thin oxide through BHF etchant systems
- Deposition of a base, spacer or sacrificial layer using phosphosilicate glass (PSG)
- Densification at 950°C for 30-60 min in wet oxygen
- Base window etching in BHF for anchors
- Deposition of structural material deposition (e.g., poly-Si using CVD method at about 600°C, 100 Pa and 125 sccm at about 150 Å/min)
- Anneal of the poly-Si at 1050°C for 1 hour to reduce stress in the structure
- Doping: in-situ, PSG sandwich and ion implantation
- Release step, selective etching of spacer layer.

The micromolding process involves use of molds to define the deposition of the structural layer. In this case, the structural material is deposited only in those areas constituting the microdevice structure. This is apparently in contrast to both bulk and surface micromachining processes. Feature blanket deposition of the structural material followed by etching to realise the final device geometry is done in one step. Once the structural layer deposition is over the mold is dissolved by using a chemical etchant. Note that the etchant does not corrugate the structural material. One of the most widely used micromolding processes is the LIGA process. LIGA is a German acronym standing for lithographie, galvanoformung and abformung, or lithography, electroplating, and molding. Photosensitive polyimides are mostly used for fabricating plating molds.

2.4 Mechanical MEMS

2.4.1 Mechanical sensors

MEMS mechanical sensors are very popular because of their easy integration procedure. The sensing mechanism utilises the following methods and principles,

- Cantilever beam sensor
- Capacitive sensing
- Accelerometers
- Microphones
- Gyroscopes
- Piezoelectricity

2.4.2 Accelerometer, Cantilever and Capacitive Measurement

Cantilever sensors can be used for the detection of physical, chemical and biological analytes with relatively good sensitivities and selectivity. The vast application areas include: acoustic applications, vibration monitoring, detection of viscosity and density, infrared and UV radiation, magnetic and electric fields, detection of chemical vapours including medical and biological agents, contaminants in water, explosive vapours such as RDX, PETN and TNT, nuclear radiation and the detection of DNA, antibodies and pathogens.

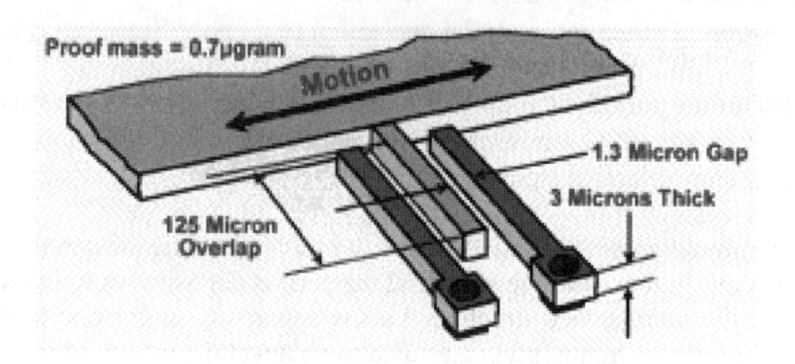

Fig. 2.2. Cantilever beam for the measurement of static and dynamic acceleration

Fig. 2.2 shows a typical cantilever sensor that can measure acceleration. The body of the sensor (proof mass) could be about 0.5-0.7 micrograms. The proof mass moves in the X- and Y-axes. Polysilicon springs suspend the MEMS structure above the substrate facilitating the proof mass to move freely. Acceleration causes deflection of the proof mass from its centre position. There could up to 32 sets of radial fingers around the four sides of the square proof mass. The fingers (middle one), shown in the figure, are positioned between two plates that are fixed to the substrate. Each finger and pair of fixed plates constitutes a differential capacitor, and the deflection of the proof mass is determined by measuring the differential capacitance. This sensing method has the ability of sensing both dynamic acceleration such as shock or vibration, as well as static acceleration such as inclination or gravity.

Many accelerometers employ piezoelectric sensing techniques. Thin film piezoelectric materials such as lead zirconate titanate (PZT) are promising materials for MEMS applications due to their high piezoelectric properties (Davis 2004: at http://www.nnf.cornell.edu/1999REU/ra/Meteer.pdf). Piezoelectric polymers are now being used for sensor applications. Piezoelectric polymeric sensors offer the advantage of strains without fatigue, low acoustic impedance and operational flexibility. The PZT converts mechanical disturbances to electrical signals. The starting material for the front-side process (FSP) is a silicon wafer that has silicon dioxide, lower metal electrode (Ti/Pt), and deposited PZT films. Industrial applications of MEMS accelerometers include airbag release mechanisms, machinery failure diagnostics, and navigational systems.

2.4.3 Microphone

Acoustic MEMS are air-coupled, and can offer a wide range of applications such as detection, analysis and recognition of sound signals. The basic component is the microphone, called micro-microphone or simply MEMS microphone. A simple definition of a microphone is that it is an electromechanical acoustic transducer that transforms acoustical energy into electrical energy. It is an ultrasonic microsensor, which takes advantage of miniaturisation and also consumes low power. When multiple microphones are arranged in an array, the device is referred to as smart system as it can offer more reliable and intelligent operations. The basic challenge that is encountered in designing the MEMS microphone is the formulation, design, and implementation of signal processing circuitry that can adapt, control and utilise the signal in noisy environments. MEMS microphones are also very suitable for outdoor acoustic surveillance on robotic vehicles, wind noise flow turbulence sensing, platform vibration sensing, and so on.

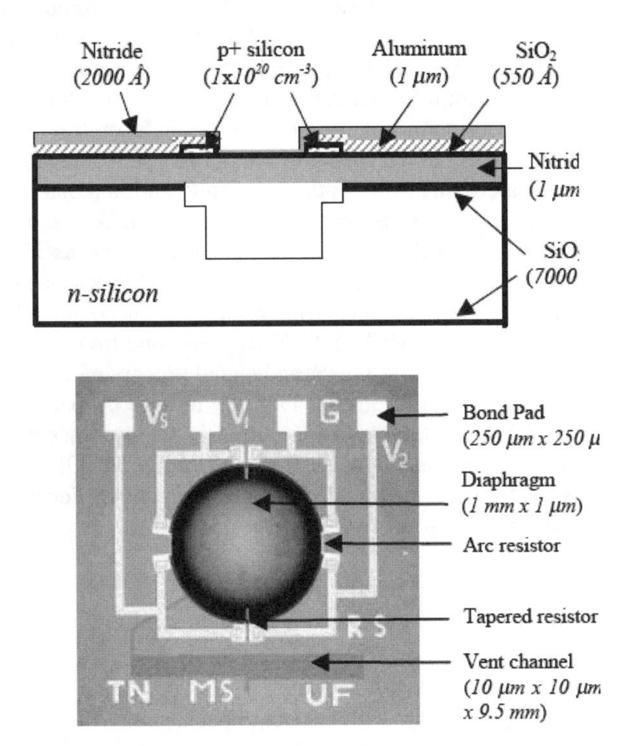

Fig. 2.3. Cross-sectional diagram of piezoelectric microphone; plan view (Arnold 2001)

The ultrasonic sensor is based on the mechanical vibration of micro membrane or diaphragm realised in the silicon platform. The diaphragm is a thin, circular membrane held in tension and clamped at the edge. Many designs use a flat free-plate that is held in proximity to the back plate by electrostatic attraction. The free-plate makes up a variable capacitor with the back plate. Its value changes during the vibration caused by the sound signal. The deformation or deviation of the membrane (free plate) from the normal values depends on the amplitude of the incident pressure. Fig. 2.3 shows the schematic as well as a micromachined SEM (Scanning Electron Microscope) picture of a MEMS microphone. The microphone can have a very low stray capacitance, is self-biasing, mass producible, arrayable, integrable with on-chip electronics, structurally simple and extremely stable over time in an ordinary environment. The typical dynamic range is from 70 to 120 dB SPL and the sensitivity can be in the order of 0.2 mV/Pa over the frequency range 100-10 kHz.

2.4.4 Gyroscope

MEMS gyroscopes (Fig. 2.4) are typically designed to measure an angular rate of rotation. A measurement of the angle is useful in many applications. A very common application is the measurement of the orientation or tilt of a vehicle running in a curved path. The MEMS gyroscope design introduces sophisticated and advanced control techniques that can lead to measure absolute angles. Some design is based on the principle of measuring the angle of free vibration of a suspended mass with respect to the casing of the gyroscope (Rajamani, 2000). The gyroscope can accurately measure both the angle and angular rate for low bandwidth applications. The measurement of orientation, for instance, is very useful in the computer-controlled steering of vehicles as well as for differential braking systems for skid control in automobiles. A typical gyroscope consists of a single mass, oscillating longitudinally with rotation induced lateral deflections being sensed capacitively. The iMEMS ADXRS gyroscope from Analog Devices, Inc., integrates both an angular rate sensor and signal processing electronics onto a single piece of silicon. Mounted inside a 7x7x3 mm BGA package, the gyro consumes 5 mA at 5 V and delivers stable output in the presence of mechanical noise up to 2000 g over a reasonably wide frequency range. A full mechanical and electronic self-test feature operates while the sensor is active. Mechanical sensors inherit many drawbacks as follows (www.sensorland.com),

- The overall silicon area is generally larger
- Multi-chip modules require additional integration steps
- Existence of larger signals from the sensor output
- Stray capacitance of the interconnections
- Comparatively larger volume with respect to packages

2.4.5 Mechanical Actuators

One of the fundamental components in MEMS technology is the actuator. A mechanical actuator is a device that usually converts electrical signals into mechanical motion. For instance, if a voltage is applied to a quartz crystal, it will change its size in a very precise and predictable way. Cantilever structures are used for actuation purposes. Cantilevers bend when pressure is applied to them and oscillate in a way similar to a spring when properly put into place. The important point that is considered in electrostatically actuated cantilever based MEMS devices is their stability.

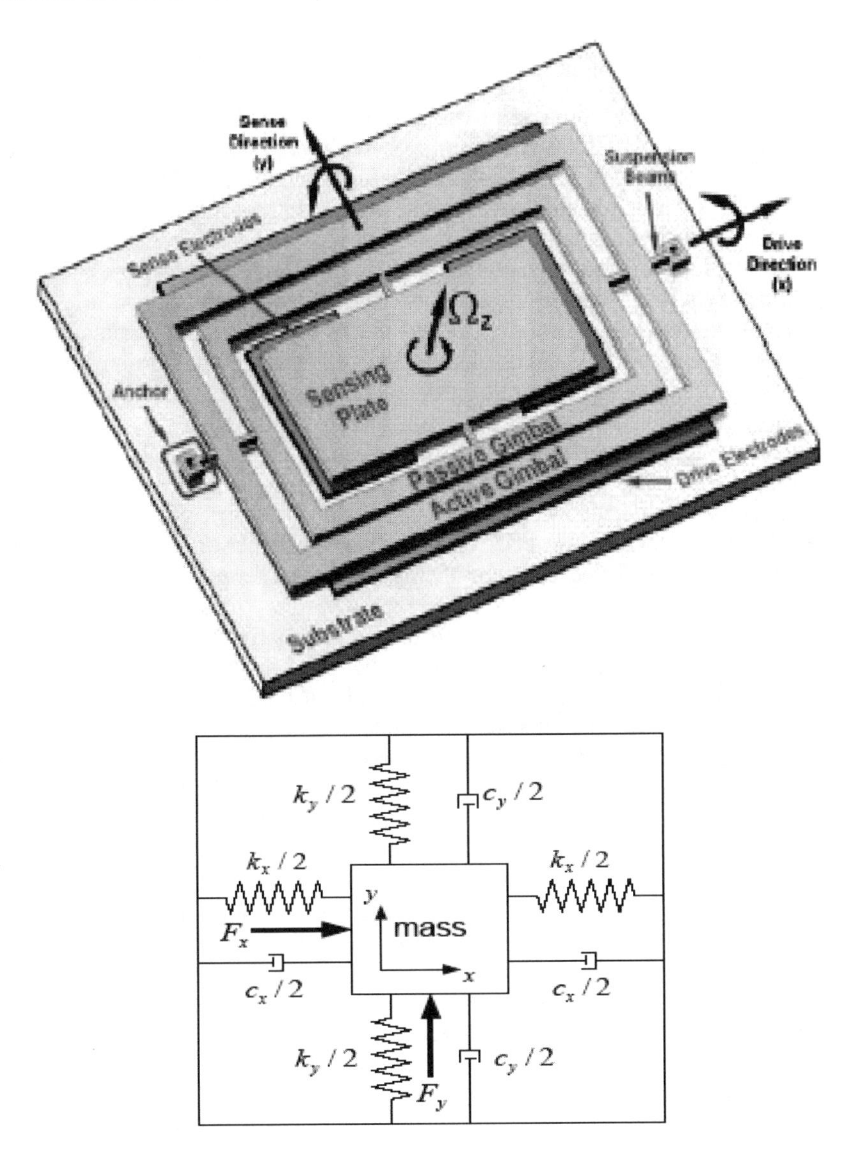

Fig. 2.4. (a) Conceptual schematic of a torsional micromachined gyroscope with non-resonant drive (Source: Acar 2004); (b) Schematic diagram of a vibratory gyroscope (Source: Piyabongkarn 2005)

2.5 Thermal MEMS

Microsystems whose functionalities rely on heat transfer are called thermal MEMS. Thermal MEMS can be sensors and actuators. When a physical signal to be measured generates a temperature profile, the principle is called thermal sensing. The physical signals could be heat radiation, non-radiant heat flux or a reaction heat. Gas pressures, masses, volume fluxes, and fluidic thermal conductivities are measured using the principle of thermal sensing. As always, sensing or actuation is a process of energy conversion, called transduction. The transduction effect is realised in three ways: the thermoelectric effect (Seebeck effect), the thermoresistive effect (bolometer effect) and the pyroelectric effect.

In 1821, Thomas Johann Seebeck discovered that a compass needle was deflected when it was placed in the vicinity of a closed loop formed from two dissimilar conductors and the junctions maintained different temperatures. This is as good as saying that a voltage (and hence the current) is developed in a loop containing two dissimilar metals, provided the two junctions are maintained at different temperatures. The magnitude of the deflection is proportional to the temperature difference and depends on the material, and does not depend on the temperature distribution along the conductors. The effect is called the thermoelectric effect, and is the basis of the thermocouple, a real temperature measuring device. The opposite of the Seebeck effect, in which current flow causes a temperature difference between the junctions of different metals, is the Peltier effect. The reversed Seebeck effect is thus the Peltier effect. A thermopile is a serially interconnected array of thermocouples. Thermopiles are used for achieving better sensitivity.

Pyroelectricity is the migration of positive and negative charges to opposite ends of a crystal's polar axis as a result of a change in temperature. The cause is referred to as electric polarisation. The polarisation phenomenon within the material further states that below a temperature, known as the Curie point, crystalline materials or ferroelectric materials exhibit a large spontaneous electrical polarisation in response to a temperature change. Materials, which possess this property, are called pyroelectric materials. The change in polarisation is observed as an electrical voltage signal and appears if electrodes are placed on opposite faces of a thin slice of such material. The design can be thought of as a typical form of a capacitor. Cooling or heating of a homogeneous conductor resulting from the flow of an electrical current in the presence of a temperature gradient is known as the Thomson effect. It is hence defined as the rate of heat generated or absorbed in a single current carrying conductor, which is subjected to a temperature gradient. Arne Olander observed that some alloys such as NiTi, CuZnAl, CuAlNi, etc. change their solid-state phase. The property is reflected as pseudo-elasticity and the shape memory effect. After alloying and basic processing, the alloy can be formed into a desired shape, a coil for example, and then set to that shape by a heat treatment. When the shape is cooled, it may be bent, stretched or deformed and then with subsequent re-heating (which should be the heat setting temperature) the deformation can be recovered. Some of the main advantages of shape memory alloys (SMA) are that they are biocompatible with good mechanical properties such as strong and corrosion resistant and can be applied to diverse actuator applications.

2.5.1 Thermometry

Rigorous thermal analyses and experiments that assist in designing MEMS structures are in progress. Some effort has been made on liquid-crystal thermometry of micromachined silicon arrays for DNA replication. This work is being carried out at Perkin-Elmer Applied Biosystems. Replication is achieved through what is known as the polymerase chain reaction (PCR) process. PCR requires accurate cycling of the liquid sample temperature with an operational range between 55 and 95°C. PCR that makes use of micromachined structures is shown in the Fig. 2.5(a). The structure assures uniformity as far as temperature and cycle timing is concerned. It also utilises less reagent and sample volumes. Thermal design requires measurement of the temperature distribution in the reacting liquid. The measurement is possible by encapsulating the liquid crystals suspended in the liquid. This in turn led to the measurement of the temperature uniformity and the time constant for about 20 vessels in a micromachined silicon array. Two separate sets of crystals are used to image temperature variations near the two processing temperature thresholds with a resolution of 0.1°C. While the thermometry technique described above is useful for characterising microfabricated PCR systems, it can also support thermal designs of a broad variety of MEMS fluidic devices.

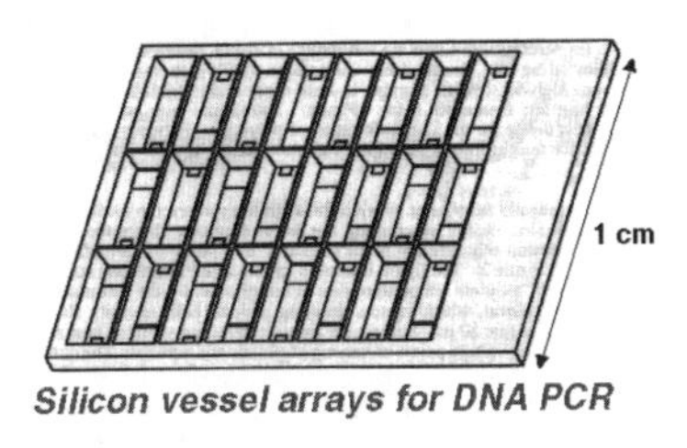

Fig. 2.5. (a) Liquid-crystal thermometry based micromachined vessel array for thermal processing of DNA using polymerase chain reaction (PCR) (Source: Woudenberg and Albin, Perkin Elmer Applied Biosystems)

Thermally activated systems are mostly employed in the bio-analytical microsystems or *lab-on-a-chip* (loac) devices. Several key issues in existing and emerging bio-analytical microsystems are directly related to thermal phenomena. Research and development on loac technology is essentially directed toward miniaturisation and the integration of chemical and biochemical analysis tools for manipulation, handling, processing and analysis of samples in a single integrated chip. In order to develop a highly sensitive and responsive thermal device, the thermal mass of the transducer element is kept as low as possible. This is only achieved by using thin film structures made by micromachining techniques.

2.5.2 Data Storage Applications

There are many other applications for thermal designs. One of the more important examples is the design of silicon cantilevers for high-density thermo-mechanical data storage applications. The design method can use the principle of the atomic force microscope (www.stanford.edu/group/microheat/seven.html). MEMS based scanning probe data storage devices are emerging as potential ultra-high-density, low-access-time, and low-power alternatives to conventional data storage devices. The implementation of a probe based storage system uses thermo-mechanical means, as described below, to store and retrieve information in thin films. The design and characterisation of a servomechanism to achieve precise positioning in a probe based storage device is extremely critical. The device includes a thermal position sensor that provides position information to the servo controller. The research work by Kenny and Chui at Stanford University in collaboration with IBM is based on the design of a microcantilever tip that exerts a constant force on a polycarbonate sample and induces localised softening and deformation during the heating phase. This is caused by a bias current along the cantilever. The resulting serrations serve as data bits, which are read by the use of another separate cantilever integrated with a piezoresistive displacement sensor (Fig. 2.5(b)) followed by measuring circuitry. The cooling time constant of the heated cantilever tip governs the rate at which it can achieve sub micrometer writing. One of the challenges in building such devices is the accuracy and the latency required in the navigation of the probe.

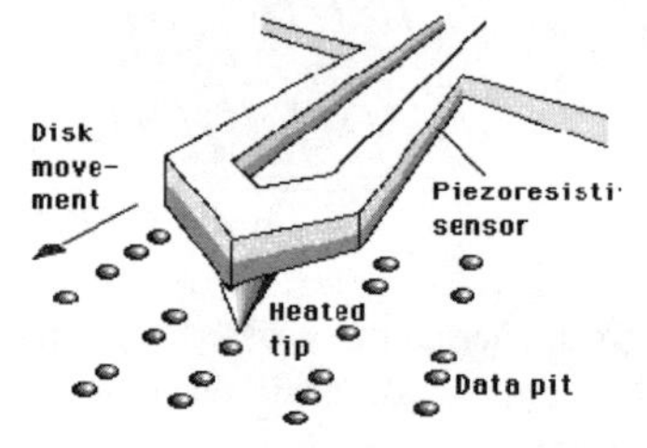

Fig. 2.5. (b) AFM based single-crystal silicon cantilevers tip for data storage (Source: JMEMS Vol.7, pp69 (1998)).

2.5.3 Microhotplate Gas Sensors

Semiconducting SnO_2 films when deposited on microhotplate platforms can detect gas species such as hydrogen and methanol (Semancik, 2001; Panchapakesan,

2001). Microhotplates are thermally isolated micromachined platforms with integrated temperature sensing and actuation for closed-loop thermal control. Typical dimensions of a thermal platform have lateral dimensions of less than 100 micrometer, and are suspended over a bulk-etched cavity for thermal isolation. The physical architecture of each microhotplate consists of a polysilicon heater, a thermoresistive film for temperature measurement, and a semiconducting SnO_2 film, which exhibits a change in conductance with the adsorption of chemical species. With aluminum as the temperature sensing film, the microhotplate can operate up to 500°C. Thermal response time for a 100-micron wide microhotplate has been measured at around 0.6 ms, with a thermal efficiency of 8 C/mW (Semancik 2001; DeVoe 2003).

2.5.4 Thermoactuators

Thermal actuation is another scenario with respect to its counterparts including electrostatic and piezoelectric types. Thermal actuation is based on electrothermal energy density transformation which is given by, $E=V^2/\rho L^2$, where V is the applied voltage, ρ is resistivity and L is the effective length of the actuator element. Microactuators based on electrothermal principles can offer significant energy density compared to electrostatic microactuators. It is worth mentioning that for some configurations, electrothermal actuation provides 1 to 2 orders of magnitude higher energy density than piezoelectric actuation, and 4 orders of magnitude higher energy density than electrostatic transduction (DeVoe 2003). Several thermal microactuator geometries have been investigated, including U-beam and V-beam geometries, respectively (Schreiber 2001; Maloney 2000). Piezoelectric microactuators usually provide a fraction of the energy density of electrothermal elements.

2.6 Magnetic MEMS

The use of magnetic materials in MEMS is a recent development in which particular emphasis is given on ferromagnetic materials. These magnetic materials could be soft or hard. The design of ferromagnetic MEMS and the methods of integrating both soft and hard magnetic materials with it are a current research and development field. Soft ferromagnetic materials have found their usefulness in microsensors, microactuators, and microsystems. However, hard magnetic materials have unique advantages that are driving their integration into other applications.

Patterns of hard magnetic materials in the micron scale are of interest for the novel design of magnetic recording media. Hard magnetic films with a thickness of several microns are grown by the sputtering technique, practically a difficult process. In order to overcome the problem, electrodeposited $Co_{80}Pt_{20}$ alloys are grown up to a several micron thickness while maintaining their hard magnetic properties. The batch fabrication process for micromachining thick films of high-performance hard magnetic materials is improving. Subtractive etching processes are needed to define the patterns of the hard materials. Micron size patterns can also be obtained by optical lithography, thus allowing a better control of the magnetic properties. Many researchers are currently trying to achieve a submicron size for the production of patterned media. Principles of magnetic MEMS are described below.

Anisotropic magnetoresistive (AMR) sensors allow for detecting the strength and direction of magnetic fields. They are used for the measurement of distance, proximity, position, angle and rotational speed. The AMR sensors undergo a change in resistance in response to an applied magnetic field vector generated by passing the current in a coil. When a magnetic field is applied, the magnetisation rotates toward the field. The variation in resistance depends on the rotation of magnetisation relative to the direction of current flow. The change in resistance of the material used in the AMR sensor is highest if the magnetisation is parallel to the current and lowest if it is perpendicular to the current. The sensitivity to the direction of the magnetic field allows for the measurement of angle also. The development of micromachined magnetic devices has relied primarily on the use of nickel-iron permalloy. Permalloy has a low magnetic anisotropy. The anisotropy field for permalloy is between 3 and 5 oersted. Permalloy is used in a number of applications since it has good soft magnetic properties, high permeability, high magnetoresistive effect, low magnetostriction, stable high frequency operation, and excellent mechanical properties. In hard disk magnetic recording heads, permalloy is widely used for magnetoresistive sensors and flux guiding elements. Devices such as magnetic separators, micropumps, magnetic micromotors, inductors, switches, and microrelays have also been fabricated using permalloy as the magnetic material as well as in moving members (Taylor 1999).

Permanent magnets are used for sensors and actuator applications that can provide the desired constant magnetic field without the consumption of electrical energy. Another important characteristic is that they do not generate heat. Furthermore, the energy stored in a permanent magnet does not deteriorate when the magnet is properly handled and micromachined. This feature is due to the fact that it does no net work on its surroundings. Moreover, permanent magnets can achieve relatively high energy density in microstructures, as compared to other energy storage devices (Cho 2002). This is why there has been a growing interest in the realisation of permanent magnets in MEMS devices recently. Apparently, magnetic MEMS devices have several advantages over electrostatic types. One of the important advantages is the generation of long-range force and deflection with low driving voltage in harsh environments. MEMS devices with a permanent magnet have benefits of low power consumption, favorable scaling and simple electronic circuitry. Lagorce et al. introduced screen-printing technology to integrate a permanent magnet in a microactuator. Actuators are capable of generating large bi-directional forces with long working lengths. These permanent magnets were screen printed on a copper cantilever beam using a magnetic paste composed of epoxy resin and strontium ferrite particles. Coercivity of 350 kA/m and residual induction of 65 mT have been reported in a disk type permanent magnet with a millimeter range diameter and 100 micron in thickness.

A number of micromachined magnetic actuators require materials with desirable magnetic properties such as high permeability and a thickness in the range of several micrometers. Investigations on the use of rare-earth magnetic powders in microsystems are in progress. The powders are essentially deposited by screen printing or other methods to fabricate strong permanent magnets with wafer level processes. Deflections of over 900 micron have been reported with surface micromachined cantilevers, magnets and pancake coils. Magnets with typical volumes measuring in the range of 0.11 mm^3 can produce forces in the mN range. A typical magnetically actuated microcantilever incorporating a screen printed magnet measures 2000x1000x100 microns. Deflection at the tip of the cantilever is about 1000 micron, subjected to a driving current of 200 mA which is passed through a 40 turn pancake coil. Large deflection causes nonlinear performance similar to electrostatic systems, however, small deflection shows a better linear current-deflection relationship.

A magnetic actuator consists of a magnetic field source and a magnetically susceptible unit. The actuation force is proportional to the magnetic field intensity, the magnetic susceptibility, and the mass of magnetic susceptible material (Son, http://sonicmems.ece.cornell.edu/Papers/Sunny_paper_mmb7.pdf). A large-force, fully integrated, electromagnetic actuator for microrelay applications is reported by Wright (Wright 1997) (Fig. 2.6). The actuator has a footprint of less than 8 mm^2 and its fabrication is potentially compatible with CMOS processing technology. It is designed for high efficiency actuation applications. The actuator integrates a cantilever beam and planar electromagnetic coil into a low-reluctance magnetic circuit using a combined surface and bulk micromachining process. Test results show that a coil current of 80 mA generates a 200 μN actuation force. Theoretical extrapolation of the data, however, indicates that a coil current of 800 mA can produce an actuation force in the millinewton range.

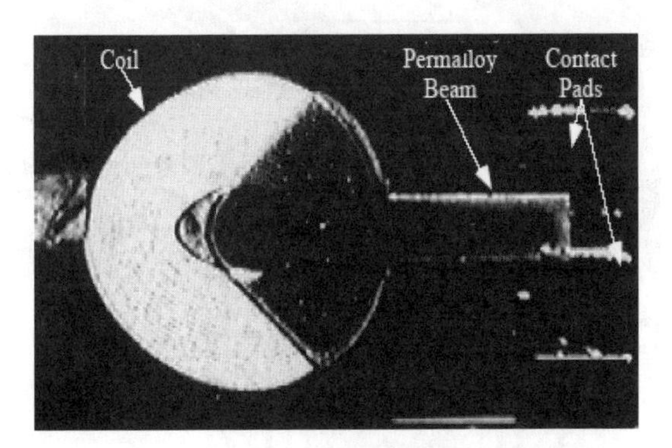

Fig. 2.6. Top view of a magnetic actuator developed by Write, et.al

An example of a bi-directional magnetic actuator used for optical scanning applications can be given. It is composed of a silicon cantilever beam and an electromagnet. At the tip of the cantilever beam, a permanent magnet array is electroplated in order to achieve the bi-directional actuation. Below the cantilever beam, the permanent magnet array is placed along the axis of the electromagnet (Cho 2002). For a large bi-directional deflection and dynamic scanning capability, Cho has designed an optical scanner supported by two serpentine torsion bars. A schematic diagram of the scanner is illustrated in Fig. 2.7. The scanner is designed to have a silicon mirror supported by two serpentine torsion bars (Fig. 2.8), carrying a permanent magnet made by bumper filling at its tip on the opposite side.

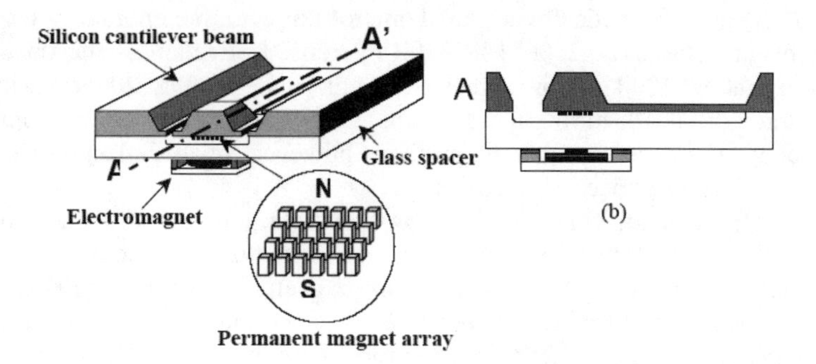

Fig. 2.7. Cantilever beam magnetic microactuator. (a) Schematic view and (b) Another view (Source: Cho 2002)

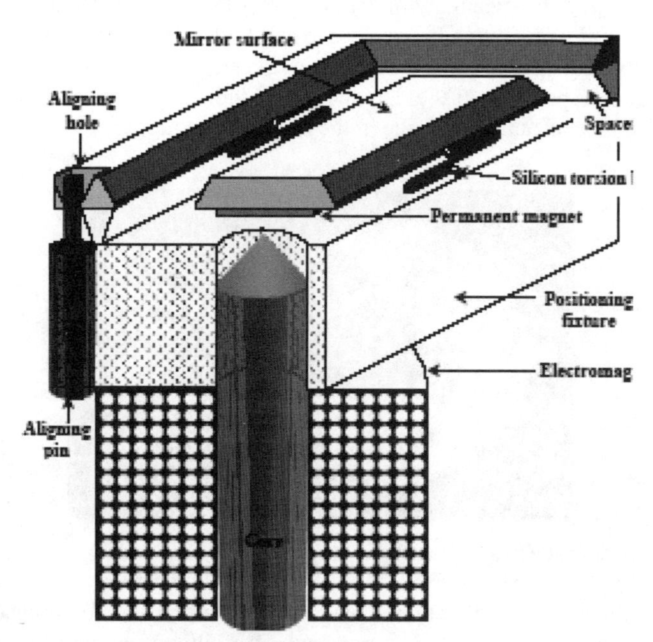

Fig. 2.8. Magnetically driven optical scanner (Source: Cho, 2002)

Son and Lal (Son and Lal 1999, 2000, 2001) reported a novel magnetic actuator that allows remote magnetic actuation with piezoresistive feedback for microsurgery applications. The actuator consists of an electromagnet, a ferromagnetic mass, and a cantilever. Conventional actuator fabrication lithography combined with electroplating is required to make high aspect ratio structures. However, high aspect ratio columns with a suspension of ferromagnetic nanoparticles and epoxy using magnetic extrusion were fabricated. The fabrication process did not require lithography. The remote magnetic actuator with feedback consists of two basic units. They are the actuation unit and feedback unit. The ferromagnetic mass is first placed on the tip of the cantilever. The electromagnet is placed above the ferromagnetic mass. If AC current is applied to the electromagnet, the magnet repeats pulling and releasing the cantilever. If the cantilever is released, the bent cantilever returns to the initial position by its own spring force. One implied factor is that if the frequency of current applied to the electromagnet equals the resonance frequency of the cantilever, the actuation is amplified by the Q factor of the mechanical resonance. The resonant frequency f_r of the cantilever with mass can be varied due to the change of its mechanical boundary conditions. The feedback unit consists of three sub-units, namely the strain gauge, amplifier and voltage controlled oscillator (VCO). When the cantilever vibrates at different frequencies, the value at the strain gauge decreases. The drop in the value of the strain gauge triggers the VCO to adjust the frequency using linear feedback to find the new f_r (Son and Lal) (Fig. 2.9).

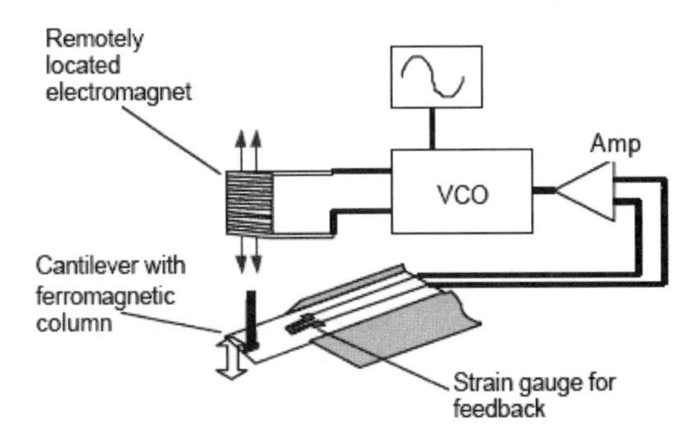

Fig. 2.9. Remotely actuated magnetic actuator (Source: Son and Lal 1999, 2000, 2001)

2.7 MOEMS

MOEMS stands for micro-opto-electromechanical systems. MOEMS is a miniaturised system combining optics, micromechanics and microelectronics, depending upon the application and device type, fabricated by the use of the collective techniques of microfabrication and micromachining (Fig. 2.10). MOEMS device technology requires a different set of rules for operation when compared with the normal MEMS devices. For example, tunable microlasers and optical switching based on MOEMS technology can improve the capability of wavelength division multiplexing (WDM) systems. MOEMS technology also manufactures structures in the micron to millimeter ranges.

Mostly, MOEMS have emerged to provide unparalleled functionality in telecommunication applications. Although the ultimate speed of these devices is unlikely to compete with solid-state electro-optic devices, the precision that can be achieved with MOEMS contributes to good performance and negligible signal degradation in the channels, thereby enabling a flexible all-optical system. System bandwidth and power consumption are the key issues. Using fiber links and optical methods, MOEMS technology demands precision interfaces and integrated components in order to achieve reliable and available quality of service (QoS). The advantages of MOEMS are:

- Miniaturisation
- Accuracy
- Insensitivity to electromagnetic interference
- Can be used in harsh environments
- Secured
- High sensitivity and selectivity
- Reasonably faster
- Reliability and availability
- Mass production capability hence low manufacturing costs

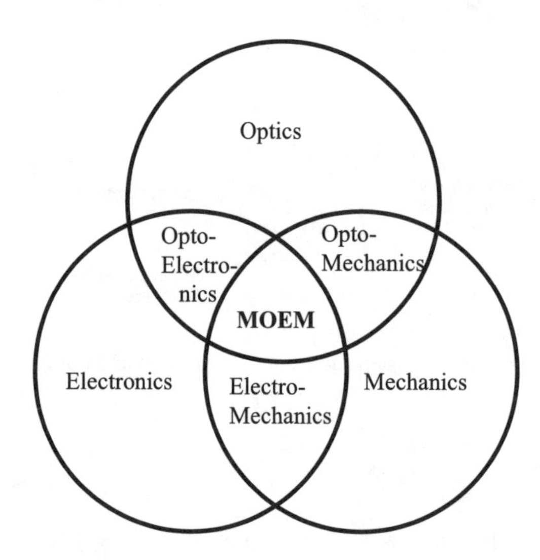

Fig. 2.10. Expression of hybrid systems

The applications of MOEMS are presented in Table 2.2. More detail is given below (http://www-leti.cea.fr/commun/europe/MOEMS/moems.htm),

- Arrays of micromirrors for digital image processing (projection and printing)
- Optical switches and routers, variable attenuators, shutters (optical communications)
- Tunable sources and filters, reflection modulators, spectral equalisers (optical communications, optical sensing, spectral analysis)
- Micro-scanners (image processing, bar code reading, obstacle detection)
- Deformable membranes for adaptive optics (astronomy, ophtalmology, FSO, defence)
- Free-space microoptics (diffractive and refractive microlenses)
- Guided optics devices
- Optical sources and photodetectors

Table 2.2. Applications of MOEMS

Broader applications	Specific applications	Communication applications
Biomedical	Scanning	Switches
Automotive	Projection	Variable attenuators
Industrial maintenance & control	Display	Equalisers
Domestics	Printing	Modulators
Space and astronomy	Sensing	Dispersion compensators
Environmental monitoring	Data storage Motors	Deformable micro mirrors
	Sensors Gyroscopes	Optical chopper
	Optical scanner	Digital or analog circuits
		Optical interconnect with micro-gratings

MOEMS devices are designed by adopting various manufacturing methods. All of these methods involve the deposition of layers of appropriate materials and etching away the sacrificial layers to produce the desired structures. One of the important combined methods used in developing the MOEMS devices is the LIGA process. LIGA is a German acronym for lithography, electroplating, and molding. The process states that each layer of the different material is deposited lithographically. These layers are different thicknesses and can overlap one another depending upon the MOEMS device being designed. One layer is usually used as a sacrificial layer to fill in and support a void area. The subsequent layer can be used to make a mold for the next layer. The succeeding layer will overlap the first and be molded into shape by the second (www.memsoptical.com/techinfo/memstut1.htm). Then the first two layers can be removed. This in turn leaves a freestanding structure.

2.8 Spatial Light Modulator

A Spatial Light Modulator (SLM) is a device that modulates light according to a fixed spatial pixel pattern. It consists of an array of electrostatic parallel plate actuators that are directly coupled with square mirror pixels by mechanical attachment posts. SLM plays a vital role in several projection and display areas where light is controlled on a pixel-by-pixel basis in order to achieve optimum system quality and performance. SLMs are primarily used to control incident light in amplitude-only, phase-only or in combination. The general idea of such devices is to modulate a constant source optical signal as a function of a modulating signal that may be optical, mechanical or electrical. The mechanical tilt of the electrostatic parallel plate actuator (integrated with mirror) takes place as a result of the electric field imposed by the potential differences between the addressable electrodes. The imposed electric field is derived from the modulating signal. The device can be fabricated using a three-layer, polysilicon, surface micromachining process. These silicon based tiltable mirrors have the potential to modulate the spatial and temporal features of constant source optical wavefronts, and have wider applications in imaging, beam-forming, and optical communication systems. SLMs are very fast, compact and reliable devices with excellent optical qualities.

2.9 Digital Micromirror Device

Digital Micromirror Device™ (DMD) is a registered trademark of Texas Instruments. Digital Light Processing™ (DLP) is another technology developed by Texas Instruments that is mainly used for display applications. It has made significant inroads in the projection and display market. DLP™ technology (Fig.2.11) is a powerful and flexible technology and has already taken momentum on data and video projectors, HDTVs, and digital cinema products. Note that at the heart of these display solutions is DMD™, a semiconductor-based *light switch array* of thousands of individually addressable and tiltable mirrors representing pixels.

DMD™ is considered to be a spatial light modulator (SLM), which reflects light that is produced externally. It is very attractive for many applications, including volumetric displays, holographic data storage, lithography, scientific instrumentation, and medical imaging, since the technology provides excellent resolution and brightness, high contrast and colour fidelity, and fast response times. These attributes are extremely important to many high-quality displays, printing and imaging applications. Liquid Crystal Displays (LCDs) are another familiar variety of SLMs.

In summary, it can be said that DMD™ is an array of semiconductor-based digital mirrors that precisely reflects a light source. Further, DMD™ enables DLP™ and displays images digitally. Rather than displaying digital signals as analog signals, DMD™ directs the digital signal directly to the screen.

The number of tiny mirrors depends on the size of the array. DMD™ are micromechanical silicon chips, which measure less than 5/8 of an inch on each side. It contains more than a quarter of a million tiny, movable aluminium mirrors and a wealth of electronic logic, memory and control circuitry. Each 16-μm^2

mirror of the DMD™ consists of three physical layers and in between them two air-gap layers so that the mirror can be tilted +10 or -10 degrees, when subjected to electrostatic voltage. That is, if voltage is applied to the addressable electrodes, the mirror can tilt ±10 degrees signifying ON or OFF in a digital fashion.

There is a light source (Fig. 2.11) and display (screen) unit within the projection system. The micromirrors are mounted in such a way that the tilts are either toward the light source (ON) or away from it (OFF), creating a light or dark pixel on the screen, respectively. The bit-streamed image code to be displayed first enters the semiconductor which then directs each mirror to switch ON and OFF up to several thousand times per second. When a mirror is switched ON more frequently than OFF, it reflects a light gray pixel and a mirror that is switched OFF more frequently, reflects a darker gray pixel.

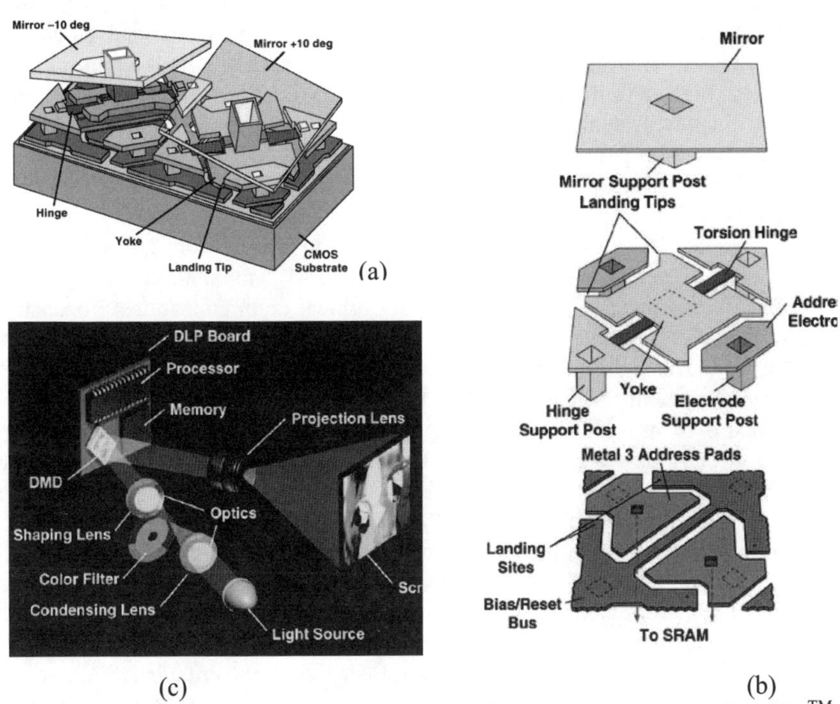

Fig. 2.11. (a) A section of DMS showing two mirrors: (b) The three layers of a DMD™ mirror; (c) The projection details of a DMD™ utilizing DLP™ technology (Courtesy: Texas Instruments Inc., Source:); http://electronics.howstuffworks. com/projection-tv5.htm, http://www.ti.com/sc/docs/products/dlp/spie-paper.pdf)

In practice, the incoming light hits the three mirror pixels. The outer mirrors that are turned ON reflect the light through the projection lens and this is passed on to the screen surface. The mirrors are responsible for producing pixel images and are geometrically square in nature. The central mirror is made to tilt so that the OFF position is achieved. This mirror is responsible for reflecting light away from the lens to a light absorber so no light can reach the screen at that particular pixel, producing a dark pixel. In the same way, the remaining mirror pixels reflect light to the screen or away from it, depending on the input coded signal that is fed to the DMD™ chip. By using a colour filter system and by varying the amount of ON time of each of the mirror pixels a full-colour digital picture can effectively be projected onto the display screen. In this way, the mirrors in a DLP™ projection system can reflect pixels, even up to 1,024 shades of gray, to convert the video signal entering the DMD™ device into a highly detailed grayscale image.

2.10 Grating Light Valve™ (GLV)

Silicon Light Machines™ and Grating Light Valve™ are trademarks of Silicon Light Machines. Silicon Light Machines (SLM™) has developed silicon chip based display architecture that can be used for many imaging applications including high-definition televisions, movie projection and high-performance desktop display systems. SLM™ was formed to commercialise a broad range of products based on patented Grating Light Valve (GLV™) technology. GLV™, an optical MEMS device built onto silicon wafers, defines itself as a tiny silicon machine, and is capable of modulating the light. After initially developing GLV™-based technology it was licensed to Sony for high-end display applications. Cypress Semiconductor acquired SLM™ in 2000.

SLM™ also focuses on components for optical communications. For instance, GLV™ devices have been used as a solution for switching, attenuating and modulating laser light. Additionally, the product includes computer-to-plate (CTP) printing equipment. When used in CTP products, this device diffracts high power laser beams to provide a high-speed, ultra-precise method of transferring digital images directly onto a printing plate. With an extremely fast switching speed and accurate diffraction control, the GLV™ device generates a writing beam on the surface of the plate enabling CTP technology to appear as products that can offer faster plate making performance and incredible image quality.

Like DMD™, GLV™ is also considered to be a digital technology. Here it is a reflective cum refractive technology in which ribbon like mirrors are physically moved to alter the path of light on the chip's surface. The difference with respect to DMD™ is that in GLV™ the pixels reflect light only in their OFF state. With regards to construction, a pixel in a GLV™ array consists of a number of aluminium-coated ribbons. The aluminium coating acts as the mirror. The dimension of the ribbon-structured mirror is about 100 μm long, 100nm thick and about 3 μm wide as shown in Fig. 2.12.

Structurally, GLV™ consists of pairs of fixed and movable ribbons located approximately a quarter wavelength above a silicon dioxide layer. The ribbons are suspended over a thin air gap. Each ribbon is pulled down a controlled distance into the air gap by means of an electrostatic voltage, called the driving voltage. The ribbons representing a pixel reflect light away from the projector's optical path. When an appropriate driving voltage is applied to alternate ribbons in a pixel, the corresponding mirror-ribbons are pulled down, forming a square-well diffraction grating. Each well is comparable with the wavelength of incident light. Light waves reflecting off adjacent up and down ribbons are now out of phase with each other. This interaction (formation of grating) causes the waves to build up in such a manner that each frequency of light will radiate at a different angle of radiation. By deliberately varying the width, separation and the amount of depth (well) of the ribbons in each pixel, in accordance with the modulating signal, the angle at which certain frequencies of the light are to be radiated can be controlled. This is turn implies that a single colour of light can be directed into the optical path. The switching between two states of a GLV^{TM} ribbon takes about 20 nanoseconds. Each pixel in the linear GLV^{TM} array is capable of reproducing precise grayscale values (Fig. 2.13) at the rate of millions of times per second, which is thousand times faster than any other light modulator technology.

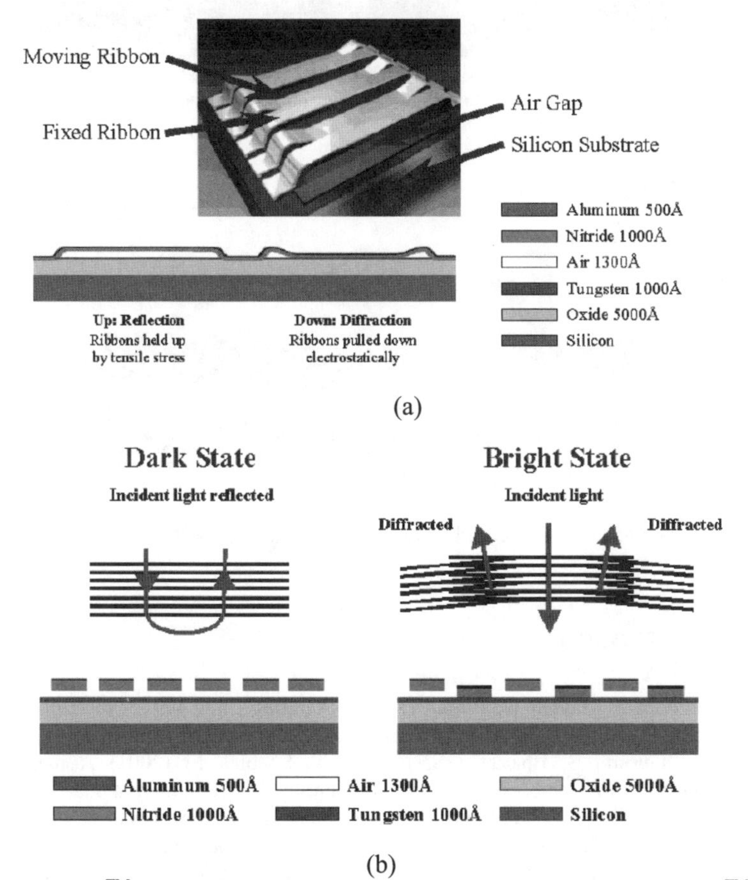

Fig. 2.12. A GLV^{TM} Valve consists of pairs of fixed and movable ribbons; (b) GLV^{TM} uses reflection and diffraction to create dark and bright image areas (Courtesy: Silicon Light Machines, Author: Bloom; Apte, 1994)

It has been reported that the response time of SLM^{TM} is faster than a DMD^{TM}'s mirror switches state. In an GLV^{TM} projector, each column of the picture is displayed sequentially, from left to right across the screen. This is done so quickly, that the device has time to display the same image three or four times during a conventional video frame. So visual resolution can be increased. The scanned GLV^{TM} architecture, which can cost-effectively create very high-resolution images, has been successfully demonstrated in a front-projection display system (http://www.siliconlight.com/htmlpgs/masterframeset/pressreleasepgs/pressrelease 3.html).

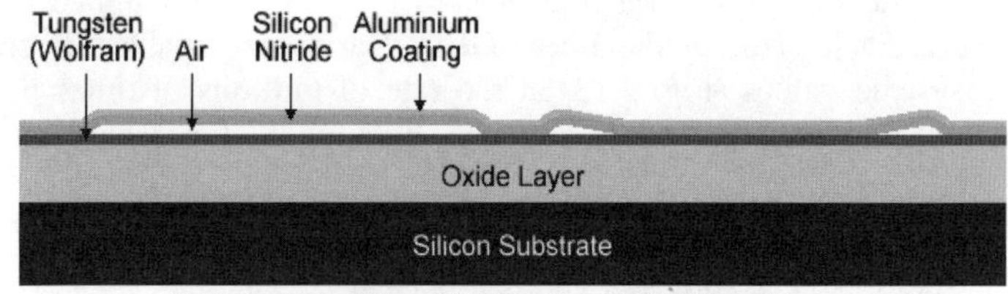

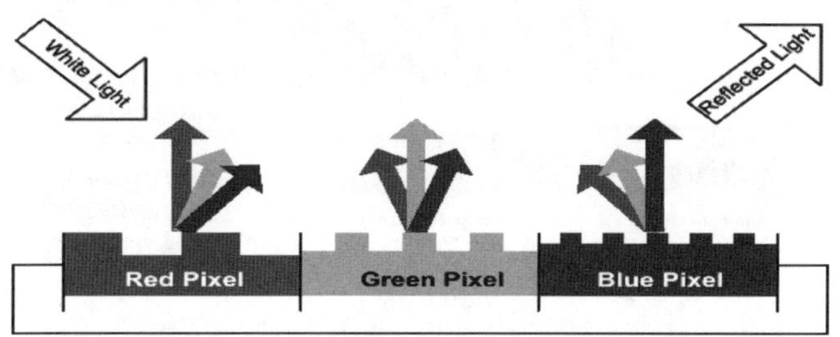

Fig. 2.13. Schematic of Grating Light Valve^{TM} technology (Courtesy: Meko Ltd.)

2.11 References

Apte R, Sandejas F, Banyai W, Bloom D (1994) Grating Light Valves for high-resolution displays. Proceedings, Solid State Sensors and Actuators Workshop

Arnold DP, Gururaj S, Bhardwaj S, Nishida T, Sheplak M (2001) A piezoresistive microphone for aeroacoustic measurements. ASME International Mechanical Engineering Congress and Exposition, November 11-16, New York

Acar C, Shkel AM (2004) Structural design and experimental characterisation of torsional micromachined gyroscopes with non-resonant drive mode, J of Micromechanics and Microengineering 14:15-25

Bernstein JJ, Dokmeci MR, Kirkos G, Osenar AB, Peanasky J, Pareek A (2004) MEMS tilt-mirror spatial light modulator for a dynamic spectral equaliser, J Microelectromechanical Systems 13:33-42

Cho HJ (2002) Micromachined Permanent Magnets And Their MEMS Applications, PhD thesis, University of Cincinnati, USA

Chui BW, Stowe TD, Ju YS, Goodson KE, Kenny TW, Mamin HJ, Terris BD, Ried RP, Rugar D (1998) Low-stiffness silicon cantilevers with integrated heaters and piezoresistive sensors for high-density AFM thermomechanical data storage, J of Microelectromechanical Systems 7:1

Damrongrit P, Rajamani R (2004) The development of a MEMS Gyroscope for absolute angle measurement, University of Minnesota, Minneapolis

Dana D, Duncan W, Slaughter J (2004) Emerging Digital Micromirror Device (DMD) applications, Texas Instruments, http://www.ti.com/sc/docs/products/dlp/spiepaper.pdf

DeVoe DL (2003) Thermal issues in MEMS and microscale systems, IEEE T on Components and packaging technologies 25:4

Haohua L, Boucinha M, Freitas PP, Gaspar J, Chu V, Conde JP (2002) MEMS microbridge deflection monitoring using integrated spin valve sensors and micromagnets, J Appl Phys 91:7774-7777

http://biz.yahoo.com/ic/105/105483.html

http://www.memsnet.org/mems/what-is.html

http://www.sensorland.com/HowPage023.html

http://www.siliconlight.com/htmlpgs/masterframeset/pressreleasepgs/pressrelease3.html

http://www.meko.co.uk/glv.shtml

http://www.coventor.com/media/papers/optical_MOEMS.pdf

Judy JW, Myung N (2004) Magnetic materials for MEMS, University of California, http://www.ee.ucla.edu/~jjudy/publications/conference/mrs_2001_judy_myung.pdf

King WP, Chaudhari AM (2004) Design of novel thermal MEMS, available at http://www.stanford.edu/group/microheat/seven.html

Kovacs GTA, Maluf NI, Peterson KE (1998) Bulk micromachining of silicon. IEEE Proceedings, 86, pp 86-88

Loom DM (2003) The Grating Light Valve: revolutionizing display technology. Technical Report, Silicon Light Machine, USA

Madou M (2004) Surface micromachining, available at http://mmadou.eng.uci.edu

Maloney J, DeVoe DL (2000) Electrothermal microactuators fabricated from deep reactive ion etching of single crystal silicon. ASME IMECE, pp 233–240

Panchapakesan B, DeVoe DL, Cavicchi R, Widmaeir M, Semancik S (2001) Nanoparticle engineering and control of tin oxide microstructures for chemical microsensor applications, Nanotechnology 12:336-349

Peter K et. al. (2003) MOEMS spatial light modulator development. In: James HS, Peter AK, Hubert KL, (eds), Proceedings of SPIE, vol. 4983, pp. 227-234

Piyabongkarn D, Rajamani R, Greminger M (2005) The Development of a MEMS Gyroscope for Absolute Angle Measurement, IEEE Trans. on Control Syst. Tech. 13:185-196

Rajamani R, Tan HS, Law B, Zhang WB (2000) Demonstration of integrated lateral and longitudinal control for the operation of automated vehicles in platoons, IEEE Trans. on Control Systems Technology 8:695-708

Robert D, Jami M (2004) Front side processing of a piezoelectric MEMS accelerometer. Technical Report, University of Notre Dame and Pennsylvania State University

Smith JH, Krulevitch PA, Lakner HK (eds) (2003) MOEMS and miniaturised systems. Proceedings of SPIE, pp 4983

Schreiber DS, Cheng WJ, Maloney JM, DeVoe DL (2001) Surface micromachined electrothermal microstepper motors, ASME IMECE
Semancik S, Cavicchi RE, Wheeler MC, Tiffany JE, Poirier GE, Walton RM, Suehle JS, Panchapakesan B, DeVoe DL (2001) Microhotplate platforms for chemical sensor research, Sensors and Actuators B77:579–91
Sakarya S, Vdovin G, Sarro PM (2004) Technology for integrated spatial light modulators based on viscoelastic layers, Delft University of Technology
Son S, Lal A (2004), A remotely actuated magnetic actuator for microsurgery with piezoresistive feedback, Univ. of Wisconsin, http://sonicmems.ece.cornell.edu/Papers/Sunny_ paper_mmb7.pdf
Son S, Lal A (1999) Silicon ultrasonic horn with integrated strain gauge, Ultrasonic International, Copenhagen
Son S, Lal A, Hubbard B, Olsen T (2001) A multifunctional silicon-based microscale surgical system, Sensors and Actuators A: Physical 91:351-356
Son S, Lal A (2000) Thin film composite for microsurgery. 22nd Int. conf. of the IEEE/EMBS, Chicago
Tang O, Xu B, Welch J, Castracane J (2003) Cantilever based biological/chemical detection sensor. Environmental quality systems. Technical Report, University at Albany
Taylor WP, Schneider M, Baltes H, Allena MG (1999) A NiFeMo electroplating bath for micromachined structures, Electrochemical and Solid-State Letters 2:24-626
Trimmer W, available at http://home.earthlink.net/~trimmerw/mems/SM_surface.html
Vittorio SA (2001) Microelectromechanical systems, Cambridge Scientific Abstract
Wen HH, Hsu TY, Tai YC (1997) A micromachined thin-film teflon electret microphone, Transducers 97, Int. Conf. on Solid-State Sensors and Actuators, vol. 2, pp 780-785
Wright JA, Tai YC, Chang SC (1997) A large-force, fully-integrated MEMS magnetic actuator, Technical Digest, Int. Conf. on Solid-State Sensors and Actuators, Transducers 97, Chicago, vol. 2, pp 793-796

3 Laser Technology in Micromanufacturing

Sungho Jeong

Department of Mechatronics, Gwangju Institute of Science and Technology, Republic of South Korea

3.1. Introduction

Laser technologies are utilised in various fields like micromanufacturing, instrumentation, imaging, medicine, communications, etc. Although each of these fields is as important as others, we will limit our discussion to the application of laser technologies for manufacturing only.

Application of lasers to micromanufacturing has several advantages such as noncontact processing, capability of remote processing, automation, no tool wear, and the possibility of machining hard and brittle materials. In this chapter, the principle of laser light generation, laser beam properties, characteristics of practical laser systems, and examples of application technologies are discussed.

3.2. Generation of Laser Light

The word 'LASER' is originally an acronym for '**L**ight **A**mplification by **S**timulated **E**mission of **R**adiation'. The generation of laser light is achieved by an amplification of light intensity through a physical process called 'stimulated emission'. Stimulated emission is a special form of light interaction with material in atomic scale.

Consider that a medium is irradiated with light, not necessarily a laser light but also sunlight or flash light. In a microscopic point of view, the interaction between the incident light and the medium can be described as an interaction between the light and the atoms consisting of the medium.

It is well known that atoms consist of a heavy, positively charged nucleus that takes most of the atomic mass and electrons of negligible mass orbiting the nucleus like the planets orbiting the sun. For the sake of simplicity, let us consider an isolated atom with only one electron orbiting the nucleus as shown in Fig.3.1(a), which is the Bohr model for a hydrogen atom (Wilson and Hawkes 1987; Milonni and Eberly 1988). The negatively charged electron will experience an attractive force Coulomb toward the positively charged nucleus, and a centrifugal force in the opposite direction due to its circling motion. Under thermal equilibrium, these two forces are balanced and the orbit along which the electron travels is determined. An atom at this condition is said to be at its ground state.

In the energy point of view, the atom possesses minimum available energy at the ground state. If external energy is supplied to a ground state atom, it will absorb the energy and reach a higher energy state, an excited state. Obviously, the radius of the electron orbit of an excited atom will be larger, (Fig. 3.1(b)), because the electron overcome attraction more easily. It is even possible for the electron to become free from the nucleus if the energy supply is sufficiently large.

However, there exists a rule that governs the excitation of an atom. This rule based on quantum mechanics, states that there exist discrete energy values permitted for an atom to occupy upon excitation such that $E_0<E_1<E_2<E_3$..., where E_0 denotes the ground state energy.

Therefore, when external energy is supplied to a ground state atom, it may be excited to a higher energy state of E_1 or E_2 but it cannot be excited to an intermediate level between E_0 and E_1 or E_1 and E_2. Specifically, absorption of external energy by an atom occurs only if the amount of supplied energy matches the energy gap $\Delta E_{ij}=E_j-E_i$ between two energy states E_i and E_j, assuming $E_i < E_j$. A more thorough discussion about the discrete energy states of atoms, molecules, and solids can be found in Milonni and Eberly (Milonni and Eberly 1988).

Let us now consider the excitation of a ground state atom by incident light as shown in Fig. 3.2(a). The amount of energy carried by a photon is equal to hf, where f is frequency of the incident light and h is the Plank's constant ($h=6.625\times10^{-34}$ J-sec). Note that 'photon' is a term describing the particle characteristics of light interacting with a medium and used to express the energy transport by light. According to above statement, excitation of the atom from E_i to E_j may occur only if $\Delta E_{ij}= hf$ is satisfied.

Otherwise, absorption of the photon would not happen. Because the energy gap ΔE_{ij} between any two states of a specific material can be known, it is possible to induce an absorption by irradiating the atom with a light of frequency f_{ij} for which,

$$\Delta E_{ij} = hf_{ij} = \frac{hc}{\lambda} \tag{3.1}$$

Where, the last term is from $c=\lambda f_{ij}$ describing the propagation speed of light in vacuum c as a function of frequency and wavelength λ.

This type of induced absorption of incident light by an atom is called 'stimulated absorption'. Although excitation of an atom is possible using stimulated absorption, the atom is not stable at the excited state and thus tends to return to the ground state without any external influence.

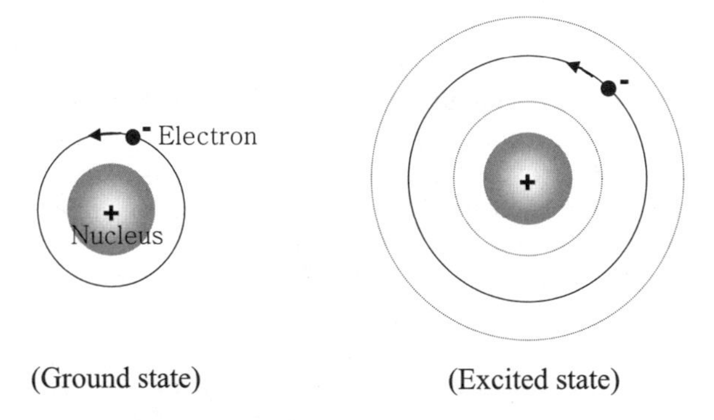

Fig. 3.1. Schematic diagram illustrating energy sates of an atom

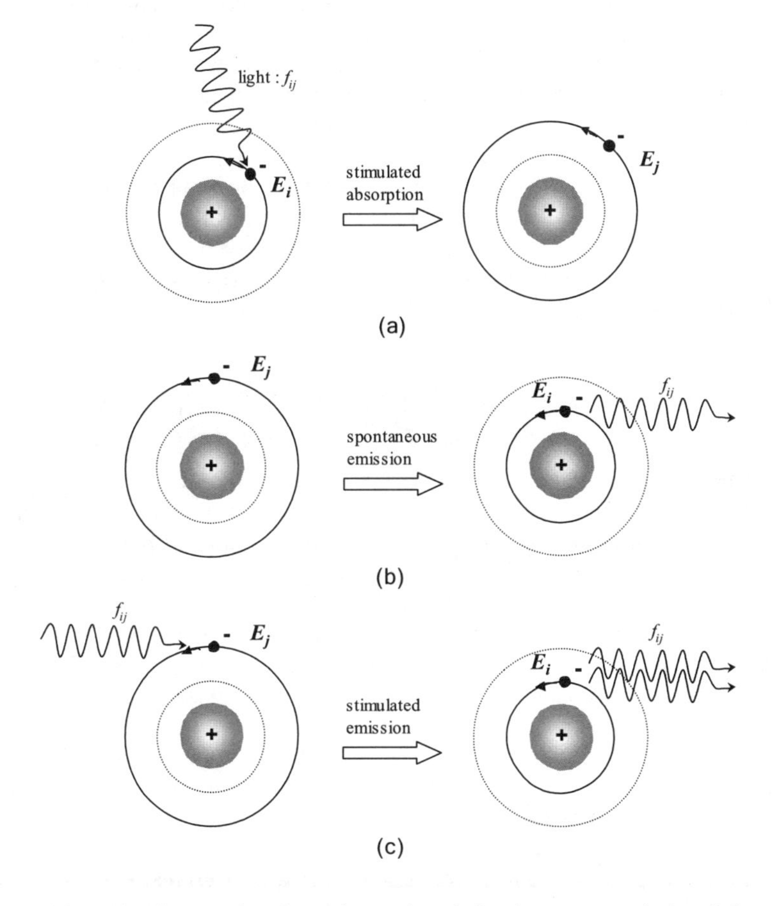

Fig. 3.2. Schematic diagram for absorption and emission by an atom during light-atom interaction

To return to the ground state, the atom needs to release excess energy, which can be done by transferring its energy to surrounding atoms through collision or by emitting light (Not to mention we are more interested in the emission of light). The duration that an atom remains at the excited state before falling to the ground state is called the lifetime and typically it is on the order of 10 nanoseconds (Hecht 1998). When the falling happens naturally and is accompanied by an emission, it is called the 'spontaneous emission', Fig.3.2(b).

Stimulated emission also represents the falling of an excited atom to a lower energy state with emission. However, in the stimulated emission, the falling is induced by an incident light of matching frequency, where matching frequency means that the incident light satisfies the relation in Eq. 3.1. Note that for the stimulated emission the lower state (E_i) does not have to be the ground state but can be any energy level below the excited state (E_j) as long as Eq. 3.1 is satisfied. The uniqueness of stimulated emission can be better understood in reference to Fig. 2(c). When a photon of frequency f_{ij} is incident on the excited atom, the atom is then stimulated not only to release its excess energy via emission but also in a fashion that the emitted photon has the same phase, polarization, and propagation direction as the incident photon. Therefore, an amplification of light intensity is achieved through the stimulated emission process as was symbolized by the word LASER.

Light amplification by stimulated emission can be achieved by shining light with proper frequency on excited atoms. However, under thermal equilibrium, most atoms are at the ground state rather than at excited states. The ratio of number of atoms between two energy states E_i and E_j ($E_i < E_j$) is expressed by the Boltzman equation,

$$\frac{N_j}{N_i} = \exp[-(E_j - E_i)/k_B T] \tag{3.2}$$

Where, N is the number of atoms (or population) at E, k_B is the Boltzman constant (1.3807×10^{-23} J/K), and T is the absolute temperature. Because the population at lower energy state is much greater than that at higher energy state, under equilibrium condition it is far more likely that stimulated absorption may take place upon light irradiation than stimulated emission. Hence, to obtain a laser light it is a prerequisite to produce a nonequilibrium condition at which more atoms exist at the excited (or upper) state rather than the lower state. This nonequilibrium distribution of atoms is known as population inversion. To see how to achieve a population inversion and thus laser light generation, let us consider the schematic diagram in Fig. 3.3. The core element of a laser is the active medium, the material undergoes stimulated absorption and emission processes for light generation, and it could be a solid, liquid or gas. Many lasers are named after the active medium; for example, ruby laser, Nd-YAG (Neodymium-doped Yttrium Aluminum Garnet) laser, CO_2 (carbon dioxide) laser, etc. Supply of external energy to the active medium, pumping, can be done electrically or optically. The schematic diagram in Fig. 3.3 shows an optical pumping in which a flashlamp generates a very bright and intense white light due to high voltage discharge. A portion of this white light whose frequency matches the absorption band is absorbed by the active medium by stimulated absorption, leading to a sharp increase of the population of excited atoms within the active medium.

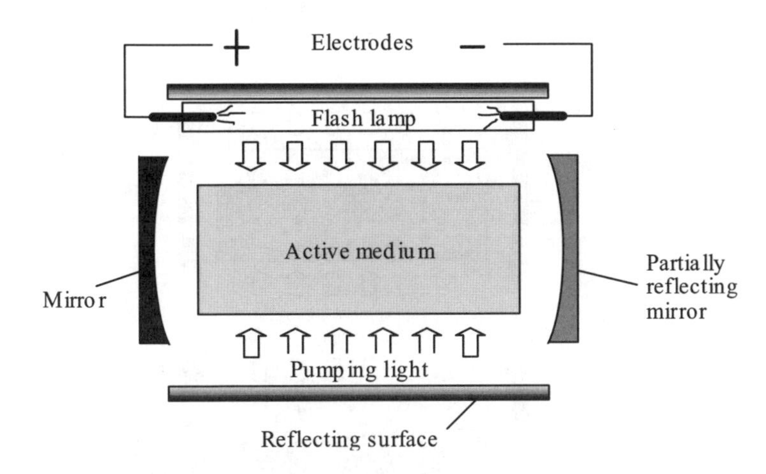

Fig. 3.3. Schematic diagram of an optically pumped laser

The excited atoms survive only for a brief moment of lifetime before falling to the ground state by spontaneous emission. Because the spontaneous emission is an uncontrolled process, the emitted light propagates in arbitrary directions (Fig. 3.4(a)). To produce a controlled and useful laser light, we need to make the emission occur into an intended direction, and this can be achieved by placing mirrors on both sides of the active medium as shown in Fig. 3.4(b).

The mirror set up like this is called an optical cavity or optical resonator. When a stray light from spontaneous emission happens to travel in the direction parallel to the mirror axis, along its course it may encounter other excited atoms and induce stimulated emission driving more photons travel in the same direction. When the photons arrive at one of the mirrors, they are reflected back into the optical cavity and amplified further. The more photons travel along the mirror axis, the stronger is the light amplification by stimulated emission. When the intensity of the laser light inside the optical cavity reaches sufficiently high level, it exits from the cavity through the partially reflecting mirror (Fig. 3.4(c)).

3.3 Properties of Laser Light

Laser light has many unique characteristics, which make lasers valuable and indispensable in many applications. The key properties of laser light include monochromacity, directionality, brightness, and coherence.

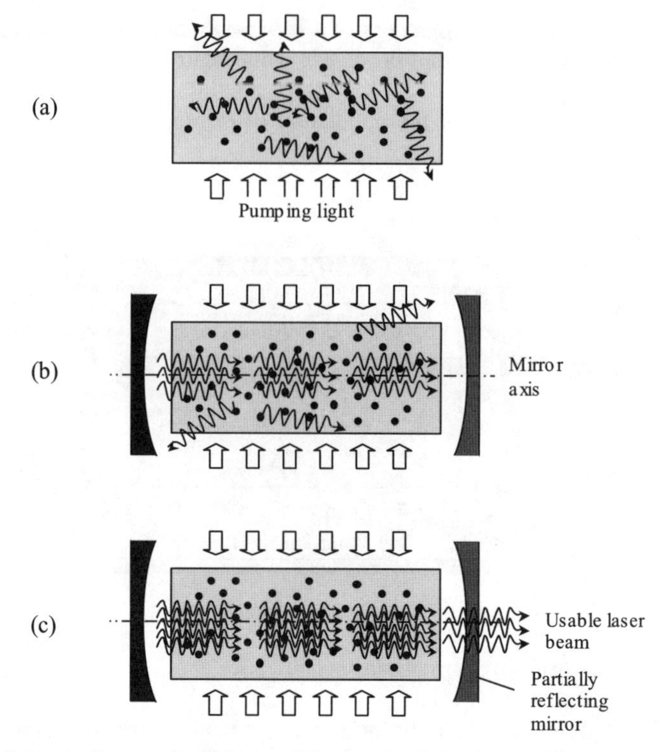

Fig. 3.4. Schematic diagram for light amplification through stimulated emission

3.3.1 Monochromacity

A monochromatic light means that all the photons have exactly one frequency. Therefore, in terms of color, a monochromatic light can be said a perfectly pure color. However, no light source is truly monochromatic but has finite band. Laser light is said to be quasi-monochromatic, meaning its linewidth is as narrow as possible. This narrow linewidth or purity of color is very useful in many application for instrumentation and micromanufacturing.

3.3.2 Directionality

Laser light propagates straight with very little divergence. This exceptional directionality of a laser beam originates from the design of optical cavity shown in Fig. 3.4. The divergence of a modern laser beam is less than 1 mrad. This small divergence benefits in many aspects for applications. For example, the laser energy is not lost during propagation and easy to focus onto a small area with a simple lens. By contrast conventional light diverges very rapidly and collection of it is very difficult.

3.3.3 Brightness

Brightness is defined as the power emitted per unit area per unit solid angle (the SI unit of brightness is W/m^2 sr). Because the divergence of a laser beam is much smaller than conventional sources, the small solid angle into which the laser beam is emitted ensures an extremely high brightness. It can be shown that a 3 mW He-Ne laser with a divergence angle of 1 mrad and beam diameter of 1 mm is 240 times brighter than the sun (Wilson and Hawkes 1987).

3.3.4 Coherence

Light is said to be coherent when the relative phase of all photons is the same. The time duration for which different photons maintain the same relative phase is defined as coherence time. Because the relative phase among photons can change due to path or material difference they travel through, coherence is an essential property for applications to instrumentation like interferometry. It is known that the coherence length, obtained by multiplying the coherence time by the speed of light, of a laser beam is as long as 3×10^5 m while that of white light is only about 3 μm (Das 1991)

3.3.5 Spatial Profile

Spatial profile defines how laser energy is distributed over the cross section of the beam. The most important and thoroughly analysed profile is the Gaussian (or TEM_{00}) mode beam. A Gaussian beam is axisymmetric and characterised by the following equation

$$I = I_0 \exp(-2r^2/w_0^2) \tag{3.3}$$

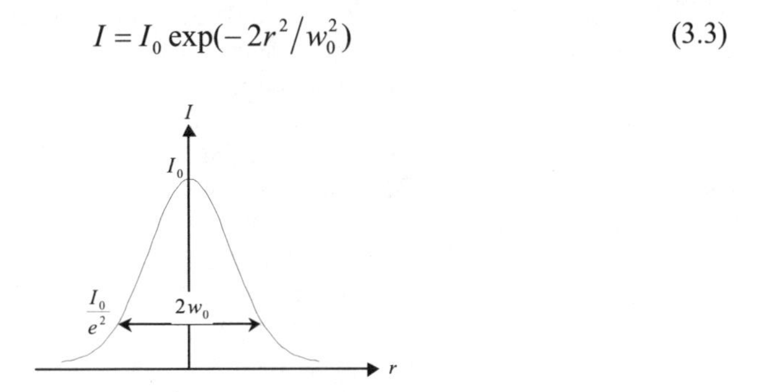

Fig. 3.5. Spatial energy distribution of a Gaussian laser beam

Where, I is irradiance (W/cm^2), I_0 is the irradiance at the center of the beam, r is radial coordinate, w_0 is the radius of the beam where I drops to I_0/e^2 (Fig.3.5). A full discussion of Gaussian beam can be found in references (Verdeyen 1995; Milonni and Eberly 1988)

3.3.6 Temporal Profile

Laser beam can be emitted from the laser continuously or intermittently. If the laser beam is continuously emitted, it is called continuous wave (CW) laser. When the laser beam is emitted with finite interval or in a single shot manner, it is called a pulse laser. The time duration for which pulse laser beam is being emitted is the pulse width or pulse duration. The term repetition rate represents how many pulses are emitted per second.

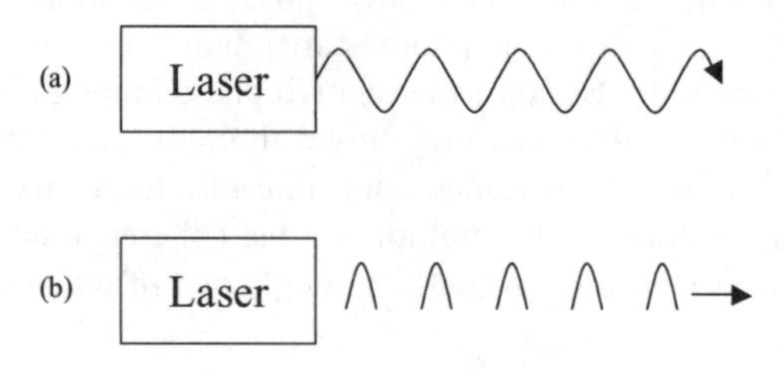

Fig. 3.6. Temporal mode of lasers: (a) CW laser, (b) pulse laser

3.4 Practical Lasers

Lasers are typically classified by active medium although classification by pulse width or other design characteristics is also possible. Nowadays, various types of lasers are commercially available and Table-3.1 shows the characteristics of most widely used lasers in industry and research. CO_2 laser is the most widely used laser in industry. Because of its unparalleled high output power (as high as 45,000 W (Ready 1997)), it has been the workhorse for conventional manufacturing such as cutting, welding, marking and drilling. Excimer laser is important for semiconductor processing. Because of the short wavelength, excimer lasers allow the fabrication of fine patterns required for high density circuit integration. Recent research reported that fabrication of a device with 50 nm feature size is viable by optical lithography with F_2 excimer lasers (Pong and Wong 2002). Semiconductor lasers (also called as diode lasers) are the most used lasers in quantity and shares the largest global market. Because of their small size, low price, lightweight, low power consumption, and the possibility of mass production, semiconductor lasers are widely used for instrumentation, telecommunications, compact disc players and data storage applications. Nd-YAG laser is another type of solid state laser with relatively high output power and wavelength (e.g., 1064 nm). Nd-YAG lasers are used as widely as CO_2 lasers in industry for the benefit of easy operation and low maintenance cost. Nd-YAG lasers produce output power as high as 2,400 W for continuous operation or up to 2 J for Q-switched pulse. Because Nd-YAG laser beam can be transmitted through an optical fiber, remote processing is possible with this laser, which is a great advantage for industry application such as automotive body welding. Using a technique known as harmonic generation, the Nd-YAG laser beam can be converted to a shorter wavelength of 532, 355, or 266 nm (Kuhn 1998).

Table 3.1. Properties of typical lasers (Ready 1997; Kuhn 1998; Carter et al. 2004)

Type	Laser	Active medium	Characteristics	Applications[3]
Gas	He-Ne	Ne	Power: 0.5-35 mW TEM_{00} mode, CW Low divergence Wavelength: 632.8 nm[1]	Interferometry Holography Alignment Velocimetry
	Ar	Ar^+	Power: upto 25 W TEM_{00} mode[2], CW Low divergence Wavelength: 514.5, 488 nm[1]	Measurements Microfabrication Entertainment Lithography
	CO_2	CO_2	Power: 100-10,000 W Multimode[2], CW & Pulse Divergence increases with output power, λ: 10.6 μm	Metal/Paper/Plastic Cutting, Drilling, Welding, Marking
	Excimer	Noble gas +Halogen gas	Energy[4]: 0.01-2 J Multimode, Pulse Medium divergence λ: 351, 308, 248, 193, 157 nm	Photolithography Micromachining Eye surgery
Solid State	Nd-YAG	Nd	Power: up to 2400 W TEM_{00} & Multimode, CW & pulse, Low divergence λ: 1064, 532, 355, 266	Welding, Marking, Micromachining, Drilling
	Ti-Saphire	Ti-Saphire	Energy: up to 2 mJ TEM_{00} mode Ultrashort pulse: 10-1000 fs [2]λ: 800 nm	Ultraprecision micromachining Nonlinear processing
	Semiconductor	Semiconducting compound	Power: 0.3-2.5 W Very small size: ~10 μm Large divergence: up to 30° [2]λ: 780-880, 1150-1650	CD & DVD player Barcode scanner Data storage Telecommunication
	Fiber	Rare earth material	Power: up to 100,000 W TEM^{00} mode High efficiency; λ: 350-2100	Telecommunication Welding, Marking Micromachining
Liquid	Dye	Organic dye	Energy[4]: up to 150 mJ Low divergence, λ: 370-900, wide range	Research Diagnostics

[1]Most common frequency, [2]Typical configuration, [3]Typical examples, [4]Pulse energy

Two other types of solid state lasers, which are relatively new but very important in emerging or future laser technologies are ultrashort pulse laser and fiber laser. Ultrashort pulse laser is a laser that outputs a pulse shorter than a picosecond (1 picosecond=10^{-12} second). Ti:Saphire laser is the most representative one in ultrashort pulse lasers and the pulsewidth ranges typically from 10's to 100's of femtoseconds (1 femtosecond=10^{-15} second). With this short pulse, extremely fine micromachining that cannot be realised with other method becomes possible (Ostendorf 2002). In a fiber laser, the optical fiber itself plays the role of an active medium and produces high quality laser beam. Fiber lasers have several advantages over other lasers in compactness, high amplification (or gain) efficiency, robustness, high thermal stability and simplicity of operation. (Schreiber et al. 2004). It is expected that fiber lasers may replace existing CO_2 or Nd-YAG lasers in many industrial applications.

3.5 Laser Technology in Micromanufacturing

Laser technology is widely applied to semiconductor processing, electronics packing, optical communication, medicine, and so on. Using lasers fabrication of microstructures on the order of micrometer (1 μm=10^{-6} m) or micromachining of hard materials like ceramic, glass, or stainless steel those are difficult to mechanically process can be easily achieved.

3.5.1 Background

The possibility of micromachining using laser beam is based on the fact that laser beam can be focused into a very small size. With high magnification objective lens, a spot diameter of 1-2 μm size is possible. Laser also plays a crucial role as an illumination source for photolithography in semiconductor manufacturing. Early 1990s, lasers have replaced halogen lamps in photolithography to reduce the critical dimension of the circuit patterns produced on silicon wafer by imaging a mask. The critical dimension in current photolithography is about 0.18 nm on the manufacturing line but the minimum pattern size of 70 nm is already achieved in research (Yim et al. 2003). All laser applications in micromanufacturing are in principle based on the absorption of the laser light by the workpiece, which leads to thermal or chemical changes in the workpiece. The development of a laser microprocessing technique and its proper application require understanding of thermal, physical, and chemical phenomena during laser beam interaction with materials and their variation with respect to the key laser parameters such as wavelength, pulse width, and energy.

3.5.2 Absorption and Reflection of Laser Light

Fig. 3.7. shows schematically the absorption and reflection of incident laser light by a solid. When the irradiance of the incident laser beam is I_0 and the reflectance of the solid surface is R, an amount of $R \cdot I_0$ is reflected at the surface while the rest is absorbed. The irradiance of the transmitted light decreases exponentially with distance in accordance with the following equation,

$$I(z) = (1 - R)I_0 \exp(-az) \tag{3.4}$$

Where, a is the absorption coefficient having a dimension of m^{-1}. If we take the inverse of the absorption coefficient, 1/a, it represents how far the light propagates before complete absorption, known as the optical penetration depth. The absorption coefficient is expressed in terms of the wavelength of the incident laser beam as,

$$a = \frac{4\pi\kappa}{\lambda} \tag{35}$$

Where, κ is the imaginary part of the index of refraction of an absorbing medium and termed as the extinction coefficient. From Eq.3.5 it is seen that the shorter the wavelength the shallow the penetration depth is. The reflectance and absorption coefficient of a medium varies significantly with wavelength and Fig. 3.8 shows theoretically determined properties of crystalline silicon, gold, and aluminum (von Allmen and Blatter 1995). In general, metals are opaque for laser light and thus light is completely absorbed within about 10 nm. The reflectance of typical metals like gold, aluminum, copper, etc. is very high for infrared light, close to 1, and thus direct processing of these metals with Nd-YAG or CO_2 laser is inefficient in energy point of view. In the case of silicon, which is the central material in semiconductor industry, infrared beam transmits over several hundreds of micrometers but ultraviolet beam is completely absorbed within about 10 nm thickness. Therefore, it is important to select the proper wavelength in accordance with the intended process.

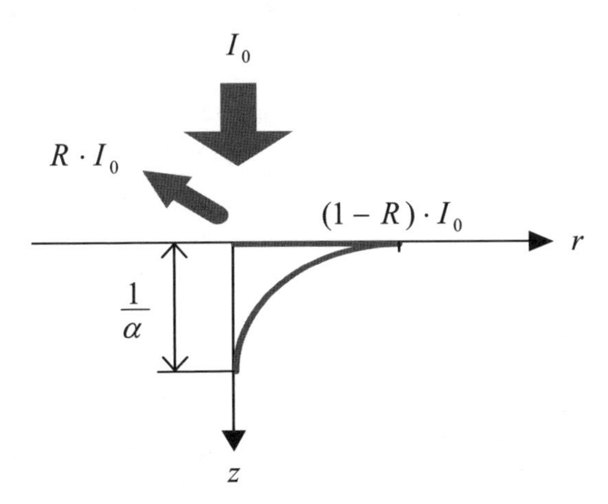

Fig. 3.7. Schematic diagram for reflection and absorption of laser energy by a solid

3.5.3 Application Technology Fundamentals

The spot diameter of a focused laser beam can be reduced to a few micrometers depending on lens' focal length, laser wavelength, and the diameter of unfocused beam. For Gaussian beam in air, the spot size d can be estimated using the following equation,

$$d = \frac{4\lambda f}{\pi D} \tag{3.6}$$

Where, λ is the wavelength, f is focal length of the lens, and D is diameter of the beam before focusing. For example, if a Nd-YAG laser beam (λ=1064 nm) with initial diameter of 2 mm is focused with a f=50 mm lens, the spot diameter will be about 34 μm.

Suppose that the laser output is from a pulse laser with pulsewidth of 5 ns and moderate pulse energy of 100 mJ. Then, the irradiance at the laser spot becomes 22×10^{15} W/m^2, which is overwhelmingly high value almost unachievable with other sources. Upon irradiation by a laser beam with this high energy density, almost all materials are immediately melted or vaporised.

This is how laser can effectively machine hard materials like diamond, glass, and ceramic. Laser irradiation of a medium may cause varying thermal effects from simple heating to melting, evaporation, or ionisation of the material. For high power laser processing, all of these phenomena occur almost simultaneous within the pulse duration, which makes the process very complicated and difficult to control. For applications like heat treatment of metal surface, annealing of polysilicon plates for liquid crystal display, texturing of computer hard disk, laser irradiation induces only heating and melting of the material with no evaporation. The substrate may undergo changes in mechanical and physical properties or surface morphology as a result of irradiation. For example, microbumps of 10-20 μm diameter and 40 nm height (Fig.3.9) can be fabricated on hard disk surface using laser melting to solve stiction problem between the hard disk surface and header (Baumgart et al. 1995).

For applications like laser trimming for fine adjustment of resistance of electronic components, drilling of printed circuit board, marking of electronic or medical components, marking of silicon wafer, fabrication of ink jet printer nozzle, and so on, laser beam removes material from the workpiece in the form of vapor, droplets or particles, or even in solid flakes. When evaporation occurs during laser microprocessing, temperature of the substrate can reach several thousand degree and the vapor pressure can be tens of atmosphere. For higher energy conditions, the vapor can absorb incident laser beam and be ionised causing significant secondary effects such as shielding of the sample surface from the laser beam. The high temperature and pressure during laser processing, and possible ionization of the vapor can lead to significant variation in the process and thus defining optimal conditions and precise control of them are crucial for real application.

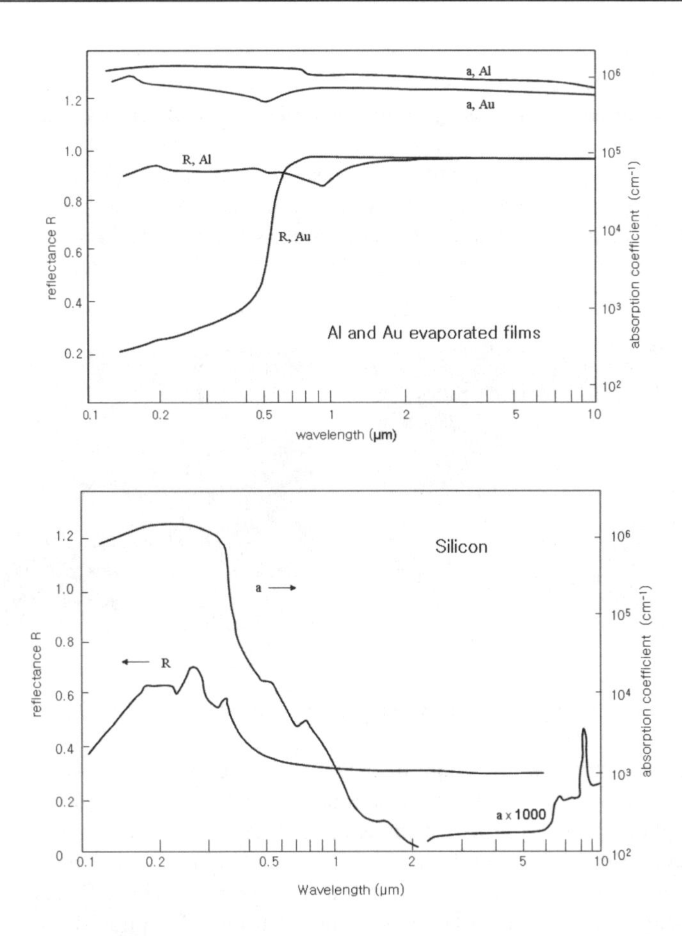

Fig. 3.8. Reflectance and absorption coefficient as a function of wavelength for different materials (von Allmen and Blatter 1995)

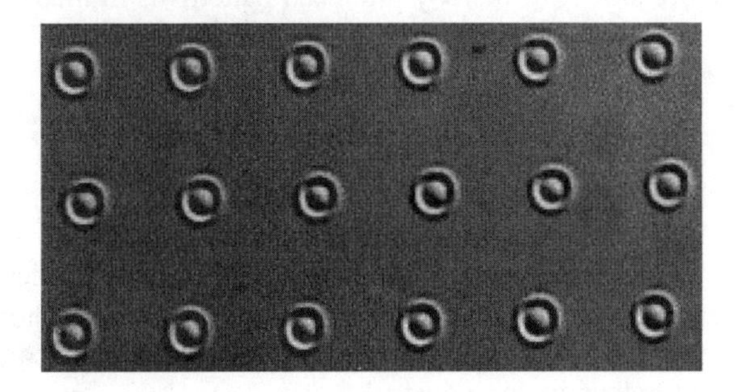

Fig. 3.9. Laser-textured microbumps on super-smooth NiP/Aluminum disk surface. The diameter and height of each bump are approximately 15 μm and 45 nm, respectively. (source: www.almaden.ibm.com)

The high vapor pressure developed during laser irradiation can be used for cleaning of contaminants on electronic components. Because no mechanical or chemical treatment is necessary, laser cleaning is free from possible mechanical damage and is environment friendly. Conventionally, laser beam is fired directly on the surface to be cleaned but in this case thermal damage of the sample may occur. Recent technique adopts a method to focus the laser beam in air above the surface and generate plasma in which cleaning is done by the high pressure of the expanding plasma. An example of plasma cleaning of intentionally contaminated solid surface is shown in Fig. 3.10.

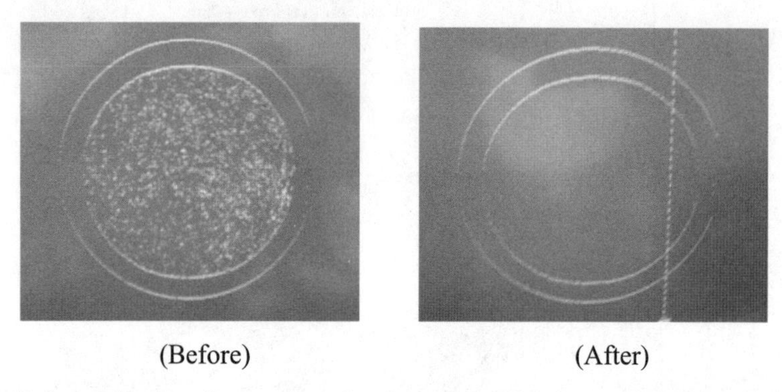

(Before) (After)

Fig. 3.10. Laser cleaning of contaminated surface. (Lee 2005, Courtesy: IMT Co. LTd.)

The heating and phase change of material during laser processing is basically thermal process. However, the precise micromachining technologies up to date is based on photochemical processes by excimer lasers whose output wavelength lies in the ultraviolet regime between 157-351 nm. The short wavelength of excimer laser beam has great importance in application point of view, especially for the photolithography in semiconductor manufacturing. In photolithography, the desired circuit pattern is first produced on a mask and this mask pattern is imaged onto a silicon wafer. Laser light is the illumination source during imaging and the minimum feature size R of thereby produced patterns is limited by diffraction by the following equation (Gower 1993),

$$R = 0.8\frac{\lambda}{NA} \tag{3.7}$$

Where, NA is the numerical aperture of the imaging optics. As it can be seen from Eq. (7), R decreases linearly with wavelength, which implies that the circuit density increases in proportion to $1/\lambda^2$. Actually, adopting a shorter wavelength excimer laser is the primary reason for the ever-increasing computer memory capacity in the semiconductor industry.

The short wavelength of excimer lasers is also a valuable property in direct micromachining. As explained earlier photon energy is inversely proportional to the wavelength as $E_p=hc/\lambda$. Photon energy of excimer lasers is much greater than that of CO_2 laser or Nd-YAG laser. When organic material like polymer is irradiated with excimer laser beam, the chemical bonding between atoms and molecules brake by incident photons and material removal in atomic or molecular form can take place. In this case, almost no heat affected zone is formed in the workpiece and micromachining with very sharp edge profile is possible (Fig. 3.11) (Gower 1999). Photon energy of various types of lasers and the bonding energies of representative chemical bonds are shown in Fig. 3.12 (Gower 1993).

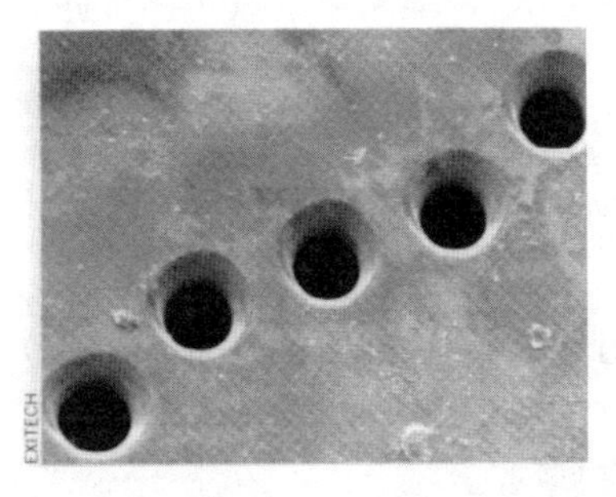

Fig. 3.11. Ink-jet printer nozzles drilled by an excimer laser. The diameter of the nozzles is about 30 μm and the substrate is polyimide (Gower 1999)

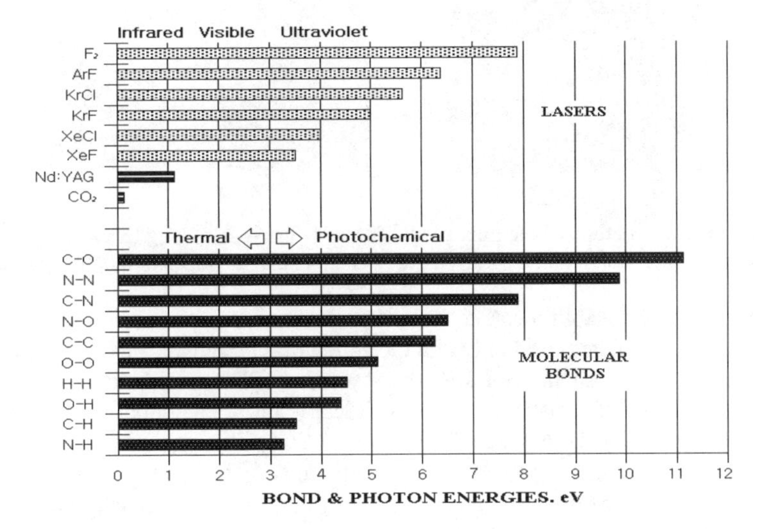

Fig. 3.12. Strengths of some common molecular chemical bonds compared with excimer laser photon energies (Gower 1993)

Since the late 1990's, there has been tremendous research and development in ultrafast laser technologies, especially for femtosecond lasers. Pulsewidth of femtosecond lasers ranges typically 10's to 100-200 fs although shorter or longer pulses are also available. To see how short this time is, let us calculate the distance light propagates within 100 fs as $L = c \times \Delta t = (3 \times 10^{14}\ \mu m/s) \times (100 \times 10^{-15}\ s) = 30\ \mu m$.

This result shows that light travels a distance equivalent only about half of a hair thickness during 100 fs. This short pulse of femtosecond lasers has significant impacts on material processing in the following aspects. First, ultrashort laser pulse easily leads to extremely high irradiance like 10^{15}-10^{20} W/m^2. At this high energy density, many nonlinear optical phenomena occur. Second, because of the ultrashort pulse, material removal occurs in thermally nonequilibrium conditions. The result of this nonequilibrium processing is a sharp machining edge with almost no heat-affected zone on the workpiece, even for metallic samples. An excellent explanation of ultrafast phenomena can be found in Craig (Craig 1998). Fig. 3.13(a) shows a striking example of femtosecond laser microfabrication in which a realistic image of a bull is manufactured in the size of a red blood cell using two photon absorption principle (Kawata et al. 2001). Other results of femtosecond laser micromachining superior to long pulse laser fabrication or non-laser methods have been demonstrated for various materials including metals (Tonshoff et al. 2000) and glasses (Minoshima et al. 2002).

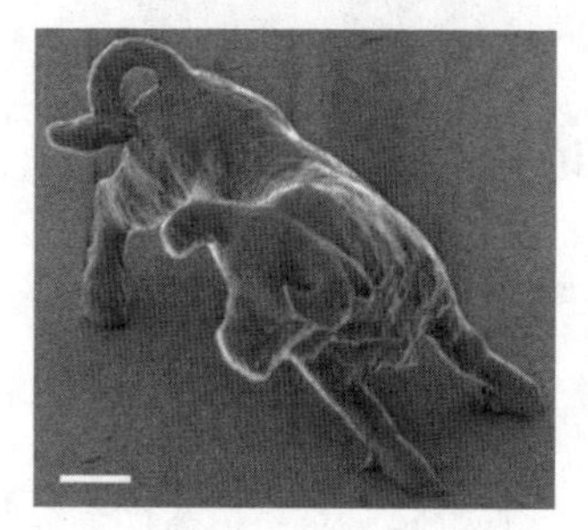

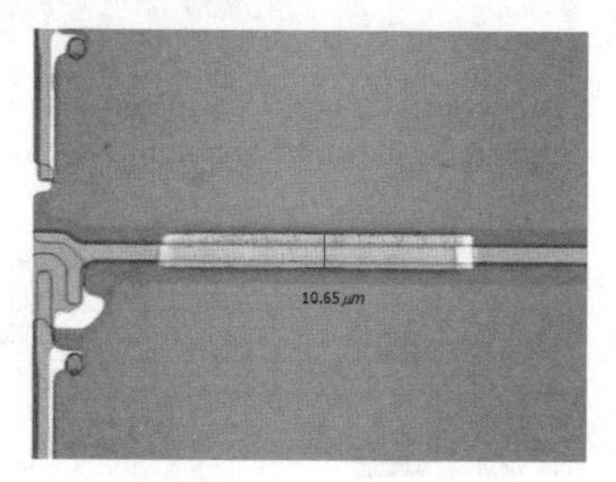

Fig.13.(a) A micrograph of the micro-bull. The scale bar is 2μm long (Kawata et al. 2001); (b) Example of circuit repair by laser-induced chemical vapor deposition on LCD glass

The high focusability of a laser beam can also be an advantage when micromachining is required only at a local area of a large workpiece. For example, flat panel displays such as plasma display panel (PDP) or liquid crystal display (LCD) are core products in electronics industry and their global market expands rapidly. Due to the large size of the PDP or LCD glasses, currently as large as 1.2×1.8 m^2, it is crucial to manufacture the circuits on these glasses free of defects because a discard of the large glass will result in a significant loss of productivity. Should there be a local defect on these glasses, it is best to fix the damage instead of discarding the entire glass. Laser can be effectively used for this type of local repair of electronic circuits. Fig. 3.13 (b) shows an example of this type of repair of an open circuit on LCD glass using laser-induce chemical vapor deposition technique (Han and Jeong 2004). A metal interconnect with a width about 10 μm is locally deposited without influencing the surrounding circuit elements. Besides deposition, laser-induced chemical etching is also possible for drilling and cutting (Shin and Jeong 2003) or fabrication of three-dimensional microstructures (Bauerle 2000)

3.6 References

Bauerle D (2000) Laser Processing and Chemistry. 3rd edn. Springer

Carter A, Tankala K, Samson B, Machewirth D, Khitrov V, Manyam U (2004) Continued advances in the designs of double clad fibers for use in high output power fiber lasers and amplifiers. Int. Symp, Laser Precision Microfabrication, Japan, SPIE Proc. 131

Craig B (1998) Ultrafast pulses promise better processing of fine structures, Laser Focus World 9:79

Das P (1991) Lasers and Optical Engineering, Springer

Gower MC (1993) Excimer lasers: current and future applications in industry and medicine in laser processing in manufacturing. In: Crafer RC, Oakley PJ (eds.), Chapman & Hall

Gower MC (1999) Excimers tackle micromachining, Industrial Laser Solutions for Manufacturing 4:14

Han SI, Jeong SH (2004) Laser-assisted chemical vapor deposition to directly write three-dimensional microstructures, J of Laser Applications 16:154-160

Hecht E (1998) Optics, 3rd edn. Addison-Wesley

Kawata S, Sun HB, Tanaka T, Takada K (2001) Finer features for functional microdevices, Nature 412, 697

Kuhn KJ (1998) Laser Engineering, Prentice

Lee JM (2004) Personal communication. Technical Report, IMT Co. LTd.

Milonni PW, Eberly JH (1988) Lasers, Wiley

Minoshima K, Kowalevicz AM, Ippen EP, Fujimoto JG (2002) Fabrication of coupled mode photonic devices in glass by nonlinear femtosecond laser materials processing, Optics Express 10:645-649

Ostendorf A (2002) Precise sturucturing using femtosecond lasers, Review of Laser Engineering 30:221

Pong WT, Wong A (2002) Feasibility of 50-nm device manufacture by 157-nm optical lithography: An initial assessment. Proceedings of IEEE conference, Electron devices meeting, Hong Kong, p 31

Ready JF (1997) Industrial Applications of Lasers, 2nd edn. Academic Press

Schreiber T, Limpert J, Liem A, Roser F, Nolte S, Zellmer H, Tunnermann A (2004) High power photonic crystal fiber laser systems. Proc., Int. Conf. on Transparent Optical Networks 1, Wroclaw, pp 131-135

Shin YS, Jeong SH (2003) Laser-assisted etching of titanium foil in phosphoric acid for direct fabrication of microstructures, J of Laser Applications 15:240-244

Tonshoff HK, Momma C, Ostendorf A, Nolte S, Kamlage G (2000) Microdrilling of metals with ultrashort laser pulses, J of Laser Applications 12:23-30

Verdeyen JT (1995) Laser Electronics. 3rd edn. Prentice

Allmen vM, Blatter A (1995) Laser-beam interactions with materials, Physical principles and applications. 2nd edn Springer

Wilson J, Hawkes JFB (1987) Lasers principles and applications. Prentice

Yim YS, Shin KS, Hur SH, Lee JD, Baik IG, Kim HS, Chai SJ, Choi EY, Park MC, Eun DS, Lee SB, Lim HJ, Youn SP, Lee SH, Kim TJ, Kim HS, Park KC, Kim KN (2003) 70 nm NAND flash technology with 0.025μm^2 cell size for 4 Gb flash memory, Technical Digest of IEEE International Electron Devices Meeting, IEDM, USA, vol. 34, pp 1-5

4 Soft Geometrical Error Compensation Methods Using Laser Interferometer

K. K. Tan and S. N. Huang

National University of Singapore, Singapore

4.1 Introduction

Micromachining is a key technology in the important fields of microimprinting, scanning microscopy, microlithography and automated alignment. A large proportion of these machines require accurate positioning of tools or probes with respect to a workpiece. Thus, much challenges behind the metrology is concerned with the accurate measurement of absolute positions and the subsequent reduction of the geometrical errors associated with this position. Compensation for geometrical errors in machines has been applied to Co-ordinate Measuring Machines (CMMs) and machine tools to minimise the relative position errors between the end-effector of the machine and the workpiece (Hocken 1980). Given an adequate machine design, a large proportion of these errors are completely repeatable and reproducible, and therefore amenable to modelling and compensation. While widespread incorporation of error compensation in machine tools has remained to be seen, the application in CMMs is tremendous and today, it is difficult to find a CMM manufacturer who does not use error compensation in one form or another (Hocken 1980). The development in error compensation is well documented by Evans (Evans 1989). Early compensation methods utilised mechanical correctors, in the form of leadscrew correctors, cams and reference straightedges. Maudslay and Donking, for example, used leadscrew correction to compensate for the errors in their scales producing machine. Compensation via mechanical correction, however, inevitably increases the complexity of the physical machine. Furthermore, mechanical corrections rapidly cease to be effective due to mechanical wear and tear. The corrective components have to be serviced or replaced on a regular basis, all of which contribute to higher machine downtime and costs. The evolution of control systems from mechanical and pneumatic-based subsystems to micromachining-based systems has opened up a wide range of new and exciting possibilities. Many operations which used to be the result of complex linkages of levers, cams, bailing wires and optical sensors can now be carried out efficiently with program codes residing in the memory of a standard computer. Soft compensation schemes thus blossomed in the 1970s. The first implementation was on a Moore N.5 CMM, a pioneering piece of work which earned Hocken the CIRP Taylor Medal in 1977. Since then, there has been an explosion of interests in soft compensation of CMM and micromachining system errors with new methods developed and implemented (Hocken et. al. 1977; Love and Scarr 1973; Kunzmann and Waldele 1984). Common to all these works and more is a model of the machine errors, which is either implicitly or explicitly used in the compensator. A common issue with these approaches is the challenge to have an adequately accurate model, which is also amenable to practical use.

4.2 Overview of Geometrical Error Calibration

Error modeling typically begins with a calibration of the errors at selected points within the operational space of the machine. For a 3D working space, the resultant geometrical errors in positioning may be decomposed into 21 underlying components. For the XY table with zero tool offsets, the error sources reduce to 7 components, including two linear errors, two angular errors, two straightness errors and the orthogonal error between the X- and Y-axis. These errors may be measured accurately using an independent metrology system such as a laser interferometer which can typically measure linear displacement to an accuracy of 1nm and angular displacement to an accuracy of 0.002 arcsec. These errors are subsequently cumulated using the overall error model to yield the overall positional error. A 3D machine has a volumetric error model described by (Zhang et al. 1985),

$$\vec{P} = \vec{X} + R^{-1}(X)\vec{Y} + R^{-1}(X)R^{-1}(Y)\vec{Z} + [R^{-1}(X)R^{-1}(Y)R^{-1}(Z)]\vec{T} \quad (4.1)$$

Where, $\vec{P}$ is the position vector of a reference point on the carriage of the machine, $\vec{X}, \vec{Y}$ and $\vec{Z}$ represent the translation along the respective axis. $R^{-1}(X), R^{-1}(Y)$ and $R^{-1}(Z)$ are the matrices representing the rotational motion about the respective axis. The offset of the tool (probe) tip with respect to the reference point is represented by $\vec{T}$. These vectors can be more explicitly expressed in terms of the various error components as,

$$\vec{X} = \begin{bmatrix} x + \delta_x(x) \\ \delta_y(x) \\ \delta_z(x) \end{bmatrix} \quad (4.2)$$

$$\vec{Y} = \begin{bmatrix} \delta_x(y) - \alpha_{xy} y \\ y + \delta_y(y) \\ \delta_z(y) \end{bmatrix} \tag{4.3}$$

$$\vec{Z} = \begin{bmatrix} \delta_x(z) - \alpha_{zx} z \\ \delta_y(z) - \alpha_{yz} z \\ z + \delta_z(z) \end{bmatrix} \tag{4.4}$$

$$\vec{T} = \begin{bmatrix} x_t \\ y_t \\ z_t \end{bmatrix} \qquad \text{4.5)}$$

$$R(u) = \begin{bmatrix} 1 & \varepsilon_z(u) & -\varepsilon_y(u) \\ -\varepsilon_z(u) & 1 & \varepsilon_x(u) \\ \varepsilon_y(u) & -\varepsilon_x(u) & 1 \end{bmatrix} \tag{4.6}$$

Where, x, y and z are the nominal positions. x_t, y_t, z_t represent the offsets of the tool tip; $\delta_u(v)$ is the translational error along the u-direction under motion in the v direction; $\varepsilon_u(v)$ is the rotation about the u axis under motion in the v direction (counterclockwise); and $\alpha_{xy}, \alpha_{yz}, \alpha_{zx}$ represent the out-of-squareness error in the XY, YZ and ZX planes respectively. Substituting Eq.4.1 with Eq.4.2-Eq.4.6, the actual displacement $\overline{x}, \overline{y}, \overline{z}$ for each axis can be expressed in terms of the individual error components.

$$\begin{gathered} \Delta x = \delta_x(x) + \delta_x(y) + \delta_x(z) - y\alpha_{xy} - z\alpha_{zx} - y\varepsilon_z(x) \\ + z[\varepsilon_y(x) + \varepsilon_y(y)] - y_t[\varepsilon_z(x) + \varepsilon_z(y) + \varepsilon_z(z)] + z_t[\varepsilon_y(x) + \varepsilon_y(y) + \varepsilon_y(z)] + x_t \end{gathered}$$

$$\begin{gathered} \Delta y = \delta_y(x) + \delta_y(y) + \delta_y(z) - z\alpha_{yz} \\ - z[\varepsilon_x(x) + \varepsilon_x(y)] + x_t[\varepsilon_z(x) + \varepsilon_z(y) + \varepsilon_z(z)] - z_t[\varepsilon_x(x) + \varepsilon_x(y) + \varepsilon_x(z)] + y_t \end{gathered}$$

$$\begin{gathered} \Delta z = \delta_z(x) + \delta_z(y) + \delta_z(z) + y\varepsilon_x(x) \\ - x_t[\varepsilon_y(x) + \varepsilon_y(y) + \varepsilon_y(z)] + y_t[\varepsilon_x(x) + \varepsilon_x(y) + \varepsilon_x(z)] + z_t \end{gathered}$$

4.2.1 Error Measurement System

For all these 21 sources of error components laser interferometer and electronic instruments can be used for the measurement. The laser measurement system facilitates machine tool manufacturers and users with the basic components necessary to measure these errors. The basic system includes a laser source, compensation electronics, optics, cables and accessories. A standard PC platform is usually necessary to control the system, and specific optic kits are required to make specific measurements of the machine geometrical properties. The complete system is able to collect and analyse different measurement data for calibration purposes. The resolution of laser measurement can reach as high as 1 nm for linear displacement and 0.002 arcsec for angular displacement. This high precision is achievable as the measurement process uses the precise wavelength of laser as a basis for the computation of distance measurement, thus achieving higher accuracy compared to other measurement systems. A two-frequency laser technique is also frequently incorporated to eliminate the problems resulting from changes in beam intensity, thus achieving better robustness and reliability. All the errors are measured by using the HP5529A laser measurement system manufactured by Hewlett Packard (HP). For illustration, a complete setup for a typical linear measurement is shown in Fig. 4.1.

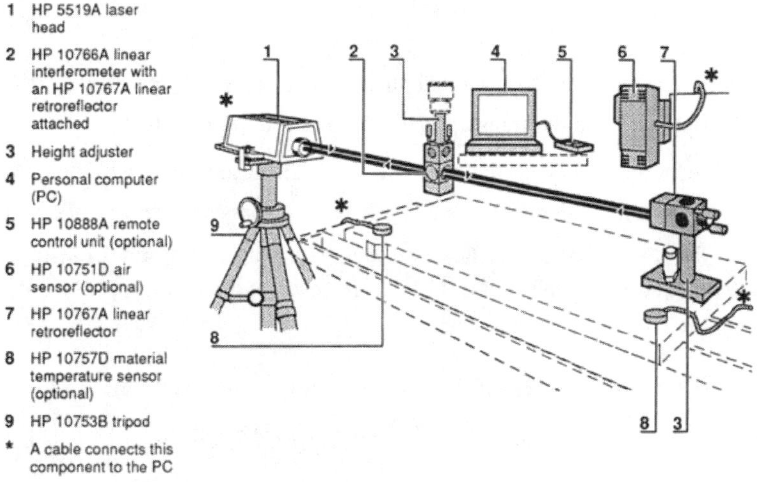

1 HP 5519A laser head
2 HP 10766A linear interferometer with an HP 10767A linear retroreflector attached
3 Height adjuster
4 Personal computer (PC)
5 HP 10888A remote control unit (optional)
6 HP 10751D air sensor (optional)
7 HP 10767A linear retroreflector
8 HP 10757D material temperature sensor (optional)
9 HP 10753B tripod
* A cable connects this component to the PC

Fig. 4.1. Setup for a typical linear measurement

An electronic measurement system can be employed to measure the roll error. The principles of operation are straightforward: it make use of a pendulum in conjunction with an electronic detection system to precisely sense the attitude of the pendulum with respect to the reference. The equipment consists of two components, namely the level unit, which is to be secured onto the moving part, and the display unit, which shows the angular deviation. An accuracy of 0.2 arcsec can readily be achieved by commercial level measurement systems. For a roll measurement, the level sensor will be fixed onto the moving part. When the axis moves to a designated position, the swing in angle on the pendulum will be reflected via the display. There is usually a PC interface provided to acquire the measurements into a PC (Fig. 4.2).

4.2.2 Accuracy Assessment

The accuracy of a machine is usually measured according to prescribed procedures (e.g., British Standards 1989). By making some assumptions for the machine of concern, a first rough assessment of achievable accuracy can be established. The following assumptions are made. Assumption-1: Repeatability error is zero; Assumption 2: The machine is calibrated at points separated by one resolution.

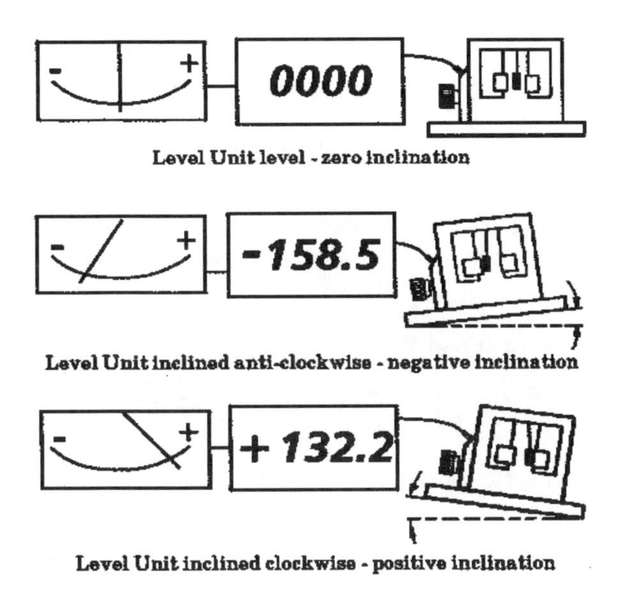

Fig. 4.2. Operation of a level-sensitive device

Consider an error model of the above 3D machine. The accuracy can be obtained by using the following formula $\Delta d = \sqrt{\Delta x^2 + \Delta y^2 + \Delta z^2}$. Suppose, $|x| \le 400, |y| \le 250, |z| \le 50$. Using a laser interferometer measurement system, linear and straightness errors can be measured accurate to 1 nm resolution, and the angular error (pitch, yaw) and squareness can be measured accurate to 0.002 arcsec resolution. Roll error cannot be directly measured using a laser interferometer. Instead an electronic sensor may be used which is accurate only to 0.2 arcsec. However, it may be possible to carry out interpolation to reach 0.002

arcsec. The system errors may be estimated over the entire working volume, assuming $x_t = y_t = z_t = 0$.

$$\Delta x \le 1+1+1+2.5\times10^8\times9.76\times10^{-9}+0.5\times10^8\times9.76\times10^{-9}$$
$$+2.5\times10^8\times9.76\times10^{-9}+0.5\times10^8(9.76\times10^{-9}+9.76\times10^{-9})$$
$$=9.34\text{ nm}$$

$$\Delta y \le 1+1+1+0.5\times10^8\times9.76\times10^{-9}+0.5\times10^8(9.76\times10^{-9}+9.76\times10^{-9})$$
$$=4.46\text{ nm}$$

$$\Delta z \le 1+1+1+2.5\times10^8\times9.76\times10^{-9}$$
$$=5.44\text{ nm}$$

The absolute diagonal error is given by: $\Delta d = \sqrt{\Delta x^2+\Delta y^2+\Delta z^2} \le 11.69\text{ nm}$. Without interpolations of roll measurements the errors increase to,

$$|\Delta x| \le 57.66\text{ nm};\ |\Delta y| \le 52.78\text{ nm};\ |\Delta z| \le 247.00\text{ nm};\ |d| \le 259.07\text{ nm}$$

If the repeatability of the linear and straightness errors is λ nm, and angular error is φ times the above angular resolution, then the errors over working volume are given by,

$$|\Delta x| \le 3\lambda+6.34\varphi;\ |\Delta y| \le 3\lambda+1.46\varphi;\ |\Delta z| \le 3\lambda+2.44\varphi$$
$$|d| \le \sqrt{27\lambda^2+48.28\varphi^2+61.44\lambda\varphi}$$

Using mechanical specifications of λ = 2000 nm, φ = 100 (repeatability is 0.2 arcsec), it follows that $|\Delta d| \le 10.99\,\mu\text{m}$. This is a conservative estimate since the error components are taken to cumulate in the worst possible manner.

4.3 Geometrical Error Compensation Schemes

Out of many types of error compensation schemes soft compensation method has been gaining momentum since two decades. However, because of the advent of new design scenarios this method is again constantly emerging. Soft compensation techniques will require that the error descriptions are stored in the computer memory. To reduce the number of points where errors must be measured, it is necessary to have a mathematical mapping to model the error behavior.

4.3.1 Look-up Table for Geometrical Errors

Traditionally, the geometrical errors to be compensated are stored in the form of a look-up table. The look-up table is built based on points collected and calibrated in the operational working space of the machine. When linear interpolation is used between the points for which errors are recorded, the data for only the six adjacent points (for 3D calibration) are recovered and interpolated.

A 1D error compensation is illustrated in Fig.4.3. Assume an axis is calibrated with a laser measurement system at equally spaced points according to encoder feedback. Denote x_1, x_2,…,x_n as the encoder measurements and e_1, e_2,…,e_n as the corresponding positioning errors derived from the laser measurement system. For a certain point x between x_{i-1} and x_i, the associated error e can be estimated via a linear interpolation process as,

$$e = (x - x_1)\frac{e_2 - e_1}{x_2 - x_1} + e_1$$

Many servo motion controllers will allow for geometrical error compensation via look-up tables. For example, the Programmable Multi-axis Controller (PMAC) from Delta Tau Data Systems Inc. is a family of high-performance servo motion controllers, which is capable of performing the look-up table compensation. PMAC has sufficient capacity to store up to eight of these compensation tables. For 2D compensation using PMAC, the amount of servo compensation for either motor will depend on the position of both motors (Fig. 4.4).

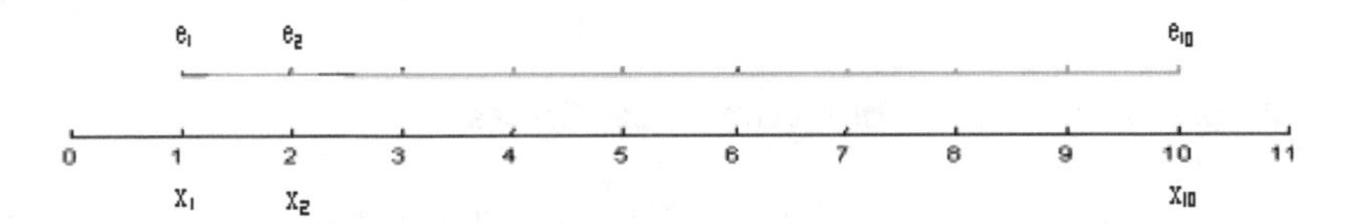

Fig. 4.3. 1D error compensation table

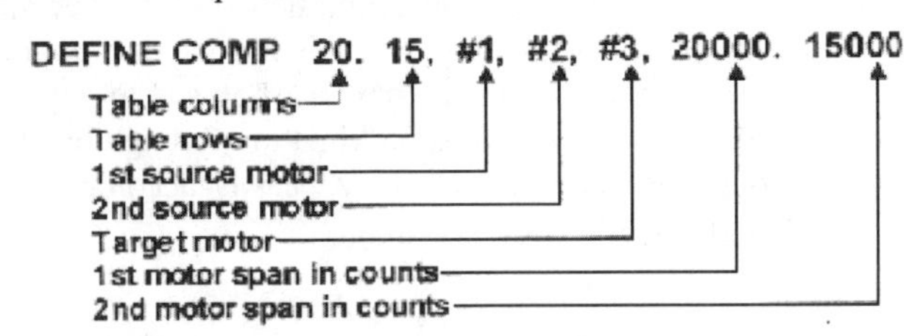

Fig. 4.4. PMAC compensation table

There are several disadvantages associated with the look-up table which become clearly significant with increasing precision requirements. First, the look-up table has extensive memory requirements. When the number of data points calibrated in a 3D workspace increases by a factor of N, the number of table entries increases by the order of N^3. This difficulty is thus especially significant for high precision machines, where a huge amount of calibration effort is necessary in order to compensate errors to within an acceptably precise threshold. Secondly, for the look-up table, the errors associated with intermediate points of the recorded data are compensated by using linear interpolation. This assumes the error to vary linearly between the calibrated points, and neighboring points are not utilised to improve the interpolation. Linear interpolation may suffice if the calibration is done at very fine intervals compared to the precision requirements. However, this will in turn imply tremendous memory requirements which may be beyond the capacity of a typical look-up table. Thirdly, the look-up table does not have a structure which is amenable to direct expansion when considering other factors affecting positioning accuracy, such as thermal and other environmental effects. When these factors are to be considered for a more precise compensation, additional tables are usually set up according to the schedules of the environmental parameters. Finally, for continuous on-line error compensation, a search through the look-up table will be necessary at every sampling interval. This is tedious when the table is large in size, especially when the calibration does not occur at regular intervals. Driven by increasingly high-precision demands, geometrical error compensation requirements are becoming more and more complex. While the traditional look-up table with limited computational power still dominates in soft compensation approaches, new function-based (as opposed to table-based) approaches are becoming more appealing in terms of the potential improvement in error compensation achievable. The following subsections describe the principles of a function-based approach based on the use of neural networks.

4.3.2 Parametric Model for Geometrical Errors

Since each error component varies with displacement in a non-linear manner, it is more naturally inclined to represent the non-linear profile using a non-linear function compared to using a look-up table. The neural networks are general tools for modelling non-linear functions. The potential of NNs for practical applications lies in the following properties they possess: i) they can be used to approximate any continuous mapping, ii) they achieve this approximation through learning, iii) parallel processing and nonlinear interpolation can be easily accomplished with NNs. Given $X \in R^N$, a three-layer NN has a net output,

$$X_{2k} = f(X;W) = \sigma[\sum_{j=1}^{N_1} W_{1jk}\sigma[\sum_{i=1}^{N} W_{ij}X_i + \theta_{1j}] + \theta_{2k}]$$

Where, $\sigma(\bullet)$ is being the activation function, W_{ij} is the first-to-second layer interconnection weights, and W_{ijk} is the second-to-third layer interconnection weights and θ_{1j} and θ_{2j} are the threshold offsets. It is usually desirable to adapt the weights and thresholds of the NN off-line or on-line in real-time to achieve the required approximation performance of the net, i.e., the NN should exhibit a learning behavior'. To obtain a good set of NN weights W, appropriate weight tuning algorithms should be adopted. A commonly used weight-tuning algorithm is the gradient algorithm based on a backpropagated error, where the NN is trained off-line to match specified exemplary pairs χ_d, γ_d with χ_d being the ideal NN input that yields the desired NN output γ_d. The learning procedure aims at driving the total approximation error to near zero via suitable adjustments of the learning parameters. This essentially constitutes a minimisation problem that the gradient techniques attempt to solve. With the appropriate choice of hidden nodes, the net can usually be driven as close to it as desired. This can be done by specifying a large number of hidden nodes in the network structure or by starting with a small number of hidden units and increasing the number until it becomes possible to drive the approximation error to within a desired threshold. The latter approach will be pursued in this chapter. The implementation of NN training may be programmed in MATLAB. We develop an M-file NN training program which can receive the user inputs relating to the error type, the raw data filename (from laser interferometer), and the training duration as shown in Fig. 4.5). The NN training is carried out and the modeling accuracy is graphically displayed by plotting the NN approximation of the error superimposed on raw calibration data.

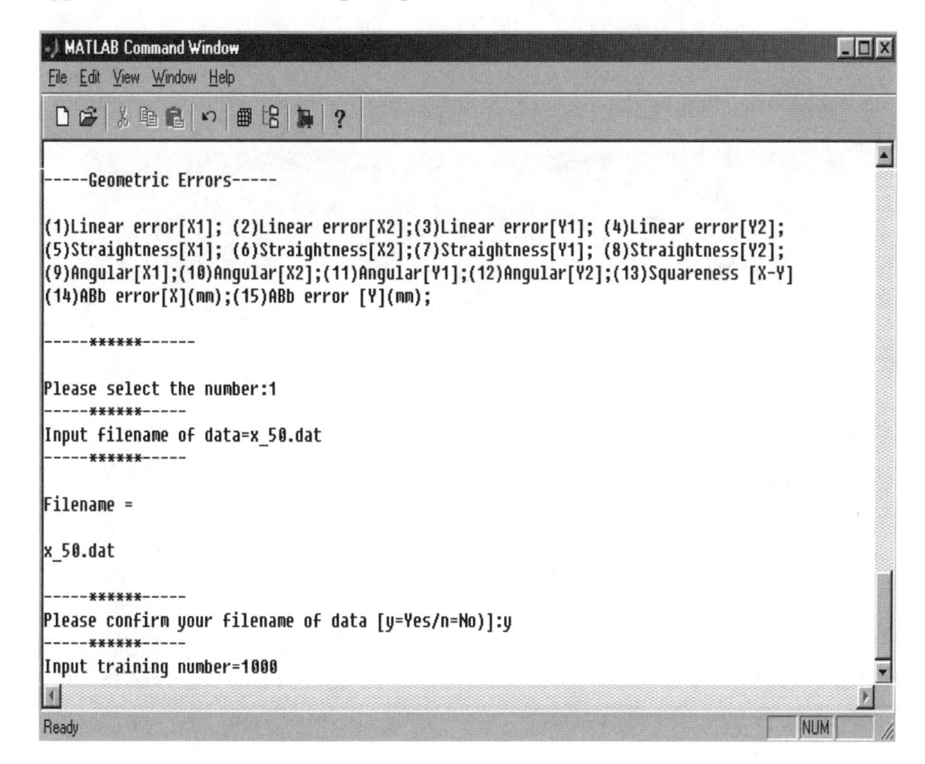

Fig. 4.5. User interface for NN training

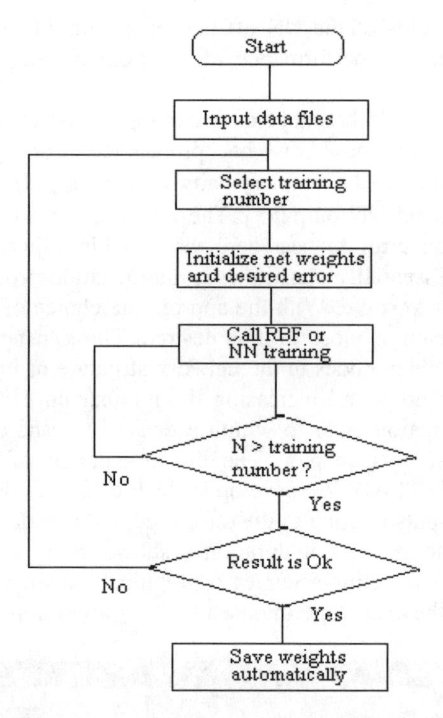

Fig. 4.6. Software flow block

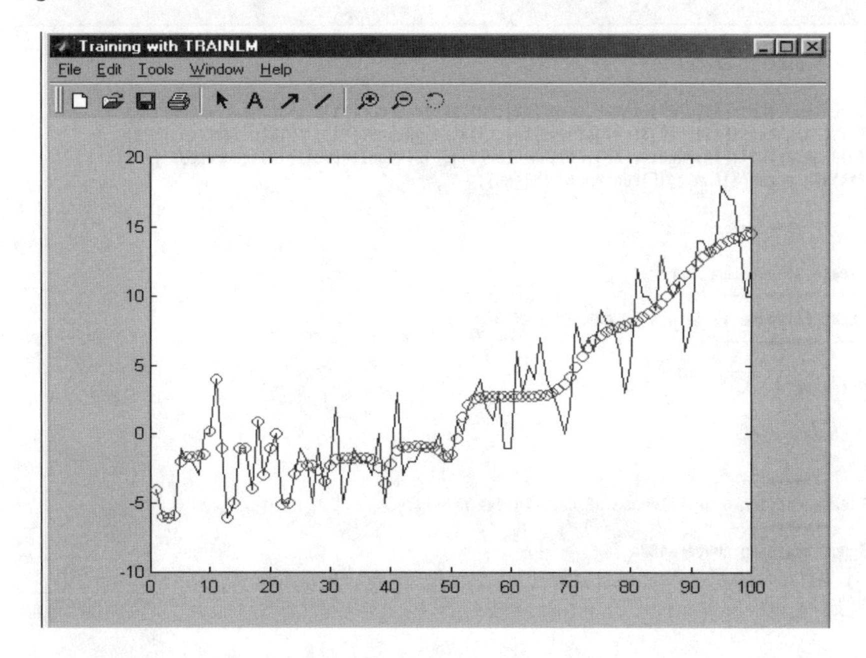

Fig. 4.7. Approximation accuracy with training number =1000

If the achievable modeling accuracy is not sufficient, the user may re-run the training with a new set of specifications until the approximation error converges to within an acceptable threshold. The flowchart of this iterative process is shown in Fig. 4.6. To illustrate, Fig. 4.7 shows the approximation accuracy achieved when duration of 1000 cycles is invoked. When re-run with a training duration of 10^6, the accuracy improves significantly as shown in Fig. 4.8. When the approximation accuracy is adequate and acceptable, the NN weights are saved in a MAT file, which will be used in the design of the S-Function in the subsequent phase.

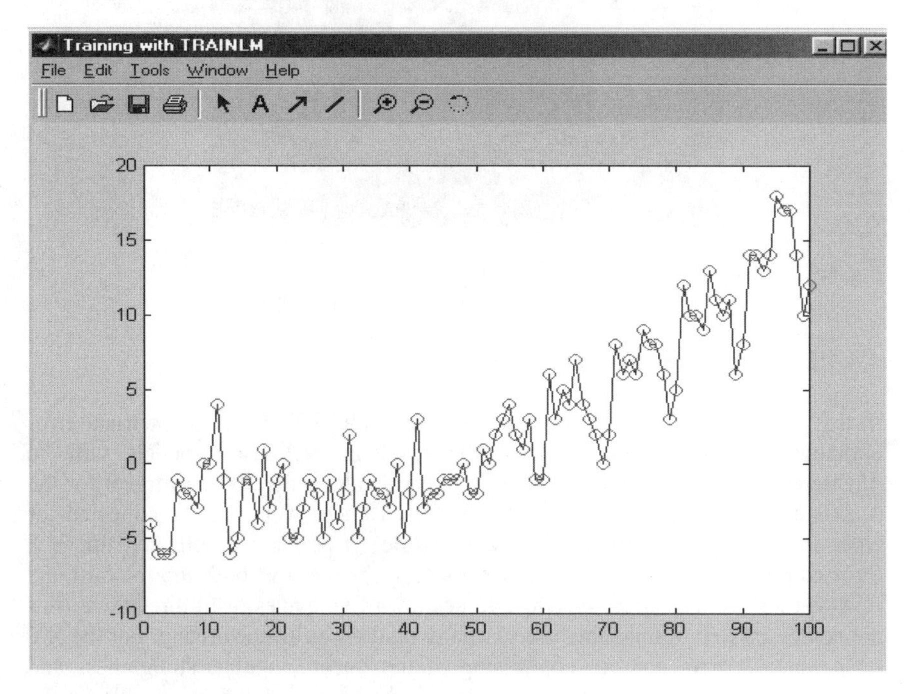

Fig. 4.8. Approximation accuracy with training number = 10^6

4.4 Experimental Results

Real-time experiments are performed on an XY table assembled using standard industrial positioning stages. Fig. 4.9 shows the XY table used in the experiments. The probe attached to the table may be moved in either the X or Y direction. Note that there is a linear offset between the probe and X-axis measurement line, giving rise to an Abbe error, which increases in proportion to the size of the offset. The Abbe error should be compensated for wherever possible.

The X and Y slides travel together to span a 200 mm x 350 mm 2D space. The Renishaw digital encoder resolution is 1 μm, which also corresponds to the minimum step size. Linear motors are used to generate the linear motion. Highly nonlinear displacement errors are thus expected, in which case, linear interpolation (in the form of look-up tables) may not be adequate and a nonlinear error model will be necessary if high-precision requirements are to be satisfied.

Fig. 4.9. XY table testbed

4.4.1 Error Approximations

In this subsection, the application of NNs to model individual geometrical error components and the adequacy of the resultant NN-based models will be illustrated. Calibration is done at 5 mm intervals along the 200 mm travel for the X axis and along the 350 mm for the Y axis. Hence, 40 points are collected for each error component along the X axis, while 70 points are collected for each error component along the Y axis. Errors corresponding to both the forward and reverse runs are measured. Average values from five cycles of bi-directional runs are computed to minimise the effects of any random influence arising. For the XY table, the six error sources comprising of the linear, angular, straightness and squareness errors are all present and they will be approximated using NNs. In the ensuing subsections, the results of error approximation using NNs will be shown.

4.4.2 Linear Errors

For each axis, the measurements must be carried out with the carriage moving in both the forward and reverse directions to determine the error associated with each direction. Fig. 4.10 shows the linear errors along the X-axis. It should be noted that the difference in the errors for the forward and reverse runs are rather insignificant in this case. Thus, for this axis, the forward and reverse errors are simply averaged.

For the modeling of the X-axis linear errors, the NN is trained using 50 samples. The results are shown in Fig. 4.11, where the output values measured using the laser system are plotted against the output values predicted by the NN. The weights are then available to commission the NN-based compensator.

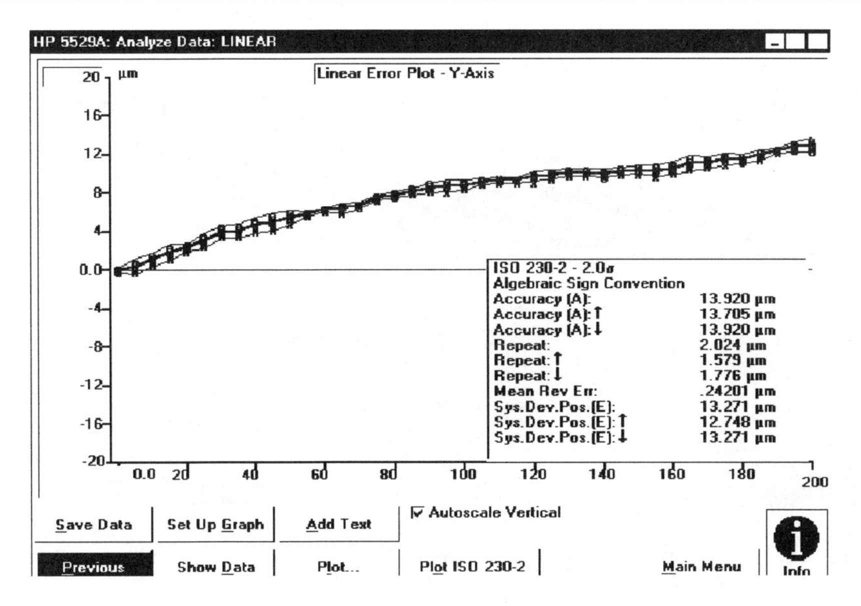

Fig. 4.10. Linear errors (X-axis): circle points represent forward run and cross points represent reverse run

For the Y-axis, the linear errors are shown in Fig. 4.12. The errors of both the forward and reverse runs are considered using separate NNs since the difference is large. For each NN, 26 hidden nodes are used. Fig. 4.13 compares the output of the NN with the actual linear error measurements (forward run) for the Y-axis, while Fig. 4.14 compares the output of the NN with the actual linear error measurements (reverse run) for the Y-axis. Although, some scattering is observed, it is observed that overall the approximation is good.

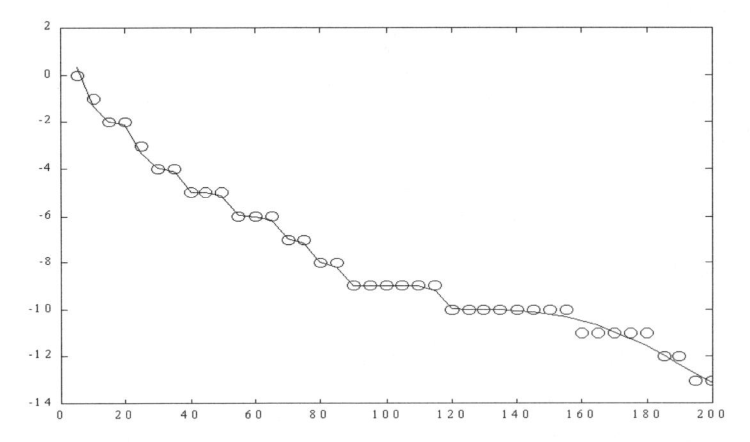

Fig. 4.11. NN approximation of the linear errors (X-axis): solid line is NN approximation and the circles represent the measured data.

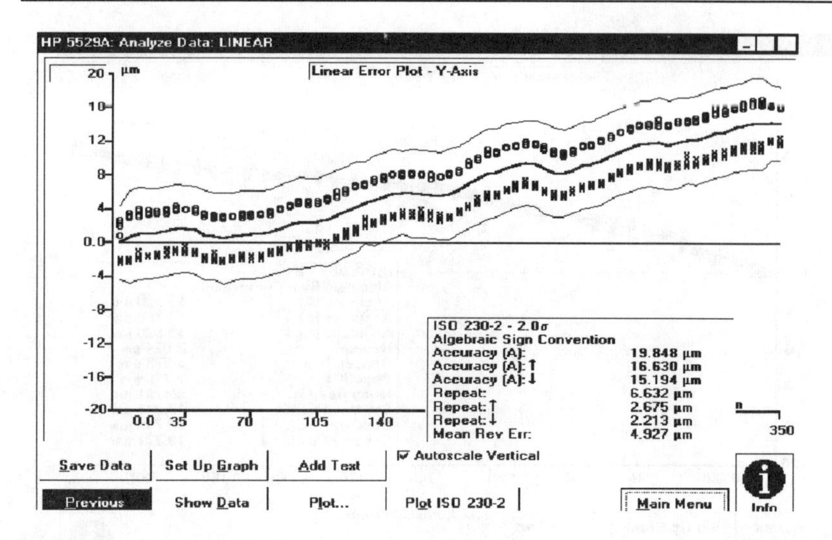

Fig. 4.12. Linear errors (Y-axis): circle points represent forward run and cross points represent reverse run

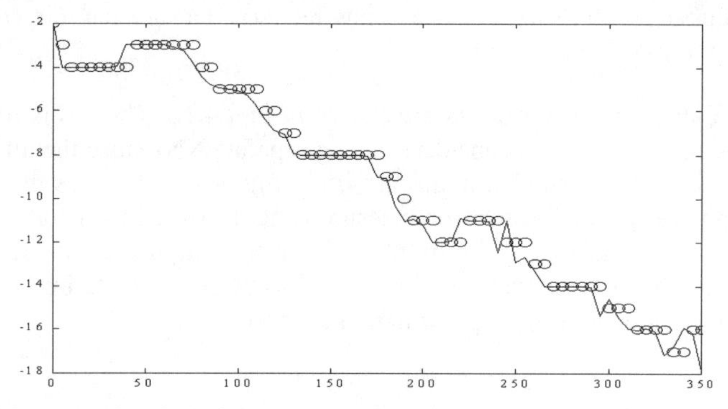

Fig. 4.13. NN approximation of linear errors (Y-axis): solid line is NN approximation and circle represents the measured data (forward run)

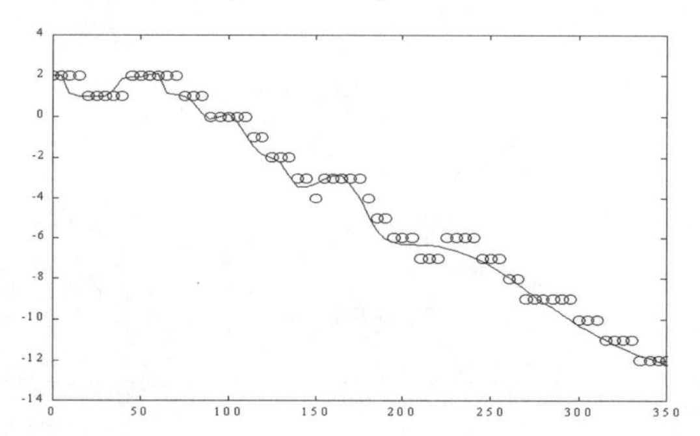

Fig. 4.14. NN approximation of linear errors (Y-axis): solid line is NN approximation and circle represents the measured data (reverse run)

4.4.3 Straightness Errors

For the XY table, there are two straightness error components to be determined: straightness of the X travel, which is concerned with deviation along the Y-axis, and the straightness of the Y travel, which is concerned with deviation along the X-axis. The good adequacy of the NN-based models in approximating the straightness errors is illustrated in Fig. 4.15 and Fig. 4.16 respectively. All the straightness error data are averaged from forward and reverse runs.

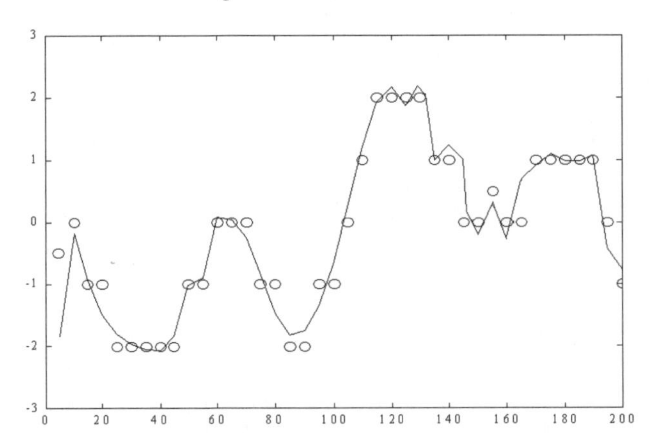

Fig. 4.15. NN approximation of straightness errors (X-axis): solid line is NN approximation and circle represents the measured data

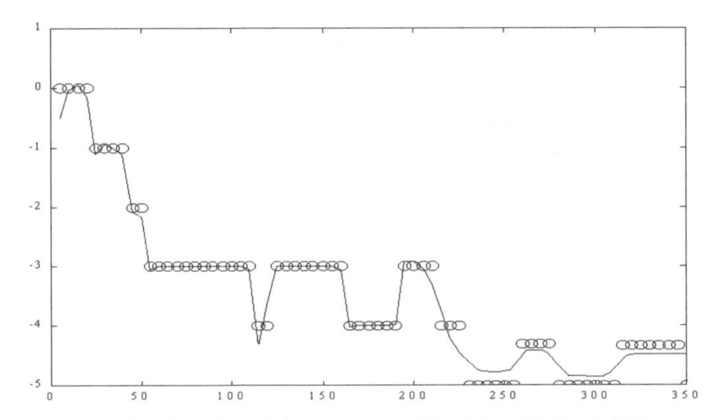

Fig. 4.16. NN approximation of straightness errors (Y-axis):solid line is NN approximation and circle represents the measured data

4.4.4 Angular Errors

For the XY table, only the yaw errors need to be measured. Pitch and roll errors are not relevant here since the XY table is a two-dimensional system. Yaw error measurements are thus made along the travel path of each axis to test for rotation about the perpendicular axis to the XY plane. With 22 hidden nodes used in each direction, the weights tuning process converges after 2.5×10^5 iterations. The outputs of the NN-based models follow the actual error measurements very closely as shown in Fig. 4.17 and Fig. 4.18.

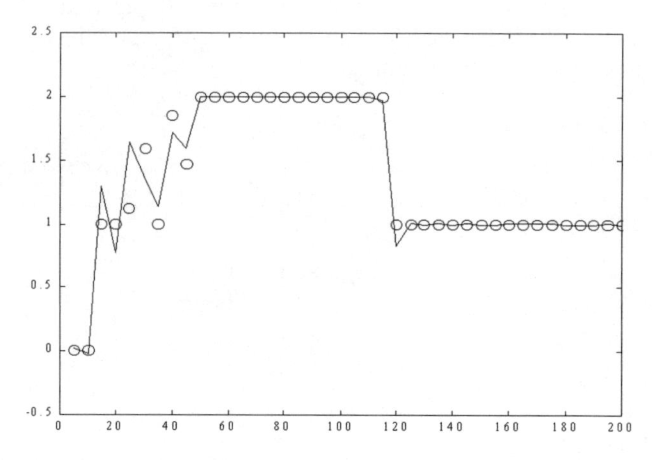

Fig. 4.17. NN approximation of yaw errors (X-axis): solid line is NN approximation and circle represents the measured data

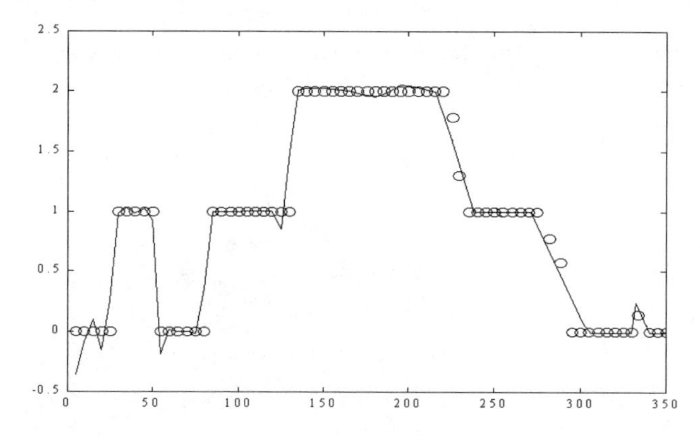

Fig. 4.18. NN approximations of yaw errors (Y-axis): solid line is NN approximation and circle represents the measured data (forward run); dotted line is NN approximation and plus line represents the measured data (reverse run)

4.4.5 Squareness Error

The squareness measurement yields a error which is single constant of 20 arcsec. In this case, the NN is not needed for squareness error. We simply substitute α = 20 arcsec directly in the overall model.

4.4.2 Assessment

To assess the compensation performance, all the error NN approximations for the individual error components are combined into the overall error model. Since bi-directional errors are used, it is necessary to consider the movement direction in the tests. For this experiment, x_p=0, y_p=52 mm since there is a probe used in this experiment. The two-axes are servo-controlled so that the carriage translated along two diagonals of the working area 200 mm x 300 mm (Fig. 4.19). This provides a fair basis to gauge the adequacy of the NN-based models. Also, this assessment follows the British Standard. The linear errors are measured across the diagonals using the HP laser interferometer system. For both measurements, five bi-directional runs are executed. At each point, there are 10 measurements taken from the forward and reverse runs. We now illustrate the actual result for the geometrical error compensation. For an X-Y diagonal (A-D), the total error before compensation is 35 μm (Fig. 4.20). After compensation, it is reduced to 11 μm (Fig. 4.21). For the other X-Y diagonal (B-C), the maximum error is 12 μm before compensation (Fig. 4.22) and 7 μm after compensation (Fig. 4.23). Thus, on the whole, a significant improvement in machine accuracy is achieved with the maximum error over the working region being reduced from more 35 μm to less than 11 μm. It is clear that the result obtained is very close to the ideal estimation.

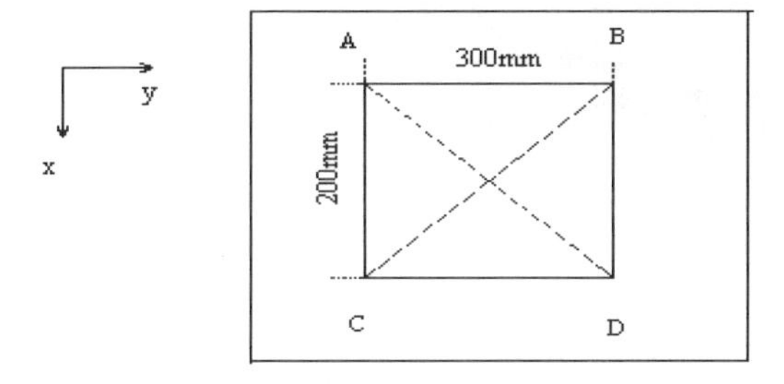

Fig. 4.19. Experimental region

4.5 Conclusions

In this chapter, we have discussed the geometrical error compensation methods, which can be applied to micromanufacturing machine systems. Specifically, NNs based models are developed and tested on a XY stage with micro level. The experimental results show that the geometrical error is reduced significantly and the developed technology can become as a tool for micromaching systems to solve their positioning, metrology, and alignment problems.

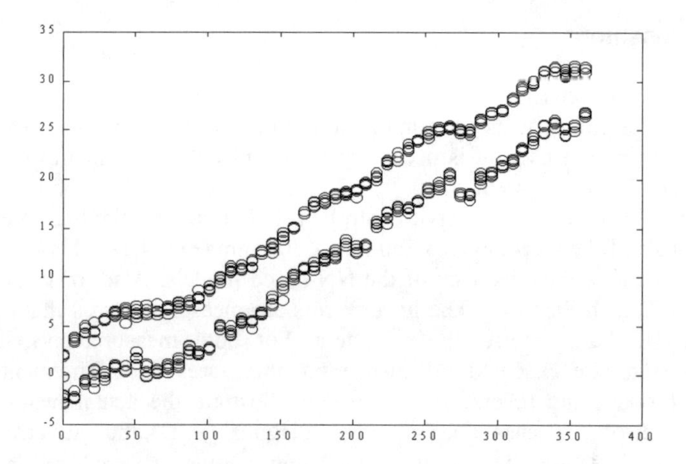

Fig. 4.20. Diagonal errors along A-D before compensation

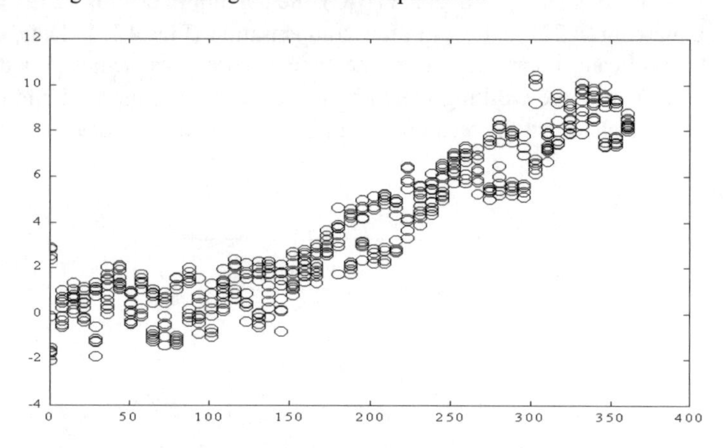

Fig.4.21. Diagonal errors along A-D after compensation (NN approximation)

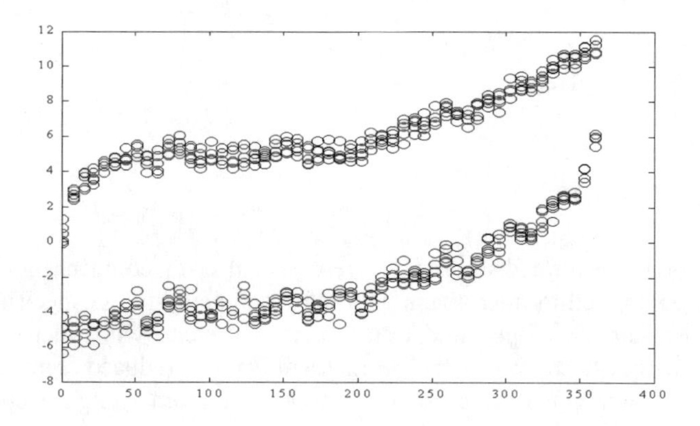

Fig. 4.22. Diagonal errors along B-C before compensation

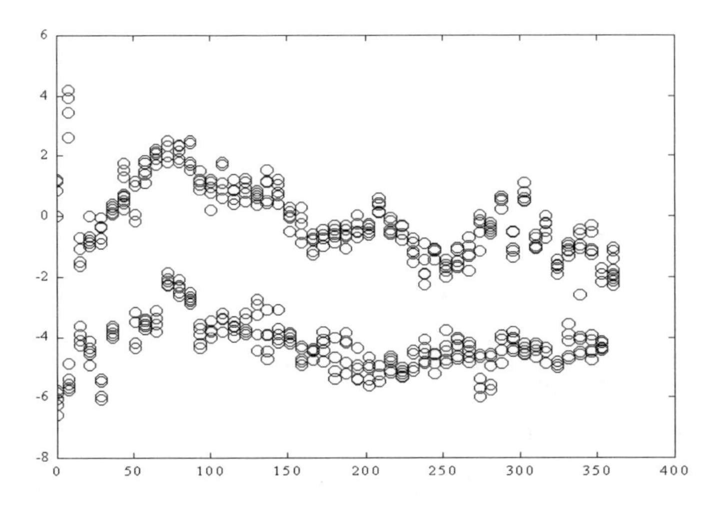

Fig. 4.23. Diagonal errors along B-C after compensation (NN approximation)

4.6 Reference

Hocken R (1980) Machine tool accuracy. Technical Report, vol. 2, Lawrence Livermore Laboratory, University of California

Hocken R, Simpson J, Borchardt B, Lazar J, Stein P (1977) Three dimensional metrology, Annals of the CIRP 26:403-408

Evans CJ (1989) Precision engineering: An evolutionary view. Cranfield Press, UK

Love WJ, Scarr Aj (1973) The determination of the volumetric accuracy of multi axis machines. Proc. 14th MTDR, pp 307-315

Bush K Kunzmann H, Waldele F (1984) Numerical error correction of co-ordinate measuring machines. Proc. Int. Symp. on Metrology for Quality Control in Production, Tokyo, pp 270-282

Tan KK, Lee TH, Lim SY, Dou HF, Seet HL (1999) Probabilistic approach towards error compensation for precision machines. Proc. of 3rd Int. ICSC Symposium on Intelligent Industrial Automation, Genova, pp 293-299

Zhang G, Veale R, Charlton T, Hocken R, Borchardt B (1985) Error compensation of coordinate measuring machines, Annals of the CIRP 34:445-448

Duffie NA, Maimberg SJ (1987) Error diagnosis and compensation using kinematic models and position error data, Annals of the CIRP 36:355-358

Tan KK, Huang SN, Seet HL (2000) Geometrical error compensation of precision motion systems using radial basis functions, IEEE Trans. on Instrument and Measurement 49:984-991

Hornik K, Stinchcombe M, White H (1989) Multilayer feedforward networks are universal approximators, Neural Networks 2:359-366

British Standard (1989) Coordinate measuring machines, Part 3. Code of practice, BS 6808

5 Characterising Etching Processes in Bulk Micromachining

S. K. Bag

Department of Electronics and Communication Engineering, Vellore Institute of Technology, Tamilnadu, India

5.1 Introduction

Machining is defined as a process of removing material from a workpiece in the form of chips in order to obtain exact shapes and sizes. It is a well-defined process method in "machine design and tooling". It includes several methods, which are usually divided into three main classes such as cutting, abrasive processes, and nontraditional machining. Machining processes design and manufacture macro components and systems. Micromachining processes bear similar meaning but the process handles microsystems. Micromachining has become a new technology, as there has been significant growth in the manufacture of components having dimensions less than a millimeter upon which micro features in the range of 1-200 μm are built. In particular, micromachining deals with microelectronics, micromechanical systems, micro-opto-electrical systems and micro-opto-mechanical systems. When these are integrated into one platform, the system is known as a MEM (Microelectromechanical) system, or MEMS. Micromachining is considered to be a process as well as a technology that is utilised to structure wafer materials or thin films in order to fabricate miniature devices such as microsensors, microactuators and passive components. The process is classified under two different process techniques: *bulk micromachining* and *surface micromachining*. If structuring is performed on wafer materials the process is called a bulk micromachining process and if it is performed on the thin film the process is called a surface micromachining process.

5.2 Wet Bulk Micromachining (WBM)

Bulk micromachining is a method of fabricating microdevices by etching the substrates and implementing the features in the bulk of the materials such as silicon, germanium, quartz, SiC, GaAs, and InP. The etching process is a special technique to cut the substrate, which can be used to form the desired shape. The process falls under three categories in terms of the state of the etchants used: wet, vapor, and plasma states. When it uses wet etchants it is called wet bulk micromachining (WBM) and the etching depends on the plane of orientation of the substrate. Orientation dependent etching is called anisotropic etching, whereas orientation independent etching is known as isotropic etching.

A vast majority of wet bulk micromachining processes are based on single crystal silicon. Some work has been done on quartz, Ge and GaAs and recently on GaP, InP and SiC. WBM is understood to be a well-established fabrication process in silicon technology. Among other techniques, surface micromachining can be named and it has been popular in recent years. The IC (integrated Circuit) industry has been using WBM for a long time, but further development has been made to adopt it into the MEMS domain. Although the technique is emerging fast, it is unlikely that the popularity of bulk micromachining will be decreased. This chapter deals with the features and properties of the etching process of the bulk micromachining method. Readers can follow the included references for further details. The next chapter deals with the important aspects of the surface micromachining process. Silicon is used as the substrate throughout the chapter.

5.3 Review

Wet etching using a mask (e.g., wax) and etchants (e.g., acid) has been found in the fifteenth century for decorating armors (Harris 1976). By the early seventeenth century, the etching process had become an established technique. Initially it was known as chemical milling. Introduction of photosensitive masks were found in 1822 but the accuracy was less then 0.05 mm (Madou and Morrison 1989). Later on, lithography and chemical milling processes were combined to achieve a new level of accuracy. PCB is the best example of an application of lithography-based chemical milling. Silicon based ICs had been fabricated by 1961. Later, many other applications like color television shadow masks, integrated circuit lead frames, light chopper and encoder discs, and decorative objects such as costume jewelry have been found that use photochemical machining (Allen 1986).

Silicon processing using isotropic etching has been available since the early 1950s. Some details of the work are available from a series of papers published by Robbins and Schwartz during 1959 to 1976, on chemical isotropic etching (Robbins and Schwartz 1960; Schwartz and Robbins 1976). They have observed the use of the chemical isotropic etchant HF and HNO_3 with or without acetic acid and water for the silicon process. Uhlir's work is based on electrochemical isotropic etching (Uhlir 1956). Isotropic etching was first applied on electrochemical cells or for electro-polishing, but the deposition of metal on the surface created a problem (Hallas 1971). Turner has found that if the critical current density is exceeded, silicon can be electro-polished in aqueous HF solutions without formation of any metal deposition (Turner 1958).

In the mid sixties, Bell Telephone Laboratories started the work on anisotropic silicon etching in the mixtures of KOH, water and alcohol and later in KOH and water. Various scientists also pursued chemical and electrochemical anisotropic etching methods during 1966 to 1977. In the mid-seventies, a new surge of activity in anisotropic etching was associated with the work on V-groove and U-groove transistors (Rodgers et al. 1977; Ammar and Rodgers 1980). Smith discovered piezoresistance using Si and Ge in 1954 (Smith 1954). Pfann et al. proposed a diffusion technique for the fabrication of Si piezoresistive sensors for stress, strain and pressure measurement (Pfann et al. 1961). Tufte et al. made the first thin Si piezoresistive diaphragms for pressure sensors using a combination of a wet isotropic etch, dry etching and oxidation processes (Tufte et al. 1962). National Semiconductor became the first to make stand-alone Si sensor products in 1972. They also broadly described Si pressure transducers in 1974 and completed a silicon pressure transducer catalogue (Editorial 1974).

Other important commercial suppliers of micromachined pressure sensor products were Foxboro/ICT, Endevco, Kulite and Honeywell's Microswitch. Other micromachined structures have been explored during the mid- to late seventies. Texas Instruments (Editorial 1977) produced a thermal print head in 1977. Hewlett Packard (O'Neill 1980) made thermally isolated diode detectors in 1980. Fiber optic alignment structures were developed at Western Electric (Boivin 1974) and IBM produced ink jet nozzle arrays (Bassous et al., 1977).

Many Silicon Valley microsensor companies played vital roles in the development of the market for Si sensor products. Druck Ltd. in the U.K. started exploiting micromachined pressure sensors in the mid-eighties (Greenwood 1984). Petersen explored the mechanical properties of a single crystalline of silicon (Petersen 1982). It is estimated that today there are more than 10,000 scientists pursuing research on the design and development of Si based microsystems. It has become a pressing need to understand the intended applications.

5.4 Crystallography and its Effects

Silicon crystalline inherits a covalent bond and diamond-cubic (DC) structure. The atoms form a tetrahedron with its four covalent bonds as shown in Fig. 5.1. These tetrahedrons make a diamond-cubic structure. The structure can be explained as two interpenetrating face-centered cubic (FCC) lattices, one displaced (1/4, 1/4, 1/4) with respect to the other, as shown in Fig. 5.2. The axes *x*, *y*, *z* are perpendicular to a plane with the three integers (h, k, l). It simplifies the illustration concerning the crystal orientations.

The lattice parameter is defined as distance between two atoms. It is approximately 5.4309 Å for the silicon structure. The DC lattice is wide-opened, with a packing density of 34%, compared to 74% for a regular FCC lattices. The plane (111) presents the highest packing density. In addition to the DC structure, silicon has many special features like high-pressure crystalline phases and stress-induced meta-stable properties.

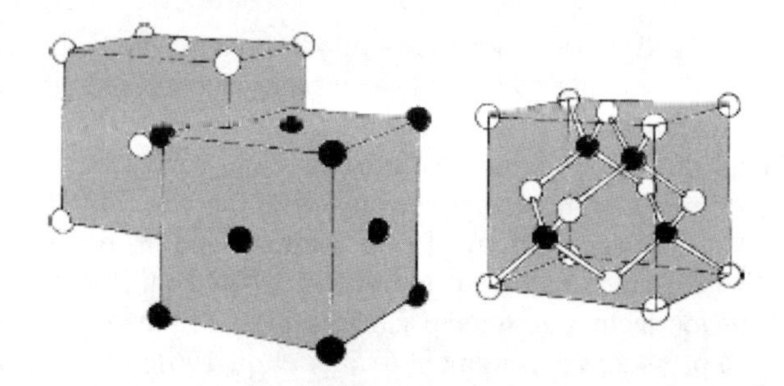

Fig. 5.1. Silicon diamond-cubic structure and covalent bonds

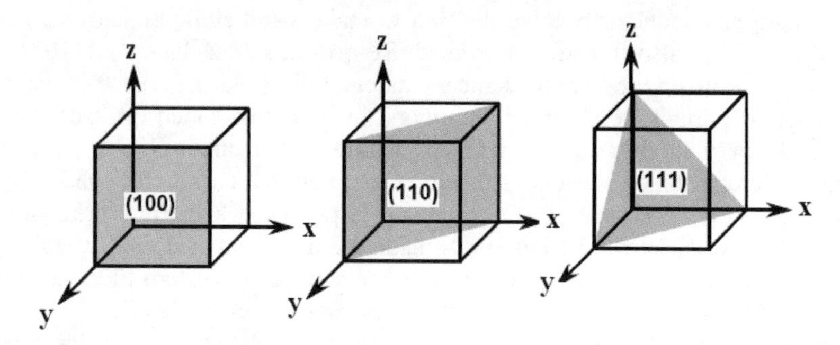

Fig. 5.2. Silicon plane orientation

Silicon wafer <100> <110> and <111> orientations are commonly used for micromachining. Wafer <100> can be machined in order to obtain (110) and (111) orientation planes on the wafer. The <110> wafer cleaves much easier than other orientations. The <111> wafers are not normally used because it is difficult to etch anisotropically. Additionally, the wafer entails a laser-assisted etching process. On a <100> silicon wafer, the <110> direction can be made easily. The flatness can be up to 3° precision. Flat areas of the wafer can be used for the determination of the orientation. The flat area also helps to place the slices on the equipment. It is important to understand the geometric relationships between the different planes within the silicon lattice. Only wafers <100> or <110> are considered to be ground planes. The wafer <111> has the highest atom packing density and cannot be etched using anisotropic etchants. Hence, wafers <100> or <110> are bound to form the sidewalls of a (111) plane. The micromachined planes are dependent on the geometry and the orientation of the mask.

5.4.1 An Example

Fig. 5.3 shows a silicon wafer <100> together with non-etching (111) planes as sidewalls. During etching, truncated pyramids or V-grooves do not widen. After etching, the (111) planes are joined to their intersection and the (100) bottom plane disappears, creating a pyramidal pit or a V-groove. The slope of the sidewall in a cross section perpendicular to the wafer surface is approximately 54.74°. The alignment of the wafer surface determines the accuracy of this angle. Due to this behavior of silicon, its mechanical properties differ from common metals. Some of the important features are discussed in the next section.

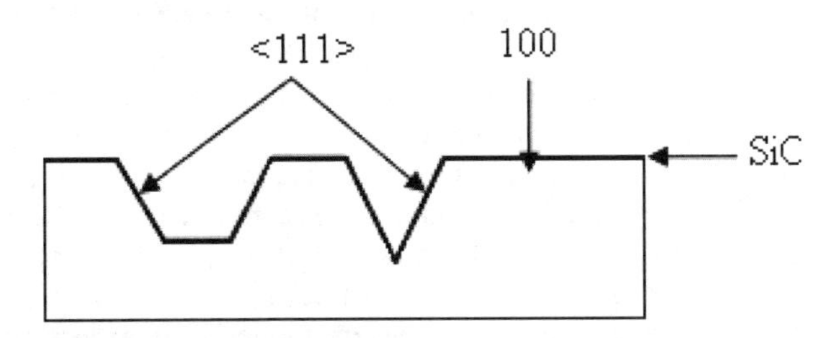

Fig. 5.3. Effect of etching on a silicon plane

5.5 Silicon as Substrate and Structural Material

Due to its intrinsic mechanical stability, sensing, integrating capabilities and electronic properties, silicon is mostly selected as the prime material for the fabrication of electromechanical microsystems. Many thin films and other microcomponent based devices use silicon as substrates or wafers.

5.5.1 Silicon as a Substrate

It is observed that the chemical sensor development in industry has moved from silicon in the 1970s and early 1980s to a hybrid thick film on ceramic approach in the late 1980s and early 1990s. However, recently, silicon has been considered to be the best choice in terms of metalisation and machinability as compared to ceramic, glass and plastic. Both ceramic and glass substrates are immune to machining processes. Plastic substrates are susceptible to metalisation. The silicon substrate is relatively costly but it can be substituted with small sizes. Some of the special features like ease of passivation layers, extreme flatness, well-established coating procedures, stability in high temperatures and formation of thin films have made the silicon substrate acceptable. Silicon has greater flexibility in design and manufacturing as compared to other substrates. Although it is much expensive, the initial capital equipment investment is not product specific, as the micromachined products require a change in *mask* but not in the equipment itself. Another factor that determines the choice for silicon is the final packaging of the device. For instance, a chemical sensor is easy to package on an insulating substrate rather than on a conductive substrate. Silicon has also some disadvantages. It is not worthwhile to use silicon for large devices and low production volumes, as well as for devices where the design of electronics circuits is not important.

5.5.2 Silicon as Structural Material

In mechanical sensors, an active structural element converts a mechanically external input signal, i.e. force, pressure and acceleration, into an output electrical signal like voltage, current or frequency. The transfer functions can be described in terms of mechanical, electromechanical and electrical conversion. In mechanical conversion, a structural active member of the microdevice is optimally loaded or stressed. An active member can be a high-aspect-ratio suspended beam or membrane. Electromechanical conversion is the transformation of a mechanical quantity into an electrical quantity such as capacitance or resistance. Sometimes, the electrical signal is amplified and further converted to voltage, current or frequency. For conversion into voltage, a Wheatstone bridge may be utilised. A charge amplifier may be provided for amplification.

To optimise the transfer functions, system modeling is necessary. One of the most important features of modeling is the method of determination of independent elasticity constants. It follows that silicon has the requisite property of elasticity constants. In the case of actuators, the same transfer function can be used in reverse way, but the mechanical and electrical properties remain same. Important mechanical properties such as stress, strain and plastic deformation and thermal properties of the silicon materials are described below.

5.5.3 Stress and Strain

Stress is defined as amount of force acting in a unit area while strain is the elongation in unit length. Another parameter which is used to define the elastic property is known as the Poisson ratio. It is defined as the amount of lateral contraction in unit diameter. Yield, tensile strength, hardness and creep of a material are all related to the elasticity properties. The stress–strain relationship curve is illustrated in Fig. 5.4. It varies from material to material, but the nature of the curve for most materials is similar. The curve follows a linear relationship in certain regions and then it becomes nonlinear. The linear part draws interest as far as the modeling of micromechanical devices is concerned. The stress is proportional to the strain of a material with a constant slope in this region. This is known as Hooke's law and the slope is defined as elastic modulus E or Young's modulus. The region marked P as in Fig. 5.4 (b), is called the elastic deformation. For isotropic media (e.g., amorphous and polycrystalline materials) the applied axial force per unit area or tensile stress σ_a, and the axial or tensile strain ε_a, are related by (Seidel et. al. 2990):

$$\sigma_a = E\varepsilon_a \tag{5.1}$$

Where, ε_a is a dimensionless ratio of $(L_2 - L_1)/L_1$, i.e., the ratio of the elongation to its original length. The elastic modulus can be considered as the resistance of the materials to deformation. Materials with high elastic modulus are stiffer. Tensile stress leads to a lateral strain, called Poisson's ratio ε_1. This is given by the dimensionless ratio of $(D_2 - D_1)/D_1 = (-\Delta D/D)$, where D_1 is the original diameter and ΔD is the change in diameter under axial stress. The negative sign means contraction rather than extension. The ratio of lateral strain to axial strain can be written as:

$$\nu = \frac{\varepsilon_1}{\varepsilon_a} \tag{5.2}$$

For most materials, ν is unchanged within the elastic limit. Small volume change is accompanied by the deformation and ν does not exceed 0.5. The value of ν changes from 10% to 50% for different materials. For an elastic isotropic medium under triaxial (*x,y,z* directions) forces, the strain in the *x* direction ε_x , can be written as:

$$\varepsilon_x = \frac{1}{E}[\sigma_x - \nu(\sigma_y + \sigma_z)] \tag{5.3}$$

Where σ_y and σ_z are stresses in the *y* and *z* directions, respectively. Similarly, ε_y and ε_z can be written using other parameters.

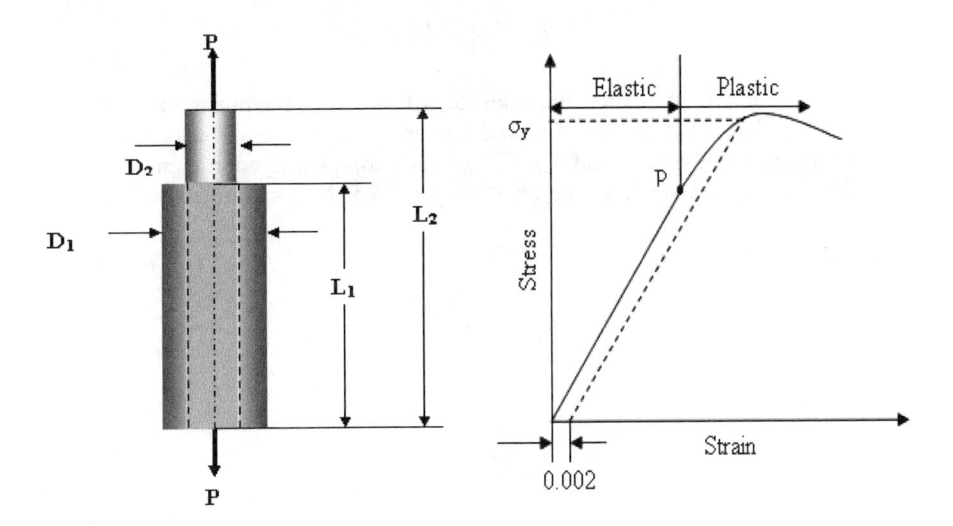

Fig. 5.4. (a) Effect of axial stress; **(b)** Elasticity curve

There exists another form of strain in the mechanical structure. This is known as the shear strain, γ. This phenomenon does not involve volume change but changes such as shape, twisting and warping appear. The corresponding stress is called shear stress. Shear strain can be written as follows:

$$\gamma = \frac{\tau}{G} \tag{5.4}$$

Where, τ is the shear stress and G is the elastic shear modulus or the modulus of rigidity. Three similar equations can also be written in the *x, y,* and *z* planes. The shear modulus, G, can be expressed in terms of Young's modulus, E, and the Poisson ratio, ν, as follows:

$$G = \frac{E}{2(1+\nu)} \tag{5.5}$$

Isotropic materials can be characterised by these two independent elastic constants. However, anisotropic materials require more than two elastic constants and the number increases with decreasing symmetry. For example, a cubic crystal such as FCC features 3 elastic constants, hexagonal crystal characterises 5 constants, and materials without symmetry require up to 21 elastic constants. The stress-strain relationship is more complex in these cases and depends on the spatial orientation of the crystallographic axes. Hooke's law can still be applied and expressed in the following form:

$$\begin{aligned} \sigma_{ij} &= E_{ijkl}.\varepsilon_{kl} \\ \varepsilon_{ij} &= S_{ijkl}.\sigma_{kl} \end{aligned} \tag{5.6}$$

Where, σ_{ij} and σ_{kl} are called stress tensors of rank 2 and expressed in N/m^2, ε_{ij} and ε_{kl} are called strain tensors of rank 2 and are dimensionless, E_{ijkl} is a stiffness coefficient tensor of rank 4, and $S_{ijkl} = [E_{ijkl}]^{-1}$ is a compliance coefficient tensor of rank 4 and expressed in m^2/N. The Eq. (5.6) above can be transformed as:

$$\varepsilon_m = \sum_{m=1,n=1}^{m=3,n=3} S_{m,n}\sigma_n \tag{5.7a}$$

$$\sigma_m = \sum_{m=1,n=1}^{m=3,n=3} E_{m,n}\varepsilon_n \tag{5.7b}$$

Where, the tensors E_{ijkl} and S_{ijkl} are replaced by $E_{m,n}$ and $S_{m,n}$, respectively. This can be obtained by substituting the indices $ij = m$ and $kl = n$. In a cubic structure it can be considered that *x, y* and *z* directions are replaced by 11, 22, 33 and the corresponding planes *xy* →12 and 21, *xz* →13 and 31, *yz* → 23 and 32 are in symmetry. With this representation the stress and strain components can be represented as:

$$\begin{aligned}\sigma_x &= E_{11}\varepsilon_x + E_{12}\varepsilon_y + E_{13}\varepsilon_z + E_{14}\gamma_{xy} + E_{15}\gamma_{xz} + E_{16}\gamma_{yz} \\ \varepsilon_x &= S_{11}\sigma_x + S_{12}\sigma_y + S_{13}\sigma_z + S_{14}\tau_{xy} + S_{15}\tau_{xz} + S_{16}\tau_{yz}\end{aligned} \tag{5.8}$$

Each of these equations has six elastic constants. The tensor matrices $E_{m,n}$ and $S_{m,n}$ are composed of 36 coefficients. A material without symmetry has 21 independent coefficients. Hence, in an isotropic crystal with symmetry the stress and strain components can be expressed by a maximum of two the elastic coefficients, mainly E and υ. In the case of a cubic crystal, the three elastic moduli are E_{11}, E_{12} and E_{44}. Furthermore, an additional relationship among them is as follows:

$$E_{44} = \alpha \frac{E_{11} - E_{12}}{2} \tag{5.9}$$

Where, $\alpha<1$ represents the anisotropic coefficient. Hence, the system can always be reduced to its independent stiffness constants. The stiffness constants for the silicon substrate <100> and <110> are functions of the plane but they are independent in the case of Si<111>. A plate made up of Si<111> can be considered as having isotropic elastic properties. Most of the elastic properties, such as the Young's modulus and Poisson ratio remain constant for both lightly and highly doped silicon. Due to the size of atomic radius, huge boron doping in many situations creates shrinkage on the lattice when compared to pure silicon, generating residual stress. A cantilever beam etched from such silicon would be expected to bend out of the plane of the wafer.

All metals deform beyond their elasticity limit when the linear relation between stress and strain does not sufficiently hold. Once metal reaches this stage, it cannot return to its original position. The point P where yielding starts (Fig. 5.4(b)), is known as the proportional limit or yielding point. At some specified strain offset, usually 0.002, a straight line is drawn parallel to the elastic portion of the stress–strain curve. The stress corresponding to the intersection of the line and the practical stress–strain curve as it bends over in the plastic region is defined as the yield strength σ_y. The yield strength of a material gives its resistance to plastic deformation. The tensile strength of a material is the maximum stress of the stress–strain curve and it corresponds to the stress that a structure can withstand by the applied load. Silicon substrates do not show any plastic deformation up to 800°C. Silicon sensors are inherently insensitive to fatigue failure even at high cyclic loads. They can be used in excess of 100 million cycles without any failures unless sticking and residual stresses come into the picture. This ability is due to the fact that the crystal has no energy absorbing or heat generating mechanism because of inter-granular movement of dislocations in its atoms at atmospheric temperatures. However, single crystal silicon is a brittle material, and shows yield catastrophically, when stressed beyond the yield limit, rather than deform plastically as metals do. SiO_2 and Si_3N_4 exhibit linear-elastic behavior at low strain and high modulus. Plastic deformation is considered as a dislocation generated in the grain boundaries due to stress. Macroscopic deformation forms inter-grain shifts within the structural material. Single crystal silicon has no grain boundaries, for which there is no plastic deformation. The deformation occurs due to two reasons i) the migration of the defects originally present in the lattice, and ii) the defects are generated at the surface. If the defects are nominal, the materials can be considered as perfect elastic material. Perfect elasticity is taken to mean that i) proportionality between stress and strain exists and ii) there is an absence of irreversibility or mechanical hysteresis. Hence, silicon has the advantage over metal as far as micromachining at relatively high temperatures is concerned.

5.5.4 Thermal Properties of Silicon

Single crystal silicon has a high temperature dependency on conductivity as compared to metals but a low thermal coefficient of expansion. The above property makes the silicon not only a better structural material but also helps microfabrication processed such as wet etching. In some processes, the devices are made by joining two wafers such as Si and Pyrex glass together as both wafers have nearly same thermal coefficient of expansion. Any mismatch will introduce thermal stress. This degrades the device performance. It is difficult to fabricate stress-free composite structures over a wide range of temperatures. Silicon on insulator (SOI) wafers can sustain high temperatures. Highly doped silicon bears a linear relationship with the temperature coefficient, resistance and sensitivity over a wide range of temperatures. A thermally isolated structure imposes difficulties in fabrication. Glass or quartz with lower thermal conductivity provides advantages.

5.6 Wet Etching Process

Chemical solutions are primarily used to etch the silicon substrate on the plane of crystallisation. Hence, it is known as wet etching. Wet etching of silicon is used mainly for cleaning, shaping, polishing and characterizing structural and compositional features (Uhlir 1956). Wet etching can provide a higher degree of selectivity than dry etching techniques. In many cases wet etching is faster and an etch rate up to 6 μm/min cam be achieved. The etching process passes through the following steps: i) reactant transportation to the surface, ii) surface reaction and iii) reaction product transportation (away from the surface). Stirring can increase the diffusion-limited etching. If (ii) is considered as the rate-determining step then etching is reaction-limited and it depends strongly on temperature, material and composition of the solution. Reaction-rate controlled processes are more active than diffusion-limited processes because of their high activation energy. The later is relatively insensitive to temperature variations. In practice, the former is preferred as it is easy to reproduce a temperature setting than a stirring. The etching apparatus should have a good temperature controller and a reliable stirring facility (Kiminsky 1985)). Wet etching can be classified into two categories depending on the direction of etch. Isotropic etching is independent of the plane of the crystal, whereas anisotropic etching is dependent on the plane of orientation.

5.6.1 Isotropic Etchants

In isotropic etching materials are removed uniformly from all directions and it is independent of the plane of orientation of the crystal lattice. Isotropic etching is used for polishing, cleaning and unidirectional etching of the materials. The etchants are a mixture of acidic solutions such as HF, HNO_3, and CH_3COOH. In single crystal silicon the etchant leads to round isotropic features. They can be used at room temperature or slightly above, but below 50°C. The mixtures have been introduced as silicon etchants by many scientists (Robbins and Schwartz 1960; Uhlir 1956; Hallas 1971; Turner 1958; Kern and Deckert 1978). Later it was discovered that some alkaline chemicals can be used to etch anisotropically. Anisotropic etchants can etch away the crystalline silicon at different rates and different directions. The pH value stays above 12. High temperatures can be maintained for slower etchants (>50°C). Isotropic etchants show diffusion limitation, while anisotropic etchants are reaction rate limited.

5.6.2 Reaction Phenomena

In acid media, the etching starts as a hole injected into the silicon covalence band. Holes are injected using an oxidant, electrical field or photons. Nitric acid and H_2O_2 in the HNA solution act as oxidants. The holes attack the covalent bond in silicon and oxidise the material. Consider the following reaction produced by HNO_3 together with water and HNO_2:

$$HNO_3 + H_2O + HNO_2 \rightarrow 2HNO_2 + 2OH^- + 2h^+ \tag{5.10}$$

In the above reaction, the holes are generated by an autocatalytic process. HNO_2 re-enters, sustaining further reaction with HNO_3 so as to produce more holes. The reaction is continued so as to produce more holes. This type of reaction takes time to introduce the oxidation process, and continues until a steady-state concentration of HNO_2 is reached. The phenomenon is observed at a low concentration of HNO_3 (Tuck 1975). After the holes are injected, OH^- groups attach to the oxidised silicon to form SiO_2 and eventually hydrogen is released.

$$Si^{4+} + 4OH^- \rightarrow SiO_2 + H_2 \tag{5.11}$$

Subsequently, hydrofluoric acid (HF) dissolves the SiO_2 and forms the water-soluble compound H_2SiF_6. The overall reaction of HNA system with silicon is given by:

$$Si + HNO_3 + 6HF \rightarrow H_2SiF_6 + HNO_2 + H_2O + H_2 \tag{5.12}$$

The above reaction model considers only the holes. The actual reaction takes place with both holes and electrons. In the isotropic model the rate-determining step in acidic etching involves hole injection into the covalence band, while in alkaline anisotropic etching it involves hole injection into the conduction band. In the first case a hole injected into the covalence band is significantly greater than the latter. The difference between the two etching processes is their reactivity.

5.6.3 Isotropic Etch Curves

Isotropic etching was introduced in late 1950s and was well characterised in the early 1960s. Schwartz and Robbins have published a series of literature in this regard (Robbins and Schwartz 1960). Isotropic etch curves are presented in two forms, one is called iso-etch curves and other one is known as the Arrhenius plot. The former characterises the HNA system for various weight percentages of the constituents as shown in Fig. 5.5 (a).

The concentration of acid plays a major role in the reaction process. In this example, concentrated acids of 49 wt% HF and 70 wt% HNO_3 (in both the cases water acts as a diluent) are shown in Fig. 5.5(a) and in Fig. 5.5(b), respectively. The curves are plotted in the sides of a triangle and the axes show the increment of concentrations from low to high. The characteristics of the HNA system can be summarised from Fig. 5.5 as follows:

- The etching is isotropic
- At high HF and low HNO_3 concentration, the iso-etch curves describe lines of constant HNO_3 concentrations
- HNO_3 concentration controls the etch rate
- Etching at those concentrations takes more time to initiate and unstable in silicon growth, hence SiO_2 forms slowly
- The etch rate is limited by amount of oxidisation as the process depends on the plane of orientation, and is affected by dopant concentration and catalyst
- At low HF and high HNO_3 concentration, iso-etch curves are lines parallel to the acid axis
- The etch rate is controlled by the ability of HF to remove the SiO_2
- In the region of maximum etch rate both the agents play important roles
- The addition of acetic acid diluents instead of water does not reduce the oxidising power of the nitric acid until large amount of diluents is added
- The rate contours remain parallel to constant nitric acid over a considerable amount of added acetic acid diluents

These plots can be used for the study of temperature effects on the HNA system. In the region around the HF vertex the surface reaction rate controls the quality of silicon surfaces and sharpness of the corners and edges (Fig. 5.6). The topology of the silicon surfaces depend strongly on the composition of the etching solution. Around the maximum etch rates the surfaces appear quite flat with rounded edges. Slow etching solutions lead to rough surfaces (Robbins and Schwartz 1960). The inner section marked '1' has a smooth surface and subsequent sections marked '2' and '3' have sharpened edges, and rough surfaces, respectively. The etch rate is highly dependent on temperature variation. The reaction rate is increased as the temperature increases. In the low temperatures, etching can be preferential and the activation energy is developed with oxidation reactions. At higher temperatures, etching creates smooth surfaces.

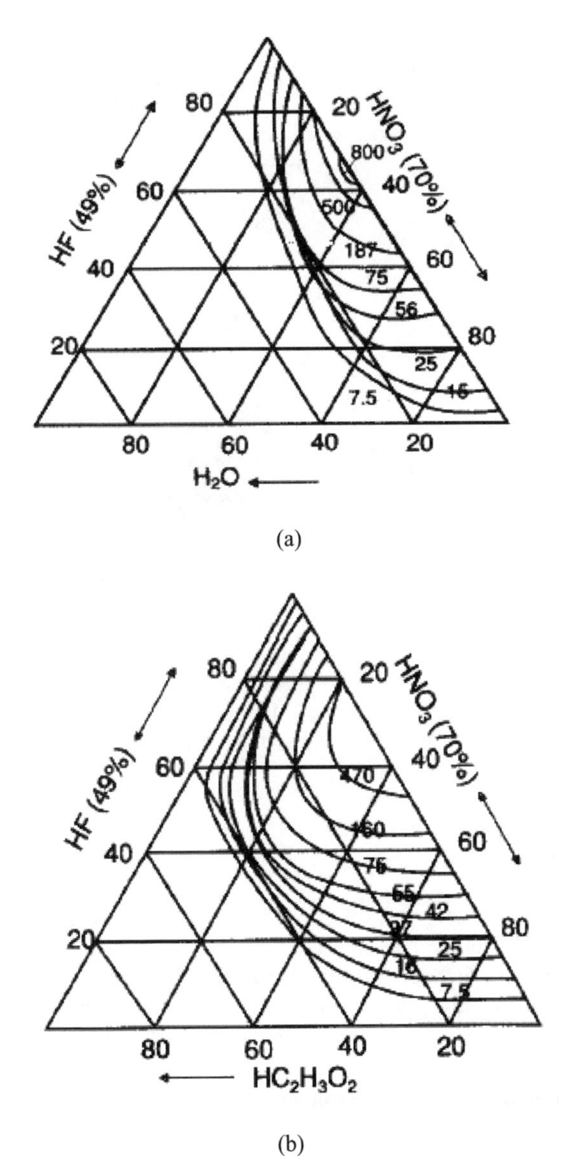

Fig. 5.5. (a) & (b) Isotropic etch curves (with permission from Dr. J-B. Lee, UTD)

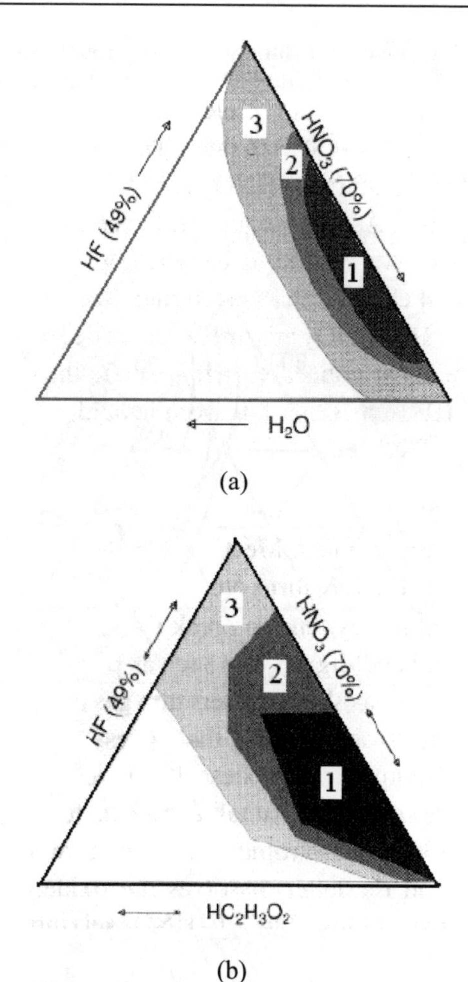

Fig. 5.6. (a) & (b) Topology or Arrhenius plots: (a) Water as diluent (b) Acetic acid as diluent

5.6.4 Masking

Isotropic etchants are very fast and do not differentiate silicon planes, all planes are etched equally. Hence, masking is essential. An etch rate of up to 50 μm/min can be obtained with 66% HNO_3 and 34% HF (Kern and Deckert 1978). Silicon oxide is used as masking material with an etch rate of 300 to 800 Å/min in the above $HF{:}HNO_3$.. One can choose a thick layer of SiO_2 for shallow etching. A mask of non-etching Au or Si_3N_4 is needed for deeper etching. Diluted HF etches Si at a higher rate because the reactions in aqueous solutions proceed by oxidation of Si by OH^- groups (Hu et al. 1986). A typically buffered HF (BHF) solution has been reported to etch Si at radiochemically measured rates of between 0.23 and

0.45 Å/min, depending on the type of doping, and the dopant concentration (Hoffmeister 1969). The doping dependence of the etch rate provides another means of patterning a Si surface. A summary of the dopant dependence (DD) of silicon isotropic etchants is given below.

5.6.5 DD Etchants

The dopant dependence of isotropic etchants is essentially a charge-transfer mechanism. Mostly, the etch rate depends on the dopant type and its concentration. A typical etch rate with an HNA system (1:3:8) for n- or p-type dopant concentration above $10^{18}/cm^3$ substrate is 1 to 3 µm/min. Reduction of the etch rate by 150 times can be obtained in n- or p-type regions with a dopant concentration of $10^{17}/cm^3$ or less (Murarka and Retajczyk 1983). Isotropic preferential etching is a kind of DD etching process. A variety of additives can be mixed into the HNA system (mainly oxidants). They can reduce and modify the etch rate, surface finishes or isotropic characteristics. This effect makes the system selective or preferential. Additives that change the viscosity of the solution can modify the etch rate. One can note that the additive only changes the diffusion coefficient of the reactants (Tuck 1975). An isotropic etchant can be transformed to anisotropic etchant using preferential additives and appropriate masking. Discussion of these effects is not within the scope of this chapter but the basic concept of the anisotropic etching is presented below.

5.7 Anisotropic Etching

Anisotropic etching presents the opposite behavior of isotropic etching. Anisotropic etchants remove the materials based on the crystal plane and do not etch uniformly in all directions. This method is used for the manufacturing of a desired shape and geometry. Anisotropic wet etching techniques were started at the Bell Laboratories in the 1960s. The process is summarised in Table 6.1.

Table 6.1. Anisotropic etching process steps

Process	Duration	Temp. (°C)
Oxidation	Variable	900 - 1200
Spinning (5000 to 6000 rpm)	25-30 sec.	Room temp.
Exposure	20-30 sec.	Room temp.
Develop	60 sec.	Room temp.
Stripping of oxide and resist	500-600 sec.	Room temp.
RCA1(NH3(25%)+ H_2O + H_2O_2:1:5:1)	600 sec.	Boiling
RCA2 (HCl + H_2O + H_2O_2:1:6:1)	600 sec.	Boiling
HF-dip (2% HF)	10 sec.	Room temp
Anisotropic etch	Several minutes	70-100

Anisotropic etching solves the problem of lateral control. The laterally masked geometry of the planar surface can be etched and a vertical profile can be easily made. Etch-stop techniques are also available for anisotropic etchants. Although it solves the problem of lateral etching, the process is not problem-free. The process is slow even in the fast etching direction of the plane (100), with an etch rate below 1 μm/min, and also consumes more time. Anisotropic etchants are temperature sensitive and insensitive to agitation agents.

5.7.1 Anisotropic Etchants

Alkaline solutions such as KOH, NaOH, LiOH, CsOH, NH4OH, and ammonium hydroxides, with the addition of alcohol are primarily used as anisotropic etchants. Alkaline organics such as ethylenediamine, choline (trimethyl-2-hydroxyethyl ammonium hydroxide) or hydrazine with additives such as pyrocathechol and pyrazine are also used. The best fit of the etch rate can be expressed using the empirical formula:

$$R = k\,[H_2O]^4\,[KOH]^{1/4} \tag{5.13}$$

Where, k is a constant. The rate is dopant insensitive over several orders of magnitude, but at certain concentration it shows bias dependence (Allongue et al. 1993). Etching of silicon can occur without external voltage. Alcohols such as propanol and isopropanol butanol slow the attack whereas pyrocathechol speeds up the etch rate. Some additives such as pyrazine and quinone are used as catalysts (Linde and Austin 1992). The etch rate increases with temperature. So etching at higher temperatures can give the best results. However, for all practical purposes, temperatures of 80 to 85°C should be avoided because the solvent starts evaporating at this temperature. The topology plot (also called Arrhenius plot) presented in Fig. 5.6 shows the effect of temperature on silicon etching (Lee et al. 1995). The effect of temperature on etch rate is high, compared to the effect of plane orientation. The slope differs for each plane. While choosing an anisotropic etchant, a variety of issues are considered. Some of them are toxicity, etch rate, etch-stop, compatibility, etch selectivity over other materials, mask material and thickness and topology of the bottom surface.

The principal anisotropic etchants are KOH and ethylene-diamine/ pyrocatechol + water (EDP). More recently, quaternary ammonium hydroxide solution such as tetraethyl ammonium hydroxide (TEAH) has become popular.

5.7.2 Masking for Anisotropic Etchants

SiO_2 cannot be used as a masking material due to its long exposure to KOH etchants. Etch rate of SiO_2 is a function of KOH concentration. The etch rate of thermally grown SiO_2 in KOH-H_2O varies and depends on the quality of the oxide, the etching container and the age of the etching solution. Thermal oxides create a strong compressive force in the oxide layer because one silicon atom takes nearly twice as much space as in single crystalline silicon. Oxides can be removed by pinholes and the etch rate in KOH remains higher than that of thermal oxides. Low pressure chemical vapor deposited (LPCVD) oxides can be used as mask materials of good quality as compared to thermal oxides. LPCVD nitride can serve as a better mask than less dense plasma deposited nitride. Nitride can easily be patterned with photoresist and etched in CF_4/O_2-based plasma. Nitride deposition does not pose any problem. In many applications it is necessary to protect the backside of the wafer from isotropic or anisotropic etchants. This can be done either by mechanical or chemical protection methods. In the former method, the wafer is held in a Teflon holder and in the latter case wax or other organic coatings spin onto the back side of the wafer. Two wafers can be glued back-to-back for faster processing.

Anisotropic etchants can leave the surface rough. This type of roughness can be found during microbridge etching. Macroscopic roughness is called notching or pillowing. Notching increases linearly with etch depth but decreases with higher concentrations of KOH. However, increases in temperature can decrease the roughness of the surface. The macroscopic smoothness can be degraded into microscopic roughness.

5.8 Etching Control: The Etch-stop

It is important to control and stop etching on the wafer when the desired shape and depth of the structure has been reached. An effective etch-stop can produce high resolutions and high yields. Practically, the wafer is not uniform and therefore, the devices cannot be etched uniformly. These effects are impossible to avoid in any microdevice. Advanced techniques in controlling, monitoring, stabilising and visualising the etch rates have been reported. The etch-stop process can be handled through etchant compositions, etchant agents, N_2 sparing, loading effects, etchant temperatures and diffusion effects. The depth of etch depends on temperature, etchant concentration, and stirring control. A stirring can lead to a smooth surface and uniform etching. Light can also affect the etch rate. Sample preparation as well as surface pre-processing can affect the etch rate. The pre-treatment and surface cleaning with 5% HF can remove the native oxides. Normally, a constant etch time method is used for the etch-stop but in the case of thin membranes it can be stopped with sufficient precision. Etch-stop can be prepared using an etch rate dependant dopant. Electrochemical etch-stop can, however, improve the accuracy. It is also easy to handle.

5.8.1 Boron Diffusion Etch-stop

The boron diffusion method is a widely used etch-stop technique. It is based on the fact that anisotropic etchants do not attack boron-doped (p+) Si layers. It can create a stress-free, dislocation-free and slow etching layer. This effect was first discovered by Greenwood. The presence of a p-n junction causes the effect (Greenwood 1969). The boron diffusion method is explained in Fig. 5.7 for micronozzle device fabrication:

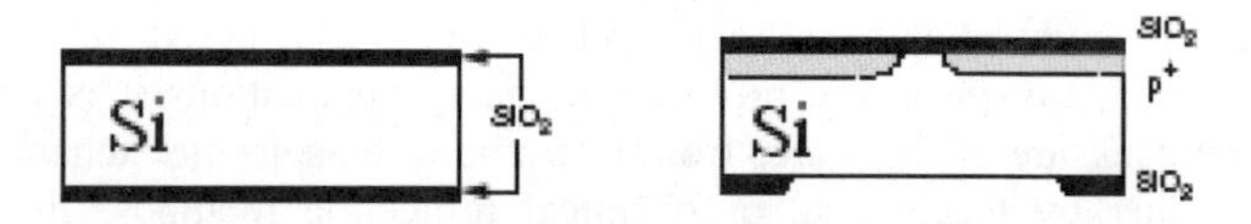

Fig. 5.7. Boron diffusion stop

The wafer is first deposited with SiO_2 followed by lithography and mask patterning. A boron p+ profile is diffused from the back of the wafer prior to the anisotropic etching. The etching process is automatically stopped at the junction. Boron atoms are smaller than silicon, hence a highly doped and freely suspended membrane or diaphragm can be stretched to create tensile stress. The microstructures can become flat and do not buckle. Boron doping causes lattice constant to decrease while germanium doping causes it to increase. A membrane doped with both boron and germanium can be etched much slower than that of undoped silicon. One disadvantage is that high concentration boron doping can not be used for bipolar type CMOS ICs.

The method can only be applied to microstructures where electronics circuits are not needed. Another limitation of this process is that it can only be used for a fixed number and angles of (111) planes. The method is not limited to anisotropic etching but can also be used for isotropic etch-stop.

5.8.2 Electrochemical Etch-stop

For microdevices such as piezoresistive pressure sensors, the doping concentration should be kept as low as $10^{19}/cm^3$ as piezoresistive coefficients may drop beyond this value. Moreover, boron doping affects the quality of the crystal and induces tensile stress and prevents the incorporation of integrated electronics. Therefore, a boron diffusion etch-stop cannot be used for all purposes. An alternative method is called an electrochemical technique that can be used as isotropic etch-stop. A lightly doped p-n junction is used as cathode and a counter electrode as anode in the etchant KOH.

A bias voltage is applied between the wafer and anode. The technique was proposed by Waggener (Waggener et al. 1967). The p-n junction forms as a large diode. A wafer is mounted on a sapphire plate, with an acid-resistant wax and immersed in the solution. An ohmic contact to the n-type epitaxial layer is connected to one pole of the voltage source and the other pole is connected via an ammeter to the solution as shown in the Fig. 5.8. In this process a p-type substrate is etched and stopped at he p-n junction.

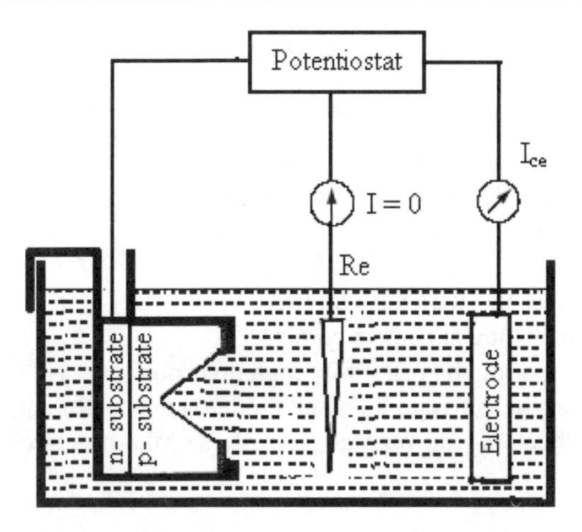

Fig. 5.8. Electrochemical etch-stop

Similarly, an n-type substrate can be etched and stopped at the n-p junction. Electrochemical etch-stop is a highly controlled method and it replaces many high temperature aggressive chemical etching processes. The method utilises a much milder solution with less complexity as far as preparation and transformation of photoresist mask is concerned (Kern and Deckert 1978). In the case of acidic etching, an electrical power supply can be employed to drive the chemical reaction by supplying holes in the silicon surface. The dissolution rate of silicon is related to the current density. Increasing the acceptor concentration can further increase the rate.

The acidic electrochemical technique in micromachining is primarily used for polishing the surfaces. Since the etching rate increases with current density, high spots on the surface are more rapidly etched and very smooth surfaces are produced. Electrochemical etch-stops with anisotropic etchants such as KOH and EDP have been performed (Waggener et al. 1967). In electrochemical anisotropic etching the p-n junction is made by the epitaxial growth of an n-type layer on a p-type substrate.

5.8.3 Thin Films and SOI Etch-stop

Another etch-stop technique is based on the principle of the silicon-on-insulator (SOI) fabrication method. Anisotropic etchant does not react with a number of materials. Hence, this idea can be employed to establish an etch-stop. An example is a Si_3N_4-based diaphragm fabrication. Silicon nitride is a strong and hard chemical and it can form stress in the film. This stress can be transformed from tensile to compressive for thin films. Another example of this category is a SiO_2 layer. A layer of SiO_2 between two layers of silicon can form an excellent etch-stop because many etchants of silicon do not react with SiO_2. SOI is a simple technique compared to the electrochemical etch-stop method.

5.9 Problems with Etching in Bulk Micromachining

Despite its simplicity in controlling the etching in bulk micromachining, wet etching is not a flexible and reliable process (Gad-el-Hak 2002). A dry etching process is an alternative choice. Unavoidable problems associated with this process are as follows:

- Extensive real estate (RE) consumption
- Difficulties in etching at convex corners
- Difficult in preparing the mask with high precision
- Etch rate is very sensitive to both agitation and temperature making it difficult to control both lateral and vertical geometries of the structure

5.9.1 RE Consumption

Bulk micromachining is involved with extensive real estate consumption because the process removes a comparable amount of materials from the bulk substrate. Fig. 5.9 illustrates the fabrication of two membranes created by etching through a <100> wafer from the backside until an etch-stop is reached. A large amount of silicon has been wasted to create this structure. The consumption of silicon can not be stopped but it can be reduced best by practice and careful design as well as appropriate etching techniques. This is defined as real estate gain by reducing the consumption of wafer materials. Two different real estate gain techniques are:

- RE gain by etching from the front
- RE gain by using silicon bonded wafers

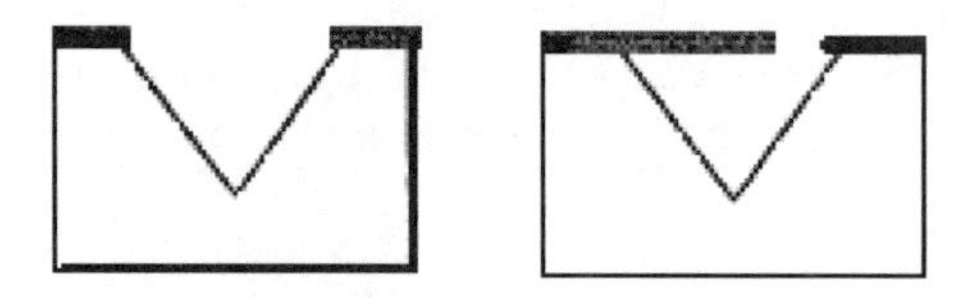

Fig. 5.9. Wastage of material

RE can be gained by removing fewer amounts of material and the use of thinner bulk instead of thick wafers. This approach is useful for wafers with a thickness more than 200 μm. An alternative solution is to etch the wafer from the front. Anisotropic etchants can remove an amount of materials depending on the orientation of the wafer with respect to the mask, and etch until a pyramidal pit is formed (Fig. 5.9). These produce an angle of 54.7° to the surface of the silicon wafer. The etching stops when the pyramid pit is completed. Hence, there is no further wastage of the wafer.

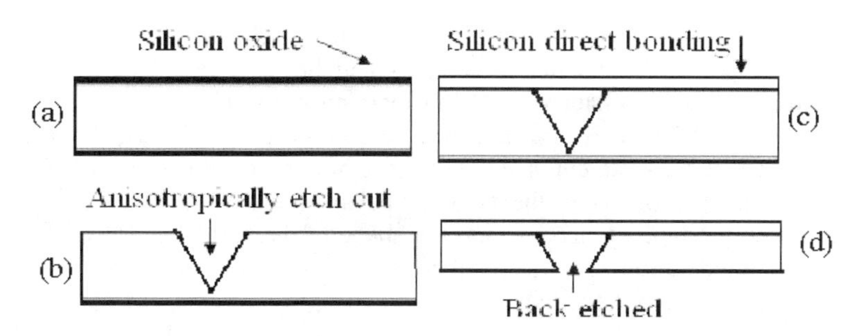

Fig. 5.10. Real estate gain by silicon diffusion bonding (SDB)

Another process to gain real estate is use of thinner wafers and join them together using the silicon direct bonding (SDB) technique. SDB is a process to bond two silicon wafers by applying pressure and annealing at a certain temperature. An example is given in Fig. 5.10. A groove (Fig. 5.10(a)) can be made using two parts such as (b) and (c). First, a thin silicon wafer <100> is taken and then it is etched to form a pyramid pit as shown in Fig. 6.10(b). In the second stage it is bonded or fused with another thin wafer to produce another desired structure (Fig. 5.10(c)). Finally, the bonded wafer is etched from the backside of the surface in order to get the final desired shape as shown in Fig. 5.10(d). The RE gain by using silicon fusion-bonded wafers can reduce the consumption of wafers, waste of etchants and time. This process saves RE and makes the design and etching simple.

5.9.2 Corner Compensation

The corners of the structure do not etch uniformly especially in anisotropic etching. In microdevices it is necessary to preserve the shape while etching. In practice, difficulty arises due to the fact that some planes etch faster than others. This results in a loss of the desired structure. The effect can be reduced through a technique called corner compensation. Convex corners can be made by adding extra structures and then removing them during etching. The effect of corner etching can lead to either underetching or undercutting of the substrate.

In general, convex corner structures and non-crystal planes are undercut during wet anisotropic etching. These characteristics have been utilised to fabricate freely suspended structures. Conversely, these effects need to be prevented in various applications. Vertical posts and vertical walls have been employed to protect convex corners and non-crystal planes from undercutting. Experimental results have shown that cavities with arbitrary shapes can be fabricated using wet anisotropic etching. In the case of etching rectangular corners, deformation occurs at the edges due to undercutting. This type of effect is unwanted where symmetry and perfect 90° corners are essential. The undercutting effect is dependent on etch time and related to the etch depth. The effect can be reduced by using saturated KOH solutions with isopropanol. This may cause a formation of hillocks and piles on the surface. Peeters has claimed that these hillocks and piles are due to carbonate contamination of the etchants and he has suggested that the etchant ingredients should be stored under an inert nitrogen atmosphere and the etching process should be carried out under an inert atmosphere (Peeters 1994). Another way to reduce the undercut is at the expense of a reduced anisotropy ratio. A mask is added to the corners in the layout. The thickness of the mask and the shape depend on the type of etchants and the shape of the corner. In the case of square corner compensation, the square of SiO_2 in the mask is enhanced by adding an extra SiO_2 square to the corner.

5.10 Conclusions

Microelectromechanical systems (MEMS) are fabricated by using three methods which include: i) bulk micromachining, ii) surface micromachining, and iii) wafer bonding. The most important and common sub-process within the above methods is etching. This chapter characterised the etching process in bulk micromachining. In particular, the chapter discussed the following topics.

- Silicon crystallography
- Silicon as the material
- The importance of properties of material with respect to etching process
- The iso-etch curves
- Thermal properties
- The difference between the isotropic and anisotropic etching and etchants
- Various types of etch-stops and their techniques
- The problems encountered in the bulk micromachining process

5.11 References

Allen DM (1986) The principles and practice of photochemical machining and photoetching. Inst of Physics Pub Inc. pp 1-18
Allongue P, Kieling VC, Gerischer H (1993) Etching of Si in NaOH solutions, Part I and II, J. Electrochem. Soc. 40:1009-1018 and 40:1018-1026
Ammar ES, Rodgers TJ (1980) UMOS transistors on (110) silicon, IEEE Trans. Electron Devices ED-27:907-914
Bassous E, Taub HH, Kuhn L (1977) Ink jet printing nozzle arrays etched in silicon, Appl. Phys. Lett. 31:135-137
Boivin LP (1974) Thin film laser-to fiber coupler, Appl. Opt. 13:391-395
Editorial (2004) Analog Devices combines micromachining with BICMOS, Semiconductor International, available at http://www.reed-electronics.com
Editorial (1974) Transducers, pressure and temperature catalog. White paper, National Semiconductor, Sunnyvale, USA
Editorial (1977) Thermal character print head. Technical Report, Texas Instruments, Austin

Gad-el-Hak M (2002) The MEMS handbook. CRC press, pp 200-211
Greenwood JC (1969) Ethylene Diamine-Cathechol-Water mixture shows preferential etching of p-n junction, J. Electrochem. Soc. 116:1325-1326
Greenwood JC (1984) Etched silicon vibrating sensor, J. Phys. E, Sci. Instrum. 17:650-652
Hallas CE (1971) Electropolishing silicon, Solid State Technology 14:30-32
Harris TW (1976) Chemical Milling. Clarendon Press, Oxford
Hoffmeister W (1969) Determination of the etch rate of silicon in buffered HF using a 31 Si tracer method, Int. J. Appl. Radiation and Isotopes 2:139
Hu JZ, Merkle LD, Menoni CS, Spain IL (1986) Crystal data for high-pressure phases of silicon, Phys. Rev. B 34:4679-4684
Kaminsky G (1985) Micromachining of silicon mechanical structures, J. Vac. Sci. Technol., B3:1015-1024
Kern W, Deckert CA (1978) Chemical etching in thin film processes. Vossen JL, Kern W (eds) Academic Press, Orlando
Lee JB, Chen Z, Allen MG, Rohatgi A, Arya R (1995) A miniaturised high-voltage solar cell array as an electrostatic MEMS power supply, J. Microelectromech. Syst. 4:102-108
Linde H, Austin L (1992) Wet silicon etching with aqueous amine gallates, J. Electrochem. Soc. 139:1170-1174
Madou MJ (1997) Fundamentals of microfabrication, CRC Press, New York
Madou MJ, Morrison SR (1989) Chemical sensing with solid state devices. Academic Press, New York
Michalicek AM (2004) Introduction to MEMS. http://mems.colorado.edu/c1.res.ppt /ppt/g.tutorial
Murarka SP, Retajczyk TFJ (1983) Effect of phosphorous doping on stress in silicon and polycrystalline silicon, J. Appl. Phys. 54:2069-2072
O'Neill P (1980) A monolithic thermal converter, Hewlett-Packard J. 31:12-13
Palik ED Faust JW, Gray HF, Green RF (1982) Study of the Etch-Stop Mechanism in Silicon, J. Electrochem. Soc. 129:2051-2059
Peeters E (1994) Process development for 3D silicon microstructures, with application to mechanical sensor design, Ph.D. thesis, Catholic University of Louvain
Petersen KE (1982) Silicon as a mechanical material, Proc. IEEE, 70:420-457
Pfann WG (1961) Improvement of semiconducting devices by elastic strain, Solid State Electron. 3:261-267
Robbins H, Schwartz B (1960) Chemical etching of silicon-II: The system HF, HNO3, H2O, and HC2C3O2, J. Electrochem. Soc. 107:108-111
Rodgers TJ Hiltpold WR, Frederick B, Barnes JJ, Jenné FB, Trotter JD (1977) VMOS memory technology, IEEE J. Solid-State Circuits SC12:515-523
Schwartz B, Robbins H (1976) Chemical etching of silicon-IV, Etching Technology, J. Electrochem. Soc. 123:1903-1909
Smith CS (1954) Piezoresistance effect in germanium and silicon, Phys. Rev., 94:42-49
Tuck B (1975) The chemical polishing of semiconductors, J. Mater. Sci. 10:321-339
Tuft ON, Chapman PW, Long D (1962) Silicon diffused-element piezoresistive diaphragms, J. Appl. Phys. 33:3322
Turner DR, (1958) Electropolishing silicon in hydrofluoric acid solutions, J. Electrochem. Soc. 105:402-408
Uhlir A (1956) Electrolytic shaping of germanium and silicon, Bell Syst. Tech. J. 35:333-347
Waggener HA, Krageness RC, Tyler AL (1967) Two-way etch, Electronics 40:274

6 Features of Surface Micromachining and Wafer Bonding Process

S. K. Bag[1], D. S. Mainardi[2] and N. P. Mahalik[3]

[1]Department of Electronics and Communication Engineering, Vellore Institute of Technology (Deemed University), Tamilnadu, India
[2]Institute for Micromanufacturing, Louisiana Tech University, USA
[3]Department of Mechatronics, Gwangju Institute of Science and Technology, Gwangju, Republic of South Korea

6.1 Introduction

Micromachining technology manufactures microdevices by implementing features in bulk materials such as silicon, quartz, SiC, GaAs, InP, Ge and glass. The features are implemented through several processes. One of the important sub-processes in micromachining is *etching*, which means the removal of selective materials. Etching is performed either on the substrate or on a preferred material layer deposited on the substrate. The very meaning of *substrate* (also called wafer) is the material upon or within which plants or animals live, however, in the context of microsystems, MEMS (Microelectromechanical Systems) components (also called *features*) and circuitry are used. In essence, a substrate is a sheet of base material in which mechanical parts, electronic components and integrated circuits (IC) are built by the process of etching. The etching process is used to remove a defined portion of the substrate in a particular manner so that the desired shape can be obtained. The process selectively removes material from the substrate. The material to be removed is determined by the etching solution (called *etchant*). Fig. 6.1 provides an insight into the etching process. Some typical structures such as plates, steps, grooves, cantilever, diaphragm, post, etc. can be fabricated or micromachined through etching. For more information on etching refer previous chapter.

The process of manufacturing of microdevices is known as *fabrication*. We will use the word "fabrication" as a synonym to micromachining, but the former one is more frequently used for IC manufacturing. As microsystem design approach is very compatible to the IC design method so they are used interchangeably. The fabrication processes of MEMS devices broadly fall under two categories: bulk micromachining and surface micromachining. There are considerable differences in the above two processing technologies, leading to differences in the fabricated structures. Both the technologies are widely used and are still being further developed. In some applications the technologies are in direct competition, whereby one dominates its counterpart. But both the methods have many common approaches as they heavily rely on the following principles and sub-processes:

- Wafer cleaning and deposition
- Photolithography and pattern transformation
- Doping (diffusion and ion implantation)
- Etching
- Metalisation with sputtered, evaporated or plated Al/Au/Ti/Pt/Cr/Ni.

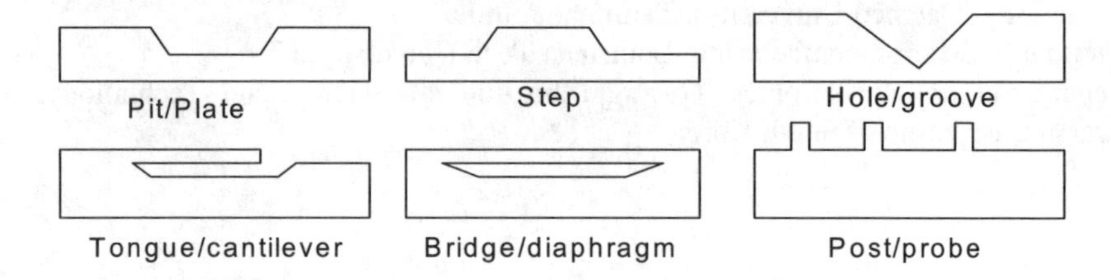

Fig. 6.1. Typical basic MEMS structures, which can be obtained by etching

6.2 Photolithography

The microcomponents and features in microdevices are created either from the substrates or from the thin layer(s) of some specific materials (e.g., the silicon dioxide layer). Assume that we are interested in fabricating a simple structure such as a *step* as shown in Fig. 6.1. An oxide layer is first deposited on the substrate (Refer to Fig. 6.2). The layer is not entirely used to create the desired component. This is due to the fact that a typical microdevice might contain many components, which could be built from the same layer. Justifiably, the components are derived from the layer. Therefore, the deposited layer has to be segmented. Subsequently, selective portions are removed or etched away so that the desired components can be produced. The segmentation process is usually achieved by a technique called *photolithography*. Photolithography is thus an essential, as well as one of the starting sub-processes, used to delineate the shape of a microstructure at the early stage. Segmentation characterises the removal of some portion from the layer. The portions of the layer, which need not be removed, are to be protected by some means. In fact this is achieved by the use of a mask, which is prepared in advance. The foremost phase of photolithography is therefore the preparation of a *photomask*. Once prepared, it is then transferred onto the thin layer (Oxide layer).

Lithography is a Latin word which means stone writing. Stone writing is a printing method that utilises flat inked surfaces to create the printed images.

Although, the early principle is not utilised in the process of developing microdevices, the meaning is similar. Photolithography is the process of using light to create a pattern, i.e., mask and subsequently transfer it onto the substrate. Photolithography is an optical means of transferring patterns.

In the meantime, the substrate (wafer) is chemically cleaned to remove particulate matter as well as any traces of organic, ionic, and metallic impurities on the surface. The wafer cleaning process and photomask preparation can go alongside. Fig. 6.2(a) shows a thin film of some selective material (e.g., silicon dioxide). Depending upon the design requirement however, other materials such as metals, alloys, etc. can also be used, which are deposited on a substrate of some other wafer material (e.g., silicon). It is required that some of the silicon dioxide is to be selectively removed so that it only remains in particular areas on the substrate.

For this we need to produce a mask. The photomask used in the photolithographic phase is a key component in the process. The mask is typically a glass plate that is transparent to ultraviolet (UV) light. The pattern of interest is generated on the glass by depositing a very thin layer of metal, usually chromium or gold. These masks are capable of producing very high quality images of micron and even sub-micron features (Fig. 6.2(c)). Photomasks are normally prepared by the help of a computer-assisted software platform (Fig. 6.2).

The next phase of the process is the coating of the photoresistive material on the wafer. This material should be sensitive to UV light. Organic polymers are usually chosen. Once the photoresist coating is completed, the photomask is placed over it as shown in Fig. 6.2(d). Then, UV light is allowed to fall on the mask. UV light is known as the *exposing energy*. The source of UV light is a mercury arc lamp, called a radiator, which has an output with spectral energy peaks at particular wavelengths. Depending upon the feature complexity, thickness and photoresist materials, the exposing process may be performed once or several times, accordingly, we have single-exposure fabrication and multi-exposure fabrication. The feature size of the bottom of the structure becomes smaller than that of its top after the development. This may cause the features to break near the bottom. Therefore, accurate doses of UV light are of paramount importance. Sometimes, additional exposure may cause large internal stress in the resist layer leading to poor adhesion between the resist layer and the substrate. The internal stress may lift-off the structure from the substrate during subsequent processes. Thus, a compromise is sought. Light may fall on the entire area of the photoresist in one go or in sequence by employing a scanning technique. Scanning is achieved by moving a small spot of light over the desired area of photoresist. When UV light falls on the photoresist (Fig. 6.2(e)) through the mask a similar pattern is developed on the photoresist layer. This phenomenon is called mask transformation or pattern transformation. The transformation is thus achieved by the combined effect of UV light and the composition of the photoresist. Photoresist can be either soluble or insoluble after being exposed to UV light (Fig. 6.2(f)&(i)). Accordingly, two types of photoresist exist, positive and negative photoresist, respectively. The soluble photoresist becomes weakened when exposed to UV light. On the other hand, the opposite occurs to insoluble photoresist. In practice, the photoresist is washed away in the region where the light was struck, conversely, the negative photoresist is not. This assures that the resist, which was not exposed to UV light, is washed away forming a negative image of the mask. The process of washing away the material is actually called etching.

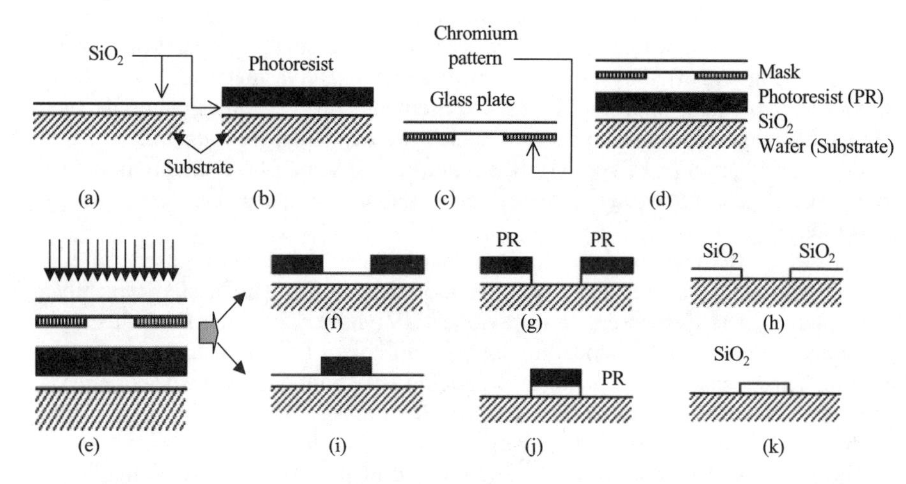

Fig. 6.2. Photolithography; (a) A substrate with SiO_2 layer, (b) Additional photoresist layer (c) The mask - glass plate with chromium pattern, (d) The mask is on the top of the photoresist-SiO_2-wafer layers, (e) UV light falling on the photoresist, (f) After exposure to UV light (positive PR), (g) Third phase of photolithography in which the opening portion of the SiO_2 layer is removed (positive PR case), (h) Final phase of photolithography in which the PR is removed (positive photoresist), (i) After exposure to UV light (negative PR), (j) Third phase of photolithography in which the opening portion of the SiO_2 layer is removed (negative PR case), (k) Final phase of photolithography in which the PR is removed (negative PR)

The photoresist is coated on the surface of the SiO_2 layer (in this typical example) by a process called *spin-coating.* In spin-coating, as it implies, the wafer is rigidly placed on a rotating base called vacuum chuck. The photoresist polymer solution is put on the solid wafer. The solvent in the solution is usually volatile. The rotor is then allowed to rotate at high speed. This rotation causes the solution to spread over the wafer surface while allowing the volatile solvent to evaporate. Depending upon the required thickness and uniformity the rotation is continued. When the solution is spun off the edges of the substrate due to centrifugal force, the rotation is stopped. The following formula is useful (Eq.6.1a) while calculating the time of spin t, for a given rotational speed w. From this equation the ratio of the thickness can also be obtained (Eq.6.1b).

$$t = \frac{3\mu}{4\rho w^2}\left(\frac{1}{h^2} - \frac{1}{h_0^2}\right) \tag{6.1a}$$

$$\frac{h}{h_0}=\frac{1}{\sqrt{1+\frac{4\rho w^2}{3\mu}h_0^2 t}} \tag{6.1b}$$

Where h_o is the initial height, h is the final height or thickness, μ and ρ are the viscosity and density of the solution, respectively.

In the third phase, the portion of the oxide layer that is now exposed through the openings of the photoresist is removed by some chemical processes (Fig. 6.2(g)&(j)). Finally the photoresist is removed, leaving the desired segmented oxide layer (Fig. 6.2(h)&(k)). The results shown in Fig.6.2(f),(g)&(h) have undergone the processes in which the positive photoresist is used and that of the Fig. 6.2(i),(j)&(k) have undergone the processes in which the negative photoresist is used. The lithography method described in the previous section is called optical lithography because it uses UV light as the exposing energy. Other exposing energies such as X-ray and electron beams can also be employed. Besides the consideration of the source of exposing energy, lithography is also characterised by hard- or soft-lithography based on the configurability of the mask.

6.3 Surface Micromachining

In bulk micromachining processes the materials are removed by different etching techniques (described in the previous chapter) while in surface micromachining processes the materials are added through thin film deposition techniques in order to form micromechanical structures. Thin films are deposited using low pressure chemical vapor deposition (LPCVD) of polycrystalline silicon, silicon nitride and silicon dioxides. The films can be deposited sequentially and removed as necessary in order to build microdevices. Surface micromachining is a relatively simple method as compared to bulk micromachining. It can produce a very thin and higher precision structure than bulk micromachining, and the process can be controlled more easily. There are many advantages in surface micromachining as compared to bulk micromachining processes. A comparison between surface and bulk micromachining is presented in the following section.

In the history of literature, surface micromachining fabrication process began in the 1960's, but it has been rapidly expanded over the past few decades. The application of surface micromachining to batch fabrication in MEMS devices was put into practice in recent years. The first example of a surface micromachined device was a resonant gate transistor (Nathanson and Wickstrom 1965). The height for this kind of device was limited to less than 10 micrometers. It was shaped like an x-y plane surface and hence the process has been named the surface micromachining process. A survey indicates that Gabriel et al. presented the first possible application of polysilicon surface micromachining in 1989 (Gabriel et al. 1989). Subsequently, microscale movable parts like pin joints, gears, springs and many other mechanical and optical components have been developed (Editor 1997; Editor 1991). The first commercial product based on surface micromachining, namely, the ADXL-50, a 50-g accelerometer for activating air-bag deployment was developed by Analog Devices (Editor 1991). Thereafter, Digital Micromirror Device™ was developed by Texas Instruments (Editor 1977). Recently, there are many commercial devices available which are manufactured using surface micromachining processes. Four major issues are dealt while developing the surface micromachining conformant devices.

- The understanding and control of the material properties of microstructure films (e.g., Polycrystalline silicon, silicon oxide and silicon nitrite)
- Design constraints such as size and layer positioning
- Adoption of precision releasing methods which consider the film stress and sticking properties of the materials
- The packaging methods

A brief overview in this regard is presented in this chapter. The following section summarises the difference between the bulk and surface micromachining processes. The properties of microstructures, bonding and packaging processes are also discussed.

6.3.1 Bulk versus Surface Micromachining

Surface micromachining is mainly active in the *x-y* plane and it is relatively difficult to process on the *z*-axis. By definition, bulk micromachining removes materials from the substrate in either direction, while surface micromachining normally adds a thin film. Bulk and surface micromachining have many sub-processes in common. Both the methods employ photolithography, oxidation, diffusion and ion implantation techniques. CVD (Chemical Vapour Deposition) method is adopted for oxide, nitride and oxy-nitride formation. Plasma etching is used in both processes. They also use many common materials like polysilicon, Al, Au, Ti, Pt, Cr and Ni. However, the processes differ in many ways like in the use of anisotropic etchants, anodic and fusion bonding, use of Si <100> as a starting material, different types of etch stops, double-sided processing and electrochemical etching in bulk micromachining, and the use of dry etching in patterning and isotropic etchants in release steps in surface micromachining. The use of polysilicon in surface micromachining avoids the fabrication difficulties associated with bulk micromachining. Polysilicon increases the freedom of the integration of complicated microdevice features. The use of a sacrificial layer in surface micromachining has advantages in the assembly of thin and tiny mechanical structures. Bulk micromachining uses larger die, waste material, and involves high production costs as compared to its counterpart. There is a $\sqrt{2}t$ effect in the bulk micromachining structure of thickness *t*, and there is no such effect on surface micromachining. Bulk micromachining does not induce any stress but this is the biggest limitation and disadvantage of surface micromachining. A brief dimensional comparison is drawn in Table 6.1.

Table 6.1. Comparison between bulk and surface micromachining (Source - J. Micromechanics and Microengineering 8 (1998))

Property	Bulk	Surface
Processing complexity	0	0/+
Lateral dimensions	3–5 mm	100–500 μm
Vertical dimensions	100–500 μm	0.5–2 μm

6.4 Characterising the Surface Micromachining Process

From the definition of the surface micromachining process, one can note that deposition of materials layers is used to create the desired structure of a typical device. Fig. 6.3 illustrates the surface micromachining process.

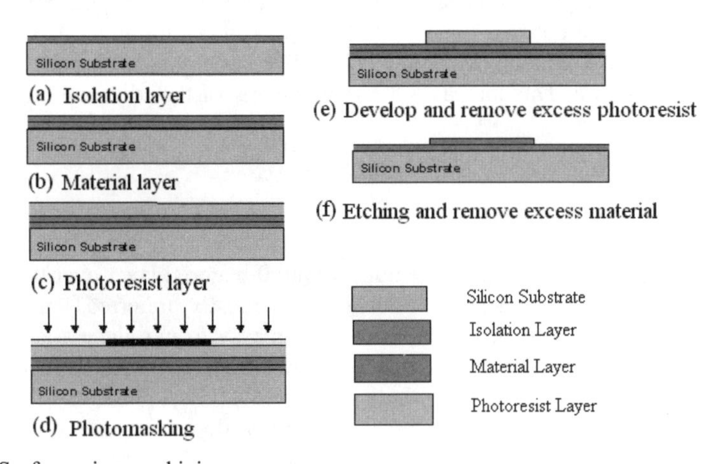

Fig. 6.3 Surface micromachining process

6.4.1 Isolation Layer

A silicon substrate is used as the ground plane. The first step in the surface micromachining process is the deposition of a thin isolation layer. This layer is deposited with a dielectric material such as silicon oxide (SiO_2) followed by a thin layer of silicon nitrite. The latter can act as an etch stop for many etchants. The isolation layer itself is configured to make a metal contact with the base substrate. A pattern is transferred to the isolation layer using isotropic etchant such as buffered HF (BHF). The ratio of NH_4F and concentrated HF is 5:1. This solution can effectively etch the SiO_2 layer. A typical etch rate is 100 nm/min. The etching may be monitored by changes of the color pattern or by supervision of the hydrophobic and/or hydrophilic behavior of the etched layer (Hermansson et al. 1991). The resist opening is the same size as the oxide thickness. After isotropic etching, it is emerges into a piranha etch bath. A strong oxidiser is then grown over the cut region. Later it is removed using diluted BHF. This is a known as cleaning. Drying follows cleaning. Once both cleaning and drying is completed, the substrate is ready for further deposition. A dry etch using CF_4H_2 is applied to cut the oxide. The process prevents the undercutting of the resist. But it has a long processing time. Etching of the isolation layer is possible by using a dry etch process followed by BHF (Tang 1990).

6.4.2 Sacrificial Layer

A phosphosilicate glass (PSG) layer can preferably be used as the sacrificial layer. Adding phosphorous and SiO_2 onto PSG can enhance the etch rate in HF (Monk et al. 1992; Tenney and Ghezzo 1973). The important benefit is that SiO_2 can behave as a solid-state diffusion dopant source making the polysilicon layer electrically conductive. The deposited phosphosilicate makes a nonuniform etch rate in HF. It is normally carried out in a furnace at 900°C to 1000°C for about two hours in a rich oxygen environment (Madou 1997). PSG undergoes a viscous flow at a certain temperature. This increases the smoothening of the edge and improves the etching condition.

6.4.3 Structural Material

The next step is the deposition of a thin structural material layer over the isolation layer. A chemical vapor deposition technique is mostly preferred, but sometimes sputtering is well suited. In the latter case, it is easy to introduce a tapered edge (Madou 1997). The commonly used structural material in microfabrication is polysilicon (poly-Si). It is essentially deposited at low pressure. The applied pressure is about 100 Pa at 500-700°C. A normal deposition rate of 200 Å/min can be achieved in an environment of temperature 500°C, pressure 200 Pa with a silane flow rate of 125 sccm. A layer of 1-µm film will take about one hour and 30 minutes. The silane is usually diluted with 70 to 80% nitrogen and deposited at temperatures from 500 to 700°C for fine-grain and from 600-800°C for coarse-grain applications. Additionally, annealing in nitrogen will reduce stress formation due to thermal mismatch of expansion coefficients. Depending on the nature of the application (e.g., sensor or actuator) other materials such as silicon nitride, silicon oxynitride, polyimide, diamond, SiC, GaAs, tungsten, α-SiH, Ni, W, and Al are also used as structural materials in surface micromachining. Silicon nitride and silicon oxide can be deposited by CVD methods but they exhibit high residual stress, which hampers their use as mechanical components (Chang et al. 1991). However, a mixed silicon oxynitride can produce substantially low or stress-free components. Other problems include oxidation at 500°C for non-passivated films.

Moreover, they do not provide good ohmic contact, and therefore are unreliable for electric contacts (Obermeier 1995).

6.4.4 Selective Etching

Surface micromachining process sequences are presented in Fig. 6.4. We like to consider silicon as the ground plane substrate because of the reasons mentioned in chapter 5.

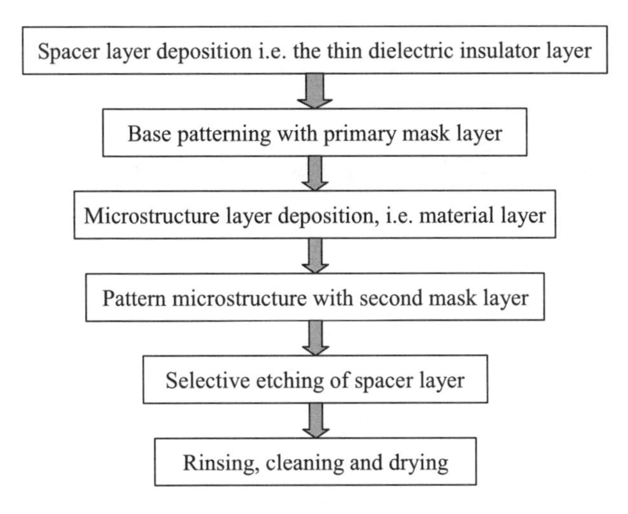

Fig. 6.4. Surface micromachining process sequence

Selective etching is employed to create movable mechanical parts in the microstructure. The structure is then freed from the spacer or sacrificial layer. Consider the design of a poly-Si based microstructure utilising RIE (Reactive Ion Etching). After patterning the poly-Si by RIE in SF_6 plasma, it has to be immersed in an HF solution to remove the underlying sacrificial layer, which makes it possible to release the structure from the substrate. A typical sacrificial layer of phosphosilicate glass, between 10 to 2000 μm long and 0.5 to 5 μm thick, is etched in concentrated buffered HF. The etch rate has to be fast in order to avoid an undesirable attack on the structural element and the insulation layer. The etch rate of PSG can be increased monotonically with concentration (Monk et al. 1994). High concentration phosphorous-doped polysilicon is prone to attack by HF. Silicon nitride deposited by LPCVD etches slower in HF than in oxide films. Kinetic and diffusion reactions are noticeable in short and long channels, respectively. The reaction therefore, shifts from kinetic- to diffusion-controlled when the channel is longer. Diffusion is observed over 300 μm of the channel in concentrated HF, affecting a large structural area (Fan et al. 1988; Mehregany et al. 1988). Rinsing and drying follow etching. Extended rinsing allows native oxide to form on the surface of the polysilicon structure. Such a layer may be desirable and can be formed more easily by a short dip in 30% H_2O_2. Surface micromachining processes, therefore, depend on the properties of the materials. Important material properties are discussed in more detail.

A simple surface micromachined cantilever beams is illustrated in Fig. 6.5(g) and (h). The polysilicon has been deposited and patterned using the RIE technique, followed by wet etching of the oxide layer under the beams in order to free them from the substrate (Fig. 6.5(h)). It shows that surface micromachining is a powerful technique for producing complicated 3D microstructures. Some of the other structures are tweesers, gear trains and micromotors.

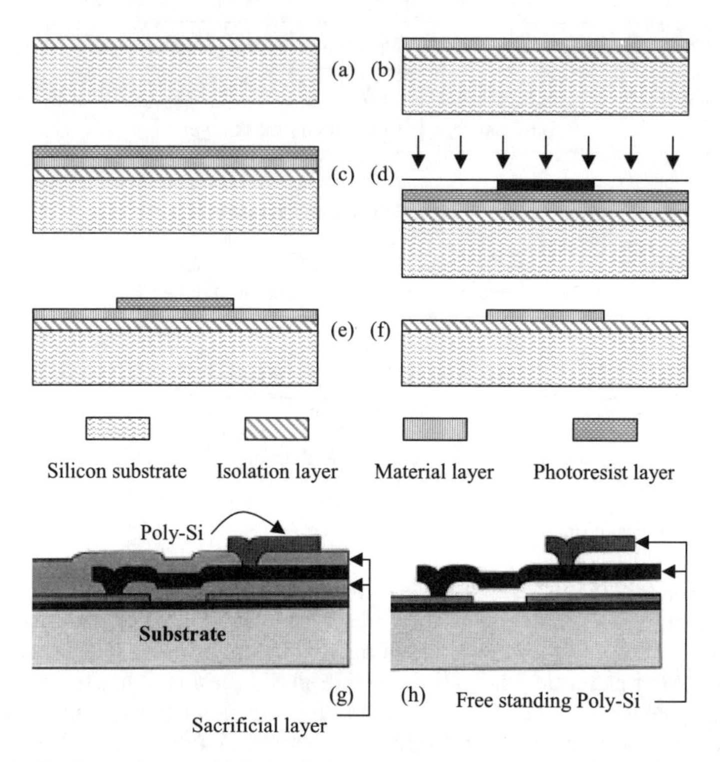

Fig. 6.5. Surface micromachining (a) Isolation layer (sacrificial or spacer layer) (b) Material layer (c) Photoresist layer (d) Photomasking (e) Removal of excess photoresist (f) Selective etching of spacer layer (g) An example of micromachining a cantilever structure (h) After micromachining (Free standing cantilever)

6.5 Properties

The fabrication process of microdevices is presented in the previous section. In order to better understand the process, there is a requirement to study the properties of the material. For example, a thin film in a surface micromachined device has to satisfy a large number of rigorous chemical, mechanical as well as electrical properties. The study can enable us to know whether or not the materials to be used are chemically, mechanically and electrically suitable. For instance, in a particular fabrication, the features such as (i) good adhesion but low in residual stress and pinhole density and (ii) good mechanical strength and chemical resistance, may be desirable. Microdevices are not manufactured for load bearing purpose, hence mechanical strength is sometimes less important. Material properties depend mainly on the deposition process and the growth conditions. Some properties depend on post-deposition thermal processing, such as annealing and surface orientation. The properties of materials change significantly from its bulk state to thin film or microstructure state (e.g., cantilever). For example, thin films display smaller grain size than bulk ones. An important reason for these differences comes from the fact that the thin films exhibit a higher surface-to-volume ratio than large chunks of material. This signifies that the thin film is strongly influenced by its surface properties (Madou 1997). With regard to physical and mechanical characteristics, thin films are also not intended for load-bearing applications. Understanding of mechanical properties is essential for improving the reliability and lifetime of thin films as well as microstructures (Vinci and Braveman 1995). The most influential considerations for surface micromachining are the understanding of stress and stickiness properties. However, others are equally important. Some of the important properties of thin film are described in the following subsections. We preferred to focus on thin films, as the manufacturing of these structures is especially crucial.

6.5.1 Adhesion

Adhesion is defined as the capability of bonding between two surfaces of the same or different materials. It depends on a number of factors including atomic structure, surface roughness and thermal condition. It is one of the most important properties of silicon micromachining materials. There are some other mechanical forces (axial and longitudinal) involved, but the effect of adhesion is more crucial in micromachining. A device may get damaged if its film is lifted from the ground substrate by a repetitive external mechanical force. Adhesion of various films and substrates is equally important as compared to the overall performance and reliability of microdevices. There are various tests done for this. Important adhesion tests include the scotch tape test, abrasion method, scratching, ultrasonic and ultracentrifuge tests, bending, and pulling.

Adhesion can be significantly improved by cleaning the substrate and using a rough bonding surface (Campbell 1970). The latter improves mechanical interlocking. Another technique to improve adhesion is by increasing the adsorption energy of the deposit and/or increasing the number of nucleation sites in the early phase of the film growth. The value of the sticking energy between the film and substrate ranges from 10-20 kcal/mole for physical absorption and chemical absorption, respectively, and the weakest form of adhesion involves van der Waals forces. van der Waals forces are weak physical forces that hold two molecules or two different parts of the same molecule together.

6.5.2 Stress

Understanding stress in thin films is the most important consideration in microstructure fabrication. Stress is developed during crystallisation and deposition. Development of stress is risky for the longevity of the component and can cause malfunctioning during normal operations. Stress on a thin film can be analysed qualitatively as well as quantitatively.

Qualitative descriptions of thin film are reflected through cracks, de-lamination and voids. Most the thin films show a state of residual stress. This is mainly due to mismatch in the coefficient of thermal expansion and non-uniform plastic deformation. Other main factors, which affect the thin films, are lattice mismatch, interstitial impurities and growth processes. The stress causing factors can be intrinsic and/or extrinsic. Intrinsic stresses, called growth stresses, are developed during film nucleation. Extrinsic stresses are introduced by external factors such as temperature gradient and packaging. Thermal stress is the most common type of extrinsic stress. It arises either in a structure (with non-homogeneous coefficients) subjected to a uniform temperature change or in a homogeneous material exposed to a thermal gradient (Krulevitch 1994). Intrinsic stresses in thin films are always larger than extrinsic stresses. They are developed due to non-uniform deposition of thin films. An example of a film deposition method is the CVD process. In a typical CVD process the atoms deposited at the beginning of the process occupy lower energy configurations than the latter ones. If the deposition rate is too high or low, atomic surface mobility gets disturbed thereby developing an intrinsic stress. Additionally, extrinsic stresses occur when parts of a material undergo a volume change during a phase transformation. Moreover, misfit stresses arise in epitaxial film due to lattice mismatches between the film and substrate. Interstitial and substitution impurities impose intrinsic residual stress. Local expansion or contraction is associated with point defects. Intrinsic stress in a thin film does not create delaminating unless the film is thick enough. High stress can result in the buckling or cracking of films.

The quantity of thin film stress can be calculated from the basic stress and strain formula. The sum of the stresses is equal to the external applied stress (σ_{extn}), an unintended external thermal stress (σ_{th}), and the intrinsic stress components (σ_{int}). The total stress in a thin film is given by:

$$\sigma_{total} = \sigma_{extn} + \sigma_{th} + \sigma_{\text{int}} \tag{6.2}$$

The three non-vanishing stress components are functions of x and y directions of the Cartesian co-ordinate plane. There is no stress in the direction normal to the substrate. In case the principal axes of the substrate coincide with the x, y axes of the plane, the shear stress τ_{xy} vanishes (Chou and Pagano 1967). With constant stress through the film thickness, the stress components can be written as:

$$\begin{aligned} \sigma_x &= \sigma_x(x,y) \\ \sigma_y &= \sigma_y(x,y) \\ \sigma_z &= 0 \\ \tau_{xy} &= \tau_{xy}(x,y) \\ \tau_{xz}(x,z) &= \tau_{yz}(y,z) = 0 \end{aligned} \tag{6.3}$$

Eq.6.2 can be reduced to the following strain–stress relationships (Seidel 1990):

$$\begin{aligned} \varepsilon_x &= (\sigma_x - \upsilon\sigma_y)/E \\ \varepsilon_y &= (\sigma_y - \upsilon\sigma_x)/E \\ \varepsilon_z &= 0 \end{aligned} \tag{6.4}$$

In the isotropic case $\varepsilon = \varepsilon_x = \varepsilon_y$, so that $\sigma_x = \sigma_y = \sigma$, or:

$$\sigma = \frac{E\varepsilon}{(1-\upsilon)} \tag{6.5}$$

Where, E is Young's modulus and υ is the Poisson's ratio of the film. The stress described in Eq.6.4 is called the biaxial modulus. Testing of uniaxial stress of thin films is difficult. The biaxial modulus, rather than Young's uniaxial modulus, is quite useful. Plane stress cannot be avoided. Employing appropriate design techniques can reduce the plane stress. Elwenspoek and Wiegerink have reported that a buckled microbridge clamped at both the ends has a critical length $L_{cr} = 2\pi\sqrt{EI/\varepsilon A}$, where, I, A and ε are the moment of inertia, cross sectional area and strain of the beam, respectively. If the microbridge is designed with the shortest length $L<L_{cr}$, then the effect of thin film stress can be reduced (Elwenspoek and Wiegerink 2000). Tensile stress can render the surface concave and compressive stress renders the surface convex.

6.5.2.1 Stress Measurement

Thin film stress can bend its base substrate by a comparable amount. The most common method for measuring the stress in a microstructure (e.g., thin film) is based on the substrate bending principle. The deformation of a thin substrate due to stress can be measured by observing the displacement at the centre of a circular disk or by using a thin cantilever beam. The radius of the curvature of the beam deflection is measured in order to calibrate the stress.

Recently, more sophisticated stress measurement systems are available and they are based upon analytical tools such as X-ray (Wong 1978), image processing, acoustics measurement, Raman spectroscopy (Nishioka et al. 1985), infrared spectroscopy (Marco et al. 1991), and electron-diffraction techniques. The relationship between the measured force, displacement and differential deformation must be modeled accurately with suitable assumptions. The deflections of suspended and pressurised micromachined membranes can be also measured by a mechanical probe (Jaccodine and Schlegel 1966) like a laser (Bromley et al. 1983), or a microscope (Allen et al. 1987). In the following section, the disk and cantilever methods of stress-measuring techniques are described.

When the thin film under stress deforms in its vertical direction the deformation does not introduce stress in the device, as it is considered to be normal movement. Thus, only stresses in the x and y directions need to be measured. A change in the radius of the curvature of the substrate is determined. Some optical or capacitive gauges can measure such deflections. The disk method is based on a measurement of the deflection at the centre of the disk before and after the application of force or load. Any change in the wafer shape can also reflect stress in the deposited film. It is relatively straightforward to calculate the stress by measuring the radius of the curvature. Hoffman (Hoffman 1976) has developed the relationship between the radius of curvature and thin film stress of the substrate as follows:

$$\sigma = \frac{1}{R}\frac{ET^2}{6(1-\upsilon)t} \tag{6.6}$$

Where, R represents the measured radius of the curvature of the bent substrate, E/ (1-υ) is the biaxial modulus of the substrate, T is the thickness of the substrate, and t is the thickness of the film. The following assumptions are made:

- The film thickness is uniform
- The disc substrate must be thin
- The film has transverse isotropic elastic properties with respect to normal direction
- The thickness of the component is much less than the substrate thickness (For most films on a silicon substrate it is assumed that $t < T$, t/T measures $\sim 10^{-3}$)
- The measurement is carried out at a uniform temperature
- Mechanical isolation between the substrate and the component exists
- Stress is biaxial and homogeneous over the entire substrate
- Film stress is constant through the thickness

Thin film residual stress can be also measured by a cantilever spiral. There are various cantilever spirals available in the literature but Fan's (Fan et al. 1990) cantilever spiral is famous in stress measurements. The spiral is anchored to the inside spring upwards and can make flexible rotations. This arrangement determines the positive strain. The arrangement in which the spiral is anchored to the outside can measure a negative strain. The positive and the negative strains produce mirror symmetry on the spiral. The strain gradient can be calculated from spiral structures by measuring the following:

- The amount of lateral contraction
- The change in height
- The amount of rotation

6.5.3 Stiction

Stiction is defined as the amount of force needed to start to move an element. There is a great deal of literature on the theoretical analysis of stiction. However, its effect on microstructures is described briefly (Lober and Howe 1988; Guckel et al. 1987; Alley et al. 1988).

The deposition of a sacrificial layer simplifies the development of movable microstructures. An important limitation of the microstructures is that they tend to deflect because of stress. It could remain attached to the substrate or isolation layer during the rinsing and drying process. This phenomenon is called stiction and it is mostly related to the effect of bonding and residual contamination. Recently, extensive study has been made to overcome stiction in microstructures (Mastrangelo et al. 1993a, 1993b). If the sacrificial layer is removed with buffered oxide enchants and rinsed in de-ionised water for long time and finally dried under an infrared lamp, stiction can be considerably prevented. One can note that as the water dries, the surface tension of the rinse water pulls the microstructure toward the substrate and a combination of forces such as van der Waals forces and bonding keeps the structure firmly attached. It is difficult to release stiction without a mechanical force, which is large enough to damage the structure. Stiction remains as a reliability issue due to contact with adjacent surfaces even after release. Stiction-free passivation can survive during packaging (Howe 1995). Making the structure stiff enough can solve the sticking problem. According to Legtenberg, if the microstructure device is designed with the critical length in Eq.6.7 then stiction can be avoided (Elwenspoek and Weigerink 2000).

$$L_{cs} > \left\{ \frac{8Ed^2h^3}{\gamma(\cos\theta_1 + \cos\theta_2)} \right\}^{\frac{1}{4}} \tag{6.7}$$

Where d and h are dimensional parameters, γ is the surface tension of the rinsing liquid and θ_1 and θ_2 are the contact angles of the liquid with the substrate. The microstructure will collapse beyond this critical length. An alternative is to increase the roughness of the surface in the microstructure because the end sticking is just a wafer bonding process (see below). Hence, the device should be designed such that its critical length is grater than L_{cs}. Another difficulty in microstructure fabrication is dimensional uncertainties, which expresses greater concern on reliability. Dimensional uncertainty exists in relatively large dimensional structures. For example, a large resonator developed using any lithography technique of quality factors of up to 100,000, the resonator frequency may vary up to 0.02 Hz over long-term (Gad-el-Hak 2002).

6.6 Wafer Bonding

Silicon wafer bonding techniques are used to join two wafers together. The physics behind the technique is simply the effect of bonding. Integrated fabrication is preferable in microdevices, but sometimes due to complex shapes and fabrication difficulties, bonding of two or more parts of the device are preferred. Many bonding methods have already been developed and applied successfully in MEMS devices. There are several techniques, which can be used for bonding the wafers. Important techniques are:

- Anodic and fusion bonds
- Glass-frit bonds

Typical process conditions used for anodic and glass-frit bonds are presented in Table 6.2.

Table 6.2. Typical wafer bonding

Bond type	Temp. (°C)	Pressure (Bar)	Voltage (V)	Surface Roughness (nm)	Precise Gaps	Hermetic Seal	Vacuum Level during Bonding (Torr)
Anodic	300-500	N/A	100 V-1 kV	20	Yes	Yes	10^{-5}
Glass-Frit	400-500	1	N/A	N/A	No	Yes	10

The wafer bonding process can be illustrated as a four-step sequence: surface preparation, fusion, pressurisation and annealing. The first step is important because the quality of the bond depends on the surface conditions. Any surface contamination due to particulate material can damage or create poor bonding. A particle of size 1 μm can have an effect on an area as large as 1 cm in diameter. In order to achieve good bonding effects, typical values of surface roughness and flatness are 5 Å and 5 μm on a 4-inch wafer, respectively.

Bond strength is critical and sometimes difficult to assess in many applications due to microscale dimensions. A number of techniques can be applied to evaluate this. One of the most common techniques is based on a surface energy measurement proposed by Maszara. This usually yields a consistent value. A blade of known thickness is inserted between the bonded elements to create a crack. The length of the crack is then measured. The surface energy can be calculated using the imperial formula as given in Eq.6.8 (Maszara et al. 1991).

$$\lambda = \frac{3}{8}\frac{Et^3 y^2}{L^4} \tag{6.8}$$

Where, λ is surface energy, E is modulus of elasticity, t is the sample thickness, 2y is the blade thickness and L is the length of the crack. The main drawback of this approach is the fourth power on the length of the crack. An error in the measurement is increased in the order of the fourth power. The Motorola automotive air bag accelerometer is the most recent example of silicon wafer bonding (Fig. 6.6). In this microdevice, a silicon top cap is bonded at the wafer level with the triple-level polysilicon surface micromachined accelerometer using a low-temperature glass-frit bond. This wafer-level silicon cap provides mechanical protection and also prevents damping and shocking. Anodic and fusion bonding techniques are described below.

Fig. 6.6. Wafer bonding application (Courtesy: Freescale Semiconductor)

6.6.1 Anodic Bonding

Anodic bonding is applied to silicon wafers and glass (pyrex borosilicate) with a high content of alkali metals. It has a sodium oxide (Na_2O) concentration of 3.5%. Anodic bonding is also known as electrostatic bonding. In this approach, silicon and glass is arranged as shown in Fig. 6.7. An electrode is connected with the glass. High negative voltage is applied between the two mechanical supports and the combination is heated up to 500°C. The high electric field in the entire area creates a strong electrostatic force, pulling the two surfaces together. The glass contains positive ions (Na^+) attached to the negative electrode area where they are neutralised by the counter charges. The transition creates a space charge at the glass-silicon interface and develops strong electrostatic bonding between them. During bonding, oxygen from the glass is transported to the glass-silicon interface.

Because of the movement, SiO_2 is formed which in turn creates a permanent bond. During the process, the electric field is high enough to allow a drift of oxygen to the positive electrode (Si), which then reacts with silicon and creates a Si-O bond. This technique can produce uniform bonds and it has been successfully applied to many microsystem applications including pressure sensors, solar cells and piezoresistive devices. However, this technique is not good for active devices such as actuators. It is also possible to bond silicon wafers together by applying gentle pressure. Other bonding methods include using an adhesive layer like wax, epoxy and SiN-SiN bonding for low temperature processes and Au-coated bonding (sometimes called eutectic bonding) using glass or photoresist. Anodic bonding can form very strong joints but they demand that the surfaces be very clean and flat. Wafer bonding techniques can be combined with basic micromachined structures to design complex microdevices like valves and pumps.

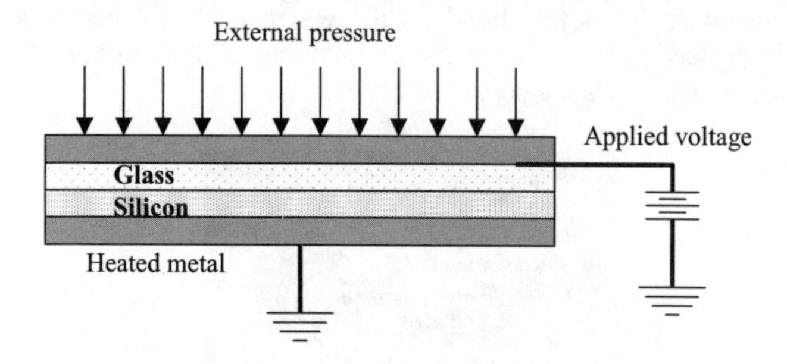

Fig. 6.7. Anodic bonding arrangement

6.6.2 Fusion Bonding

Fusion bonding is known as Silicon-Direct-Bonding (SDB). The SDB technique has been used in a wide range of applications in microelectronics and microsystems technology. It is a suitable method for manufacturing devices including p-n-junctions. Two silicon wafers are bonded through the formation of oxides on their surfaces. This technique is commonly used in bulk micromachined MEMS packaging. SDB is a low cost packaging technology for silicon and silicon compounds such as SiO_2 or Si_3N_4. In comparison to other bonding techniques, SDB has the following advantages:

- Flexible and low cost
- Thermal mismatch between the bond materials is low
- Bond partners can be featured easily
- High strength because of homogeneity
- Can be bonded at high or low temperature depending on the material used

Bonding relies on the need for both very smooth and flat surfaces to adhere. Maximum bond strength is obtained at temperatures between 700-1100°C.

Thermally sensitive devices can be bonded with a sufficient strength at temperatures between 200-400°C by using chemical surface activation methods. The first step of low temperature SDB of two wafers is the cleaning and hydrophilisation in an acid mixture (H_2SO_4 and H_2O_2), followed by rinsing and spin-drying. The wafers are then wetted with silicate solutions like sodium silicate (NaSi) or tetraethylorthosilicate (TEOS). The wafers are rinsed and dried for a second time. Then, they are joined by applying an external force or load. Bonding is initiated with a single point contact at the middle, and subsequently a uniform pressure is applied.

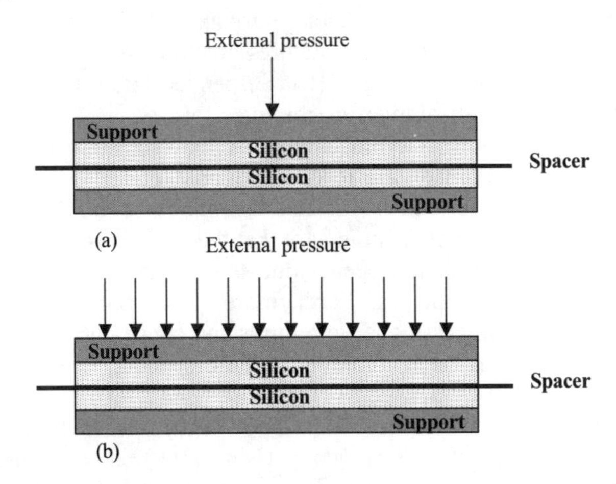

Fig. 6.8. Fusion wafer bonding, (a) Single point contact (b) Uniform surface contact

Bonding propagates from the centre to both sides and it is influenced by a change in viscosity and pressure of the ambient gas, as well as the surface energy. The mechanical spacers, as shown in Fig. 6.8, are usually provided in order to facilitate mechanical and electrical contacts. Sometimes gases and moisture in between them make the bond weak. For this reason the wafers are heated in order to dehydrate the surface. The wafer molecules diffuse from Si-to-Si bond as the process continues. At higher temperatures, oxygen is also diffused into the crystal lattice. Bonding at temperatures greater than a thousand degree centigrade, the strength approaches that of silicon itself. But this process has shown some drawback in some applications, especially concerning the formation of cavities.

6.7 Summary

The term micromachining refers to the fabrication of 3D MEMS structures with the aid of advanced lithography and etching techniques. Lithography patterns the structural material whereas etching removes the selective portion of the substrate or thin film already deposited. In general, the micromachining process either can use the material to form microstructures by etching directly into the material or can use structural sacrificial layer to produce the same. A sacrificial layer is etched away in order to obtain a freestanding 3D structure. Broadly, the fabrication processes fall into three categories:

- Bulk micromachining
- Surface micromachining
- Wafer bonding

Bulk micromachining refers to etching through both the sides (front and back) to form the desired structure. The structures are formed either by wet chemical etching or by reactive ion etching (RIE). Suspended microstructures are usually fabricated using wet chemical micromachining.

The advantage of bulk micromachining is that substrate materials such as quartz or single crystal silicon are readily available, and reasonably high aspect-ratio structures can be fabricated. It is also compatible with IC technologies. The disadvantages of bulk micromachining are that the process is pattern and structure sensitive and pattern distortion occurs due to different selective etch rates on different crystallographic planes. Furthermore, since both the front side and backside are used for processing, severe limits and constraints are encountered on the minimum feature size and minimum feature spacing. Bulk micromachining utilises the etch selectivity between {111} planes and {100} and/or {110} planes in aqueous alkaline etchants. Si wafers with (100) and (110) orientations are essentially used in bulk micromachining. Using (100) silicon, simple structures such as diaphragms, V-grooves, nozzles, and more complex structures such as corner cubes and rectangular masses, can be fabricated. Vertical microstructures can be fabricated on Si (110) wafers. This is because Si (110) has four {111} planes that intersect the wafer surface vertically.

Surface micromachining is another method for the fabrication of MEMS structures out of deposited thin films. It involves the creation of mechanical structures in thin films formed on the surface of the wafer. The thin film may be composed of three layers of materials: low pressure chemical vapor deposition (LPCVD) polycrystalline silicon, silicon nitride, and silicon dioxides. Silicon has excellent mechanical properties making it an ideal material for machining. The layers are deposited in sequence and subsequently some selective portions of the layer are removed to build up a 3D mechanical structure. Polysilicon, silicon nitride and silicon dioxide are commonly used as the structural, insulation and sacrificial layer, respectively. Polysilicon material based surface micromachining has been the backbone of the fabrication technology for many microsensors and actuators. Hydrofluoric acid can dissolve the sacrificial layer. Once etched, the structure can be freed from the planar substrate. This is called the release process.

The surface micromachining process is a simple method and entails the study of the important properties of the material. Note that the properties of material vary at the microstructure level. In particular, the following issues require careful attention:

- Basic understanding and control of the material properties of structural films
- Releasing method for the microstructure
- Fabrication features for hinged structures and high-aspect ratio devices
- Packaging methods

In wafer bonding, the silicon wafer and glass substrate are brought together and heated to a high temperature. Then an electric field is applied across the joint, which develops a strong bond between the two materials.

This chapter presented the features of surface micromachining and wafer bonding. In particular, the properties of materials such as adhesion, stiction and various types of stresses were the matter of discussion. References are provided for further study.

6.8 References

Allen DM (1986) The principles and practice of photochemical machining and photoetching. Inst of Physics Pub Inc., pp 1-18

Allen MG, Mehregany M, Howe RT, Senturia SD (1987) Microfabricated structures for the in situ measurement of residual stress, Young's modulus and ultimate strain of thin films, Appl. Phys. Lett. 51:241–243

Alley RL, Cuan GJ, Howe RT, Komvopoulos K (1988) The effect of release etch processing on surface microstructure stiction. echnical Digest of the Solid-State Sensor and Actuator Workshop, Hilton Head Island, pp 202–207

Allongue P, Kieling VC, Gerischer H (1993) Etching of Si in NaOH solutions, Part I and II, J. Electrochem. Soc. 40:1009-1018 and 40:1018-1026

Ammar ES, Rodgers TJ (1980) UMOS transistors on (110) silicon, IEEE Trans. Electron Devices ED-27:907-914

Bassous E, Taub HH, Kuhn L (1977) Ink jet printing nozzle arrays etched in silicon, Appl. Phys. Lett. 31:135-137

Boivin LP (1974) Thin film laser-to fiber coupler, Appl. Opt. 13:391-395

Bromley EI, Randall JN, Flanders DC, Mountain RW (1983) A technique for the determination of stress in thin films, J. Vac. Sci. Technol. B1:1364–1366.

Campbell DS, (1970) Mechanical properties of thin films. In: Maissel L, Glang R (eds.) , Handbook of Thin Film Technology, McGraw-Hill

Chang S, Eaton W, Fulmer J, Gonzalez C, Underwood B, Wong J, Smith RL (1991) Micromechanical structures in amorphous silicon. Proc. of the 6th Int. Conf. on Solid-State Sensors and Actuators (Transducers 91), San Francisco, pp 751–754

Chou PC, Pagano NJ (1967) Elasticity: tensor, dyadic, and engineering approaches. Dover Publications Inc., New York

Core TA, Tsang WK, Sherman SJ (1993) Fabrication technology for an integrated surface micromachined sensor, Solid State Technol. 36:39–47

Diem B, Delaye MT, Michel F, Renard S, Delapierre G (1993) SOI(SIMOX) as a substrate for surface micromachining of single crystalline silicon sensors and actuators. Proc. of the Int. Conf. on Solid-State Sensors and Actuators, Japan, pp 233–236

Editorial (2004) Analog Devices combines micromachining with BICMOS, Semiconductor International, available at http://www.reed-electronics.com

Editorial (1974) Transducers, pressure and temperature catalog. White paper, National Semiconductor, Sunnyvale, USA
Editorial (1977) Thermal character print head. Technical Report, Texas Instruments, Austin
Elwenspoek M, Weigerink R (2000) Mechanical microsensors. Springer, Germany
Fan LS, Muller RS, Yun W, Huang J, Howe RT (1990) Spiral microstructures for the measurement of average strain gradients in thin films, Proc. of the IEEE Micro Electro Mechanical Systems, (MEMS 90), pp 177–181
Fan LS, Tai YC, Mulller RS (1988) Integrated movable micromechanical structures for sensors and actuators, IEEE Trans. Electron Devices 35:724–730
Gabriel K, Jarvis J, Trimmer W (1989) Small machines, large opportunities. A report on the emerging field of microdynamics, National Science Foundation
Gad-el-Hak M (2002) The MEMS handbook. CRC press, pp 200-211
Greenwood JC (1969) Ethylene Diamine-Cathechol-Water mixture shows preferential etching of p-n junction, J. Electrochem. Soc, 116:1325-1326
Greenwood JC (1984) Etched silicon vibrating sensor, J. Phys. E, Sci. Instrum. 17:650-652
Guckel H, Burns DW, Nesler KC, Rutigliano CR (1987) Fine grained polysilicon and its application to planar pressure transducers. Proc. of the 4th Int. Conf. on Solid-State Sensors and Actuators (Transducers 87), Tokyo, pp 277–282
Hallas CE (1971) Electropolishing silicon, Solid State Technology 14:30-32
Harris TW (1976) Chemical milling. Clarendon Press, Oxford
Hermansson K, Lindberg U, Hok B, Palmskog G (1991) Wetting properties of silicon surfaces. Proc. of the 6th Int. Conf. on Solid-State Sensors and Actuators (Transducers 91), San Francisco, pp 193–196
Hoffman RW (1976) Mechanical properties of non-metallic thin films: Physics of nonmetallic thin films (NATO Advanced Study Institutes Series: Series B, Physics). In: Dupuy CHS, Cachard A (eds), Plenum Press, pp 273–353
Hoffmeister W (1969) Determination of the etch rate of silicon in buffered HF using a 31 Si tracer method, Int. J. Appl. Radiation and Isotopes 2:139-144
Howe RT, (1995b) Recent advances in surface micromachining. Technical Digest of the 13th Sensor Symposium, Tokyo, pp 1–8
Hu JZ, Merkle LD, Menoni CS, Spain IL (1986) Crystal data for high-pressure phases of silicon, Phys. Rev. B 34:4679-4684
Jaccodine RJ, Schlegel WA (1966) Measurements of strains at Si-SiO_2 interface, J. Appl. Phys. 37:2429–34
Jian L, Destaa YM (2003) SU-8 based deep x-ray lithography/LIGA: Micromachining and microfabrication process technology. Proceedings of SPIE, pp 4979-4983
Kaminsky G (1985) Micromachining of silicon mechanical structures, J. Vac. Sci. Technol. B3:1015-1024
Keller CG (1998) Microfabricated high aspect ratio silicon flexures: Hexsil, RIE, and KOH etched design & fabrication, MEMS Precision Instruments Pub.
Kern W, Deckert CA (1978) Chemical etching in thin film processes. In: Vossen JL, Kern W (eds), Academic Press, Orlando
Krulevitch PA (1994) Micromechanical investigations of silicon and Ni–Ti–Cu thin films. Ph.D. thesis, University of California at Berkeley
Lee JB, Chen Z, Allen MG, Rohatgi A, Arya R (1995) A miniaturised high-voltage solar cell array as an electrostatic MEMS power supply, J. Microelectromech. Syst. 4:102-108
Legtenberg R, Elders J, Elwenspoek M (1993) Stiction of surface micromachined structures after rinsing and drying: model and investigation of adhesion mechanisms. Proc. Int. Conf. on Solid-State Sensors and Actuators (Transducers 93), pp 198–201

Linde H, Austin L (1992) Wet silicon etching with aqueous amine gallates, J. Electrochem. Soc. 139:1170-1174
Lober TA, Howe RT (1988) Surface micromachining for electrostatic microactuator fabrication. Technical Digest of the Solid State Sensor and Actuator Workshop, Hilton Head Island, pp 59–62
Luder E (1986) Polcrystalline silicon-based sensors, Sensors and Actuators A 10:9–23
Madou MJ (1997) Fundamentals of microfabrication. CRC Press, New York
Madou MJ, Morrison SR (1989) Chemical sensing with solid state devices. Academic Press, New York
Marco S, Samitier, Ruiz O, Morante JR, Esteve-Tinto J, Bausells J (1991) Stress measurements of SiO_2-Polycrystalline silicon structures for micromechanical devices by means of infrared spectroscopy technique. Proc. of the 6th Int. Conf. on Solid-State Sensors and Actuators (Transducers 91), San Francisco, pp 209–212
Mastrangelo CH, Hsu Ch (1993) Mechanical stability and adhesion of microstructures under capillary forces-Part II: Experiments, J. Microelectromech. Syst. 2:44–55
Mastrangelo CH, Hsu CH (1993a) Mechanical stability and adhesion of microstructures under capillary Forces-Part I: Basic Theory, J. Microelectromech. Syst. 2:33–43
Maszara W, Jiang BL, Yamada A, Rozgonyi GA, Baumgart H, Kock, AJR (1991) Role of surface morphologyin wafer bonding, J. Appl. Phys. 69:257
Mehregany M, Senturia SD (1998) Anisotropic etching of silicon in hydrazine, Sensors and Actuators A 13:375–390
Michalicek AM (2004) Introduction to MEMS, http://mems.colorado.edu/c1.res.ppt /ppt/g.tutorial
Middlehoek S, Dauderstadt U (1994) Haben mikrosensoren aus silizium eine zukunft? technische rundschau, Technical Report, pp 102-105
Monk DJ, Soane DS, Howe RT (1992) LPCVD silicon dioxide sacrificial layer etching for surface micromachining. Proc. of the National Conf. on smart materials fabrication and materials for microelectromechanical systems, San Francisco, pp 303–310
Monk DJ, Soane DS, Howe RT (1994a) Hydrofluoric acid etching of silicon dioxide sacrificial layers. Part II. Modeling, J. Electrochem. Soc. 14:270–274
Murarka SP, Retajczyk TFJ (1983) Effect of phosphorous doping on stress in silicon and polycrystalline silicon, J. Appl. Phys. 54:2069-2072
Nathanson HC, Wickstrom RA (1965) A resonant-gate silicon surface transistor with high-Q band-pass properties, Applied Physics Letters 7:84-86
National Materials Advisory Board (1997) Microelectromechanical systems: Advanced materials and fabrication methods, National Academic Press Washington, pp 1-61
Nishioka T, Shinoda Y, Ohmachi Y (1985) Raman microprobe analysis of stress in Ge and GaAs/Ge on SiO_2-coated silicon substrates, J. Appl. Phys. 57:276–81
Noworolski JM, Klaassen ME, Logan J, Petersen K, Maluf N (1995) Fabrication of SOI wafers with buried cavities using silicon fusion bonding and electro-chemical etchback. Proc. of the Int. Conf. on Solid-State Sensors and Actuators (Transducers 95), Stockholm, pp 71–74
Obermeier E (1995) High temperature icrosensors based on polycrystalline diamond thin films. Proc. of the 8th International Conference on Solid-State Sensors and Actuators (Transducers 95), Stockholm, pp 178–181
Obermeier E, Kopystynski P (1992) Polysilion as a material for microsensor applications, Sensors and Actuators A 30, 149–155
O'Neill P (1980) A monolithic thermal converter, Hewlett-Packard J. 31:12-13
Palik ED Faust JW, Gray HF, Green RF (1982) Study of the Etch-Stop Mechanism in Silicon, J. Electrochem. Soc. 129:2051-2059

Park J, Allen M (1998) Development of magnetic materials and processing techniques applicable to integrated micromagnetic devices, J. Micromech. Microeng. 8:307–316

PeetersE (1994) Process development for 3D silicon microstructures, with application to mechanical sensor design. Ph.D. thesis, Catholic University of Louvain

Petersen KE (1982) Silicon as a mechanical material. Proc. IEEE, 70:420-457

Pfann WG (1961) Improvement of semiconducting devices by elastic strain, Solid State Electron. 3:261-267

Robbins H, Schwartz B (1960) Chemical etching of silicon-II: The system HF, HNO3, H2O, and HC2C3O2, J. Electrochem. Soc. 107:108-111

Rodgers TJ Hiltpold WR, Frederick B, Barnes JJ, Jenné FB, Trotter JD (1977) VMOS memory technology, IEEE Transaction on Solid-State Circuits SC12:515-523

Schwartz B, Robbins H (1976) Chemical etching of silicon-IV, Etching Technology, J. Electrochem. Soc. 123:1903-1909

Singer P (1992) Film stress and how to measure it, Semicond. Int. 15:54–58

Smith CS (1954) Piezoresistance effect in germanium and silicon, Phys. Rev. 94:42-49

Tang WCK (1990) Electrostatic comb drive for resonant Sensor and actuator applications. Ph.D. thesis, University of California at Berkeley

Tenney AS, Ghezzo M (1973) Etch rates of doped oxides in solutions of buffered HF, J. Electrochem. Soc. 120:1091–1095

Tuck B (1975) The chemical polishing of semiconductors, J. Mater. Sci. 10:321-339

Tuft ON, Chapman PW, Long D (1962) Silicon diffused-element piezoresistive diaphragms, J. Appl. Phys. 33:3322

Turner DR, (1958) Electropolishing silicon in hydrofluoric acid solutions, J. Electrochem. Soc. 105:402-408

Uhlir A (1956) Electrolytic shaping of germanium and silicon, Bell Syst. Tech. J. 35:333-347

Vinci RP, Zielinski EM, Bravman JC (1995) Thermal strain and stress in copper thin films

Voronin VA, Druzhinin AA, Marjamora II Kostur VG, Pankov JM (1992) Laser–recrystallised polysilicon layers in sensors, Sensors and Actuators A 30:143-147

Waggener HA, Krageness RC, Tyler AL (1967) Two-way etch, Electronics 40:274

Wong SM (1978) Residual stress measurements on chromium films by X-ray diffraction using the sin2 Y method, Thin Solid Films 53:65–71

Xu YB, Okubo T, Sadakata M (1997) Synthesis of ultrafine ß-Sic particles from siox (X=0,1, 2) powders and C2h2, J. of Chem. Eng. of Japan 30:662-668

7 Micromanufacturing for Document Security: Optically Variable Devices

R. A. Lee

CSIRO Manufacturing and Infrastructure Technology, Victoria, Australia

7.1 Preamble

By its very nature document security microtechnology is a subject much less able to be discussed in an open manner than many other areas of research and application. As well as the normal commercial in confidence issues, all companies, research groups and end users working in this area are mindful of the need to restrict the disclosure of information which might compromise the integrity of particular security devices being applied to protect official documents from counterfeiting. Not only are counterfeiting groups now much better organised in their activities, but they are also much more technology focussed in seeking to overcome new security measures. Holographic technology, foil embossing equipment, application processes and information associated with security device design is now being adopted by counterfeiters in their efforts to overcome the new security measures being introduced by the industry. However it is also important for the ongoing development of the industry that document security microstructure issues remain accessible to the research community. Only in this way can new ideas for improved security microstructures be developed and applied in order to stay ahead of the counterfeiters. With these issues and context in mind, this review will have a particular focus on the contribution of electron beam lithography to the development of particular optical security microstructures. It is this type of origination process, which is perceived by the industry to offer a higher level of counterfeit deterrence than the alternative hologram technologies.

7.2 Introduction

The counterfeiting of banknotes and other financial transaction documents as well as visas, passports, ID cards, other official Government security papers and the counterfeiting of software, pharmaceuticals and brand name products is now becoming one of the world's fastest growing areas of criminal activity. With the advent of colour photocopiers and digital scanning devices the traditional security printer's skill and craftsmanship is no longer sufficient to protect the integrity of high value documents (Fagan 1990). According to the International Chamber of Commerce Counterfeiting Intelligence Bureau over 5% of world trade consists of counterfeit product. The cost of this illegal activity amounts to over US$350 billion annually and is growing. To counteract this rapidly evolving threat security printers and central bank issuing authorities have turned to new technology in the form of the diffractive Optically Variable Device (OVD) (Renesse 1994). Mass produced by embossing into hot stamping foil, these diffractive surface relief microstructures have proven to be a highly effective solution to the counterfeiting problem. The key security feature underpinning all optically variable devices is, as the name implies, the variability of the image with changing angle of view or changing angle of illumination by a light source. This image variability can even take the form of a switch between one artwork pattern to a totally different piece of artwork; from the image of a face to the number 50, for example. Because of this image change effect these devices cannot be copied by the usual photographic, computer scanning or printing techniques, and therein lies their attractiveness to security printers who have seen their traditional anti-counterfeiting technologies of intaglio print and watermarking undermined by the counterfeiters using the new computer scanning and copying technologies developed since the 1980's. In developing this discussion the intention is to first describe the origins of OVD technology as applied to financial transaction documents and then explore the history and influence of two particular high security developments in the fabrication of optically variable device (OVD) anti-counterfeiting technology for high security document applications. The focus will be on those features of two technologically competitive technologies, which are perceived by the security printing industry to be significant in reducing the counterfeiting threat to official documents. By comparing the security attributes of these competitive OVD technologies and their influence on different sections of the security printing industry from the early 1970's to the late 1990's, we will be able to draw some general conclusions about the suitability and uniqueness of particular OVD features and their suitability for future extension and development. The security value of the different OVD effects will be classified here according to the degree of difficulty in originating the device, the recognisability and uniqueness of their corresponding optical effects and finally by the degree of availability of these effects from alternative and less secure origination sources. In subsequent sections we will consider in more detail particular applications of the electron beam lithography origination process; the origination process regarded by the security printing industry as providing the most secure and controlled set of optical effects. By control here we mean the ability of the industry and its component suppliers to restrict the availability of these high security effects to high security printing industry applications only and in so doing prevent unscrupulous operators from gaining access to the technology. Finally we will consider the limitations of this current high-end origination process and its corresponding foil based replication process and consider a new paradigm for addressing these limitations in a way that provides a future path for extended developments of the underlying core technology.

7.3 OVD Foil Microstructures

The first application of OVD technology as an anti-counterfeiting feature was the result of the adaptation of existing holographic technology to high volume manufacture by direct embossing onto foil and the application of that foil onto the security document by a hot stamping process. The security hologram was developed initially by the American Bank Note Company through its subsidiary, American Bank Note Holographics (ABNH) and was first adopted by MASTERCARD in 1982 in the form of a 2D/3D holographic foil image. Two years later VISA used a 3-D variation of the ABNH technology to protect its new credit card series (Fagan 1990). In both cases the use of the hologram resulted in a marked decrease in the rate of credit card counterfeiting and fraud.

7.3.1 The Security Hologram

In the years that followed many other variations of holographic foil images have appeared on various types of financial transaction cards. During the same period attempts were made to apply similar technology to paper substrates, including prototype banknote applications. It then became apparent that the hologram suffered from several serious problems, which reduced its effectiveness as a security feature when applied to these less robust substrates.

Some of these problems already existed in the case of the card applications. The key disadvantages included: (1) image blurring (and therefore reduced recognition) under extended light source illumination, (2) lack of image brightness, (3) very poor brightness and recognition after crumpling of the paper substrate and finally (4) lack of control over the origination technology which sometimes allowed unscrupulous operators to re-originate the security image or simulate the effects from commercially available foils with similar optical effects (Antes 1986).

While some of these problems were eliminated or reduced by the development of new origination processes such as holographic stereography and dot matrix technology (Souparis 1990), the results were usually perceived to be less than satisfactory from a security printer's point of view. This unfavourable perception was mainly due to the lack of a clear optical effects design philosophy as well as an inability to guarantee complete security with respect to the availability of specialised optical effects. In the following sections we shall show how these critical issues were resolved by the development of the first OVD technologies with optical effects exclusively reserved for security printing applications (Lee 1998).

7.3.2 The Kinegram™

Developed by the Landis & Gyr Corporation in the early 1980's, the Kinegram™ was the first OVD technology to be designed explicitly as a no compromise anti-counterfeiting measure exclusively reserved for banknote and other high security document applications. The Kinegram™ technology (Antes 1982, 1988; Moser 1996; Renesse 1994, 2004) represents probably the most advanced form of OVD microstructure written by laser interference techniques. An example of such a Kinegram™ microstructure is shown in Fig. 7.1. The photomicrograph shows some fundamental microstructure units of a Kinegram™ OVD. The small circular 60-micron diameter grating pixels have groove spacings and angles which vary throughout the device, according to the input picture information, thereby providing the means by which the macroscopic artwork of the Kinegram is made optically variable. The photomicrograph is taken from the Kinegram Saudia Arabia passport application of 1987. The initial stimulus for the Kinegram™ OVD development came from the Swiss National Bank and followed earlier work by Landis & Gyr on the development of machine-readable optical codes for banknotes. This work in turn drew on Landis & Gyr's still earlier work on optically coded pre-paid telephone cards. The expertise developed on these earlier projects, particularly in relation to the mathematical analysis of inverse scattering problems, provided the critical mathematical tools necessary for calculating the Kinegram™ microstructure configuration for particular optical effect designs. In conjunction with Gregor Antes' application of mathematical physics to the design of these unique optical effects, Landis &Gyr also developed a novel system for originating the required Kinegram™ microstructures. In order to satisfy the security requirements of the banknote printing industry, the Kinegram™ origination technology was (and still is) maintained as a single site process exclusively reserved for high security anti-counterfeiting applications. Fig. 7.2 shows some applications of the Kinegram OVD technology, including the Swiss 50 Franc and Finnish 500 Mark banknotes.

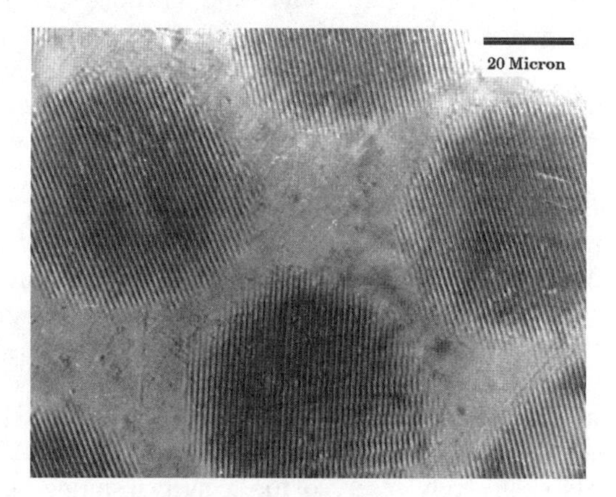

Fig. 7.1. Example of a Kinegram OVD microstructure

Fig. 7.2. Kinegram OVD applications, including the OK marketing Kinegram

Underlying this origination development work was a well-defined optical effects design philosophy, which drew heavily on a particular interpretation of the existing knowledge of the psychophysical aspects of visual perception. The two principal psycho-physical Kinegram™ design rules can be stated as follows: (1) use crisp sharp lines in an image to efficiently help memorisation, as shown for example in the 20 mm diameter Kinegram shown in Fig. 7.3 and (2) best memorisation of the variability of an image is achieved when the image change is accompanied by sudden and brisk changes of intensities. As well as these rules of perception, Kinegram™ images are also designed to have high brightness, be easy to observe under a wide range of illumination conditions and to be resistant to image degradation due to crumpling of the foil on the banknote. To complete this unique package of high security design and origination attributes, Landis & Gyr also restricted the availability of the Kinegram™ to banknotes and other official documents. This restriction on availability was intended to ensure that Kinegram optical effects would not be compromised in the way holographic optical effects had been compromised by their appearance on many commercial applications outside of security printing.

With such a powerful and tightly controlled set of security attributes, it is not surprising that the Kinegram™ quickly became the market leader in OVD protection for high security documents. Appearing first on the Saudia Arabia passport in 1987 and the Austrian 5,000 schilling banknote in 1990, the Kinegram™ later appeared on Finnish banknotes, the new Swiss currency in 1995, German banknotes in 1997 and finally the low denomination euro banknotes of 2002. Fig. 7.4.(a) shows the behaviour of the Kinegram on the Finnish 500 mark banknote. It should be noted however that not all of these projects conformed to the classical Kinegram™ design philosophy of sparkling opto-kinetic effects based on optically variable guilloche line artwork. The “moving 50” Swiss banknote Kinegram as shown in Fig.7.4(b) for example, is composed of arrays of diffractive tracks more reminiscent of an Exelgram™ microstructure than the juxtaposed micrograting elements of the other applications. The acceptance of the

OVD as an essential feature of modern banknote design (Lancaster and Mitchell 2004) has seen the emergence of competitors to the Kinegram™. This is demonstrated by the selection of Hologram Industries' Moviegram™ for the higher denomination banknotes of the new European currency. Some of the competing products also show clear evidence of the adoption of the highly successful Kinegram design and optical effects philosophy. Companies such as ABNH of the US, AOT and De La Rue Holographics of the UK , Hologram Industries of France and the Toppan printing company of Japan are able to offer sometimes similar effects via alternative origination scenarios (Newswager; Aubrecht et al. 2004). While these companies have yet to create opto-kinetic effects of similar brightness to the Kinegram™ OVD it is clear that such improvements are only a matter of time.

With the increased availability and reduced control status of these Kinegram™ effects, there has developed an increasing interest in OVD technologies which offered additional optical effects unable to be produced by either holographic or opto-kinetic origination systems and which are able to be exclusively restricted to security printing applications. A particularly effective example of an alternative approach is provided by the electron beam lithography origination process as exemplified by the Exelgram™ OVD technology developed by the Commonwealth Scientific and Industrial Research Organisation (CSIRO) of Australia.

Fig. 7.3. Characteristic opto-kinetic Kinegram OVD effects at three angles of view

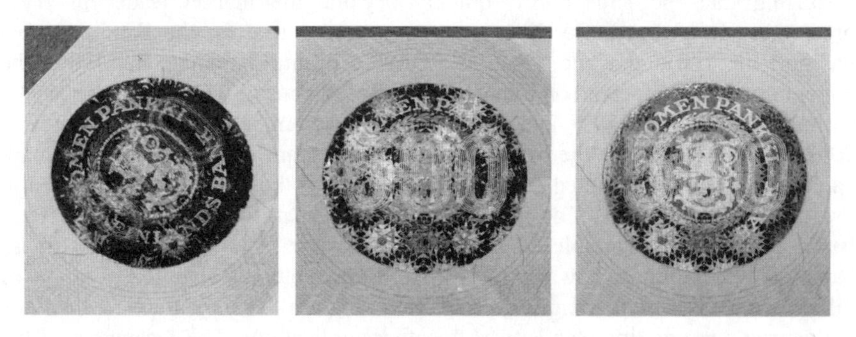

Fig. 7.4(a). Optical behaviour of the Kinegram on Finnish 500 mark banknote

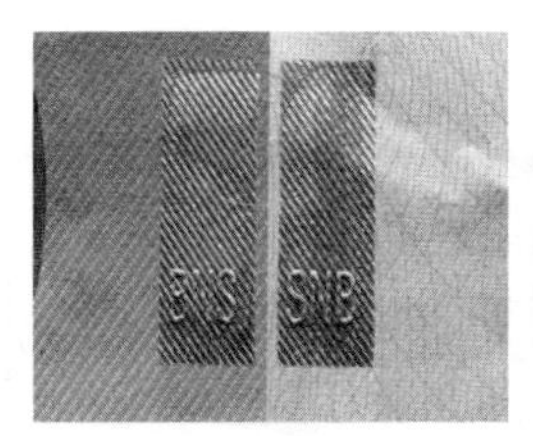

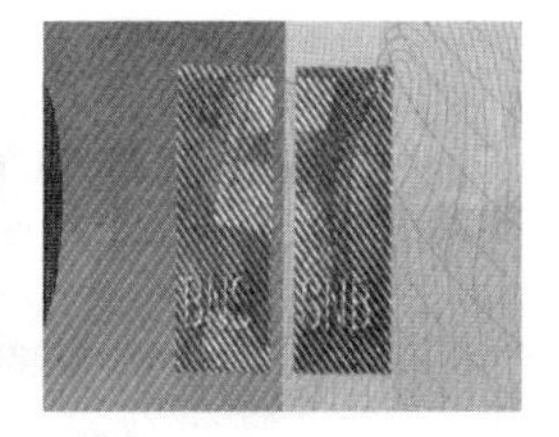

Fig. 7.4(b). The Swiss 50 Franc banknote Kinegram at three angles of view

7.3.3 The Catpix™ Electron Beam Lithography Microstructure

While the first examples of Exelgram™ technology did not appear until late 1993, the origins of the security philosophy which influenced the development of this OVD technology go back almost twenty years earlier to what was called the CNRD (Currency Note Research and Development) project in the early 1970's. The CNRD project was a joint collaboration between the CSIRO and Reserve Bank of Australia (RBA) and had the aim of developing a new high security banknote containing features that could not be copied by the coming generation of colour photocopiers (Hardwick 1990; Wilson 1998). In contrast to the optically variable guilloche philosophy, which drove the development program of the Kinegram™, the Exelgram™ traces its principal design influence back to the optically variable portraiture requirements of the RBA and similar influences from Bank of England studies during the same early 1970's period. Just as the opto-kinetic line art effects generated by Kinegrams can be regarded as optically variable versions of the complex guilloche effects used in traditional banknote design, so too can the positive to negative image switch of the American Express Centurion Exelgram™ incorporated into the new series of travellers cheques (Wood 1999; McHugh 1999) issued in 1997 be regarded as an optically variable example of the traditional portraits used on the majority of the world's banknotes. In this sense the Kinegram™ and the Exelgram can be regarded as alternative and complementary solutions to the same problem of protecting their common banknote heritage from counterfeiting. While portraiture was a critically preferred feature of the CNRD OVD, other design objectives also had to be met. These included the addition of optically variable graphics as well as a requirement that the diffractive properties of the device should be resistant to crumpling of the banknote surface. The final result of this work was the CATPIX™ grating concept developed by CSIRO in 1985 –1987 (RA Lee, EP 0449 893 B1). This Catpix™ groove structure was used on the Captain Cook portrait OVD on the Australian ten-dollar plastic banknote issued in January 1988. This was the first use of an OVD on a banknote anywhere in the world. The OVD on the 1988 polymer banknote took the form of a 20 mm high schematic portrait of Captain Cook, two angles of view of which are shown in Fig. 7.5, and the underlying diffractive microstructure was written by a modified JEOL 5A electron beam lithography system. Interestingly this was the first electron beam lithography system to incorporate an oblique line writing capability as developed by CSIRO and incorporated into the JEOL 5A system installed at the Reserve Bank of Australia note printing department in 1980. The OVD microstructure developed on this machine was designed in the form of a network of continuously connected curvilinear regions (catastrophe pixels) in order that the resultant diffraction pattern from the OVD would be less sensitive to banknote surface crumpling effects. Fig. 7.6 shows a small section of a curvilinear region of this type of OVD.

7.3.4 Structural Stability

As briefly discussed above, the crinkling and flexing of the banknote surface was an issue that arose very early on in the process of developing an effective security microstructure for anti-counterfeiting applications. Curving the rulings of the grating microstructure in an undulatory fashion was found, through earlier work on the mathematical and optical properties of catastrophe gratings (Lee 1983, 1985), to be an effective technique for generating expanded diffraction beams from small surface regions of the banknote OVD. This increased the observational stability of each diffracted beam because each expanded beam could be observed from a broader range of viewing angles than in the case of straight line grating structures, which produced unexpanded diffracted beams. Fig. 7.7 shows the first order Catpix grating Fourier plane diffraction pattern corresponding to the catastrophe grating (Lee 1983) ruling pattern shown in Fig. 7.6. While the Australian Reserve Bank Catpix™ technology shared a similar security philosophy to that of Landis & Gyr's Kinegram technology (unique optical effects, single source origination and controlled availability), there was a fundamental problem with the Catpix™ technology in that its portraiture capabilities were still very limited. Because optically variable greytone line art portraiture had still not been achieved, CSIRO initiated a new research project in late 1988 aimed at securing the optically variable portrait objective by using a radically different underlying grating structure to that used in Catpix™. In the course of this work it was found that the modulated curvature geometry used to improve the observational structural stability of the Catpix OVD could also be adapted to solve the problem of creating a realistic OVD portraiture capability.

Fig. 7.5. Two angles of view of the Captain Cook Catpix OVD; the first banknote OVD

Fig. 7.6. Photomicrograph of a small section of the Captain Cook Catpix grating OVD

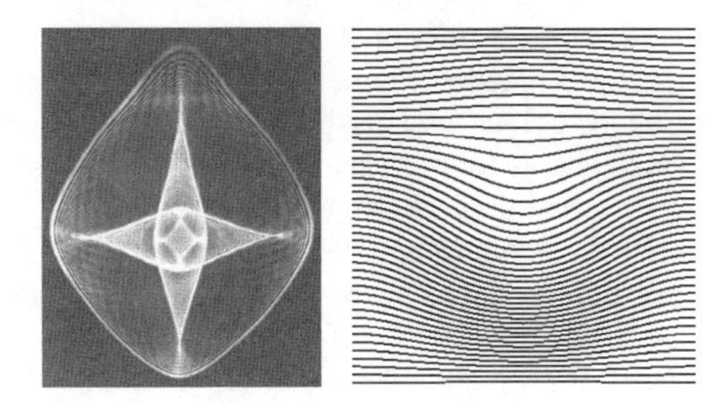

Fig. 7.7. Fourier plane diffraction pattern from a component of a Catpix grating OVD

7.3.5 The Pixelgram™ Palette Concept

An alternative and much more flexible approach to the design of OVD microstructures is based on the use of a palette of independent pixel grating functions of different line curvatures to represent the palette of grayscale or intensity values within a portrait image. The artwork to microstructure mapping process can then be thought of as part of a broader and more flexible mapping concept designed to allow for the conversion of any piece of artwork into optically variable form. The development of this diffractive microstructure palette technology was first implemented in 1989 via a new OVD technology (Lee 1991) developed by CSIRO called Pixelgram™. Each element of a Pixelgram palette is designed to be in one to one correspondence with the RGB palette used in the construction of the input artwork, e.g. as represented on a computer screen. The particular advantage of this microstructure palette concept is that it enables the electron beam lithography data file size to be minimised. In 1989 the data explosion problem was particularly acute for diffractive devices of areas of several square centimetres because in this case the electron beam lithography pattern file consisted of millions of micro trapezoids of different angles, widths and lengths. Converting this pattern file to binary on a Vax computer for the Leica EBMF series of machine took many hours of computer time and resulted in very large data files of several billion e-beam shapes. Disc storage devices at this time also could not very easily store the resultant multi-gegabyte binary pattern file.

Redefining the OVD image in terms of a palette of 10 or twenty micro grating files solved these problems. It was then only necessary to convert each micro grating file to binary and the resultant binary file for the palette was normally no larger than ten or twenty megabytes. In this new electron beam lithography writing format the image is defined by a special Job control data file consisting of a list of stage positions and exposure conditions for positioning and exposing each micro grating palette element according to the image layout in the original artwork file. A second key advantage of this palette-writing format is the ability to separate the macroscopic properties of the pixellated microstructure array from the microscopic properties of the system. In the case of the diffractive optically variable device this means that many different optically variable images can be constructed from the same microstructure palette. Each image simply corresponds to a different distribution of 30 micron X 30 micron or 60 micron X 60 micron diffractive microstructure palette elements. An example of a four-element diffractive microstructure palette and its corresponding RGB artwork palette is shown in Fig. 7.8 which shows four elements of a 16-element greyscale Pixelgram palette and its corresponding RGB greyscale palette. The diffractive microstructure pixels are 60 microns X 60 microns in size. An example of an optically variable portrait produced from such a palette is also shown at two angles of view. The characteristic OVD effect is a positive to negative tone image switch. Element 16 of the palette also includes a super high-resolution text message as an additional covert security feature. Fig. 7.9 shows a photomicrograph of a small section of an typical Pixelgram OVD.

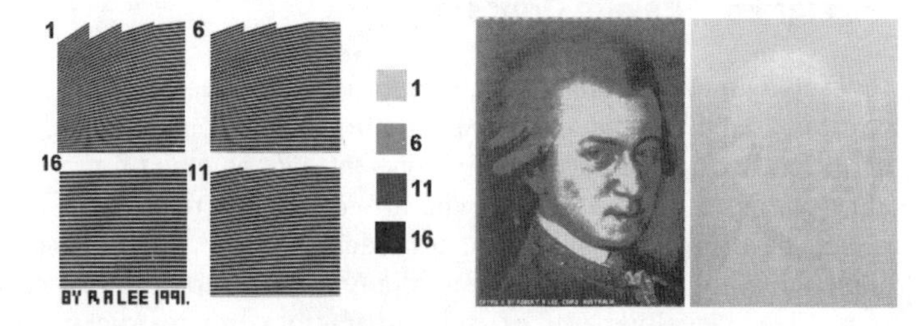

Fig. 7.8. Four elements of a greyscale Pixelgram palette and corresponding RGB palette

Fig. 7.9. Small section of a Pixelgram OVD showing pixels of different groove structure

The Pixelgram^{TM} technology developed by CSIRO required the use of this new e-beam writing format, which was developed in collaboration with Leica Microsystems of the UK using data file structures developed by CSIRO Australia. The resulting specialised software package, called PATJOB, was implemented on a Leica EBMF 10.5 using a slight modification to some of the operating system routines. The implementation of this new e-beam writing configuration led to the rapid development of the Pixelgram^{TM} OVD technology. All later OVD and security microstructure technologies developed by CSIRO have made use of this proprietary palette based electron beam lithography-writing format. More recently (in 2002) CSIRO and the Central Microstructure Laboratories of the Rutherford Appleton Laboratories in the UK developed an equivalent version of the palette based writing format for the Leica Vectorbeam VB6 series of electron beam lithography machines. This VB6 palette based data format, called PALTEK is a significant improvement over PATJOB as it has three times the resolution and ten times the writing speed of the earlier EBMF 10.5 version.

7.3.6 The Exelgram^{TM} Track based OVD Microstructure

By early 1993 the size of the individual pixel gratings of the Pixelgram^{TM} technology had been reduced to only 30 microns X 30 microns, providing greatly increased image resolution. However this reduction in pixel size also caused some loss in image brightness due to diffuse scattering effects from the edges of the pixel gratings. This diffuse scattering problem was solved by dispensing with the pixel based configuration of the OVD microstructure and replacing it with a surface relief structure based on a multiplicity of very thin diffractive tracks with continuous and smoothly varying groove angles and spacings (Lee, US patent nos. 5825547 and 6088161, 1997). This new alternative technology to Pixelgram^{TM} was named EXELGRAM^{TM}; exel being the common term used to describe the fundamental exposure elements of the electron beam writing process. An example of how the diffractive microstructure palette concept can be used to generate application specific specialised optical effects is given for the case of optically variable greyscale portraiture. In this example the 16 greyscales of the input OVD artwork are mapped to 16 different microstructures of different grating groove curvature. Fig.7.10 shows an example of such a microstructure region. American Express US$ and Euro Travellers cheques used Exelgram OVD microstructures of this type. In Fig. 7.10 the portrait dimensions are approximately 8 mm X 10 mm. Horizontal groove regions diffract the negative version of the portrait while sloping groove regions diffract light into angles where the positive version of the portrait can be observed, as shown above.

Fig.7.11 shows two views of the American Express Euro 50 Exelgram OVD at two different angles of observation. The application of the Exelgram to the Euro denomination cheques followed the highly successful application of the technology on the earlier US cheques (McHugh 1999; Wood 1999). These images illustrate the two channel switching capabilities of the technology as well as the highly accurate portraiture representations, which can be achieved. These images are enlarged views. The actual long axis dimension of the foil on the cheque is 19mm. While optically variable portraiture, for example as illustrated in Fig.7.12, was the driving force behind the development of Exelgram, this effect is just one of many unique optical effects available within the Exelgram™ optical effects package. Most of these new effects relate to Exelgram's highly flexible optically variable graphics capability, which is due to the diffractive palette basis of the technology and the electron beam lithography basis of the origination process.

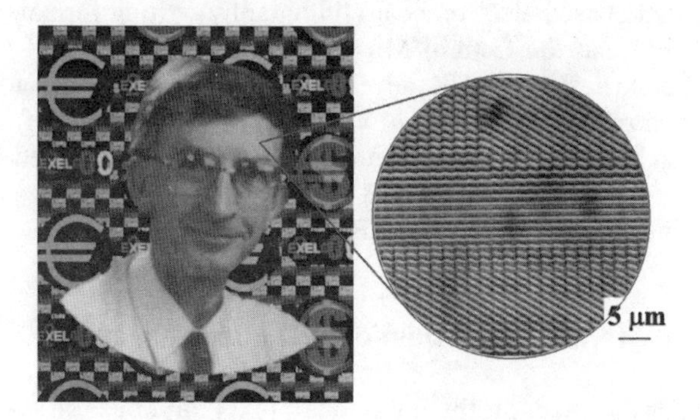

Fig. 7.10. Exelgram portrait OVD and corresponding diffraction grating microstructure

Fig. 7.11. American Express Euro Exelgram OVD at two angles of view

Fig. 7.12. Exelgram OVD showing two-channel switching and portraiture effects

For example the same curved line and variably spaced groove elements used for the portraiture effects can also used to generate very fast switching two-channel effects. In this case two interlaced micro diffraction grating structures are produced in which one group of grating structures corresponds to the RGB mapping from one input picture while the second interlaced group corresponds to the RGB mapping from a second picture.

By arranging the groove angles and/or spacings of the two sets of microstructures to be quite different, the resultant diffractive images generated by the OVD can be made to appear at quite different angles of view. Fig. 7.13(a) and Fig. 7.13(b) shows examples of the use of multi-channel microstructures (Lee 2000). In Fig. 7.13(a) the separation of the images is due to the different grating groove frequencies. This causes the different images to be diffracted at different angles of view to the OVD. By incorporating a slight degree of curvature into the individual groove elements much faster channel switching speeds can be obtained as well as less cross channel overlap. The Exelgram foils used on the Vietnamese bank cheques issued in 1996 used multi-channel effects of this type.

Examples of other high security Exelgram graphic effects can be seen on the new series of Hungarian banknotes issued in mid 1997. These banknotes contain an Exelgram^{TM} foil stripe, which generates an image switch upon rotation of the banknote by ninety degrees.

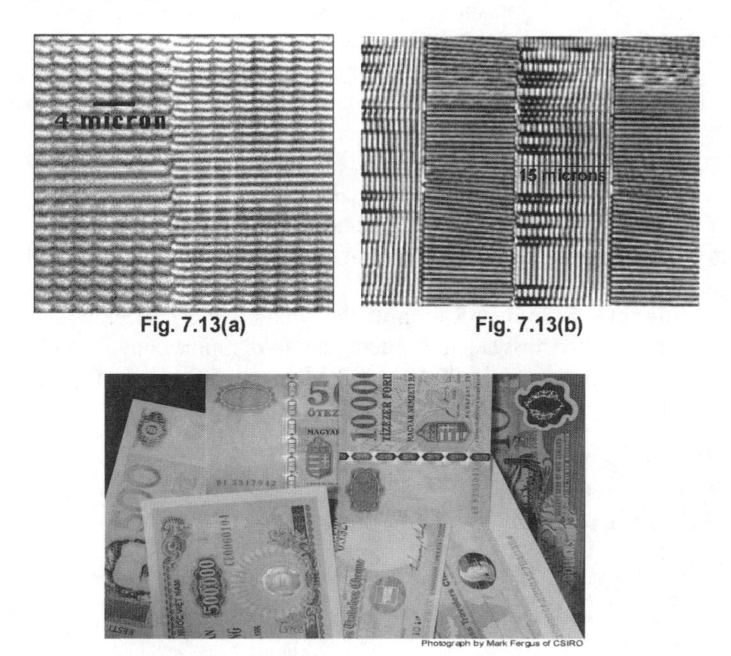

Fig. 7.14. Currency applications; Exelgram microstructures as an anti-counterfeiting feature

In this case the OVD image switches from the crown of Hungary to the letters MNB when the banknote is rotated by 90 degrees with respect to the normal. A small part of the OVD microstructure used to achieve this right angle switching effect is shown in Fig. 7. 13(b). Fig. 7.14 shows some sample Exelgram projects, including two Hungarian banknote denominations where the ninety-degree rotation image changes from the crown of Hungary to "MNB" on the foil stripe can be clearly seen. The effectiveness of Exelgram technology as an anti-counterfeiting feature is well documented.

7.3.7 Covert Image Micrographic Security Features

A unique advantage of the electron beam lithography origination process is its ability to fabricate complex non-diffractive anti-copy micrographic features and to integrate these features within the diffractive microstructure of the image. In this case the OVD image is designed to incorporate very small scale graphic elements consisting of combinations of alpha-numeric characters and other graphic elements such as logos and line drawings containing a range of feature sizes from 1 to 30 microns which act to diffusely scatter incoming light so that a palette of such micrographic elements can be used to form optically invariable macroscopic images of the greyscale or line-art form (Lee and Quint, PCT/AU98/00821). The Exelgrams used to protect the new Ukrainian passport/visa and the 2000 special edition New Zealand plastic $10 millenium banknote include micrographic features of this type for enhanced protection against counterfeiting. Fig. 7.15(a) and Fig. 7.15(b) show examples of two types of micrographic elements based on a euro theme test. In Fig. 7.15(b) the micrographic doves forming the background wallpaper pattern are each 20 microns across. The first three letters of the word EURO shown here are comprised of the second type of micrographic element, an enlarged view of which is shown in Fig. 7.15(a). A typical Exelgram OVD of this type may contain up to 100,000 individual micrographic elements embedded within a background diffractive microstructure. Particular advantages of such hybrid diffractive/ micrographic OVD microstructures include; (1) easy microscopic forensic authentication of the smallest piece of embossed substrate, (2) extremely high security against attempted holographic copying and (3) reduced metallic appearance of the OVD foil due to diffuse scattering from the micrographic elements.

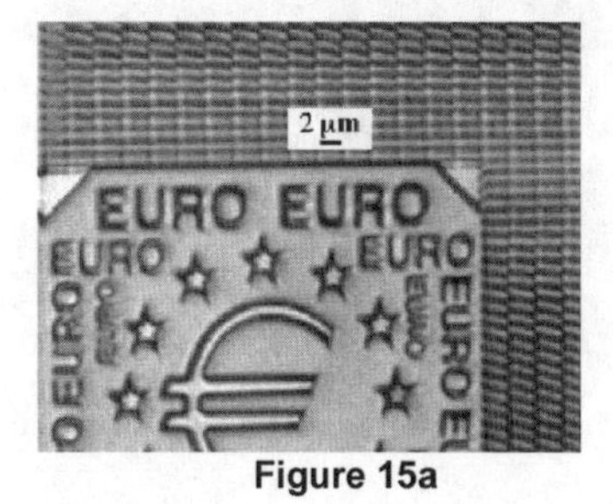

Figure 15a

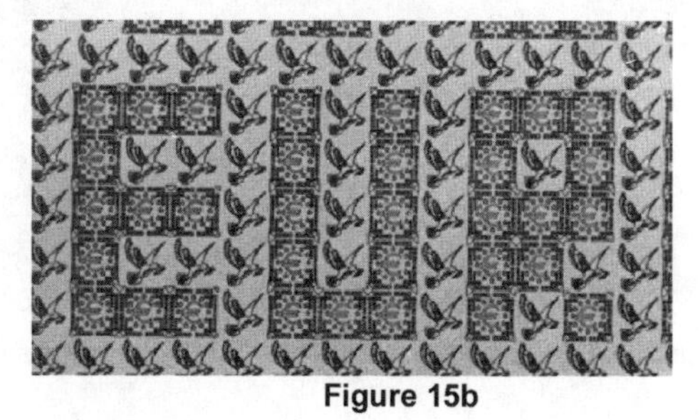
Figure 15b

Fig. 7.15. Micrographic doves forming the background

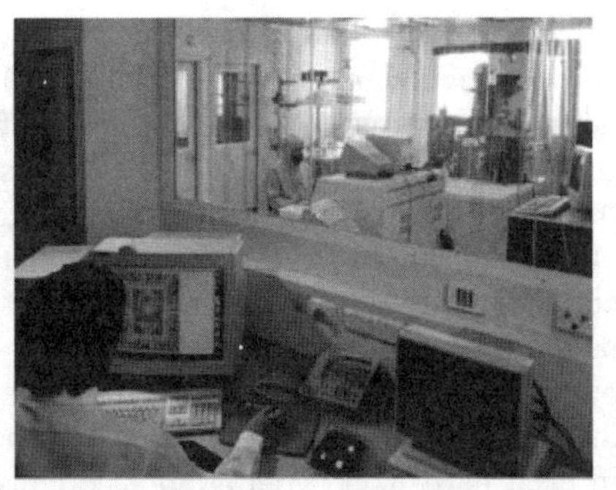

Fig. 7.16. CSIRO high security electron beam lithography laboratory established in 1993

7.3.8 Kinegram™ and Exelgram™: Comparison

The foregoing comparison of the attributes of the Exelgram™ and Kinegram™ technologies shows a very high degree of agreement with respect to satisfying security printing industry requirements. Both technologies are based on high cost single source origination systems and each technology generates its own unique set of tightly controlled optical effects. A summary of Kinegram and Exelgram optical effects and security attributes is shown in Table 7.1. The above discussion relating to the particular advantages of using electron beam lithography in the fabrication of specialised diffractive microstructures for anti-counterfeiting applications shows that this technique has much greater flexibility, resolution and geometric precision than previous techniques based on laser interference or holographic imaging. This is particularly true in the case of fast switching graphic effects, optically variable greyscale and line art portraiture and complex anti-copy micrographic effects. The origination of these effects is not possible using the earlier optical techniques. The mathematically defined input structure functions combined with the polygon based sequential writing mode of the EBL origination technique offers an almost unlimited range of possibilities for the generation of security microstructures that can be applied to security documents as specialised "inks", by direct embossing or as embossed hot stamping foils.

7.3.9 Vectorgram™ Image Multiplexing

As a further example of the versatility inherent in the electron beam lithography method for the micromanufacturing of optical security devices we will now discuss the technique of image multiplexing. The Vectorgram™ multiplexing concept (Lee, US patent 6,342,969) is best understood by considering a Pixelgram™ or Exelgram™ microstructure palette as a set of optical field generators. For example a typical palette of 26 elements might consist of two main groups. By element here we mean a 30 micron X 30 micron pixel containing a miniature diffraction grating of curved or straight grooves. Exelgram pixels are constrained by the requirement to smoothly connect to other pixels vertically above or below in the image. Elements 1 to 10 in a typical palette might consist of horizontal grooves of different spatial frequency; element 1 being of the lowest frequency and element 10 of the highest frequency. These elements are used for the generation of kinematic effects within the image. Elements 11 to 26 on the other hand might have a fixed spatial frequency, but varying degrees of groove curvature between the different elements. This group of 16 elements can therefore be used for positive to negative switch portraiture effects. Two channel effects can be produced by putting pixels of different spatial frequency next to each other in the image or by splitting the pixels in half and using different spatial frequencies in each half pixel. Micrographic effects can be produced with other palettes by putting little pictures into each 30-micron pixel. Different degrees of diffuse scatter between the different micro pictures can be used to generate optically invariable image effects. A two channel Pixelgram or Exelgram palette using split pixels can be regarded as a multi-component pixel palette because each pixel is a generator of two different optical fields.

Table 7.1. A summary of Kinegram and Exelgram security elements and optical effects

Security elements and optical effects	Kinegram™	Exelgram™
Single source origination	Yes	Yes
Origination technology	Analogue by proprietary laser exposure system	Digital by vector scan electron beam lithography
Technology availability	Tightly controlled	Tightly controlled
IP protection	Yes, various patents	Yes , various patents
Origination costs	High	High
Microstructure geometry	Arrays of 60 micron diameter pixels or 60 μm wide lines	Based on arrays of 15 or 30 micron wide diffractive tracks with internal groove angle modulations
Image resolution	60 micron	30 micron
Image brightness	Very high	High
Kinematic effects	Yes, observed by rotating OVD about axis perpendicular to the plane of the OVD.	Yes, observed by rotating the OVD about an axis within the plane of the OVD.
+/- greyscale portraiture	No	Yes
Greyscale effects	No	Yes
Sharp colour effects	Limited	Yes, sharp and highly visible
Multichannel effects	Yes	Yes
Channel switching speed	Slow	Very fast
Diffuse scattering efects	No	Yes
Microtext	Yes	Yes
Micrographic effects	No	Yes
Covert image effects	Yes. Fourier plane covert image observed under laser illumination	Yes. OVD plane image observed through coded transparent screen.

While the total image field generated by the two channels Exelgram may be regarded as a vector object, the local image field (i.e. the image field observed at a particular angle of view) is not a vector object because the two channels are not overlapping at the observer's eye. This is a preferred requirement as it is usually desirable to avoid cross talk between channels so as to generate a fast and clean switch from one image in channel 1 to another image in channel 2. However sometimes it is desirable to deliberately maximise cross talk between channels in order to create new types of optical effects. This multiplexing of optical fields to generate new types of optical effects is the key idea behind the Vectorgram™ process.

The process of arranging different types of optical microstructure elements within a Vectorgram™ pixel in order to cause a combined optical effect at the observer's eye is analogous to the process of combining the elementary directional aspects (Vector components) of a physical field to produce a resultant directional effect (Vector); hence the name Vectorgram™. Two examples of optical effects, which can be generated by this process, are described.

By splitting each pixel of a pixellated diffractive device into three areas corresponding to the red, green and blue components of a full colour image of a face or scene it is possible to create a device, which under point source illumination, generates a reproduction of that face or scene at one particular angle of view. The input artwork here involves three pictures corresponding to the red, green and blue components of the image. Fig. 7.17 shows a small section of a Vectorgram OVD with this type of structure. The area inside the black square represents a pixel of dimensions 30 microns X 30 microns. Each triplet of tracks contains diffraction grating structures of three slightly different spatial frequencies so that a red, green and blue triplet of spectra is allowed to overlap at one angle of view of the observer. The intensity of the red, green and blue components is controlled by varying the width of the pixel tracks according to the intensity required by the input artwork pixel.

In the optically variable device known under the trademark Pixelgram portraiture effects are achieved by mapping the greyscale palette of an image into a palette of miniature diffraction gratings (e.g. each of size 30 microns X 30 microns) where the grating rulings are curved in such a way as to modulate the intensity of the diffracted light in accordance with the level of greyness in the original portrait artwork. Lighter shades of grey correspond to greater degrees of groove curvature and more expansion of the diffracted beam. In the case of the Exelgram technology the modulation of the diffracted light intensity is achieved by varying the angle of the grooves within each grating track.

By combining both groove modulation techniques it is possible to generate optically variable portrait effects with increased brightness and a higher degree of variability in the transition from positive tone image to negative tone as the angle of view is changed. Fig. 7.18 shows a small section of the microstructure of such a Vectorgram portrait structure. A photograph of a corresponding example Vectorgram OVD at one angle of view is also shown. As can be seen from the micrograph, the combination process is a result of interlacing columns of Pixelgram type grating structures with Exelgram type microstructures. Fig. 7.19

shows a comparison of Exelgram positive to negative switching portraiture effects with the corresponding Vectorgram equivalent. Note the increased brightness of the Vectorgram positive tone image when compared to the Exelgram version. In each case the OVD size is 22mm X 22mm.

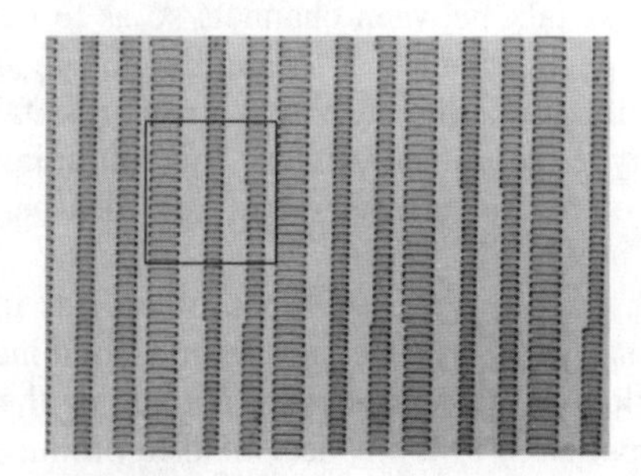

Fig. 7.17. Micrograph of a small section of a Vectorgram OVD

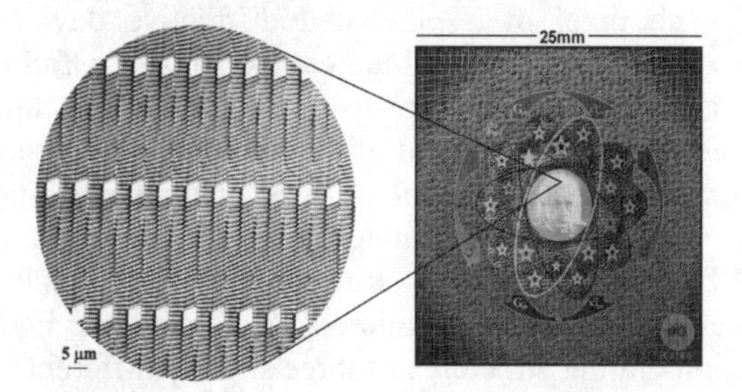

Fig. 7.18. Vectorgram microstructure for producing high brightness dynamic portraiture

Fig. 7.19. Comparison of Exelgram images (A , B) and Vectorgram images (C , D)

7.3.10 Interstitial Groove Element Modulation

In the Pixelgram, Exelgram and Vectorgram scenarios, the techniques for image modulation of the input artwork include groove frequency modulation, groove angle modulation, curvature modulation and combinations thereof. In all of these cases the fundamental modulation unit is the pixel palette element consisting of a fixed number of diffraction grating grooves of particular angles, spacings and curvatures. However it is also possible to extend the resolution of the modulation unit by considering modifications to the diffracted wavefront due to the "doping" of individual pixel palette elements with particular groove elements. The principle idea (Lee PCT/AU99/00520) here is the concept of using a continuous background groove pattern to generate a diffracted carrier wave and then to modulate this carrier wave with optically variable information by "doping" the background groove pattern with a multiplicity of interstitial groove elements to modulate both the spatial frequency and amplitude from each interstitial groove region according to the requirements defined by the optically invariable input picture information. In particular, the angles of diffraction from any particular interstitial groove are given by the local spatial frequency in that region and this is related to the number of parallel interstitial groove elements in that region. The intensity of diffracted light from the particular interstitial groove region under consideration is proportional to the length, angle, shape and degree of curvature of the interstitial groove elements. Hence the interstitial groove element concept provides a possible mechanism for modulating both diffracted light intensity and diffraction angle at a much higher degree of resolution. Fig. 7.20 shows a small section of an OVD microstructure of this type.

Fig. 7.20. Small section of an interstitial groove element optical security microstructure

7.4 Generic OVD Microstructures

The high security foil microstructures described in the previous section can be characterised as belonging to a class of optically variable devices ideally suited to high volume banknote type applications. High volume means volumes greater than 40 million units; all formed using the same OVD image. For these application volumes the high origination costs of the master plate, typically of the order of US$40,000, can be amortised over the production run resulting in a unit foil price of much less than 1 US cent. For lower volume applications the Kinegram or Exelgram solution is much less cost effective and there is therefore a need for much lower cost solutions outside of the security compromised dot matrix hologram technologies. Three different approaches to this low volume and high security requirement are described.

7.4.1 Optically Variable Ink Technology

One method for increasing flexibility, while retaining security through optical variability, involves the development of liquid media or inks containing microscopic flakes or pigment particles, which are in themselves individually optically variable.

7.4.1.1 Multilayer Thin Film Pigments

The US company Flex Products has a long history in the development of products of this type; in particular products based on microscopic thin film flakes which change colour (e.g. from red to green) with changing angle of view due to the cancellation of particular wavelengths and the reinforcement of other wavelengths as the incoming white light is partially reflected and partially transmitted at the boundaries of different layers of different refractive index within each flake (Phillips and Bleikolm 1996). Optically variable pigments of this thin film interference type can be distributed within a transparent liquid to produce a range of optically variable inks. Optically variable inks of this type are used on a wide range of banknotes throughout the world. While the optically variable effect is much simpler (being just a colour change), less pronounced and much more subtle than for conventional security foil patches, the ability of the ink to be used within standard offset printing processes with complete in-house artwork freedom and at relatively low cost (when compared to security foil technology) is a big attraction of the technology for printers with both high and low volume security printing requirements.

7.4.1.2 Diffractive Ink Pigments and Micrographic Powders

An alternative optically variable ink technology to the above multilayer thin film pigment can be produced by replacing the multilayer flakes with microscopic flakes in the form of miniature surface relief diffraction gratings. In this case the optical properties of the ink are determined by the diffraction of light from the groove patterns contained within the ink flakes. Arranging the groove spacings within the flakes to have predefined spacings, and also arranging the groove spacings to vary according to some modulation function can control the optical properties of the ink controlled to a much higher degree than with thin film devices (Lee, US patent no. 5912767; Argoitia and Chu 2004).

This higher degree of flexibility is due to the fact that the ink flake microstructures can be varied to an almost unlimited degree by using electron beam lithography to originate the master embossing plate for production of the diffractive flakes. The concept of a security ink based on diffractive flakes or pigments incorporating novel diffraction grating groove patterns originated by electron beam lithography was first proposed several years ago by CSIRO in Australia. More recently the idea has been further studied in the context of printable holograms (Argoitia and Chu 2004).

On a broader front it can be seen that not only are diffractive ink pigments able to be constructed in this way, but also ink flakes or powders based on the incorporation of microscopic micrographic elements as described earlier. This provides a basis for the production of low cost multi-element micrographic powders and pigments suitable for a range of security and track and trace applications. Mixing together two or more single element micrographic or diffractive powders selected from a palette of 26 or more individual single element powders can be used to create a range of personalised powders for scattering on documents or products to link the item to the owner or user. For example, imagine a micrographic powder palette consisting of 26 elements, where each micrographic element is a letter of the alphabet approximately 3 microns in height. For each letter (e.g. “A”) a master embossing shim (comprising a repeating pattern of the letter, e.g. “A”, over a fixed area) is produced by electron beam lithography and electroplating. The embossing shim is then used to emboss 6 or 7-micron diameter polymer beads in a roll or stamping type embossing press with an area of polymer powder particles being impressed with the letter (e.g. “A”) via the rolling or stamping process. When this process is repeated with each of the 26 letters of the alphabet we will have 26 powders to use as the basis for the construction of the multi-element powders. The number of possible multi-element powder combinations is very large. In fact a little mathematical analysis shows that a micrographic powder palette of N elements can generate $2^N - 1$ combination powders. The Table 7.2 shows examples of combination powders for the alphabet micrographic palette. Hence a 26 element micrographic powder palette can produce more than 67 million individual combination powders; enough to service over 67 million customers with their own micro ID signature for protecting their documentation and assets. Optical pigments constructed in this way can also be used to trace particular types of dangerous materials such as explosives.

Table 7.2. Micrographic powder combinations for the alphabet micrographic palette

Palette elements	Micrographic elements	Powder combinations	No. of powder types
1	A,	A	1
2	A, B	A, B, AB	3
3	A, B, C	A, B, C, AB, AC, BC, ABC	7
4	A, B, C, D	A, B, C, D, ABCD, AB, AC, AD, BC, BD, CD, ABC, BCD, CDA, ABD	15
5	A, B, C, D, E	.	31
.	.	.	.
26	A, B, C, ..., Z	.	$2^{26} - 1 = 67{,}108{,}863$

7.4.2 Diffractive Data Foils

Foil based data storage systems for financial transaction and other types of applications have been in use around the world for many years now. Every credit card carries on its reverse side a 1 cm by 9 cm magnetic foil stripe containing information pertaining to the user so that financial transactions can be carried out within the retail sector and for financial transactions at automatic teller machines (ATM) and electronic funds transfer point of sale (EFTPOS) terminals. The problem is that unscrupulous operators using hand held "skimming" devices or other more sophisticated data retrieval systems easily copy the information on the magnetic stripe. Once the information is collected it can be used to construct counterfeit credit cards for use in electronic theft and ID fraud applications. Because of this data-counterfeiting problem there is a need for more secure personal data storage technologies, which are both user friendly and open to network applications. Various innovative concepts have been proposed to increase the level of counterfeit difficulty. Some of these new technologies, including Valuguard™ and Xsec™ aim to increase security by making use of the inherent properties, such as noise and data jitter, within the mag stripe. Other proposed solutions make use of the security characteristics offered by OVD microstructures. Two of these OVD based data protection solutions are described.

7.4.2.1 Holomagnetic Data Storage

One such OVD technology based solution is the holomagnetic stripe technology developed by the company American Bank Note Holographics of the US and Leonhard Kurz GmbH. Holomagnetics uses a special machine readable holographic overlay on top of a magnetic stripe. The distribution and registration of the holographic pattern is unique to each card and is read and represented by a characteristic number stored on the magnetic stripe. A special reader can read the holographic pattern and calculate a corresponding pattern and compare with the number stored. Only when a match is confirmed is the financial transaction allowed to proceed via the normal magnetic stripe reading and comparison process.

7.4.2.2 Pixelgram Optical Pulse Sampling (POPS)

POPS is a secure data storage technology that unlike the magnetic stripe component technologies discussed above, uses diffractive microstructures for both the data storage process and the authentication process (Lee, US patent no. 5811775). POPS uses the concept of continuous foil patterns of Pixelgram and Exelgram diffractive microstructures which, under laser illumination, generate unique diffractive signature beams that can be machine read by photodetectors programmed to recognise and accept only beam intensity profiles of a unique predetermined envelope structure. Data is encoded into a POPS foil stripe by erasure (e.g. by laser ablation) of the diffraction grating microstructure within small cellular areas across and down the strip. This has the effect of sampling the diffracted light pulses in a manner characteristic of the information stored within a pulse. The highly unusual diffraction signatures determine the security value of the POPS system or pulse shapes which can be generated by novel diffraction grating microstructures fabricated by the electron beam lithography process. For example a typical POPS data storage palette might consist of a set of 8 Exelgram type-grating microstructures of different groove slopes. Three elements of this type of palette are shown in Fig. 7.21. Micrograting structures such as those shown in Fig. 7.21 can be used to control the intensity of the diffracted light. Intensity is measured by a photodetector positioned to collect light from a straight line grating of the same underlying spatial frequency as the micrograting palette elements.

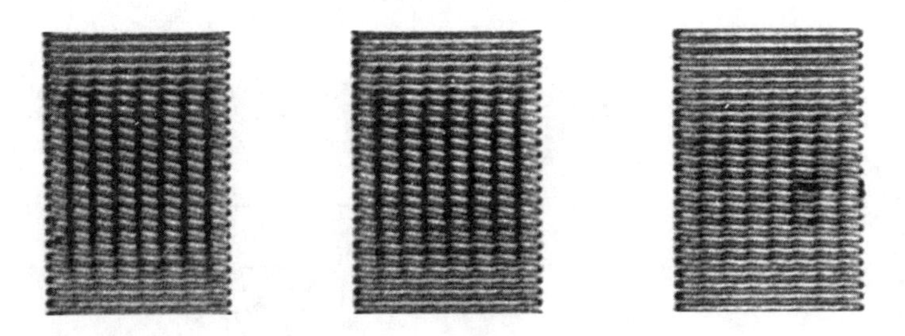

Fig. 7.21. Examples of POPS palette elements

Fig. 7.22 shows how a pallete of such microgratings can be used to design 8 bit pulse shapes into which binary data can be encoded by the erasure (by laser ablation) of particular pulse palette elements. The data readout system for the POPS system is shown schematically in Fig. 7.23. Greyscale areas represent microgratings of particular groove curvature. Fig. 7.23(b) is an example of a pulse signature grating structure. Micrograting areas can be as large as required and the data storage capacity is a function of POPS stripe length and data redundancy required.

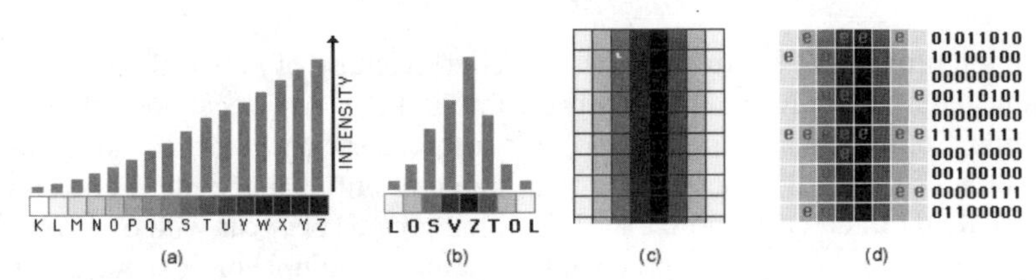

Fig. 7.22. (a) POPS palette (b) Example of a pulse signature grating structure (c) Section of unencoded POPS strip. (d) POPS strip and 8 bit byte structure after data encoding by laser erasure in cells labelled "e"

In this arrangement the POPS stripe replaces the magnetic stripe on the back of the credit card and the laser scanning system is built into the Automatic Teller Machine (ATM) or the point of sale card. This carding reading requirement creates a problem, which applies, not just to the POPS system, but also to all alternative technologies to the current magnetic stripe system. This is the problem of embedded infrastructure; that is, the current installed base ATM machines, etc. Changing the data storage technology on the card also requires a change in the card reading infrastructure and this is a significant cost barrier to the introduction of new technology. The inherent security characteristics of this optical data storage system is determined by the unique pulse shapes or signatures which can be constructed using micrograting palette elements incorporating grating groove structures of very unusual configuration. These grating structures can only be originated using electron beam lithography origination systems and this origination technology by nature of its complexity and expense acts as a further barrier to the counterfeiter wishing to emulate the characteristics of the POPS system.

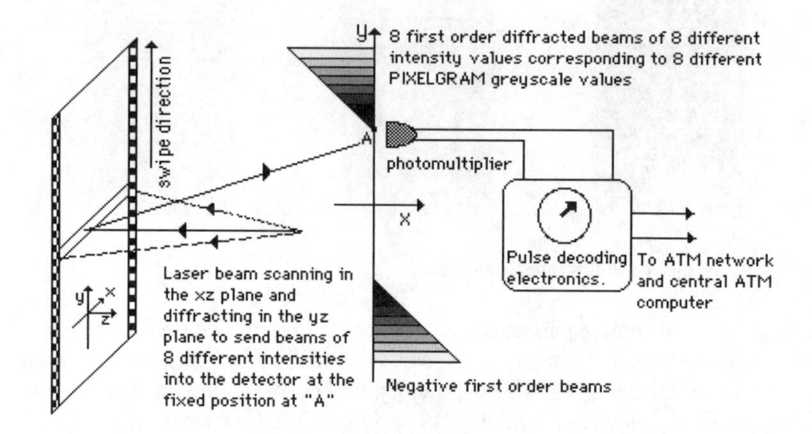

Fig. 7.23. POPS data readout system

7.4.3 Biometric OVD Technology

While the problem of ID fraud due to the counterfeiting of personal identification documentation has always been with us, this issue is now receiving much more attention from document issuers and law enforcement agencies due to the rapid escalation of financial losses related to this area of criminal activity. For example, in the UK, ID fraud costs more than 1.3 billion pounds per year with over 100,000 people directly affected in the process. Furthermore, multiple or false identities are used in more than a third of terrorist related activity and in organised crime and money laundering (www.standards.com.au, 2003).

In Australia, the situation is similar with financial losses due to ID fraud in 2002-2003 amounting to over $1.1 billion. There is therefore a need for improved levels of security for documents related to the legitimate identification of individuals for travel, access control and other purposes. Current ID document protective technologies include the traditional features of intaglio printing, watermarked paper, screen angle modulation printing as well as some relatively recent technologies such as the use of optically variable devices in the form of hot stamping foil patches or transparent overlays. While the use of OVDs in these two forms has certainly improved the security of ID documents, there is still a risk associated with altering the identity information without changing the other security features (e.g. by temporarily peeling away the transparent OVD overlay and replacing the photograph of the ID document holder with that of another individual).

The linking personal biological information associated with the ID document holder (e.g. iris scan information, fingerprints, etc. (Ashbourn 2003) to the document in a much more secure fashion is of course possible using electronic data storage methods such as chip cards. However this method requires machine authentication of the document and this requirement limits the utility of this method for protecting the document; e.g. on the spot visual authentication by field personnel is not possible without field portable equipment. As well as issues associated with expensive authentication infrastructure and equipment, there are also issues here related to individual privacy and the safeguarding of the document holder's biometric information and ensuring that this information is not made available for other purposes outside of its intended ID document use.

With the above background in mind, it can be seen that there is a need for an ID document solution, which links the visual image of the document holder to an OVD solution, which is unique to that individual. While this is possible using current OVD technologies such as Exelgram, the cost of such an application would be prohibitive in the extreme because an OVD origination would need to be produced for each document holder. However there has recently been developed an alternative and much lower cost method for achieving a similar result.

In its simplest form the biometric OVD concept (Lee 2004; McCarthy 2004) involves firstly originating a generic diffractive microstructure consisting of only two different groove spatial frequencies. This is illustrated in Fig. 7.24 for the case of a track based OVD such as Exelgram. In this case every even numbered track contains a grating microstructure of a fixed spatial frequency, while every odd numbered track contains a grating microstructure of a different spatial frequency to the even numbered tracks. This generic micrograting array forms the blank OVD canvas. After production of the corresponding embossed hot stamping foil, the security documents are prepared by hot stamping individual generic foil patches onto these documents. In this way all ID documents can be prepared prior to the inclusion of the ID information. The encoding of the image data into the blank OVD canvas to produce a two channel OVD now takes place via the following procedure. First, a portrait of the document holder is obtained and converted into a gridded and binary dithered image as shown by the example in Fig. 7.25 (A). Then a similar procedure is used to prepare the image chosen for the second channel of the OVD as shown by way of example in Fig. 7.25(B).

Finally the two binary dithered images are interlaced into a set of tracks where the even numbered tracks contain the dithered image of the document holder and the odd numbered tracks contain the image data of the second channel image. The resultant image pattern is shown in Fig. 7.26. The preparation of this two channel image data is carried out by special software and is effectively instantaneous following the image capture of the ID document holder's face.

The encoded two channel image data file is now used to control the operation of a laser exposure or ink jet printing system which scans the blank generic OVD canvas on the ID document and modifies the reflectivity of the grating pixels at selected locations according to the form of the image data file. The resultant OVD produces an image of the ID document holder at one angle of view and also replays the second channel image at a different angle of view corresponding to diffraction from the second set of micro grating pixels. These two images are shown in Fig. 7.27. Fig. 7.28 shows an example of an experimental biometric OVD produced in the manner described above. Note that the method described above for producing a diffractive biometric OVD structure involves an in-register modification of the diffraction properties from pre-selected micro grating pixels aligned in parallel tracks.

Fig. 7.24 Biometric OVD canvas

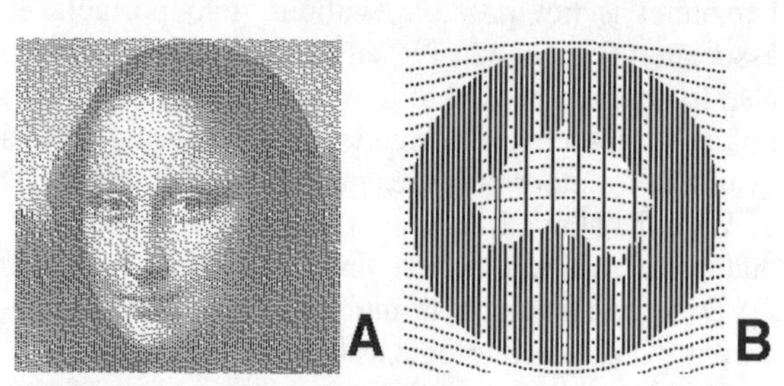

Fig. 7.25 Channel 1 input image (A) and channel 2 input image (B)

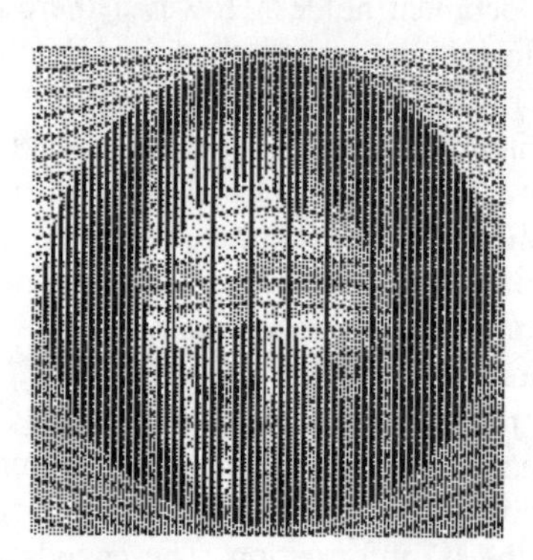

Fig.7. 26. Combined channels input data file for biometric OVD

Fig. 7.27. Biometric OVD image at two different angles of view

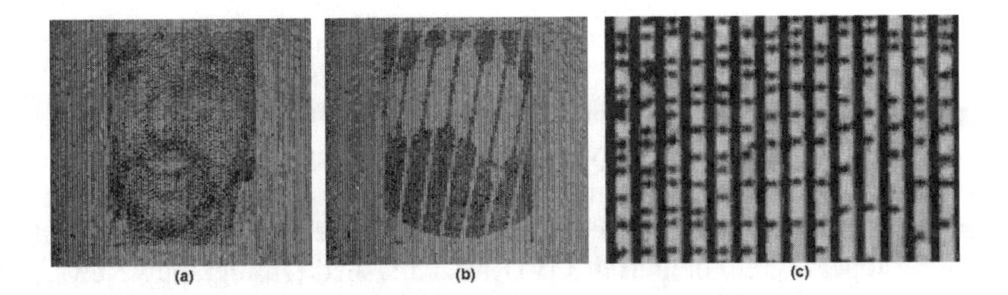

Fig. 7.28. Experimental biometric OVD image (of size approx. 6 mm X 10 mm) at two different angles of view (a), (b) and a magnified view of the laser ablated Exelgram track structure (black spots) for one of the channels (c). The black stripes in (c) correspond to the second channel, which diffracts at a different angle of view.

Using this structure the interlaced double image information is initially scrambled by the data encoding process and then decoded or unscrambled by the two angle micro grating array. The method can be regarded as a subtractive OVD origination method in contrast to current diffractive origination methods where both the OVD microstructure and the project specific artwork or image information is required to be originated within a single process. By separating the optically variable microstructure information from the project specific artwork information into two processes involving a generic microstructure origination step and a low cost data-encoding step a personalised OVD imaging process can be achieved. Personalizing the microstructure by subtractive modification of individual micro grating pixels by laser ablation or ink jet printing is the low cost per unit part of this process. Origination of the generic two channel OVD canvas and the production of the corresponding embossed hot stamping foil is of a similar cost to conventional OVDs.

7.5 NanoCODES

All of the optically variable device technologies described in sections 1 to 3 above have one thing in common. They all use embossed hot stamping foil as the transfer medium for attaching the optical microstructure to the security document requiring protection from counterfeiting. However foil based OVD technology, while in standard use throughout the world today on more than 70 different types of banknotes, is not regarded by security printers as the most preferred protection measure for security documents.

Firstly the foil basis of the technology requires a separate foil production facility and this reduces security, due to the risks associated with the transporting of the security foil to the security printing plant, and adds to production costs because of the need for specialised equipment for hot stamping the foils onto the security document. Secondly, and more recently, a new problem has arisen due to the inherently uncontrolled nature of holographic and diffractive OVD technology and some of its optical effects. Counterfeiters can now buy sophisticated dot matrix (Lee CK et al. 2000) master plate origination technology off-the-shelf so to speak and can also buy the necessary foil embossing and hot stamping technology from other sources. Thirdly the diffraction grating basis of OVD foils has an inherent problem due to the periodic nature of the grating groove structures in most of these devices. The more periodic the grating structure the greater the possibility of holographic copying (McGrew 1990) of the microstructure using holographic contact copying techniques. Counterfeit banknotes are already starting to appear with counterfeit OVD foils attached (Holography News, 2004; Renesse 2004). While the quality of these counterfeits is at present very poor, it is only a matter of time before more sophisticated versions appear. These new threats have given impetus to the search for much more sophisticated diffractive optical security microstructures and more sophisticated origination technologies. Electron beam lithography origination is now the increasingly preferred fabrication process (Tethal 2004) because of the more complex and secure specialised optical effects able to be produced by this process and the increased resistance to holographic contact copying due to the ability of these systems to produce curved and variable spaced grating groove structures. Holographic systems are unable to create microstructures, which generate these more specialised optical effects preferred by the security printing industry. However there is a view within the industry that a more fundamental change is required in order to provide a much greater degree of document security. This view includes a preference for a new OVD technology which eliminates the need for foil, is well differentiated in optical effects from the diffractive foil technologies and has sufficient microstructure depth to allow for more cost effective direct printing of the OVD microstructure onto the security document using an in-line printing process. Importantly, associated with the above required paradigm shift in thinking is a further requirement that any new origination process must involve technology which is much less accessible to the counterfeiters.

In this section we will address these issues by describing a new approach to the design of optically variable devices (OVDs) for anti-counterfeiting applications. In particular we shall address the range of problems related to improving the security and differentiation of OVD optical effects, in-line replication of the resultant image and lowering the overall cost of the OVD replication process while at the same time creating an origination process much less accessible to organised counterfeiting groups. In order to accommodate the broadest possible range of optical microstructure types, including the pre-existing diffractive types discussed above, as well as new reflective and refractive mechanisms and combinations of reflective and diffractive versions thereof, we will introduce a new umbrella term which also defines the principal area of application of these devices as relating to document security. The term NanoCODES (Nanofabrication of Cellular Optical Device Encryption Surfaces) is the proposed generic descriptor for this new approach as it emphasises that the main objective of this type of research is to find mechanisms for the coding of easily recognisable images into an optically variable form in a manner which is both secure from copying and simulation and easily recognisable by the general public as an authentication feature. One particular subset of NanoCODES devices of course relates to the diffractive devices discussed in previous sections. The second subset, which we will discuss, is based on devices consisting of arrays of micromirror optical elements.

7.5.1 The Micromirror OVD

The most important step change underpinning the new OVD approach is the replacement of micro diffraction grating elements within the OVD architecture by arrays of micro mirror elements (Lee, patent application no. PCT/AU02/00551). Optical micromirror array devices will be shown to offer the best hope of creating OVDs with sufficient microstructure depth to allow for direct printing onto the document substrate via the use of specialised reflective inks. The objective here is to do away with the current practice involving the costly intermediate step of producing an embossed hot stamping foil as the carrier of the OVD microstructure.

In the case of diffractive OVDs the diffraction grating groove spacings must be of the order of the wavelength of light in order for diffraction effects to take place. The depth of the grooves is related to the spacing by nature of the origination process, whether by electron beam lithography or laser interference techniques, and this limited groove depth capability (usually much less than 1 micron) is the main reason diffractive OVD microstructures are unable to be printed by normal printing techniques. The roughness of the substrate surface is normally of much greater depth than the surface relief of the diffractive microstructure. Micro mirror devices on the other hand are not so limited in their microstructure depth characteristics. For example a micro mirror array of 60 micron square micro mirror pixels inclined at an average angle of 30 degrees to the horizontal would have an average microstructure depth of 60Sin(30) = 30 microns. It is this greatly increased depth characteristic that holds out the promise of being able to produce OVD microstructures which can be directly printed. The development of passive micro mirror arrays with the above dimensional requirements can be achieved by combining the palette based cellular origination process used for diffractive OVDs with the well-known techniques of greytone lithography.

Greytone lithography is a technique for producing high-resolution microstructures with quite complex surface profiles and significant depth dimension. The process starts with the production of a variable transparency mask by electron beam lithography techniques. This mask is then typically used in an optical projection system to transfer the variable transparency distribution into a variable exposure distribution on a thick optical resist layer. Development of the exposed resist then yields a surface relief microstructure with a depth distribution corresponding to the distribution of transparency on the mask modified by the transfer function of the optical projection system and the exposure characteristics of the optical resist. The greytone lithography technique has been successfully applied to the fabrication of many different types of complex 3D high-resolution microstructures (Purdy 1994; Reimer et.al. 1997; Gimkiewiczet al. 1999; Kuddryashov et al. 2003). Examples include Fresnel lens array structures, blazed gratings, microprism arrays for optical interconnects and greyscale structure formations in general.

The graytone lithography technique can also be applied within a cellular array format to create palette based imaging structures. Because the set of fundamental microstructure elements of the cellular array is finite in number it can be regarded as a palette of greytone elements in an analagous manner to the RGB palette of a computer graphic image. Because of this analogy with the elements of a conventional colour palette this extended greytone lithography technique has been given the name "Colourtone Lithography" (Lee 2002). Further, investigations into this new colourtone lithography technique have been carried out by Leech and Zeidler (Leech and Zeidler 2003) with the aim of understanding the optical lithography exposure conditions required for increasing the depth of the microstructures. Recently further work has been undertaken to investigate the optical reflection and imaging properties of cellular greytone structures and to investigate the direct printing capabilities of these novel cellular array microstructures (Lee 2004).

7.5.2 Origination of a Micromirror OVD

A typical micro mirror array OVD microstructure may contain up to one million micro mirrors of dimensions 60 microns X 60 microns or less with each mirror arranged at a predetermined angle according to the input picture information. The procedure for producing a micromirror OVD involves two main phases – production of an image specific mask by electron beam lithography and the use of that mask within a photolithography exposure system to produce the required micromirror surface relief structure within a thick optical resist layer. The micromirror printing die is then formed by the electroplating of this optical resist layer to produce a hard nickel tool. The details regarding this origination process may be described as follows.

7.5.2.1 Greytone Palettes

The simplest type of greytone palette is one where each palette element corresponds to a fixed "shade of grey". The mask element corresponding to each palette element in this case consists of a uniform distribution of micro apertures with the micro aperture sizes varying between the greytone palette elements to give the different exposures. However a much more versatile lithography concept can be conceived by allowing each greytone palette element itself to have a variable microaperture distribution within the greytone pixel area. This is the basis of the "Colourtone Lithography" concept mentioned previously. In analogy to the standard greytone technique this new process involves the fabrication of a variable transparency mask by electron beam lithography and wet or dry etching followed by the use of this mask to profile a photoresist surface by means of a photolithography process. However in contrast to a greytone mask the "Colourtone Mask" is a pixellated array of individual mask pixel elements, where each mask pixel is typically of size 30 microns X 30 microns or 60 microns X 60 microns, and is a member of a finite set of mask pixels ("Mask Palette") of different transparency distribution. Secondly the distribution of mask pixels within the pixellated array is controlled by a colour coded input "Picture File" where each colour coded pixel or "RGB Pixel" of the picture file is a member of an "RGB Pixel Palette" and each member of the RGB palette is in one to one correspondence with each mask pixel of the mask palette. An example of an RGB picture palette and corresponding 3D microstructure palette is shown in Fig. 7.29. Not shown in Fig. 7.29 is the intermediate phase variable transparency mask palette, which generates the 3D pixel elements through exposure of the mask elements in an optical lithography system. This process modulates the optical transparency distribution of each mask palette element by both the transfer function of the optical system and the exposure properties of the optical resist. Hence calculation of the required mask palette for a given 3D pixel palette involves a similar functional relationship in the optics and resist technology to that used in conventional or non-palette greytone lithography. The colourtone lithography concept can therefore be seen to allow for the fabrication of complex microsurface structures containing a finite set of elemental 3D microsurface structures. An example of a four-element artwork palette, mask palette and corresponding microstructure palette, which can be used to construct a micro mirror device, is shown in Fig. 7.30.

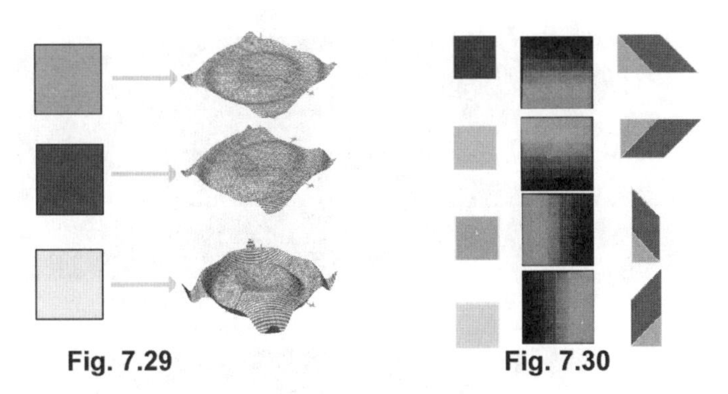

Fig. 7.29 Fig. 7.30

7.5.2.2 Micromirror Two Channel Switch Effect

As an explicit example of the use of the colourtone lithography technique to construct an optically variable micro mirror array, consider the artwork sample shown in Fig. 7.31. To create an optically variable microstructure version of this artwork we simply map the interlaced rows of light and dark pixels, representing the euro and "50" symbols respectively, into corresponding interlaced rows of micro mirrors of different slopes. Then when the surface relief microstructure is rotated about an axis parallel to the rows of interlaced micro mirrors the euro or "50" symbol will alternatively reflect light to the eye depending on which set of micro mirrors is aligned for maximum reflection.

To construct the device the artwork of Fig. 7.31 is converted into a corresponding pixellated greytone mask element array by palette based electron beam lithography techniques whereby each light and dark artwork pixel is replaced by the corresponding mask palette elements shown in Fig. 7.30. For the background area in Fig. 7.30 the artwork pixels are mapped into a mask pixel element (not shown in Fig.7.30) corresponding to a diffuse scattering microstructure element. ZEP 7000 is used as the electron beam resists and the pattern is exposed at 35 microcoulomb/cm^2 in a Leica EBMF 10.5 electron beam lithography system. After development of the plate a mask is produced by reactive ion etching in a Cl_2/O_2 plasma.

The variable transparency colourtone mask corresponding to Fig. 7.31 is then placed in a one to one mask aligner and used to expose a 30-micron thick layer of optical resist to the variable exposure pattern created by the mask. After development the optical resist surface is electroplated to produce a highly reflecting hard nickel surface replica of the device. Fig. 7.32 shows an optical micrograph of a small section of an experimental micro mirror microstructure produced in this way. Rows of micromirrors of opposing slopes can be seen here. In this particular case each micromirror has a dimension of 30 microns X 30 microns and therefore the total width of the micrograph is approximately 165 microns. Fig. 7.33 shows experimental images of the optically variable switch effect produced by the artwork shown in Fig. 7.31.

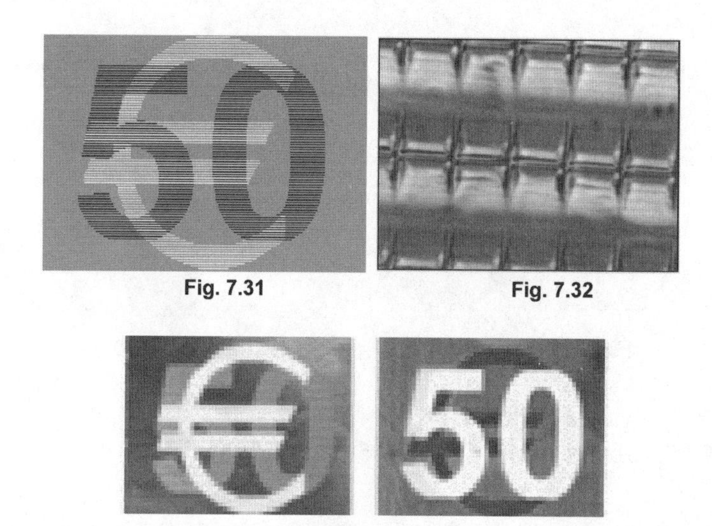

Fig. 7.33. Images of an experimental micromirror optically variable device at two angles of view showing the corresponding images embedded into each channel using the artwork configuration of Fig. 7.30 and the micromirror palette shown in Fig. 7.31

This example shows that two channel image switching is readily achieved within a micromirror framework by forming the individual channel images from micromirrors of different slopes and then interlacing these columns or rows of differently sloping micromirrors so that both images overlap each other within the same OVD area. Image separation is then achieved by the different reflectances of the two groups of micromirrors as the angle of view or the angle of illumination of the device is changed.

7.5.2.3 Optically Variable Portraiture using Micromirror Arrays

In the case of portraiture effects a micromirror mechanism is needed which modulates the degree of brightness of a micromirror pixel within a portrait image as a function of position within the portrait in accordance with the greyness value of the corresponding pixel within the input artwork. There are two possible mechanisms for achieving this variable intensity pixel behaviour. The first possibility is to fix the average slope of the micromirror pixels corresponding to the portrait image while also varying the curvature of the micromirror pixel surfaces so that the light is reflected into an expanding cone of light, only part of which is intercepted by the eye at any one angle of view. In this scenario variable degrees of surface curvature represent the different shades of grey within a portrait image. In practice this type of intensity modulation mechanism is very difficult to achieve because of the inherent difficulty in the micro sculpturing of such small surface areas using greytone lithography techniques. An alternative, and much simpler technique is to modulate the reflected light intensity by varying the area of the micromirror surface. Fig. 7.34 shows 4 elements of a 16-element portraiture mask palette corresponding to a 16-element grayscale portrait palette. White areas correspond to the areas of greatest light transmission within the mask elements. Therefore the top parts of the mask pixels will generate the greatest intensity of exposure during the photolithography part of the origination process. The corresponding areas on the photoresist will therefore have the greatest depth after development and the resulting micromirror pixels will consist of micromirrors sloping downwards from bottom to top (using the orientation of the pixels shown in Fig. 7.34). By interlacing these portrait type micromirror pixels with a different image comprised of micromirrors sloping in the opposite direction we can create a two channel micromirror OVD with a portrait image in one channel. Fig. 7.35 shows a photograph of an example of such a micromirror OVD.

Fig. 7.34. 4 mask elements of a 16-element micromirror portrait palette used for modulating the reflected light intensity within micromirror portrait image OVD

Fig. 7.35. Photographs of the images generated by a nickel shim containing a two-channel micromirror OVD with a grayscale portrait image in one channel. The diameter of the OVD is 36 mm

7.5.3 Summary of Micromirror OVD Optical Effects

From the above examples it can be seen that micromirror optically variable devices can generate various other optical effects. In fact in can be shown that there exists an isomorphic relationship between diffraction and reflection effects generated by diffractive and reflective OVDs respectively. In both cases the diffracted or reflected optical fields are described by gradient mapping equations with respect to the optical phase function generated by the microstructure. The main differences between the two systems relates to the presence of the wavelength and order number parameters in the diffractive case. Diffractive OVDs differ from reflective OVDs through the presence of colour effects and multiple diffraction orders. Except for these differences we can map all of the usual optical effects associated with diffractive OVDs into an equivalent set of optical effects produced by reflective OVDs. Table 7.3 compares and summarises the different mechanisms for achieving particular optical effects within the two OVD scenarios. A particular advantage of the micromirror OVD concept relates to the relatively large angular separation between channels (sometimes of the order of 120 degrees). Since there is only one propagating order for each channel in the micro mirror case, micro mirror OVDs are likely to be far less sensitive to crumpling of the paper surface in comparison to diffractive OVDs where the degree of cross talk between different channels due to both multi order propagation and relatively small angular differences (e.g. 30 degrees) between channels often causes the individual channels to be separately unobservable after crumpling of the paper surface. Another interesting aspect of this type of security microstructure is its ability to be used as an embossing tool for transparent substrates; such as polymer banknotes. In this case the micro mirror structures become micro prism arrays and the image becomes an optically variable watermark in transmission; that is, looking through the transparent substrate at a light source we observe an image switch as we move the transparent device from above the light source to below the light source.

It is also clear from the above discussion and the isomorphic relationship between the optics of diffractive and reflective pixellated arrays that the micromirror OVD concept easily lends itself to the creation of generic OVD substrates for application in the area of biometric OVD technology. A diffractive version of this technology was described in section 7.4.3 above. If we replace the two channel Exelgram structure by a generic two channel reflective micromirror structure then all other aspects of the biometric OVD concept can be carried over to the micromirror case (including the laser written data encoding process) with the result that a personalised direct print version of the technology can be created.

It can be seen from the example described in the previous section that the biometric OVD concept can also be applied to the storing of bar code data in a very secure manner. Since a bar code is simply a picture of black and white bars of different widths organised into a particular sequence, we can imagine using two different bar code images as channel 1 and channel 2 images encoded into an interlaced sequence according to the method described earlier. The bar code reader is then presented with two bar codes to read at two different angles of view and modified software within the reader correlates the two pieces of bar code data to determine the authenticity of the object carrying the two channel bar code. For example, an ID card or passport could include both a biometric OVD image of the document holder for visual authentication and an OVD bar code containing other data pertaining to that individual. In both cases the data is written onto the document at the time of issue using the subtractive data encoding method described in the previous section. In a similar manner to the biometric OVD case it can be seen that the micromirror origination process can also be used to produce a reflective version of the POPS data storage system discussed in section 7.4.2. In the case of micromirrors the data palette is constructed from micromirror pixels of variable micromirror width and the signature pulse reflected from the printed micromirror strip is constructed from sets of 8 micromirror areas of different reflectivity corresponding to different micromirror widths. The data encoding process in this case is similar to the diffractive case in that destroying the reflectivity of particular micromirror areas by laser ablation creates a binary “1”, the unerased areas being binary “0”s.

In concluding this section it should be mentioned that, while we have focused the discussion on the use of a generic micro mirror array as the underlying OVD microstructure, similar biometric OVD and personalised data encoding techniques can be applied when the underlying generic OVD microstructure takes the form of a transparent film embossed with a generic micro mirror surface relief structure to create arrays of micro prisms corresponding to different angles of refraction of the transmitted light. In this case the images are viewable in transmission at two different angles of view, thereby creating an optically variable watermark effect. In this case of course no reflective ink need to be applied to the substrate to view the images. It can be seen that the encoding of the data onto this film substrate can be carried out using similar in-register laser ablation techniques in order to modify the refractive properties of individual prism pixels. In fact, in this case of course, the laser ablation process changes the individual micro prism pixels into diffusely transmitting pixels.

Table 7.3. Comparison of diffractive and reflective OVD optical effects

Optical property	Diffractive pixel OVD	Reflective pixel OVD
Resolution dependence.	Determined by diffractive pixel dimensions	Determined by micro mirror pixel dimensions
Palette structure	Comprised of a finite set of diffractive pixels of different grating groove spacings, modulations, curvature and groove orientations.	Comprised of a finite set of micro mirror pixels of different mirror angles, curvatures and orientations.
Multi-channel image switching	Produced by using adjacent diffractive pixels of different spatial frequency	Produced by using adjacent micro mirrors of different mirror slope angles.
Microtext	Produced by constructing letters out of diffractive pixels of the same type	Produced by constructing letters out of micro mirror pixels of the same angle slope and orientation.
Right angle effects	Produced by using adjacent diffractive pixels with grooves orientated at 90 degrees to each other.	Produced by using adjacent micro mirror pixels with mirror slopes orientated at 90 degrees to each other.
Covert image effects	Produced by interlacing a positive and negative version of the same image, where the positive image diffractive pixels have different groove curvature to the negative image diffractive pixels.	Produced by interlacing a positive and negative version of the same image, where the positive image micro mirror pixels have different mirror curvature or width to the negative image micro mirror pixels.
Movement effects	Produced by filling adjacent areas of the OVD with diffractive pixels of different special frequency.	Produced by filling adjacent areas of the OVD with micro mirror pixels of different micro mirror angle.

7.6 Conclusions

In this chapter we have reviewed the specialised area of micromanufacturing and nanotechnology as applied to the fabrication of optical security microstructures. In particular we have discussed in detail the specific area of optically variable devices and the advantages offered by the electron beam lithography method of manufacturing these devices. While the growth of development and adoption of foil based OVDs during the 1990's has certainly resulted in improved protection against counterfeiting, the need for more application specificity, particularly in the

ID document area, reduced costs in both origination and replication and the ongoing need to stay ahead of the counterfeiters as they attempt to undermine the latest developments, has led to a demand by security printers for more innovative technologies producing better differentiated optical effects, lower costs and better control on the availability of the underlying core technologies. In CSIRO we have responded to these industry challenges by initiating new research activities to develop biometric specific OVD microstructures as well as optical microstructures, which can be directly printed onto the security document without the costly intermediate foil phase. As the counterfeiters become better organised and more technically focussed we can expect security printers and their associated technology suppliers to become much more R&D focussed in meeting this evolving threat from the counterfeiters. Current thinking within the industry now accepts the need to upgrade document security on a regular five-year basis and therefore we can expect that micromanufacturing and nanotechnology R&D for document security will become a growing area.

7.7 References

Antes G (1986) Holograms and Kinegrams as visual and machine readable security features on securities and plastic cards and the stability and conservation of photographic images: Chemical, electronic and mechanical. Proc. Int. Smp., Thailand, pp 24-27

Antes G (1982) Document with diffraction grating: European patent no. 0 105 099

Antes G, Saxer C (1988) Document with microlettering: European patent no. 0 330 738

Antes G (1988) Multichannel surface patterns: European patent no. 0 375 83

Argoitia A, Chu S (2004) The concept of printable holograms. In: Renesse RL (ed) Optical Security and Counterfeit Deterrence Techniques. Proc. SPIE, vol. 5310, pp 275–288

Ashbourn J (2003) Biometrics: A contemporary view of existing biometric technologies, Keesing's Journal of Documents 2:3-8

Aubrecht I et al. (2004) The use of mix-matrix techniques to create holograms, Keesing's Journal of Documents 5:24-26

Coombs PG et al. (2004) Integration of contrasting technologies into advanced optical security devices. In: Renesse RL (ed) Optical Security and Counterfeit Deterrence Techniques. Proc. SPIE, vol. 5310, pp 299-311

Gimkiewicz C et al. (1999) Fabrication of microprisms for planar optical interconnections by use of analog grayscale lithography with high-energy-beam-sensitive glass, Appl. Opt 38:4

Hardwick, Bruce A (1990) Performance of the diffraction grating on a banknote– the experience with the Australian commemorative note. In: Fagan WF (ed) Optical security and anticounterfeiting systems. Proc. SPIE, vol. 1210, pp 20-26

Heslop J (1990) The need for industry to take a pro-active role against pirates. In: Fagan WF (ed) Optical security and anticounterfeiting systems. Proc. SPIE, vol. 1210, pp 129-132

Holography News (1992) vol. 6, No. 9, p 1

Holography News (2002) vol. 16, No. 5, p 1

Holography News (2004) vol. 18, p 1

Keesing's Journal of Documents (2004) Editorial 7:33-34

Kontnik LT (1990) Survey of holographic security systems. In: Fagan WF (ed) Optical security and anti-counterfeiting systems. Proc. SPIE, Vol. 1210, pp 122-128

Kuddryashov V, et al. (2003) Grey scale structures formation in SU8 with E-beam and UV, Microelectronic Engineering 67/68:306-311

Lancaster M, Mitchell A (2004) The growth of optically variable features on banknotes. In: Renesse RL (ed) Optical Security and Counterfeit Deterrence Techniques. Proc. SPIE, vol. 5310, pp 34-45

Lee CK, et al. (2000) Optical configuration and colour-representation range of a variable-pitch dot matrix holographic printer, Applied Optics 39:40–53

Lee RA et al. Diffraction grating: European patent no. 0449 893 B1

Lee RA, Quint G, Micrographic device: Int. patent application PCT/AU98/00821

Lee RA, A diffractive device: US patent no. 5,825,547; A diffractive device at least partially comprised of a system of tracks: US patent no. 6,088,161; Diffraction grating and method of manufacture: European patent no. EP 0 490 923 B1; Diffractive structure with interstitial elements: International patent application PCT/AU99/00520; Diffractive indicia for a surface: US patent no. 5,912,767; Optical data element: US patent no. 5, 811,775; An optical device and methods of manufacture: International patent application no. PCT/AU02/00551; Multiple image diffractive device: US patent no. 6,342,969

Lee RA (1983) Generalised curvilinear diffraction gratings I: Image Diffraction Patterns, Optica Acta 30:267-289

Lee RA (1983) Generalised Curvilinear Diffraction Gratings II: Generalised Ray Equations and Caustics, Optica Acta 30:291-303

Lee RA (1983) Generalised Curvilinear Diffraction Gratings V: Diffraction Catastrophes Optica Acta 30:449-464

Lee RA et al (1985) Diffraction Patterns of Generalised Curvilinear Diffraction Gratings, Optica Acta 32:573-593

Lee RA (1991) The Pixelgram: An application of electron beam lithography for the security printing industry. In: Fagan W (ed) Holographic Optical Security Systems. Proc. SPIE, Vol. 1509, pp 48-54

Lee RA (1997) A new high security OVD technology for anti-counterfeiting applications, 9th Interpol conference on currency counterfeiting, Helsinki

Lee RA (1998) The optically variable device as an anti-counterfeiting eature. Interpol 75th Anniversary Publication, Kensington Publications, pp 98-101

Lee RA (2000) Microtechnology for anti-counterfeiting, Microelectronic Engineering 53:513-516

Lee RA (2001) OVD microstructures for direct printing applications. Intergraf, XVIIIth Int. Security Printers Conf., Sorrento, Italy

Lee RA (2002) Colourtone lithography, Microelectronic Engineering 61:105-111

Lee RA, et al. (2003) A method of forming a diffractive authentication device: Australian provisional patent no. 903502

Lee RA (2004) Micro mirror array nanostructures for anti-counterfeiting applications. In: Renesse RL (ed) Optical Security and Counterfeit Deterrence Techniques. Proc. SPIE, Vol 5310, pp 350-368

Leech PW, Zeidler H (2003) Microrelief structures for anti-counterfeiting applications, Microelectronic Engineering 65:439-446

McCarthy LD, et al. (2004) Modulated digital images for biometric and other security applications. In: Renesse RL (ed) Optical Security and Counterfeit Deterrence Techniques. Proc. SPIE, vol. 5310, pp 103-116

McGrew SP (1990) Hologram counterfeiting. In: Fagan WF (ed) Optical security and anticounterfeiting systems. Proc. SPIE, vol. 1210, pp 66-76

McHugh J (1999) American express cheque redesign, Aanti-counterfeiting initiatives conference on document counterfeiting protection, pp 28-29, USA

Moser JF et al. (1996) Perceptual information from OVD diffraction security devices. Renesse RL (ed) Proc. SPIE, vol 2659, pp 53-58

Moser JF (1994) Document protection by optically variable graphics (Kinegram). In: Renesse RL (ed) Optical Document Security. Artech House Publishing, pp 169-185

Newswanger C (2004) Holographic diffraction grating patterns and methods for creating the same: European patent application, 0467601A2

Phillips RW (1990) Optically variable films, pigments and inks, SPIE – Optical thin films III, New developments 1323:98-109

Phillips RW, Bleikolm AF (1996) Optical coatings for document security, Applied Optics 35:28

Purdy D (1994) Fabrication of complex microoptic components using photo-sculpturing through halftone transmission masks, Appl. Opt. 3:167-175

Reimer K, et al. (1997) Fabrication of microrelief surfaces using a one-step lithography process, Proc. SPIE, vol. 3226, p 6

Renesse RL (1994) Iridescent optically variable devices: A survey. In: Renesse RL (ed) Optical Document Security. Artech House Publishing, pp 207-225

Renesse R (2004) The security power of DOVIDS, Holography News, 18:9

Souparis H (1990) Embossed stereogram developments and security applications, In: Fagan WF (ed) Optical security and anticounterfeiting systems. Proc. SPIE, vol. 1210, pp 59-65

Tethal T (2004) Using electron beam lithography to create high quality DOVD, Keesing's Journal of Documents 5:1

Wilson GJ (1998) Australian polymer banknote. In: Renesse RL (ed) Proc. SPIE, vol. 3314, pp 2-6

Wood G (1999) Holograms in the protection of commercial documents of value, Proc. 1st world Product and Image Security Conf., Barcelona, p 1435

www.standards.com.au, (2003) Securities industry research centre of Australia (SIRCA), Identity fraud in Australia; An evaluation of its nature, cost and extent

8 Nanofinishing Techniques

S. Jha and V. K. Jain

Department of Mechanical Engineering, Indian Institute of Technology Kanpur, 208016, India

8.1 Introduction

Final finishing operations in manufacturing of precise parts are always of concern owing to their most critical, labour intensive and least controllable nature. In the era of nanotechnology, deterministic high precision finishing methods are of utmost importance and are the need of present manufacturing scenario. The need for high precision in manufacturing was felt by manufacturers worldwide to improve interchangeability of components, improve quality control and longer wear/fatigue life (Keown 1987). Taniguchi (Taniguchi 1983) reviewed the historical progress of achievable machining accuracy during the last century. He had also extrapolated the probable further developments in microtechnology and nanotechnology, Fig. 8.1. The machining processes were classifieds into three categories on the basis of achievable accuracy viz. conventional machining, precision machining and ultraprecision machining. Ultraprecision machining are the processes by which the highest possible dimensional accuracy can be achieved at a given point of time. This is a relative definition, which varies with time. It has been predicted that by 2000 AD, machining accuracies in conventional processes would reach 1 μm, while in precision and ultraprecision machining would reach 0.01μm (10 nm) and 0.001μm (1 nm), respectively (Taniguchi 1983). His predictions made around two decades before are in line with the current advances in manufacturing technology. These accuracy targets for today's ultraprecision machining can't be achieved by simple extension of conventional machining processes and techniques.

Nanotechnology (Taniguchi 1983) was first used to classify integrated manufacturing technologies and machine systems, which provide ultraprecision machining capabilities in the order of 1 nm. Since then ultraprecision technologies have grown rapidly over recent years and have tremendous impact on the development of new products and materials. Nanotechnology is the target of ultraprecision machining because the theoretical limit of accuracy in machining of any substance is the size of an atom or molecule of that substance. With the advent of new materials, manufacturing scientists are facing challenge in machining them to meet their functional requirements. As the demand moves from the microtechnology (1 μm accuracy) to the nanotechnology (1 nm accuracy) region, the systems engineering demands rapid increase in stringency and complexity (Taniguchi 1983). The traditional finishing processes alone are therefore incapable of producing required surface characteristics to meet the demand of nanotechnology. Even in certain cases these processes can be used but then they require expensive equipment and large skilled labour, finally leave them economically incompetent.

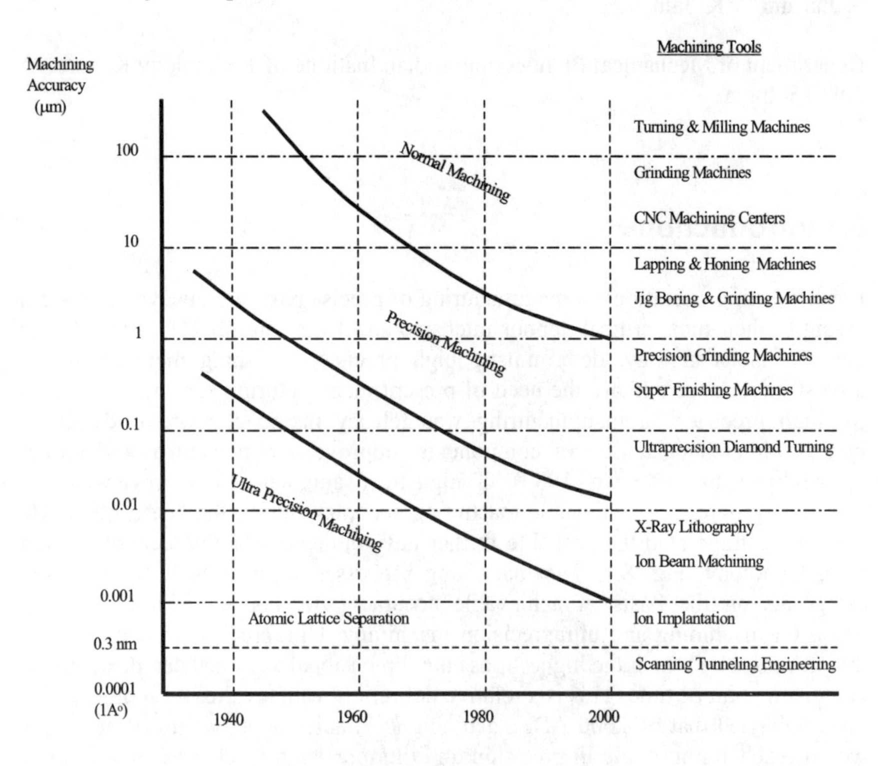

Fig. 8.1. Achievable Machining Accuracy (Taniguchi 1983)

New advanced finishing processes were developed in last few decades to overcome limitations of traditional finishing processes in terms of higher tool hardness requirement and precise control of finishing forces during operation. This helped in finishing harder materials and exercising better in process control over final surface characteristics. Another limitation relaxed by some advanced finishing processes using loose abrasives is to finish complicated geometries by enhancing reach of abrasive particles to difficult-to-access regions of the workpiece surface. In this way, newly developed finishing processes are, to a large extent, helpful in meeting the requirements of 21st century manufacturing.

8.2 Traditional Finishing Processes

Before continuing discussion on advanced ultra precision finishing processes, it is useful to understand the principle of working of commonly used traditional finishing processes; grinding, lapping and honing. All these processes use multipoint cutting edges in the form of abrasives, which may or may not be bonded, to perform cutting action. These processes have been in use from the earliest times because of their capability to produce smooth surface with close tolerances. Higher hardness of abrasive particles compared to workpiece hardness is an important prerequisite of these processes. If properly conducted, these abrasive machining processes can produce a surface of high quality with a controlled surface roughness combined with a designed residual stress distribution and freedom from surface and sub-surface damages (Schey 1987).

8.2.1 Grinding

Grinding is the most widely used abrasive finishing process among all traditional finishing processes used in production. In grinding, the material is removed from the workpiece surface by relative motion between the workpiece and the grinding wheel having abrasive particles embedded on its periphery. The abrasive particles are bonded together to form porous revolving body (Shaw 1996) which when come in contact with workpiece results in material removal. Grinding in a broad perspective is divided into two regimes – stock removal grinding (SRG), and form and finish grinding (FFG). In SRG, the main objective is to remove the superfluous material from the surface, while in FFG, the surface quality is a main concern. The abrasives on a grinding wheel are firmly bonded with an appropriate binder and at the same time also have possibility to allow grain fracture to generate new cutting edges. Abrasive grain wears rapidly on grinding harder materials so a less strongly bonded wheel is preferred for operation. Porosity on the grinding wheel is a controllable factor to provide room to accommodate chips. Wheels bonded with glass are strongest and hardest while organic bonds are of lower strength. The size and distribution of grits along with wheel structure play an important role in grinding performance. A proper selection of wheel according to finishing requirements is very important. Grinding is mainly employed for simple geometries like cylindrical or plane surfaces where size is limited by grinding wheel movement.

8.2.2 Lapping

Lapping uses loose abrasives to finish the surface. It works on three body abrasive wear principle in which finishing action takes place through abrasion by hard particles trapped between workpiece surface and a relatively soft counter formal surface called lap. After introducing abrasive slurry between workpiece and lap surface, the workpiece is held against lap and moved in random paths under pressure. Simple three-dimensional shapes and curved surfaces (concave, convex etc.) can be finished by designing a compliant lap. As this process is generally employed for improving surface finish and accuracy, the amount of material removed is insignificant.

8.2.3 Honing

Honing is another abrasive finishing process generally used to finish internal cylindrical surfaces. The abrasives in the form of stones or sticks carried in an expanding and oscillating mandrel are used to generate random cross-marked surface with good finish. The stick pressure on workpiece surface is comparatively more than lapping. The surface produced after honing has self-lubricating property due to oil retaining capability in crosshatched pattern.

8.3 Advanced Finishing Processes (AFPs)

There are many advances taking place in the area of finishing of materials with fine abrasives, including the processes, the abrasives and their bonding, making them capable of producing surface finish of the order of nanometer. Earlier there has been a limit on the fineness of size of abrasives (~a few μm) but today, new advances in materials syntheses have enabled production of ultra fine abrasives in the nanometer range. Abrasives are used in a variety of forms including loose abrasives (polishing, lapping), bonded abrasives (grinding wheels), and coated abrasives for producing components of various shapes, sizes, accuracy, finish and surface integrity.

The electronics and computer industries are always in demand of higher and higher precision for large devices and high data packing densities. The ultimate precision obtainable through finishing is when chip size approaches atomic size (~ 0.3 nm) (Taniguchi 1983). To finish surfaces in nanometer range, it is required to remove material in the form of atoms or molecules individually or in groups. Some processes like Elastic Emission Machining (EEM) and Ion beam Machining (IBM) work directly by removing atoms and molecules from the surface. Other processes based on finishing by abrasives, remove them in clusters. On the basis of energy used, the advanced finishing processes (AFPs) can be broadly categorised into mechanical, thermoelectric, electrochemical and chemical processes (Jain 2002). The performance and use of a certain specific process depend on workpiece material properties and functional requirement of the component. In mechanical AFPs, very precise control over finishing forces is required. Many newly developed AFPs make use of magnetic/electric field to externally control finishing forces on abrasive particles. To name a few, these magnetic field assisted finishing processes include Magnetic Abrasive Finishing (MAF), Magnetic Float Polishing (MFP), Magnetorheological Finishing (MRF), and Magnetorheological Abrasive Flow Finishing (MRAFF). Chemo Mechanical polishing (CMP) utilises both mechanical wear and chemical etching to achieve surface finish of nanometer range and planarisation. CMP is the most preferred process used in semiconductor industry for silicon wafer finishing and planarisation. Since the material removal in fine abrasive finishing processes is extremely small, they can be successfully used to obtain nanometer surface finish, and very low value of dimensional tolerances. Advanced abrasive finishing processes belong to a subset of ultra precision finishing processes which are developed for obtaining surface finish in the range of nanometer. A comparison of surface finish obtainable from different finishing processes is given in Table 8.1. This chapter discusses the principle of working and potential applications of such processes in following paragraphs.

Table 8.1. Comparison of surface finish obtainable by different finishing processes

Finishing Process	Workpiece	Ra value (nm)
Grinding	-	25 - 6250
Honing	-	25 - 1500
Lapping	-	13 - 750
Abrasive Flow Machining (AFM) with SiC abrasives	Hardened steel	50
Magnetic Abrasive finishing (MAF)	Stainless steel	7.6
Magnetic Float Polishing (MFP) with CeO_2	Si_3N_4	4.0
Magnetorheological Finishing (MRF) with CeO_2	Flat BK7 Glass	0.8
Elastic Emission Machining (EEM) with ZrO_2 abrasives	Silicon	<0.5
Ion Beam Machining (IBM)	Cemented carbide	0.1

8.3.1 Abrasive Flow Machining (AFM)

Abrasive Flow Machining (AFM) was identified in 1960s as a method to deburr, polish, and radius difficult-to-reach surfaces and edges by flowing abrasive laden viscoplastic polymer over them. It uses two vertically opposed cylinders, which extrude abrasive medium back and forth through passage formed by the workpiece and tooling. Abrasion occurs wherever the medium passes through the restrictive passages. The key components of AFM are machine, tooling, types of abrasives, medium composition, and process settings (Rhoades 1988). Extrusion pressure, number of cycles, grit composition and type, and fixture design are the process parameters that have the largest impact on AFM results.

Abrasive action accelerates by change in the rheological properties of the medium when it enters and passes through the restrictive passages (Rhoades 1988). The viscosity of polymeric medium plays an important role in finishing operation (Jain et al., 2001). This allows it to selectively and controllably abrade surfaces that it flows across. The work piece held by fixture is placed between two medium cylinders, which are clamped together to seal so that medium does not leak during finishing process. Three major elements of the process are; (a) The *tooling*, which confines and directs the abrasive medium flow to the areas where deburring, radiusing and surface improvements are desired, (b) The *machine* to control the process variables like extrusion pressure, medium flow volume, and flow rate, and (c) The abrasive laden *polymeric medium* whose rheological properties determine the pattern and aggressiveness of the abrasive action. To formulate the AFM medium, the abrasive particles are blended into special viscoelastic polymer, which shows a change in viscosity when forced to flow through restrictive passages.

AFM can process many selected passages on a single workpiece or multiple parts simultaneously. Inaccessible areas and complex internal passages can be finished economically and productively. It reduces surface roughness by 75 to 90% on cast, machined or EDM'd surfaces (Kohut 1989). The same AFM set up can be used to do a variety of jobs just by changing tooling, process settings and if necessary, abrasive medium composition. Because of these unique capabilities and advantages, AFM can be applied to an impressive range of finishing operations that require uniform, repeatable and predictable results (Kohut 1989).

AFM offers precision, consistency, and flexibility. Its ability to process multiple parts simultaneously, and finishing inaccessible areas and complex internal passages economically and effectively led to its applications in a wide range of industries. Aerospace, aircraft, medical components, electronics, automotive parts, and precision dies and moulds manufacturing industries are extensively using abrasive flow machining process as a part of their manufacturing activities. Recently, AFM has been applied to the improvement in air and fluid flow for automotive engine components, which was proved as an effective method for lowering emissions as well as increasing performance. Some of these potential areas of AFM application are discussed below:

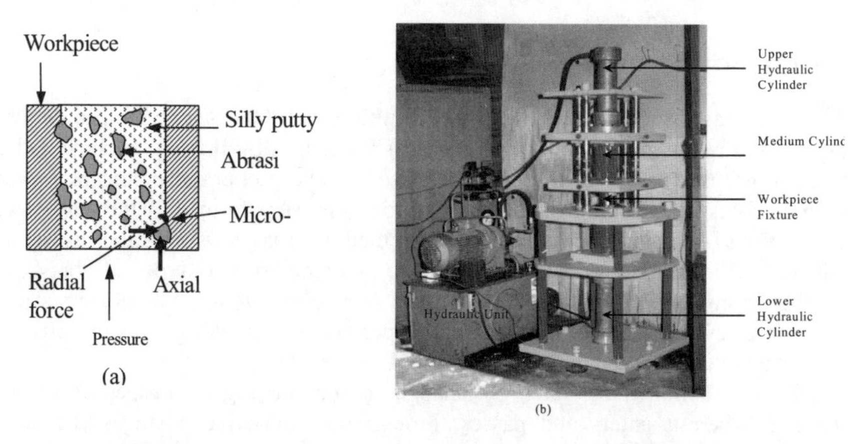

Fig. 8.2. Abrasive Flow Machining, (a) forces acting on abrasive particle in AFM process; (b) Experimental setup

AFM process was originally identified for deburring and finishing critical hydraulic and fuel system components of aircraft in aerospace industries. With its unique advantages, the purview of its application has been expanded to include:

adjusting air flow resistance of blades, vanes, combustion liners, nozzles and diffusers; improving airfoil surface conditions on compressors and turbine section components following profile milling, casting, EDM or ECM operations; edge finishing of holes to improve the mechanical fatigue strength of blades, disks, hubs, and shafts with controlled polished, true radius edges; and finishing auxiliary parts such as fuel spray nozzles, fuel control bodies and bearing components. The surface finish on the cast blades is improved from original 1.75 μm to 0.4 μm R_a. Cooling air holes are deburred and radiused in one operation on turbine disks as large as 760 mm in diameter. It is also used to remove milling marks and improve finish on the complex airfoil profiles of impellers and blades. Intricate intersections can also be finished easily. Since in the AFM process, abrading medium conforms to the passage geometry, complex shapes can be processed as easily as simple ones. Dies are the ideal workpiece for the AFM process as they provide the restriction for medium flow, typically eliminating fixturing requirements. AFM process has revolutionised the finishing of precision dies by polishing the die surfaces in the direction of material flow, producing a better quality and longer lasting dies with a uniform surface and gently radiused edges. The uniformity of stock removal permits accurate 'sizing' of undersized die passages. Precision dies are typically polished in 5 to 15 minutes unattended operation.

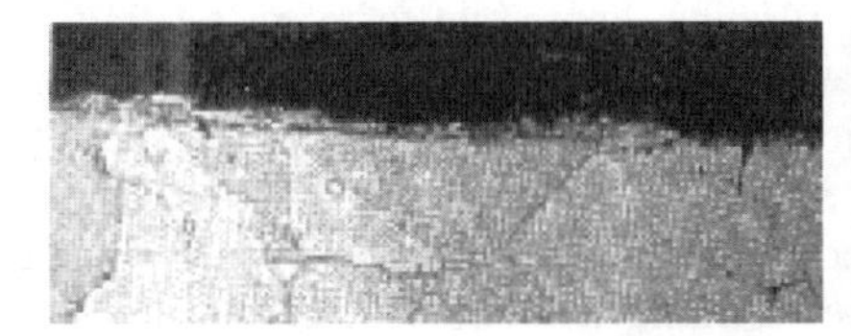

Fig. 8.3. 500X photomicrograph showing complete removal of EDM recast layer, (a) before AFM, (b) after AFM [courtesy: Extrude Hone corporation, USA]

The original 2 μm R_a (EDM finish) is improved to 0.2 μm with a stock removal of 25 μm per surface. Fig. 8.3 shows the complete removal of EDM recast layer. In some cases, the tolerances have been achieved as good as 13 μm (Rhoades 1991). AFM process is used to enhance the performance of high-speed automotive engines. It is a well-known fact in automobile engineering that smoother and larger intake ports produce more horsepower with better fuel efficiency. But it is very difficult to obtain good surface finish on the internal passageway of intake ports because of its complex shape (Gorana et al. 2004). The demand for this process is increasing rapidly among car and two wheeler manufacturers as it abrades, smoothes and polishes the surfaces of 2-stroke cylinders and 4-stroke engine heads for improved air flow and better performance. The AFM process can polish anywhere that air, liquid or fuel flows. Rough, power robbing cast surfaces are improved by 80-90 % regardless of surface complexities.

Advanced high-pressure injection system components in diesel engines are subjected to repeated pulses of very high pressure that can generate fatigue failure at high stressed areas. Smoothing and removing surface stress risers, cracks, as well as making uniform radius of sharp edges by AFM process can significantly extend component life. Flow tuned spray orifices of fuel injector nozzles can reduce particulate emission and improve the fuel efficiency of diesel engines.

8.3.2 Magnetic Abrasive Finishing (MAF)

Magnetic abrasives are emerging as important finishing methods for metals and ceramics. Magnetic abrasive finishing is one such unconventional finishing process developed recently to produce efficiently and economically good quality finish on the internal and external surfaces of tubes as well as flat surfaces made of magnetic or non-magnetic materials. In this process, usually ferromagnetic particles are sintered with fine abrasive particles (Al_2O_3, SiC, CBN or diamond) and such particles are called ferromagnetic abrasive particles (or magnetic abrasive particles). However, homogeneously mixed loose ferromagnetic and abrasive particles are also used in certain applications. Fig. 8.4 shows a plane MAF process in which finishing action is performed by the application of magnetic field across the gap between workpiece surface and rotating electromagnet pole. This gap is known as machining gap or working gap. The enlarged view of finishing zone in Fig. 8.4 shows the forces acting on the work surface to remove material in the form of chips. Force due to magnetic field is responsible for normal force (F_n) causing abrasive penetration inside the workpiece while rotation of the magnetic abrasive brush (and North pole) results in tangential force (F_t) leading to the material removal in the form of chips.

The magnetic abrasive grains join each other magnetically between magnetic poles along the lines of magnetic force, forming a flexible magnetic abrasive brush (usually 1 to 3 mm thick). MAF uses this magnetic abrasive brush for surface and edge finishing. The magnetic field retains the powder in the gap, and acts as a binder causing the powder to be pressed against the surface to be finished (Kremen 1994). 3-D minute and intricately curved shape can also be finished by this process. Controlling the exciting current of the magnetic coil precisely controls the machining force of the magnetic abrasives acting on the work piece (Jain et al. 2001).

Since the magnitude of machining force caused by the magnetic field is very low but controllable, a mirror like surface finish (R_a value in the range of nano-meter) is obtained. In MAF, mirror finishing is realised and burrs are removed without lowering the accuracy of the shape. These fine finishing technologies using magnetic abrasives have a wide range of applications. The surface finishing, deburring and precision rounding off the edges of the workpiece can be done simultaneously. MAF can be used to perform operations as polishing and removal of thin oxide films from high speed rotating shafts. Shinmura (Shinmura et al 1987) have applied MAF to the internal surface of work pieces such as vacuum tubes and sanitary tubes.

Fig. 8.5 shows the magnetic abrasive jet finishing of internal surface of a hollow cylindrical workpiece. It's a variant of MAF process in which working fluid mixed with magnetic abrasive particles is jetted into the internal surface of the tube, with magnetic poles being provided on the external surface of the tube (Kim 1997). The magnetic abrasives in the jet mixed with fluid are moved towards the internal surface by the magnetic force. Because of relative motion between the abrasive particles and workpiece surface, the internal surface of the workpiece is finished effectively and precisely.

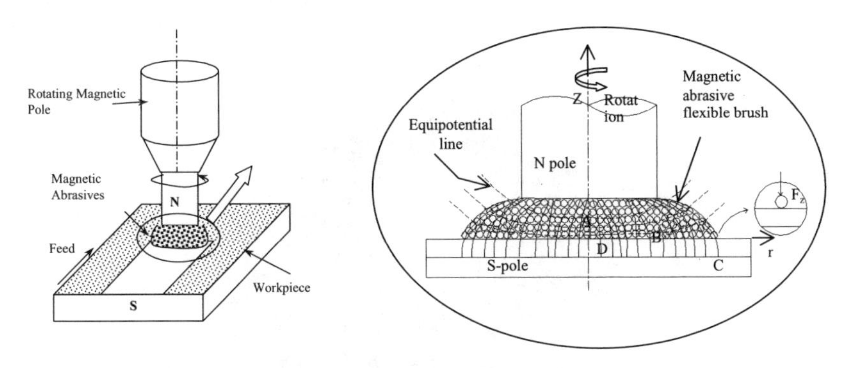

Fig. 8.4. Plane Magnetic abrasive finishing (Kim 1997)

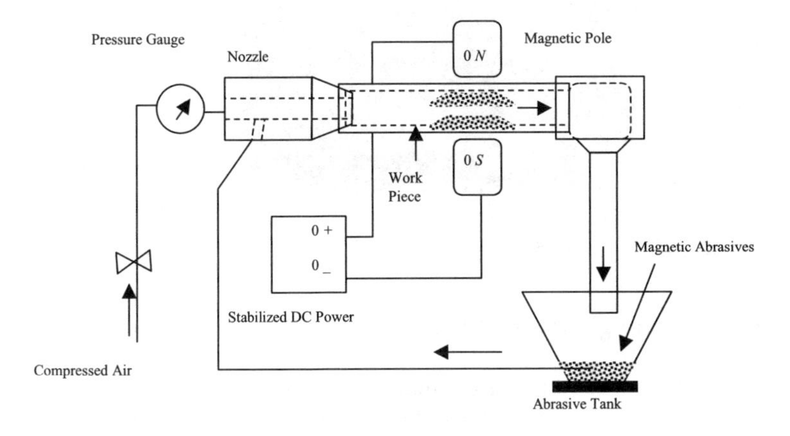

Fig. 8.5. Magnetic abrasive jet finishing (Kim 1997)

Fig. 8.6. shows a schematic of a typical MAF process in which a cylindrical workpiece to be finished is located between two magnetic poles. The working gap is filled with magnetic abrasive powder which forms a flexible magnetic abrasive brush (1 - 3 mm thick). MAF uses this magnetic abrasive brush for surface and edge finishing. The magnetic field retains powder in the gap, and acts as a binder causing the powder to be pressed against the surface to be finished (Kremen 1994). A rotary motion is provided to cylindrical workpiece, such as ceramic bearing rollers between magnetic poles. Also axial vibratory motion is introduced in the magnetic field by the oscillating motion of the magnetic poles to accomplish surface and edge finishing at a faster rate and with better quality. The process is highly efficient, and the material removal rate and finishing rate depend on the workpiece circumferential speed, magnetic flux density, working clearance, workpiece material, and size, type and volume fraction of abrasives. The exciting current of the magnetic coil precisely controls the machining force transferred through magnetic abrasives on the work piece. Results were reported in the literature of finishing on stainless steel rollers using MAF process to obtain final Ra of 7.6 nm from an initial Ra of 0.22 μm (=220 nm) in 30 seconds (Fox 1994) have been reported.

Fig. 8.6. Magnetic Abrasive Finishing of cylindrical surface

8.3.3 Magnetorheological Finishing (MRF)

Traditional methods of finishing high precision lenses, ceramics and semiconductor wafers are very expensive and labor intensive. The primary obstacle in manufacturing high precision lenses is that lenses are usually made of brittle materials such as glass, which tends to crack while it is machined. Even a single microscopic crack can drastically hinder a lens's performance, making it completely ineffective for its intended application. Every device that uses either lasers or fiber optics requires at least one high precision lens, increasing its demand higher than ever. Lens manufacturing consists of two main processes: grinding and finishing. Grinding gets the lens close to the desired size, while finishing removes the cracks and tiny surface imperfections that the grinding process either overlooked or created. The lens manufacturer generally uses its in-house opticians for the finishing process, which makes it an arduous, labor-intensive process. Perhaps the biggest disadvantage to manual grinding and finishing is that it is non-deterministic. To overcome these difficulties, Center for Optics Manufacturing (COM) in Rochester, N.Y. has developed a technology to automate the lens finishing process known as Magnetorheological Finishing (MRF) (Kordonski 1996).

The MRF process relies on a unique "smart fluid", known as Magnetorheological (MR) fluid. MR-Fluids are suspensions of micron sized magnetisable particles such as carbonyl iron, dispersed in a non-magnetic carrier medium like silicone oil, mineral oil or water. In the absence of magnetic field, an ideal MR-fluid exhibits Newtonian behaviour. On the application of external magnetic field to the MR-fluid, a phenomenon known as Magnetorheological effect, shown in Fig. 8.7, is observed. Fig. 8.7(a) shows the random distribution of the particles in the absence of external magnetic field. In Fig. 8.7(b), particles magnetise and form columns when external magnetic field is applied. The particles acquire dipole moments proportional to magnetic field strength and when the dipolar interaction between particles exceeds their thermal energy, the particles aggregate into chains of dipoles aligned in the field direction. Because energy is required to deform and rupture the chains, this micro-structural transition is responsible for the onset of a large "controllable" finite yield stress (Furst 2000). Fig. 8.7(c) shows an increasing resistance to an applied shear strain, γ due to this yield stress. When the field is removed, the particles return to their random state (Fig. 8.7(a)) and the fluid again exhibits its original Newtonian behaviour.

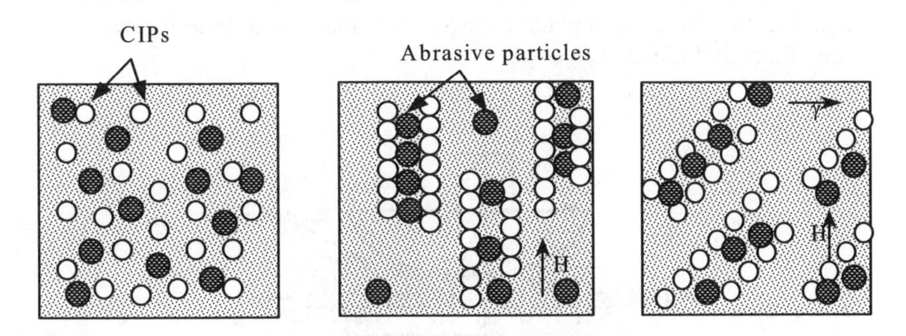

Fig. 8.7. Magnetorheological effect, (a) MR-fluid at no magnetic field, (b) at magnetic field strength H, and (c) at magnetic field H & applied shear strain γ

Rheologically, the behaviour of MR-fluid in presence of magnetic field is described by Bingham plastic model (Kordonski et al. 2001): $\tau = \tau_o + \eta(d\gamma/dt)$; where, τ is the fluid shear stress, τ_o is magnetic field induced yield shear stress, η is dynamic viscosity of MR-fluid and $d\gamma/dt$ is shear strain rate. The dynamic viscosity is mostly determined by the base fluid. The field induced shear stress τ_o depends on the magnetic field strength, H. The strength of the fluid (i.e. the value of static yield shear stress) increases as the applied magnetic field increases. However, this increase is non-linear since the particles are ferromagnetic and magnetisation in different parts of the particles occur non-uniformly (Ginder and Davis 1994). MR-fluids exhibit dynamic yield strength of 50-100 kPa for applied magnetic fields of 150-250 kA/m (~2-3 kOe) (Carlson 1996). The ultimate strength of MR-fluid is limited by magnetic saturation. The ability of electrically manipulating the rheological properties of MR fluid attracts attention from a wide range of industries and numerous applications are explored (Klingenberg 2001). These applications include use of MR-fluid in shock absorbers and damping devices, clutches, brakes, actuators, and artificial joints. The magnetic field applied to the fluid creates a temporary finishing surface, which can be controlled in real time by varying the field's strength and direction. The standard MR-fluid composition is effective for finishing optical glasses, glass ceramics, plastics and some non-magnetic metals (Lambropoulo et al. 1996). In the Magnetorheological finishing process as shown in Fig. 8.8, a convex, flat, or concave workpiece is positioned above a reference surface. A MR-fluid ribbon is deposited on the rotating wheel rim, Fig. 8.9. By applying magnetic field in the gap, the stiffened region forms a transient work zone or finishing spot. Surface smoothing, removal of sub-surface damage, and figure correction are accomplished by rotating the lens mounted on a spindle, at a constant speed while sweeping the lens about its radius of curvature through the stiffened finishing zone (COM 1998). Material removal takes place through the shear stress created as the magnetorheological polishing ribbon is dragged into the converging gap between the part and carrier surface. The zone of contact is restricted to a spot, which conforms perfectly to the local topography of the part. Deterministic finishing of flats, spheres, and aspheres can be accomplished by mounting the part on a rotating spindle and sweeping it through the spot under computer control, such that dwell time determines the amount of material removal.

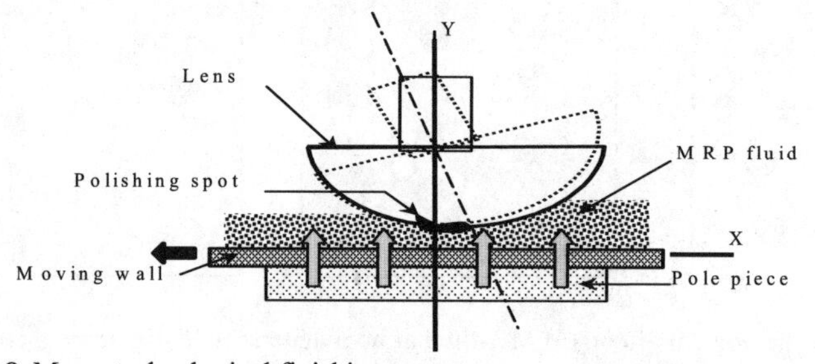

Fig. 8.8. Magnetorheological finishing process

The MR-polishing fluid lap has following merits over traditional lap:-

- Its compliance is adjustable through the magnetic field.
- It carries heat and debris away from the polishing zone.
- It does not load up as a grinding wheel.
- It is flexible and adapts the shape of the part of the workpiece.

The computer controlled magnetorheological finishing process has demonstrated the ability to produce the surface accuracy of the order of 10-100 nm peak to valley by overcoming many fundamental limitations inherent to traditional finishing techniques (Kordonski 1996). These unique characteristics made magnetorheological finishing as the most efficient and noble process for high precision finishing of optics. MRF makes finishing of free form shapes possible for first time.

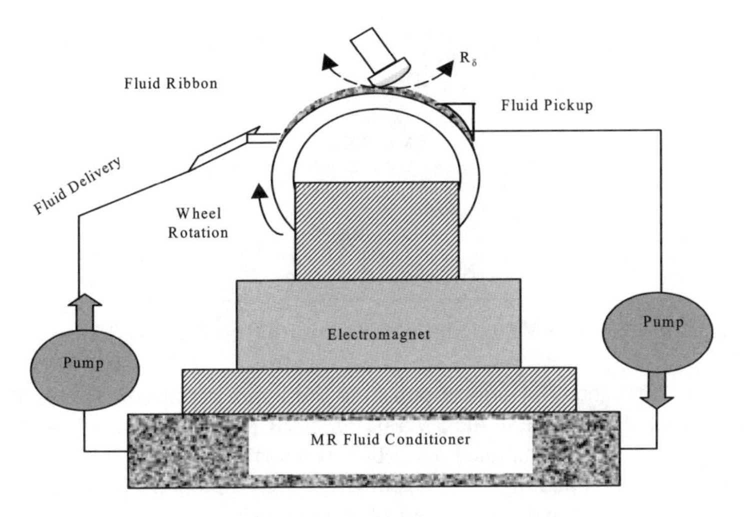

Fig. 8.9. A Vertical MRF Machine (COM 1998)

Applications that use high precision lenses include medical equipment such as endoscopes, collision-avoidance devices for the transportation industries, scientific testing devices and military's night vision equipment like infrared binoculars. Missiles are equipped with a wide variety of high precision lenses for navigation, target location, and other functions. The nano diamond doped MR-fluid removes edge chips, cracks, and scratches in sapphire bend bars.

8.3.4 Magnetorheological Abrasive Flow Finishing (MRAFF)

In AFM, the polishing medium acts as compliant lap and overcomes shape limitation inherent in almost all traditional finishing processes. As abrading forces in AFM process mainly depend on rheological behaviour of polymeric medium, which is least controllable by external means, hence lacks determinism. The process named as magnetorheological finishing described above, uses magnetically stiffened ribbon to deterministically finish optical flats, spheres and aspheres. In order to maintain the versatility of abrasive flow machining process and at the same time introducing determinism and controllability of rheological properties of abrasive laden medium, a new hybrid process termed as "*Magnetorheological Abrasive Flow Finishing (MRAFF)*" is used, Fig. 8.10 (Jha and Jain 2004). This process relies on smart behaviour of magnetorheological fluids whose rheological properties are controllable by means of magnetic field.

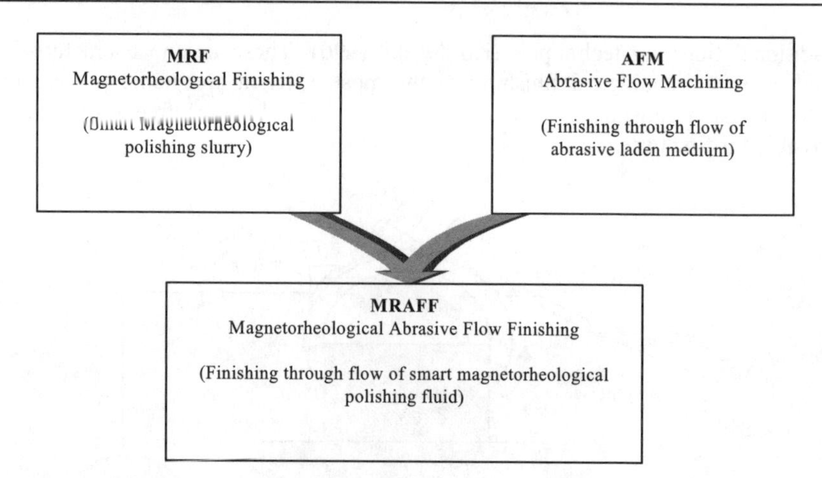

Fig. 8.10. Development of Magnetorheological Abrasive Flow Finishing process

The use of magnetorheological polishing fluid with cerium oxide abrasives for finishing optical lenses up to the level of 0.8 nm R.M.S. (root mean square) value has already been demonstrated by MRF process (Kordonski 1996). MRAFF process has the capability of finishing complex internal geometries up to nanometer level. It imparts better control on the process performance as compared to AFM process due to in-process control over abrading medium's rheological behaviour through magnetic field. Magnetorheological (MR) fluids, invented by Rabinow (Rabinow 1948) in late 1940s, belong to a class of smart controllable materials whose rheological behaviour can be manipulated externally by the application of some energy field. Magnetorheological polishing fluid comprises of MR-fluid with fine abrasive particles dispersed in it. On the application of magnetic field the carbonyl iron particles (CIP) form a chain like columnar structure with abrasives embedded in between. Figs. 8.11(a) & (b) show actual photographs taken by optical microscope for the case when no magnetic field is applied and the structure formed in the presence of magnetic field, respectively. The magnetic force between iron particles encompassing abrasive grains provides bonding strength to it and its magnitude is a function of iron concentration, applied magnetic field intensity, magnetic permeability of particles and particle size. In MRAFF process, a magnetically stiffened slug of magnetorheological polishing fluid is extruded back and forth through or across the passage formed by workpiece and fixture. Abrasion occurs selectively only where the magnetic field is applied across the workpiece surface, keeping the other areas unaffected. The mechanism of material removal of the process is shown in Fig. 8.12. The rheological behaviour of polishing fluid changes from nearly Newtonian to Bingham plastic upon entering, and Bingham to Newtonian upon exiting the finishing zone. The abrasive (cutting edges) held by carbonyl iron chains rub the workpiece surface and shear the peaks from it, Fig. 8.13. The amount of material sheared from the peaks of the workpiece surface by abrasive grains depends on the bonding strength provided by field-induced structure of MR-polishing fluid and the extrusion pressure applied through piston. In this way magnetic field strength controls the extent of abrasion of peaks by abrasives.

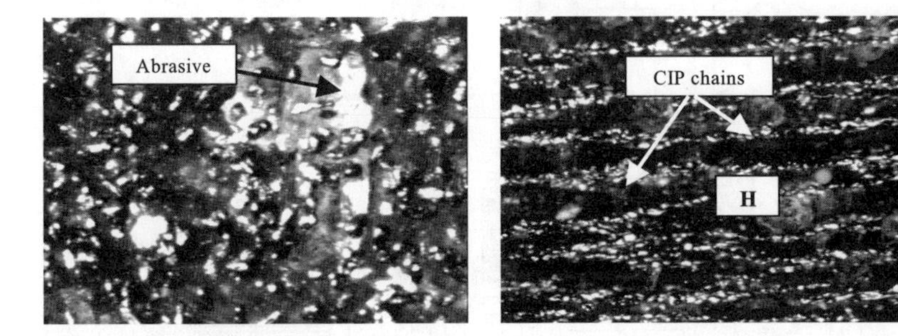

Fig. 8.11. Chain formation in Magnetorheological polishing fluid, (a) abrasives & carbonyl iron particles at zero magnetic field, (b) abrasive particles embedded in carbonyl iron particles chains on the application of external magnetic field

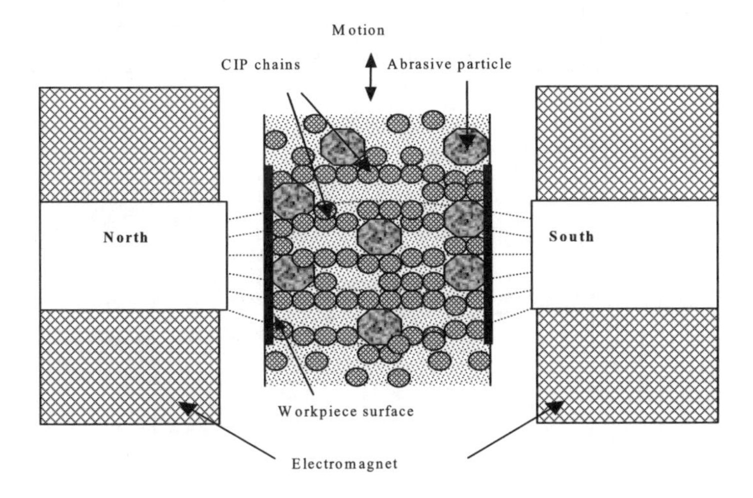

Fig. 8.12. Mechanism of MRAFF process

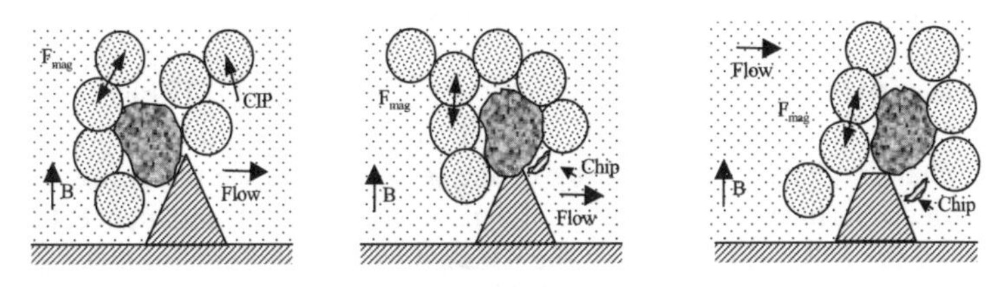

Fig. 8.13. Finishing action on a single profile in presence of external magnetic field (a) Abrasive grain along with CIP chains approaching roughness peak, (b) Abrasive grain takes a small cut on roughness peak in presence of bonding forces, (c) Abrasive grain crossing the roughness peak after removing a microchip during cutting action

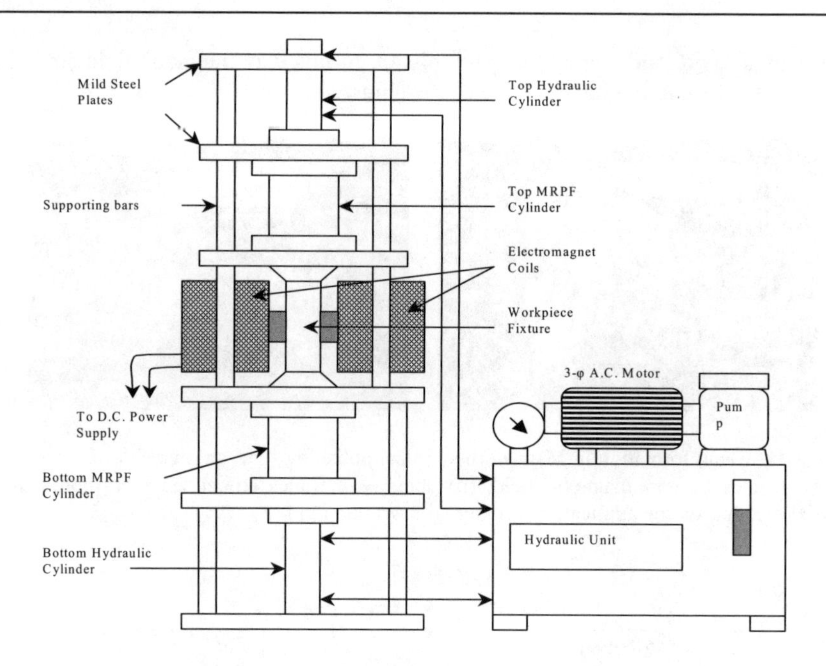

Fig. 8.14. Schematic of MRAFF Machine

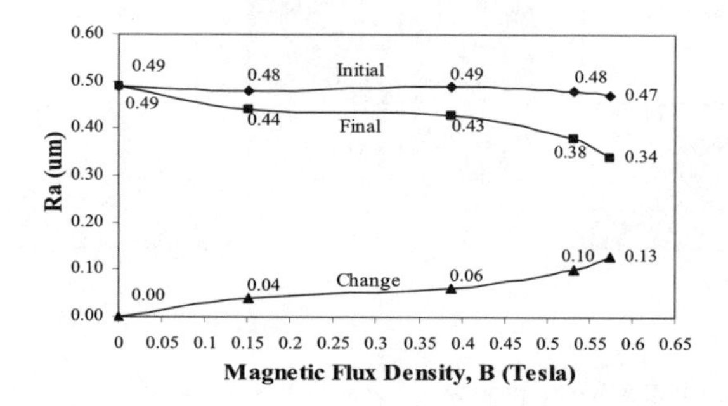

Fig. 8.15. Effect of magnetic flux density on Ra value

The viscosity of smart magnetorheological polishing fluid (MRPF) is a function of applied magnetic field strength, and it is varied according to the desired finishing characteristics. The shearing of the Bingham plastic polishing fluid near the workpiece surface contributes to the material removal and hence finishing. Extrusion of the MRP-fluid through the passage formed in the workpiece fixture is accomplished by driving two opposed pistons in MRPF cylinders using hydraulic actuators operated in the desired manner with the help of designed hydraulic circuit, Fig. 8.14. The role of magnetic field strength in MRAFF process is clearly distinguished in Fig. 8.15, as at no field conditions no improvement in surface finish is observed and the improvement is significant at high magnetic field strength. This is because, in the absence of magnetic field the carbonyl iron particles and abrasive particles flow over the workpiece surface without any finishing action due to the absence of bonding strength of CIPs. As the magnetic field strength is increased by increasing magnetising current, chains of carbonyl iron particles keep on holding abrasives more firmly and thereby result in increased finishing action. Even 0.1521 Tesla field is capable of removing to some extent, loosely held ploughed material left after grinding process and expose the actual grinding marks made by abrasives. Depths of initial grinding marks were reduced progressively by reducing asperities as the experiments were performed at higher flux density. Fig. 8.16(a) shows the initial closely spaced grinding marks before MRAFF, which were sparsely located as shown in Fig. 8.16(b) after finishing for 200 cycles at 0.574 Tesla (Jha and Jain 2004). The other process variables that affect performance of MRAFF are number of finishing cycles, extrusion pressure and MRP-fluid composition. The relative particle size of abrasives and CIPs plays an important role in finishing action.

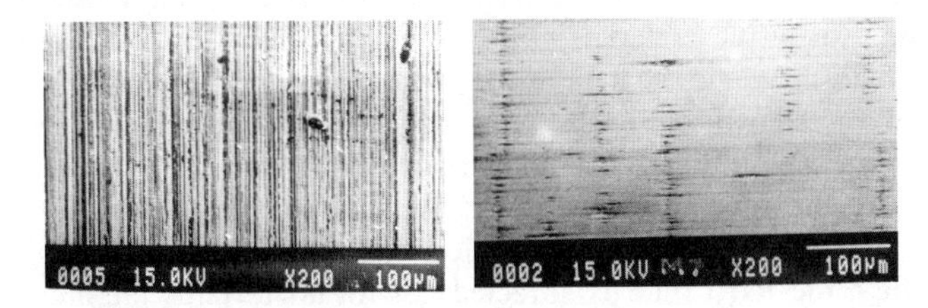

Fig. 8.16. Comparison of surface before and after MRAFF, (a) Initial surface before MRAFF, (b) Final surface after MRAFF for 200 cycles at B = 0.574 T

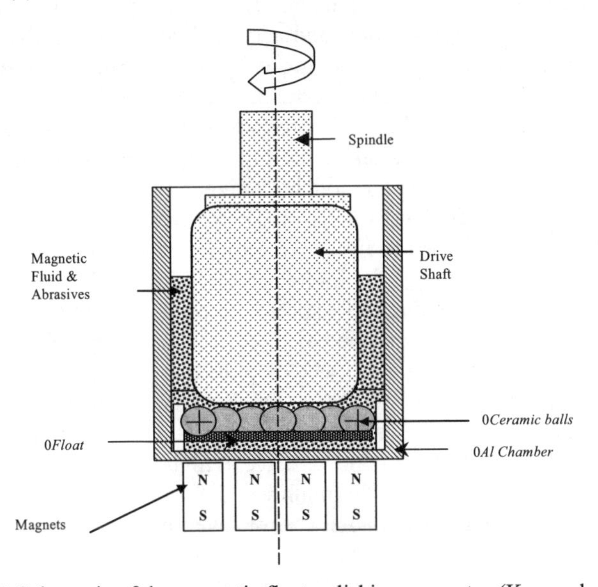

Fig. 8.17. Schematic of the magnetic float polishing apparatus (Komanduri 1996)

8.3.5 Magnetic Float Polishing (MFP)

The applications of advanced ceramics are limited because of their poor machinability and difficulties involved in processing them into useful shapes. Ceramics are extremely sensitive to surface defects resulting from grinding and polishing processes. Since fatigue failure of ceramics is driven by surface imperfections, it is of utmost importance that the quality and finish of the elements of ceramic bearings be superior with minimal defects. For this gentle and flexible polishing conditions like low level of controlled forces and use of abrasives softer than workmaterial are required. A recent advancement in fine abrasive finishing processes involves the use of magnetic field to support abrasive slurries in finishing ceramic balls and bearing rollers, and the process is known as Magnetic Float Polishing (MFP). The magnetic float polishing technique is based on the ferro-hydrodynamic behaviour of magnetic fluids that can levitate a non-magnetic float and abrasives suspended in it by magnetic field. The levitation force applied by the abrasives is proportional to the magnetic field gradient and is extremely small and highly controllable. It can be a very cost-effective and viable method for superfinishing of brittle materials with flat and spherical shapes. The schematic diagram of the magnetic float polishing apparatus used for finishing advanced ceramic balls is shown in Fig. 8.17. A magnetic fluid containing fine abrasive grains and extremely fine ferromagnetic particles in a carrier fluid such as water or kerosene fills the aluminium chamber. A bank of strong electromagnets is arranged alternately north and south below the chamber. On the application of magnetic field the ferro fluid is attracted downward towards the area of higher magnetic field and an upward buoyant force is exerted on non-magnetic material to push them to the area of lower magnetic field (Rosenweig 1966). The buoyant force acts on a non-magnetic body in magnetic fluid in the presence of magnetic field. The abrasive grains, ceramic balls, and acrylic float inside the chamber are of non-magnetic materials, and all are levitated by the magnetic buoyant force. The drive shaft is fed downward to contact the balls and press them downward to reach the desired force level. The balls are polished by the relative motion between the balls and the abrasives under the influence of levitation force (Tani and Kawata 1984). In this polishing method, both higher material removal rate and smoother surface are attained by stronger magnetic field and finer abrasives. Magnetic float polishing was used to finish 9.5 mm diameter Si_3N_4 balls. Si_3N_4 is considered as a candidate material for high-speed hybrid bearing in ultra high-speed precision spindles of machine tools and jet turbines of aircraft (Komanduri 2000). Conventional polishing of ceramic balls generally uses low polishing speed (a few hundred rpm) and diamond abrasive as a polishing medium. In practice, it takes considerable time (say, 12-15 weeks) to finish a batch of ceramic balls. The long processing time and use of expensive diamond abrasive result in high processing cost. Also, the use of diamond abrasive at high loads can result in deep pits, scratches, and micro cracks. To minimise the surface damage, 'gentle' polishing conditions are required, namely, low level of controlled force and abrasives which are not much harder than the work material. The magnetic float polishing (MFP) process easily accomplishes this. The surface finish obtained was

4 nm Ra and 40 nm Rmax. The best sphericity obtained of the Si_3N_4 balls was 0.15 to 0.2 μm. Finished surfaces relatively free of scratches, pits, etc. were obtained.

8.3.6 Elastic Emission Machining (EEM)

Though this process was developed as early as in 1976 (Mori et al. 1976), it attracts the attention because of its ability to remove material at the atomic level by mechanical methods and to give completely mirrored, crystallographically and physically undisturbed finished surface.

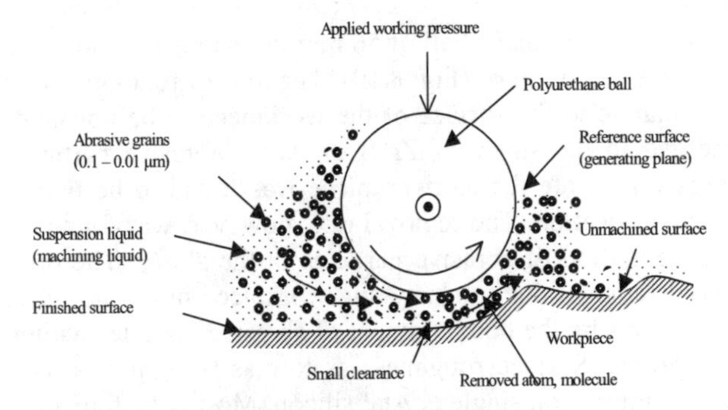

Fig. 8.18. Rotating sphere and workpiece interface in EEM (Mori 1976).

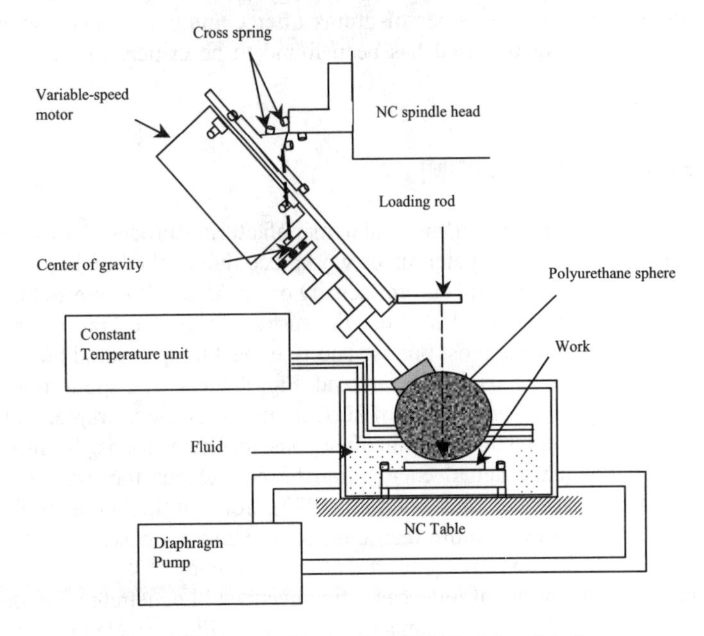

Fig. 8.19. Schematic of EEM assembly used on NC Machine (Mori)

The material removal in conventional machining is due to the deformation or fracture based on migration or multiplication of pre-existing dislocations, or by the enlargement of cracks originating from pre-existing micro cracks. If the material removal can occur at atomic level, then the finish generated can be close to the order of atomic dimensions (0.2 nm to 0.4 nm). Using ultra fine particles to collide with the workpiece surface, it may be possible to finish the surface by the atomic scale elastic fracture without plastic deformation (Mori 1987). This new process is termed as Elastic Emission Machining (EEM). Mori et al. established theoretically and experimentally that atomic scale fracture can be induced elastically and the finished surface can be undisturbed crystallographically and physically (Mori). The rotating sphere and workpiece interface in elastic emission machining is shown in Fig. 8.18.

In the EEM process, a polyurethane ball of 56 mm in diameter is mounted on a shaft driven by a variable speed motor (Fig. 8.19). The axis of rotation is oriented at an angle of ~ 45° relative to the surface of the workpiece to be polished. The workpiece is submerged in the slurry of ZrO_2 or Al_2O_3 abrasive particles and water. The material removal rate for the workpiece was found to be linear with dwell time at a particular location. The removal rate, however, was found to vary non-linearly with concentration of abrasive particles in the slurry. The proposed mechanism of material removal due to slurry and workpiece interaction involves erosion of the surface atoms by the bombardment of abrasive particles without the introduction of dislocations. Surface roughness as low as 0.5 nm rms has been reported on glass and 1 nm rms on single crystal silicon. Mori et al. found out that the material removal process is a surface energy phenomenon in which each abrasive particle removes a number of atoms after coming in contact with the surface. The type of abrasive used has been found to be critical to the removal efficiency.

8.3.7 Ion Beam Machining (IBM)

Ion Beam Machining (IBM) is a molecular manufacturing process based on the sputtering off phenomenon of materials of workpiece due to the bombardment of energised ions of 1 to 10 keV and current density of $1 mA/cm^2$. It removes material in the form of 'atoms or molecules' from the surface of the workpiece. The Ion beam machining is one of the most precise and fine machining method because of its basis of action on atomic size stock removal, Fig. 8.20(a). The sputtering off is basically a knocking out phenomenon of surface atoms of the workpiece due to the kinetic momentum transfer from incident ions to target atoms. Removal of atoms will occur when the actual energy transferred exceeds the usual binding energy of 5-10 eV (Spencer and Schmidt 1972). Ions of higher energy may transfer enough momentum to more than one atom to cause a cascading effect in the layers near the surface, removing several atoms at a time.

At sufficiently high energy of ions, cascading events will penetrate more deeply into the solid. As a result, several atoms or molecules will be ejected out and the bombarding ion will get implanted deep within the material, Fig. 8.20(b). Many microscopic damage centers will result from the energetic displacements of the atoms. Clearly it is not desirable from surface quality point of view.

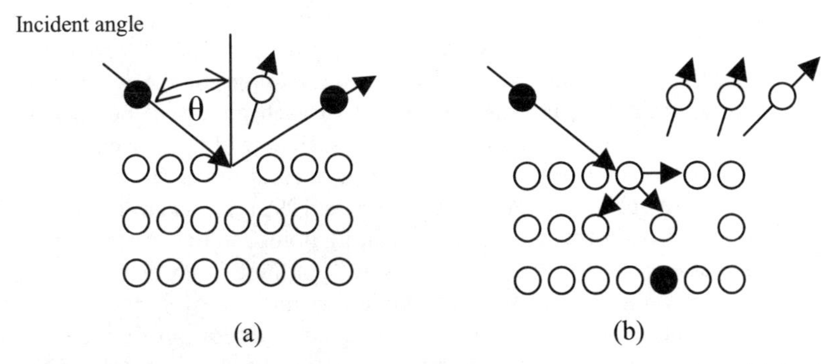

Fig. 8.20. Schematic illustration of the mechanism of material removal in IBM (a) Low energy Ideal case, (b) High energy Ion implantation

The sputtering yield is the most important machining characteristic of ion beam machining. The sputtering yield S is defined as the mean number of atoms sputtered from the target surface per incident ion. The sputtering yield S depends on the material being machined, the bombarding ions and their energy, ion incidence angle to the target surface and the crystal face of the target. In order to describe the ion beam machining characteristic more practically, the concept of the specific sputter machine rate V(θ)[(μm/h)/(mA/cm^2)] has been introduced. The relationbetween the sputtering yield S and the sputter machine rate V(θ) is represented as (Somekh 1976),

$$V(\theta) = 576 \times 10^9 \frac{S(\theta)}{n} \times \cos\theta \ \ (\mu\text{m/h})/(\text{mA/cm}^2)$$

Where, n is the atomic density of the target material in atoms/cm^2, and cosθ term accounts for the reduced current density at an angle θ to the normal, Fig. 8.20a. Therefore if the sputter machine rate is obtained experimentally, the sputtering yield can be calculated.

Workpiece material, ion etching gas, angle of incidence, ion energy and current density are identified as the main factors affecting machining characteristics in IBM. Sputtering yield is a function of atomic number, binding energy, grain size, number of electron shell (Smith 1976) etc. of the workpiece material and atomic weight of the incident ions. Ions having high atomic number will yield high MRR. Since the sputtering yield is related to the binding energy of the material being etched it is possible to vary its value by introducing reactive gas. The reactive gas will react with the work surface, vary its binding energy and consequently vary the etch rate. The ion incidence angle at which the maximum sputtering yield occurs increases when the bombarding ion energy or its mass increases. The reason for this is that as the ion incidence angle increases, more atoms of the work piece can be knocked out or sputtered away easily from the two or three atomic layers of the work surface. However, when the ion incidence angle further increases, the machining rate begins to decrease because the ion current density for unit working area decreases by cosθ and the number of ions reflected from the surface of the workpiece without sputtering off atoms increases. Success of ion beam polishing depends crucially on the grain size and initial morphology of the work surface. With very small grain size, the machining rate of each grain will be almost the same, and therefore uniform machining over the surface will take place. For large grain size, the difference between the machining rates of the grains results in the increase in value of surface roughness (Miyamoto 1993). Ion beam machining is an ideal process for nano-finishing of high melting point and hard brittle materials such as ceramics, semiconductors, diamonds etc. As there is no mechanical load on the workpiece while finishing, it is suitable for finishing of very thin objects, optics and soft materials such as CaF_2. Diamond styli for profilometer were sharpened using Kaufman type ion source. Argon ion beam of energy E=10keV and ion current density of 0.5 mA/cm^2 was used to sharpen the styli to the tip radius of 10 nm (Miyamoto and Ezawa 1991).

8.3.8 Chemical Mechanical Polishing (CMP)

Chemical Mechanical Polishing (CMP) is the fastest growing process technology in the semiconductor manufacturing. CMP is a planarisation process, which involves a combination of chemical and mechanical actions. The importance of each contribution depends on the polished material. It was developed at IBM in the mid-1980s as an enabling technology to planarise SiO_2 interlevel dielectric (ILD) layers so that three or more levels of metal could be integrated into a high-density interconnect process. This technology was adapted from silicon wafer polishing. By employing CMP, subsequent structures could he fabricated on a nearly planar surface. The initial application of CMP was ILD planarisation, but its application to other areas in the overall semiconductor fabrication sequence followed quickly.

A similar variant is Chemo-Mechanical polishing in which driving factor for material removal is chemical action between abrasive particles and workmaterial followed by mechanical action for the removal of reaction products (Vora et al. 1982). In chemical mechanical polishing, the reaction is between the fluid and the workmaterial, and the abrasive removes the reaction product by mechanical action (Nanz and Camilletti 1995). Chemo-mechanical polishing is expected to overcome many problems of surface damage associated with hard abrasives, including pitting due to brittle fracture, dislodgement of grains, scratching due to abrasion, etc. , resulting in smooth, damage free surfaces (Komanduri et al. 1997). A schematic of CMP planarisation process is shown in Fig. 8.21. The wafers are pressed downward by carriers and rotated against the polishing pad covered with a layer of silica slurry.

The polishing pad on rotating plate is used to hold the slurry particles, transmit load to the particle-wafer interface, and conform precisely to the wafer being polished. Aqueous colloidal silica suspension is widely used for polishing. Initial chemical reaction of silicon with the aqueous solution form a thin silica layer and this is then mechanically removed by the polishing slurry. The thin layer of silica reduces the friction force, provides uniform distribution of slurry particles and remove eroded material and heat generated. During CMP, the kinetic energy of the slurry particles moving between the wafer and the pad, erodes the surface of the wafer. CMP has been used for a long time in manufacturing of silicon wafers. CMP results in defect free surfaces in contrast to mechanically polished surfaces (Jhansson et al. 1989). Simultaneous double side polishing is also feasible (Wenski et al. 2002). Double side polishing gives a better parallelism compared to single side polishing and less adherence of particles since both sides are smooth. Another use of CMP substrate is in thin film transistor (TFT) technology (Chang et. al 1996) and polishing of IC wafers (Venkatesh et al. 2004). In TFT, used for instance for making flat displays, polycrystalline silicon is deposited on glass substrates. To obtain more global planarisation in contrast to local, CMP was introduced to planarise the deposited interlevel dielectric (ILD) oxide layers in metal interconnection layers. CMP for ILD planarisation is used in the aluminum wiring – tungsten plug interconnection technology.

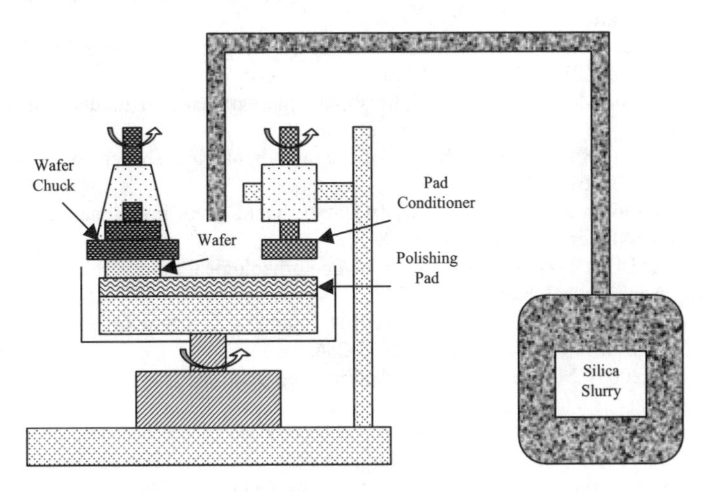

Fig. 8.21. Schematic of CMP planarisation process equipment (Hayashi et al.)

8.4 References

Keown PA (1987) The role of precision engineering in manufacturing of the future, Annals of CIRP 36:495-501

Taniguchi N (1983) Current status in, and future trends of ultraprecision machining and ultrafine material processing, Annals of CIRP 2:573-582

Schey JA (1987) Introduction to manufacturing processes. 2nd edn., McGraw-Hill Book Company

Shaw MC (1996) Principles of abrasive processing. Clarendon press, Oxford
Jain VK (2002) Advanced machining processes. Allied publishers, New Delhi (India)
Rhoades LJ (1988) Abrasive flow machining, Manufacturing Engineering 1:75-78
Jain VK Ranganath C, Murlidhar K (2001) Evaluation of rheological properties of medium for AFM process, Machining Science and Technology 5:151-170
Kohut T (1989) Surface finishing with abrasive flow machining, SME technical paper, pp 35-43
Rhoades LJ (1991) Abrasive flow machining: a case study, J of materials processing Technology 28:107-116
Gorana VK, Jain VK, Lal GK (2004) Experimental Investigation into cutting forces and active grain density during abrasive flow machining, Institution of Engrs. 44:201-211
Kremen GZ (1994) Machining time estimation for magnetic abrasive processes, International J of Production Research 32:2817-2825
Jain VK, Kumar P, Behera PK, Jayswal SC, (2001) Effect of working gap and circumferential speed on the performance of magnetic abrasive finishing process, Wear 250:384-390
Shinmura T (1987) Study on magnetic abrasive finishing-characteristics of finished surface, J of Japan Society of Precision Engineering 53:1791-1793.
Kim JD (1997) Development of a magnetic abrasive jet machining system for internal polishing of circular tubes, J. of material processing technology 71:384-393
Kremen GZ (1994) Machining time estimation for magnetic abrasive processes, Int. J of Production Research 32:2817-2825
Fox M, Agrawal K, Shinmura T, Komanduri R (1994) Magnetic Abrasive Finishing of Rollers. Annals of CIRP 43:1
Kordonski WI (1996) Magnetorheological finishing, International J of modern physics B 10:2837-2849
Furst EM, Gast AP (2000) Micromechanics of magnetorheological suspensions, Physical Review E 61:6732-6739
Kordonski W, Gordokin S, Zhuravski N (2001) Static yield stress in magnetorheological fluid. Int. J. of Modern Physics B 15:1078-1084
Ginder J M, Davis LC (1994) Shear stresses in magnetorheological fluids: Role of magnetic saturation. Appl. Phys. Letter 65:3410-3412
Carlson JD, Catanzarite DM, Clair KA (1996) Commercial magnetorheological fluid devices, Int. J of Modern Physics B 10:2857-2865
Klingenberg DJ (2001) Magnetorheology: Applications and challenges, AIChE Journal 47:246-249
Lambropoulo SJ, Yang F, Jacob SD (1996) Optical fabrication and testing, Technical digest series (Optical Society of America, Washington DC), vol. 7, pp 150-153
COM (1998) Magnetorheological finishing, Article by Center for Optics Manufacturing (http://www.opticam.rochester.edu)
Jha S, Jain VK (2004) Design and development of magnetorheological abrasive flow finishing process, Int. J of Machine Tool and Manufacture 44:1019-1029
Rabinow J (1948) The magnetic fluid clutch, AIEE Trans 67:1308
Klingenberg DJ (2001) Magnetorheology: Applications and challenges, AIChE Journal 47:246-249
Rosenweig, RE (1966) Fluid magnetic buoyancy, AIAA Journal 4:1751
Tani Y, Kawata K (1984) Development of high-efficient fine finishing process using magnetic fluid, Annals of CIRP 33:1-5
Komanduri R (1996) On material removal mechanisms in finishing of advanced ceramics and glasses. Annals of CIRP 45:509-514

Komanduri R (2000) Finishing of silicon nitride balls, Technical Report, Oklahama State University
Mori Y, Ikawa N, Okuda T, Yamagata K (1976) Numerically controlled elastic emission machining, Technology reports of the Osaka University, vol. 26, pp 283-294
Mori, Y;Yamauchi, K (1987), Elastic Emission Machining, Precision Engineering vol. 9, pp. 123-128
Mori Y, Ikawa N, Sugiyama K (2000) Elastic emission machining-stress field and fracture mechanism, Technology Reports of the Osaka University, vol. 28, pp 525-534
Mori Y, Ikawa N, Sugiyama K Okuda T, Yamauchi K (2002) Elastic emission machining – 2nd Report: Stress field and feasibility of introduction and activation of lattice defects, Japan Society of Precision Engineers, vol. 51, pp 1187-1194
Spencer EG, Schmidt PH (1972) Ion beam techniques for device fabrication, J of Vacuum science and technology 8:S52-S70
Somekh S (1976) Introduction of the ion and plasma etching, J of Vacuum Science and Technology 13:1003-1007
Smith M (1976) Ion etching for pattern delineation, J of Vacuum Science and Technology, 13:1008-1022
Miyamoto I (1993) Ion beam fabrication of ultra fine patterns on cemented carbide chips with ultra-fine grain, Annals of CIRP 42:1
Miyamoto I, Ezawa TV (1991) Ion beam fabrication of diamond probes for a scanning tunneling microscope, Nanotechnology 2:52-56
Vora H, Orent TW, Stokes RJ (1982) Mechano-chemical polishing of silicon nitrid, J of American Ceramic Society 65:140-141
Nanz G, Camilletti L.E. (1995), Modeling of chemical-mechanical polishing: A review, IEEE Trans. on Semiconductor Manufacturing 8:382-389
Komanduri R, Lucca DA, Tani Y (1997) Technological advances in fine brasive processes, Annals of CIRP 46:2
Hayashi Y, Nakajima T, Kunio T (2001) Ultrauniform chemical mechanical polishing (CMP) using a hydro chuck, featured by wafer mounting on a quartz glass plate with fully flat water supported surface, Japanese J of Applied Physics 35:1054-1059
Jhansson S, Schweitz JA, Lagerlof KPD (1989) Surface defects in polished silicon studied by cross-sectional transmission electron microscopy, J of the American Ceramic Society 72:1136-1139
Wenski G, Altmann T, Winkler W, Heier G, Holker G (2002) Double side polishing- A technology mandatory for 300 mm wafer manufacturing, Materials Science in Semiconductor Processing 5:375-380
Chang C Y, Lin HY, Lei TF, Cheng JY, Chen LP, Dai BT (1996) Fabrication of thin film transistors by chemical mechanical polished polycrystalline silicon films, IEEE Electron Device Letters 17:100-102
Venkatesh VC, .Izman S, Mahadevan SC (2004) Electrochemical mechanical polishing of copper and chemical mechanical polishing of glass, J of Materials Processing Technology 149:493-498

9 Micro and Nanotechnology Applications for Space Micropropulsion

Giulio Manzoni

Microspace, Italy

9.1 Introduction

Microfabrication technologies (MST) allow the realisation of highly sophisticated products integrated with mechanical, optical and electronic functions completely. Such products are already exploiting very big markets in automotive and bio-medical applications. Other sectors including aerospace are still observing the developments and are becoming aware of the potential applications of MEMS (Microelectromechanical Systems), MOEMS (Micro-opto-electromechanical Systems) and MST. The products will be accepted if the specifications are met. Companies, universities and research groups are more open to experimentation and many research projects in this field are financed from Space Agencies in the advanced countries (Europe, USA, Japan, Korea, etc.) (Liu 2003).

The field of nanotechnology is now explored worldwide by academia, research centers and industries. Applications and products are appearing on the market progressively. Nanotechnology is often merged with microtechnology to enable particular functionalities. It is applied on macroscopic products to enhance functionality. Many disciplines such as physics, chemistry, engineering, biology, and informatics are now converging into multi-disciplinary department in order to develop micro- and nanotechnology based products. One can note that the application of micro and nanotechnology to the aerospace field requires a very interdisciplinary engineering approach with a very wide field of view. In fact, from the consideration of material property at molecular level which affects the manufacturing at nanoscale (10 nm = 10E-9 m) to the comprehension of space mission requirements related to positioning of the satellite in orbit the engineer in charge of the mission study must have a general knowledge on a field spread over more than 15 orders of magnitude (Manzoni 2003) and a very specific knowledge on the 6 to 9 orders of magnitude (Fig. 9.1) covering the nanometer to millimeter size. Devices for aerospace application very often require environmental ranges, which are more extended than similar terrestrial devices. This complicates the design or the selection of materials and requires sometimes a dedicated manufacturing process. The system designers should consider the following aspects prior to the conception of component and architecture.

- The availability of reliable technologies
- The active and passive components of MEMS device
- Manufacturability
- Precision and standard
- Dimension and tolerances
- Performance

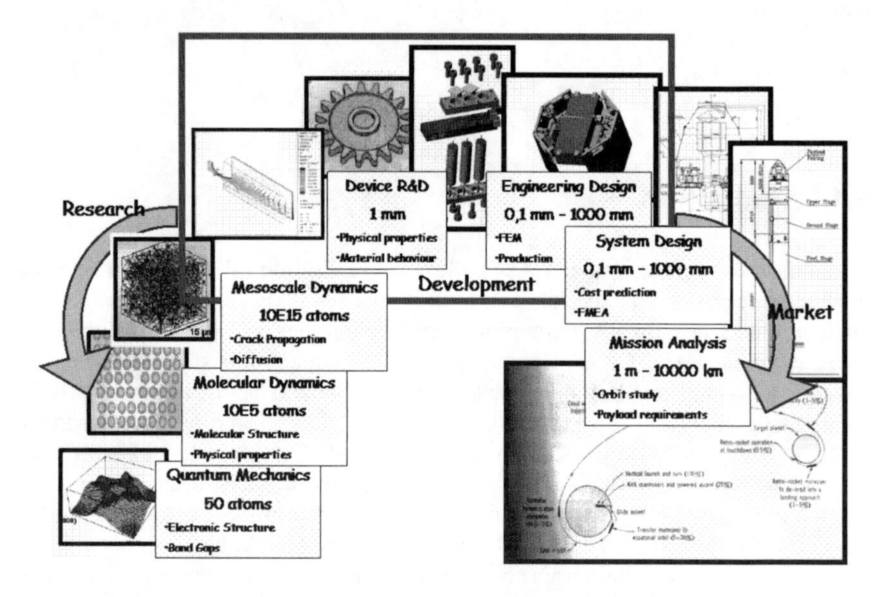

Fig. 9.1. Order of magnitudes

The aerospace field is characterised from the fundamental need of vehicle mass reduction in comparison with the payload mass and the reduction of the payload size and mass in relation to the mission requirements. Therefore, miniaturisation of the payload of the transporting vehicle is of primary interest. Space applications require also a higher level of intelligence of the systems since in the majority of the cases the human intervention is impossible or not fast enough. For such reasons, the miniaturisation and autonomy of the system are among the first targets of aerospace engineering and research. Such targets can be achieved by means of integrated systems involving microtechnology methods and using materials with enhanced performances. Several architectures are possible for the different systems, but in any case subsystems usually composed by hundreds of assembled elements, which are realised on a silicon wafer. A satellite is a very expensive product, composed by non-standard components, qualified with very high reliability and with a mission, from the initial phase, through realisation, launch and operation. It is also characterised by a very complex and expensive management. The concept is not any more fully justified for the enormous costs involved and therefore different possibilities can be explored. The Fig. 9.2 represents the mass of several satellites as a function of application domains. We observe how the mass can range from less than 1 Kg as in the PicoSat of DARPA to some tons like for the Hubble Space Telescope. Further classification follows.

- Mini satellite: above 100 kg
- Micro satellite: 10 – 100 kg
- Nano satellite: 1 – 10 kg
- Pico satellite: below 1 kg

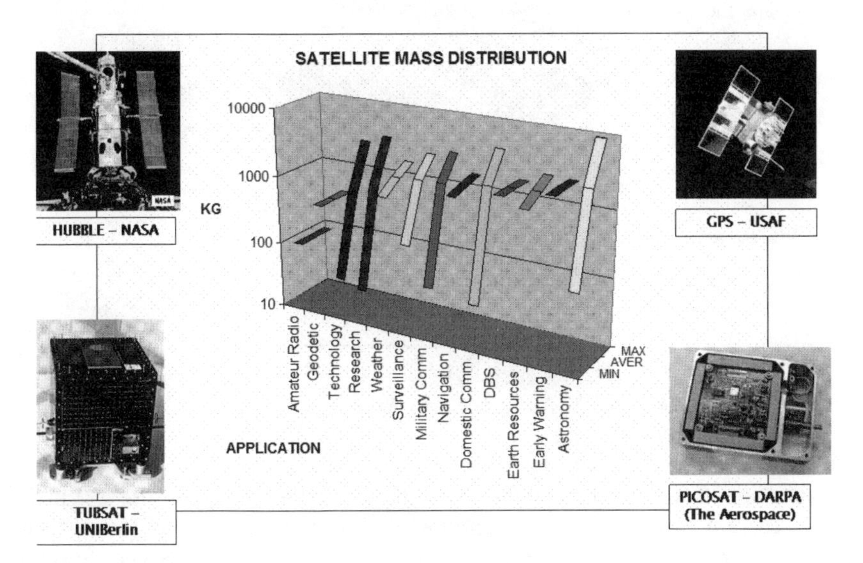

Fig. 9.2. Satellite trends

Considering the launch cost and the structure of the satellite we clearly recognise the benefit of miniaturisation as the payload is directly related to the mass. Different principle and technology can separately influence the volume and mass of the spacecraft. Since the first concept of nanosatellite, represented in Fig. 9.3, fully manufactured on silicon wafers, has been proposed by "The Aerospace corp." in 1995 (Janson 1995), many research groups are on the way to produce and demonstrate such promising idea.

The reasons to build a nanosatellite are obviously related to the cost reduction in terms of launch and to the possibility to realise totally new class of missions. In practice the space community is still waiting the boom of the nanosatellite constellations, mainly because the enabling technologies are still not fully available or maybe because a different approach is necessary. The economical reasons to accomplish a mission only with nanosatellites are not fully demonstrated, especially due the high investments necessary to first demonstrate the reliability of such concept. The following four types of space based systems will be commercially developed in the next 10 to 15 years.

- Global positioning and navigation services
- Global communication services
- Information transfer services
- Global reconnaissance services

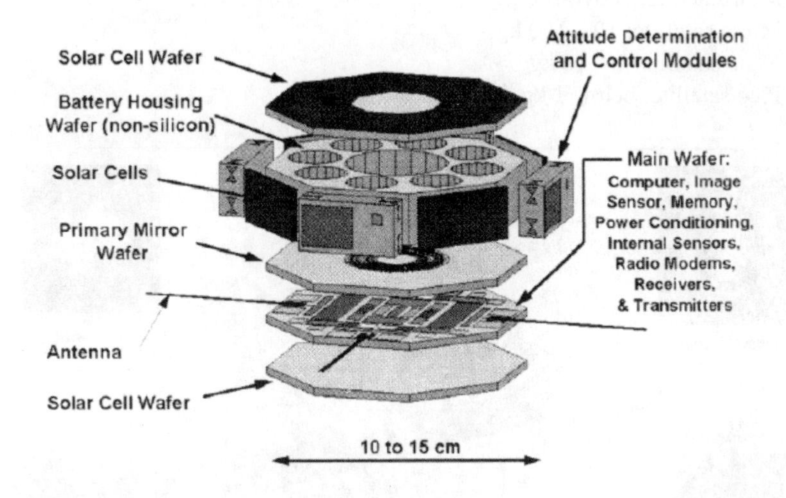

Fig. 9.3. The nanosatellite of the Aerospace Corporation

Each of these systems will be part of the *infosphere*, the volume where most of the information will transit. Other exciting concepts, like the use of many nanospacecrafts for planetary exploration and solar sail steering (http://solar-thruster-sailor.info/sts/sts.htm) are under study at Intelligent Space Systems Laboratory of the University of Tokyo (Furoshiki Satelllite). In that case the nanospacecrafts can have the function to provide attitude control to the sail and to realise the necessary tension by spinning the payload for the exploration. When such kind of collaborating sailing spacecraft fleet will reach the target, the sail can be removed and the spacecrafts can form a constellation around the planet, some of them can land on it and perform the exploration.

Each such nanospacecraft will perform one specific function and therefore will be equipped with one single instrument integrated with sensors and COTS (Component Off The Shelf). The whole concept of microspacecraft makes sense if it is designed based on a high modularity and standardisation. Several possible payloads can be easily connected to the spacecraft bus. Given the size of available sensing devices like imaging CCD or CMOS, accelerometers and gyros and magnetic sensors. It is reasonable to design a fully performing nanosatellite of the size of 10 cm. On this class of spacecrafts (Fig. 9.4) there is a big research activity worldwide, especially at university level, like the CubeSat XI developed at the University of Tokyo (Nakasuka et al. 2003), which was successfully launched on the 30th of June 2003. While CubeSat does not have all the functionality of conventional spacecrafts, it is expected that in near future it will be fully competitive. As can be seen in the Fig. 9.5 there is no much difference from a well packed PC or other electronic devices and a nanosatellite at the present status of development. Employing advanced technology and integrating miniaturised sensory systems the nanosatellite design would make the difference.

Fig. 9.4. CubeSat “XI” of the University of Tokyo

Fig. 9.5. CubeSat XI structure

9.2 Subsystems and Devices for Miniaturised Spacecrafts Micropropulsion

The improvement of mission performances has two sides. On one side we would like to increase the volume of the payload to get more amount of information, cover a bigger data collection surface and to use less power for the antenna. This can be achieved through a payload distribution. In the same time we try to reduce cost by means of mass reduction and therefore we miniaturise the payload without loosing the performance and increasing the resolution. For the reasons of cost and feasibility the slight increase in spacecraft size can be ignored and a compromise may be sought. It is beneficial to have a constellation of smaller satellites with miniaturised payloads equipped with sensors based on nano and micro technologies. In order to have precise correlation between the data obtained from the different spacecrafts, high performance Attitude Determination and Control System (ADCS) is needed. The components related to payload and to ADCS are of primary importance. The ADCS is basically composed by sensors and actuators. Optical sensors can be used for the most precise attitude determination. These sensors can obtain the image of the earth horizon, sun position or a star map and from such image the attitude of the satellite can be calculated. An example of such sensor is the bolometer which is an IR (Infrared) optical sensor used to determine the position of the horizon giving the values of pitch and roll. The fabrication process of a typical microbolometer is shown in section in the Fig. 9.6 (Sedky 1997).

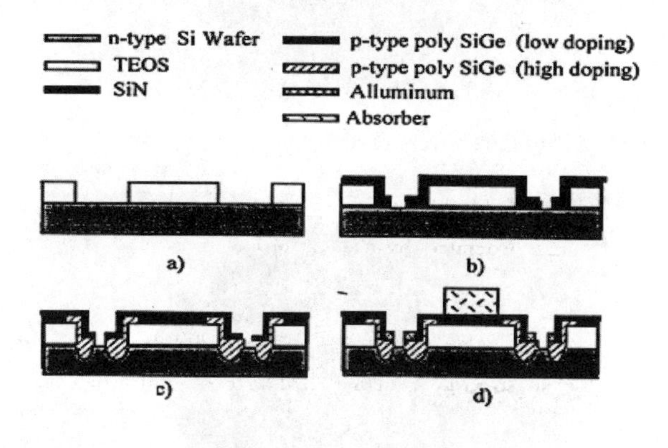

Some steps of the bolometer fabrication. a)Teos deposition and patterning. b)Deposition and doping of polySiGe, deposition of SiN, patterning of the two layers. c) Implantation of the supports with a high boron dose. d)Deposition and patterning of metal contacts and absorber, removal of the sacrificial layer.

Fig. 9.6. Microbolometer fabrication

A microbolometer is a resistor with a high thermal coefficient of resistivity (TCR), such that a small temperature variation will produce a high resistance change. For such reason the sensor is thermally insulated. The active element (metal or vanadium oxide) is deposited on a membrane (nitride or silicon oxide), which is thermally and electrically insulated. This way is not compatible with the microelectronic process. The alternative approach consists of a sensible element (Polysilicon or poly-Silicon-Germanium for a better tuning of the thermal conductivity) in the shape of a microbridge. Such microbridge, as shown in the figure is obtained by successive depositions of semi conductive layers on a sacrificial layer, which is removed at the end of the process. The layers are realised following typical patterns and doped in order to obtain the desired thermal and electric characteristics.

The effect of inertia is reduced with the volume while the main disturbances, sun radiation and atmospheric drag are related to the surface of the satellite. The best system to realise a full control of the attitude of the satellite is based on micropropulsion devices. Micropropulsion devices are necessary to execute orbit manoeuvres like plane change and spinning of the satellite. Micropropulsion is considered as a propulsion system which is applicable to a microspacecraft (Micci and Ketsdever). It produces a thrust in the order of μN and has a mass of few grams or tens of grams. Micropropulsion systems can be electrical or chemical. Fig. 9.7 represents a Field Emission Electrostatic Propulsion (FEEP) where metallic ions are accelerated from an electrical field in order to give the thrust (Helvajan 1995).

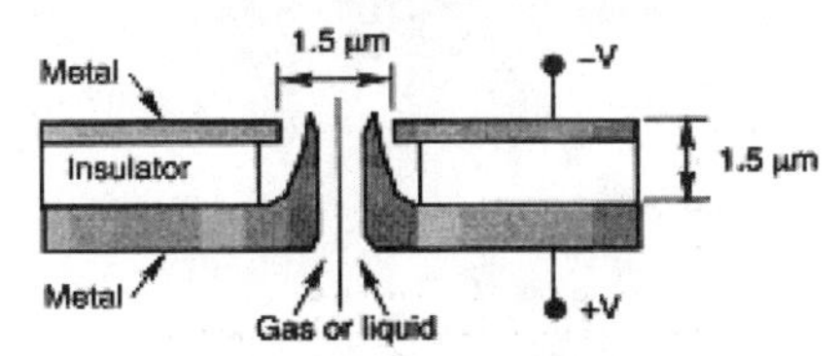

Fig. 17.25. Schematic cross section of the Spindt microvolcano field emission ionizer.

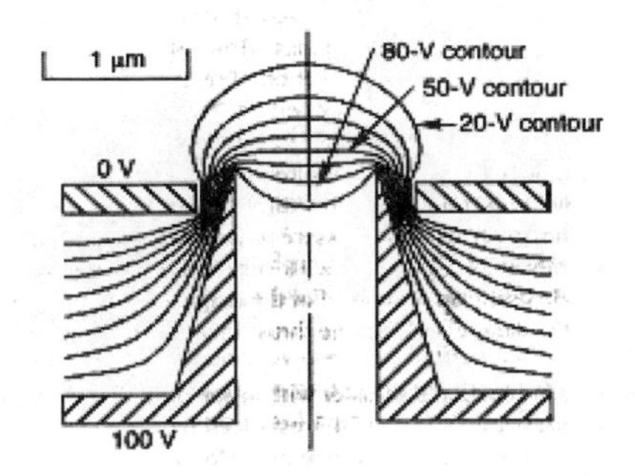

Fig. 9.7. Micro FEEP

A more classic principle is used in the microrocket arrays. In this case an array of small chambers is filled with a solid propellant material and sealed with a substrate which contains microheaters deposited on the membranes which closes the chambers. On the other side of the membrane a rough nozzle is obtained in the substrate as shown in the Fig. 9.8. High voltage and relatively high power are necessary requiring special electronics.

In the ideas of the developers (Rossi 1999), such microthrusters can be distributed on the satellite surface and activated sequentially to produce the required impulse. When the selected resistor is heated, the propellant is ignited and the charge burns producing the gas. The temperature and the rise of pressure break the membrane and the gas exits from the nozzle producing thrust. This integration has the disadvantage of a variable misalignment from the thrust vector to the centre of gravity at each firing. The impulse is not very precise and only a limited number of pulses can be produced. Also the thrust cannot be adjusted. Nanotechnology plays an important role in such kind of micropropulsion systems in the fabrication of nanosized particles, which can provide higher uniformity in the charge, reliable ignition and higher combustion efficiency (Jigatch et al. 2002).

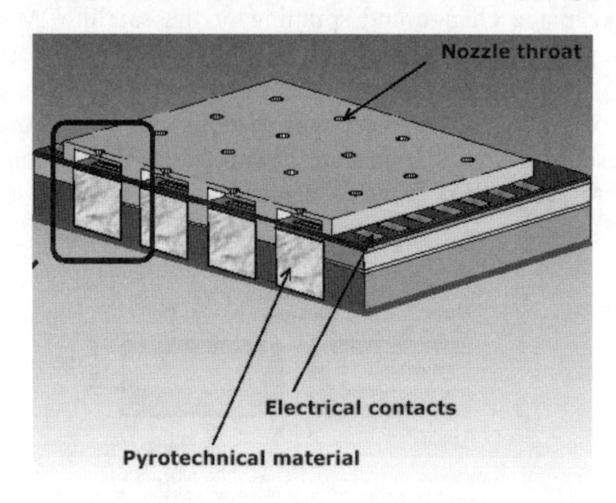

Fig. 9.8. Solid propellant microthruster

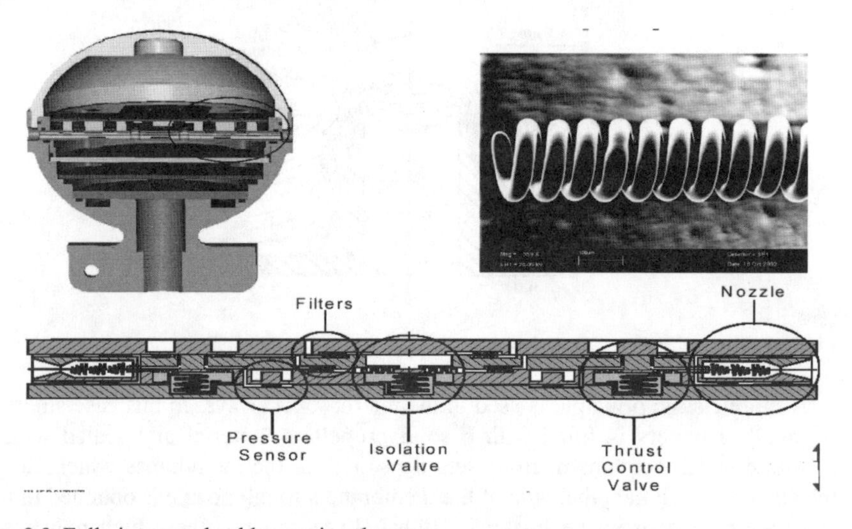

Fig. 9.9. Fully integrated cold-gas microthruster

The most refined solution for the attitude control uses single Cold-Gas thrusters to produce very precise thrust vectors and torques around the three axis of the satellite. The miniaturisation of thrusters is a very challenging topic, which requires knowledge of thermodynamics, chemistry, fluid dynamics and technology. Several groups are working on Cold-Gas microthrusters, typically based on silicon technologies and with a different level of integration. Fig. 9.9 represents a section of the microthruster on chip developed at the Angstrom Centre in Sweden. Two central wafers realise the nozzle structure, a heating chamber and the filtration. The external wafer contains the fluidic connections and the deformable element of the proportional valve. The initial idea of a piezoelectric actuation of the valve has been abandoned in favour of a thermally actuated microvalve (Stenmark 2000).

Another field where microsystems could contribute in the realisation of mass economies is in the power generation subsystem. Miniaturised turbine generators are under development in many institutes like the famous project of the MIT (Epstein and Senturia 1977). A section of a microturbogas is represented in Fig. 9.10. This is probably the most challenging project in the field of MEMS power devices.

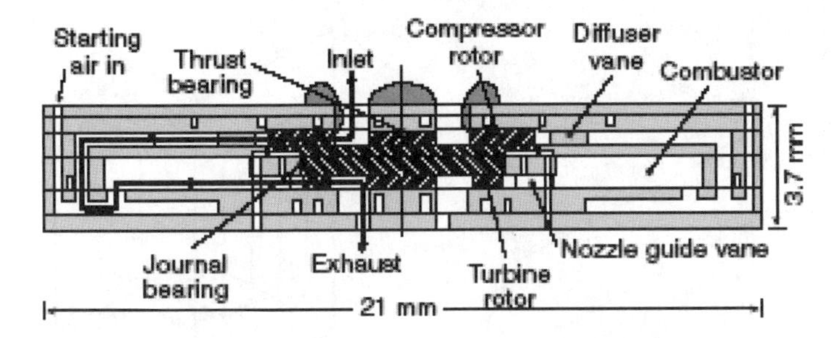

Fig. 9.10. MIT Microturbogas

The valve that separates the tank from the piping is opened once the satellite is in space. The valve functionality is based on cutting of a metallic pipe by means of a pyrotechnic charge. A miniaturised version of pyrotechnic actuator, which enables the construction of a microvalve, was developed by the author for the European Space Agency (Manzoni et al. 2000). As can be seen in the Fig. 9.11 a microcharge is exploded and the energy released in form of gas to actuate the movement of the valve.

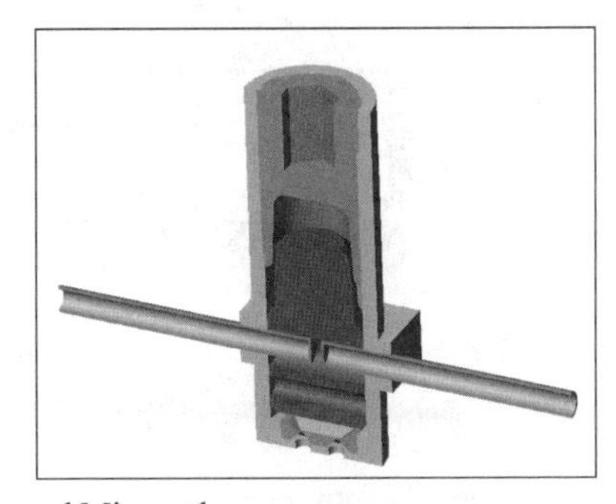

Fig. 9.11. Miniaturised Pyrotechnic Actuator and Microvalve

Satellites are subjected to the heating from the sun and the energy reflected by the planets surfaces. The internal electronic constitutes another source of heating which can accumulate and making some parts reaching a critical temperature. On the other hand the parts facing the space emit energy and the temperature could reach critical negative values. A thermal control is required in order to transfer energy from different satellite points and reject or release the thermal excess to the space. This function can be achieved by means of microfluidic devices like the microheatpipes and controlled by thermal microswitches or microlouvers (Gilmore 1997). The Fig. 9.12 represents a concept for the realisation of microheatpipes and a microthermal switch.

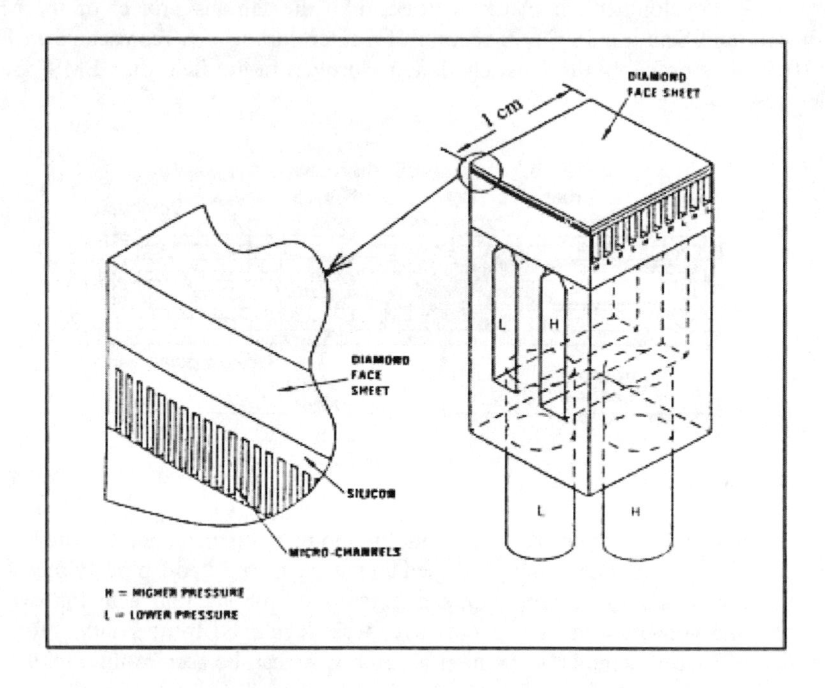

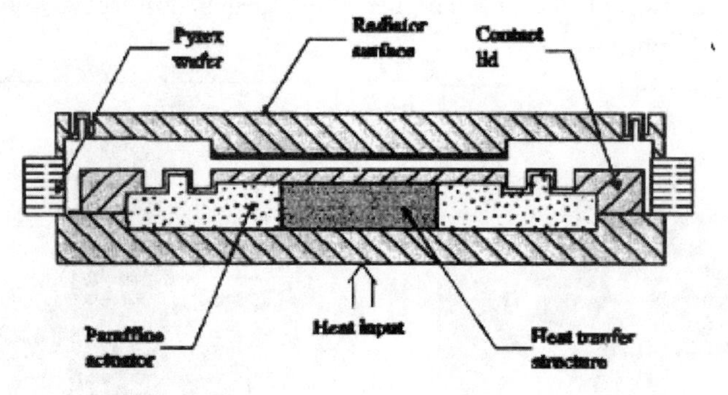

Fig. 9.12. Thermal microswitch and microheatpipes

9.3 Propulsion Systems

As micro and nanosatellites promise revolutionary access to space, several concepts have already been proposed for a standardised architecture. The important functional subsystems are,

- Guidance and navigation
- Attitude determination and control
- Telemetry, tracking and commanding
- Propulsion, power unit and thermal control
- Payload, structure and structural elements

Fig. 9.13 shows a basic diagram of a cold-gas system for control of attitude around one axis of the spacecraft (Manzoni 2001). Among the several subsystems, propulsion and attitude control are the most difficult to be miniaturised, due to the complex interaction between physical principles, technologies and engineering problems. The important micropropulsion approaches are i) Chemical micropropulsion (Liquid-gas propellant (Cold-gas, Monopropellant, Bi-propellant and Turbo-pump fed systems) and Solid propellant, and ii) Electric micropropulsion. Brief descriptions on solid propellant, cold- and warm-gas, monopropellant and bipropellant systems, colloid thrusters and regenerative-pressurisation cycles are presented below.

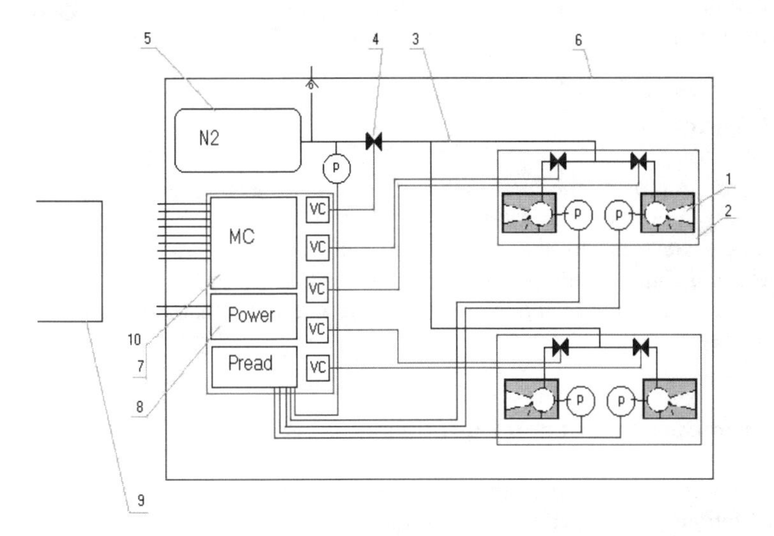

Fig. 9.13. Cold Gas main parts [1. Microthruster (Micronozzle, Pressure sensor, Microvalve, Mechanical module); 2. Thrusters group; 3. Fluidic connections; 4. Safety valve; 5. Gas tank (i.e. N_2); 6. Mechanical interface; 7. Control unit (Microcontroller (MC), Valve controller (VC), Pressure read circuit (Pread)); 8. Power; 10. Satellite structure; 11. Control Software]

9.3.1 Solid Propellant

In a solid propellant each charge is considered as a single shot. It is possible to obtain an array of charges. Each shot generates a well-quantified impulse and patterns of charges can be fired to generate specific torques as well as thrust vectors. Issues include achieving consistency of propellant cavity filling, reliable ignition, and performance. Initiators as opposed to propellants have been considered; hence analysis of detonation rather than combustion is required. A major question is how the area required for the thrusters array will impact on photovoltaic panel mountings in the smallest satellites, reducing available power levels.

9.3.2 Cold-Gas

Cold-gas is a micro fluidic system. The main development issues are the demonstration of low power and high-pressure valves, the use of high-density impulse propellants (e.g. Xe, Butane, nitrous oxide) and integration of thrust chamber components with ancillary components in a small total package.

9.3.3 Colloid Thrusters

The colloid thrusters are vaporising micro liquid system. Presently there is no activity in Europe but they are studied strongly in USA.

9.3.4 Warm-Gas

Warm-gas improves the performance of a cold gas system by heating the propellant, up to the temperature limits of MST materials. Full integration of all the fluidic devices on the same chip will require a trade-off in terms of temperatures limits, and performance. Deposition of appropriate heaters, heat transfer to the propellant and retention of heat within the chamber are major issues to be resolved.

9.3.5 Monopropellant and Bipropellant Systems

The scaling of a combustion chamber is possible but completely new configurations are expected due to the different thermal behaviour and the underlying scaling effects. Issues such as effectiveness of catalytic decomposition (monopropellant), injection, mixing and combustion (bipropellant) on the microscale are very poorly understood. It requires considerable amount of modelling effort. The main operational challenges to be resolved are operations at high pressure and high temperature. Before development of a bipropellant system considerable research into methods of cooling will need to be carried out.

9.3.6 Regenerative-Pressurisation Cycles

The time needed to efficiently combust propellants is dependent on chemical reaction times, combustion pressure and chamber size. For the small chamber sizes dictated by micropropulsion systems, high-pressure operation is likely to be a prerequisite. Therefore, the selected pressurisation method may mandate pump feed. Pump feed is of particular interest for micropropulsion, as small pumps may compensate for the inability of MEMS valves to seal at high pressures.

9.3.7 ADCS

The purpose of an ADCS is to counteract the effects of disturbances which are present in space. By carrying out the calculation of the disturbances (Manzoni 2000) we obtain that the thrust required to control the attitude of a micro or nanosatellite ranges from few μN to few mN. It is also very important to evaluate how often the thrusters must "fire" to keep the attitude under the requested range and accuracy (Manzoni 2004). The total number of firing is the ratio between the total mission time and the time necessary to correct the maximum allowed attitude error.

$$n = \frac{t_{mission}}{t_{\max}} = \frac{t_{mission}}{4\sqrt{\frac{2\theta_{\max}}{\dot{\omega}}}} = \frac{t_{mission}}{4\sqrt{\frac{2\theta_{\max}}{T/I}}}$$

Fig. 9.14 shows how the number of pulses per orbit changes with the spacecraft size and the maximum allowed excursion in order to balance only the effects of the drag. It is clearly recognised that, in the field of chemical microthrusters, which are today the first choice for the fine attitude control of small satellites, the solid propellant thrusters arrays are not useful due to the very limited number of shots, while cold-gas systems are the only reasonable possibility due to the unlimited number of shots.

9.4 Realisation of a Cold-Gas Microthruster

Among the several micropropulsion systems a Cold-Gas based microthruster has been selected due to the reasonable maturity of the required technology and the possibility to realise a working prototype in the time frame of the present study. A

Cold-Gas system has been therefore designed, prototyped, tested and integrated in a nanosatellite. Further improvements on the system, namely the heating of the gas and the conception of a system to provide such heating in an efficient way, have been studied as basis for further research steps.

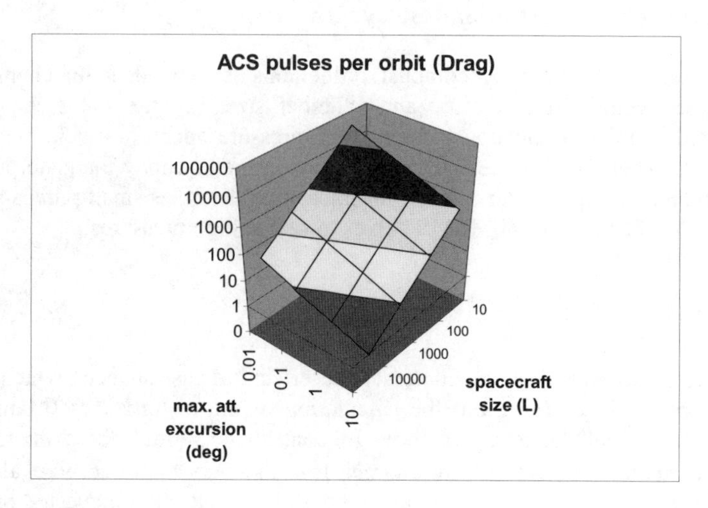

Fig. 9.14. Pulses vs. size and ACS accuracy

9.4.1 Gas- and Fluid Dynamics

The preliminary design of the micronozzle has been done using a simple one-dimensional model for the adiabatic flow of the compressible fluid with viscous friction (Oates 2004). With this model the first estimation of the nozzle thrust has been possible. The formulae used to take in to account of the losses are certainly very simplified and have been originally developed for macroscopic systems (Manzoni 2000). The coefficients of the formulae maintain their validity partially in the microscopic scale. In order to have a better idea of the flow in the designed nozzle a CFD simulation is performed. In order to use the Navier-Stokes equations the characteristics of the flow has to be checked (Gad-el-Hak 1999). In fact fluid flows in small devices differ from those in macroscopic applications because of:

- Higher surfaces effects
- The mean free path is close to the characteristic dimensions
- There could be a relative movement between the fluid and the walls (the no slip-condition looses availability)

As depicted in the Fig. 9.15 where the above-mentioned limits are traced and the working range of the microthruster is represented in the preliminary design, we observe that the Navier-Stokes equation and the continuum approach are still applicable.

The CFD simulation shows that a certain selection of nozzle length introduces losses higher than expected and the exit pressure is reached well before the end of the nozzle. With the help of such information, the length of the nozzle can be reduced in order to obtain an optimised design for better efficiency.

Flow character		Throat
Reynolds	Re	659
Mach	Ma	1
Knudsen	Kn	0,002
Normalised gas density	r/r0	0,6

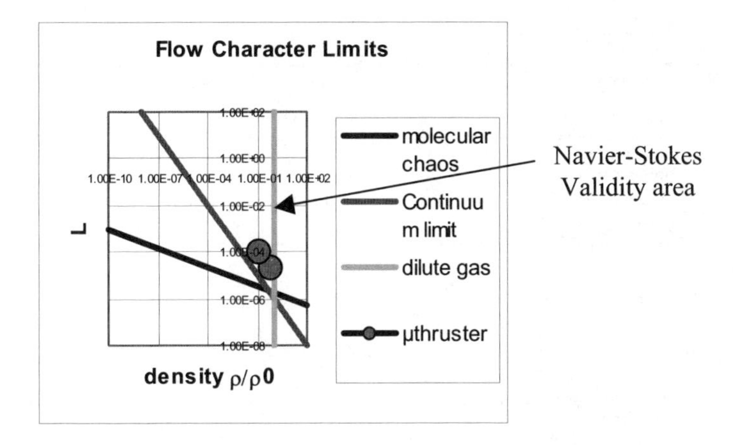

Fig. 9.15. Flow Character

9.4.2 Prototyping

The manufacturing of the micro components has been based on silicon and glass assembled by means of anodic bonding and on some packaging techniques for the assembly with the other COTS, electronic and mechanical components. An investigation on the components for micropropulsion available on the market has been conducted and the results are represented in the Fig. 9.16.

We have observed that ancillary components like microvalves or pressure sensors are available on the market also in MEMS configuration while the heart of the propulsion system, the micronozzle, has been only demonstrated by some universities. The microfabrication focus has been concentrated on the microthruster, while the other necessary MEMS components have been selected from devices available on the market. The preparation of the nanolithography mask and the lithography of the micronozzle have been performed initially from the Italian Institute for the Physics of the Matter in Trieste in cooperation with the company Microspace. Such process is an adaptation of the state of the art,

Reactive-Ion Etching (RIE) and has been optimised for the smoothness of the surface. Fig. 9.17 and Fig. 9.18 represents the recipe and sequence of the steps for the realisation of the microlithography mask and the lithography on silicon to realise the micronozzle (Santoni et al. 2002). Fig. 9.19 shows a view of the micronozzle in comparison with a human hair (about 100 micrometers). The Fig. 9.20 shows a detail of the throat area after a cut with microsaw; the round profile of the channel, due to the particular isotropic etching given by the RIE, can be noticed.

Chemical Micropropulsion Technology Readiness Evaluation		
FEED LINE	MEMS	Miniaturised
pressure reduction system	not available	under development
Gas Generation System		under development
miniaturized NC valve	concept and basic element available	commercially available
MICROROCKET-MICROTHRUSTER		
pressure sensors	commercially available	commercially available
microvalves	available but leaking and slow actuation	commercially available
microcombustion chamber	not available	under development
micronozzle	demonstrated	available
microturbine	under development	demonstrated
micropump	available for low pressure	available for medium pressure
microelectrical generator	demonstrated	available

Fig. 9.16. Technology readiness evaluation

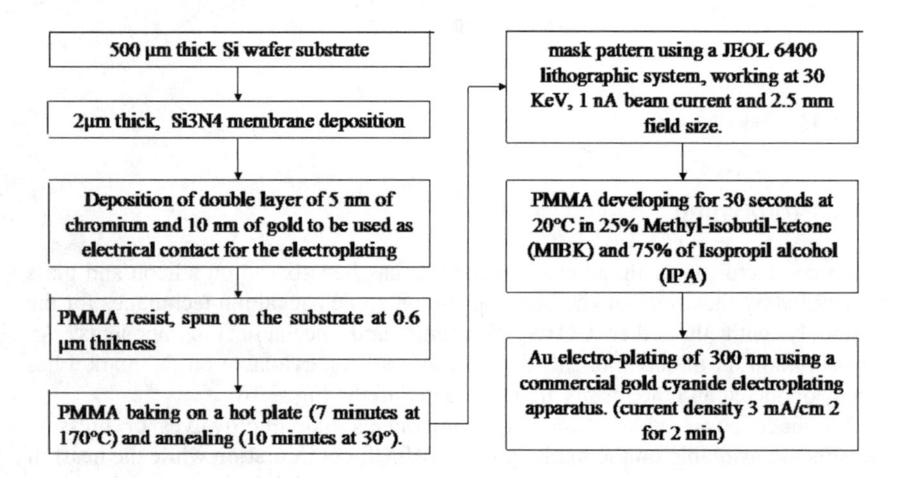

Fig. 9.17. Mask nanolithography process

Fig. 9.21 shows a detail of the throat by focussing the microscope to the upper surface of the silicon and the bottom surface of the channel. In the first figure we notice the smoothness of the channel geometry. The narrowest part of the throat is 40 micrometers wide. The small irregularities in the round geometry are due to the errors in the optical transfer process from the optical mask to the RIE metallic temporary mask attached to the silicon surface. A geometry analysis can confirm the accuracy. The analysis shows differences of 1 or 5 micrometers from the several thrusters produced on the same wafer.

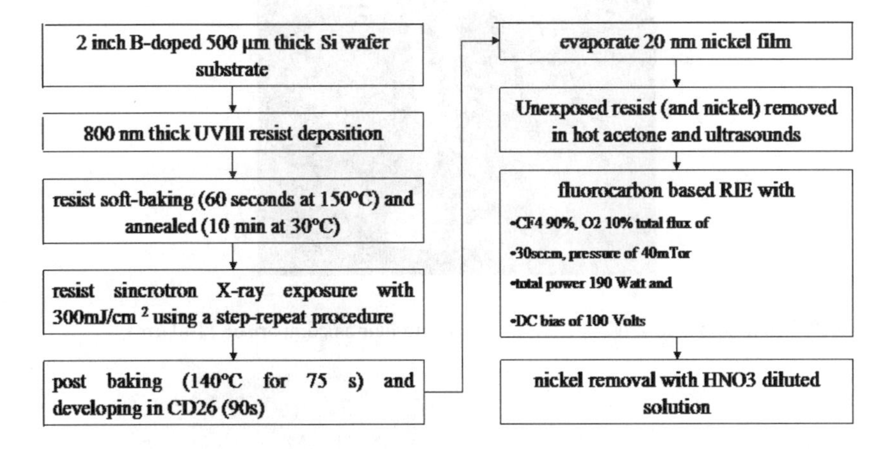

Fig. 9.18. Micronozzle nanolithography process

After the wafer with the thrusters has been produced it will be bonded to the support glass, previously prepared (with microholes) for each thruster. Fig. 9.23 shows the final result of the process. It is primarily a cube of 4x4x4 mm with one thruster, ready to be assembled in the micropropulsion system. An improved version of the micronozzle, shown in Fig. 9.24 is now under development at the National Institute for Advanced Industrial and Science Technology (AIST) in Japan in collaboration with the Microspace, Italy (Manzoni 2004). The improvements concern the use of a more repeatable process, better characterisation of the performances and realisation of a reliable packaging. The assembly of the main thruster's components comprehend essentially 4 elements.

- Micronozzle
- Microvalve
- Pressure sensor
- Mechanical structure

Several architectures are possible with respect to the interfaces of the single selected components. Fig. 9.25 shows a typical assembly structure to be used (Graziani et al. 2000) for commercial microvalve and pressure sensor by the company Redwood and Fraunhofer, respectively. Such assembly has also been mounted on the Italian microsatellite UNISAT-2 of the University of Roma "La Sapienza", launched in December 2002. Due to the big size of the packaging of the Redwood microvalves and the Fraunhofer pressure sensors, a very big volume is introduced between the valves and the thrusters. This kind of assembly does not allow a fast dynamic of the thrust variation. Furthermore the Redwood microvalves, being thermally actuated, are not functioning over 70 degrees Celsius.

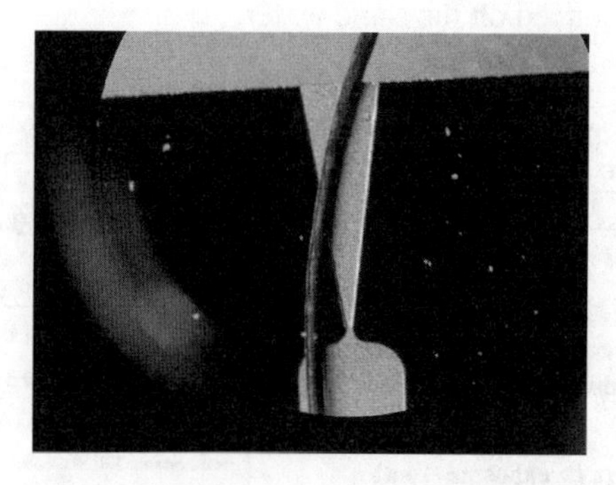

Fig. 9.19. Micronozzle microscope photo (25x) with hair as comparison (100 μm);

Fig. 9.20. Micronozzle section cut photo

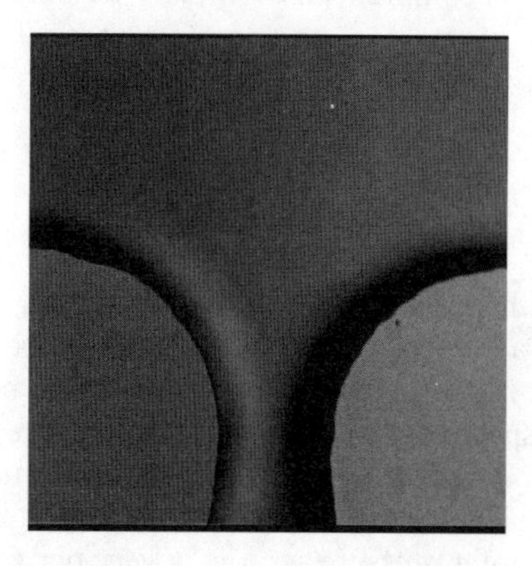

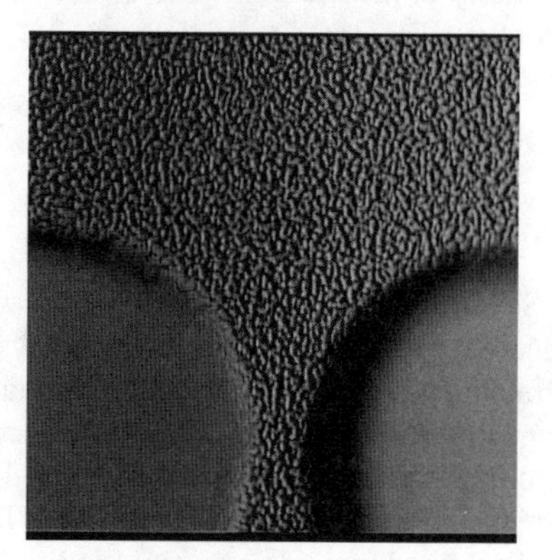

Fig. 9.21. Micronozzle throat microscope photo

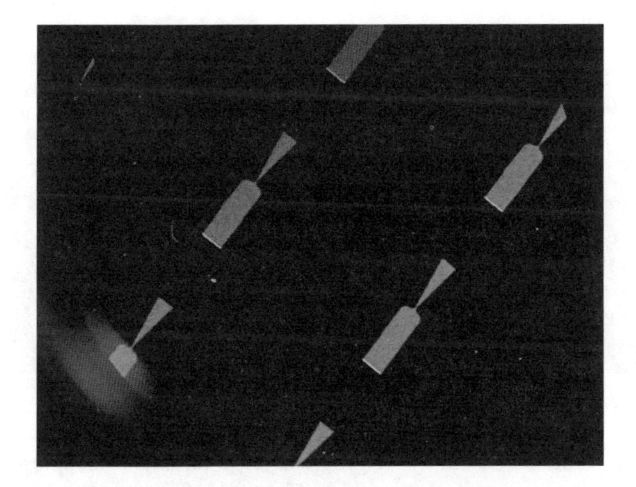

Fig. 9.22. Micronozzle production

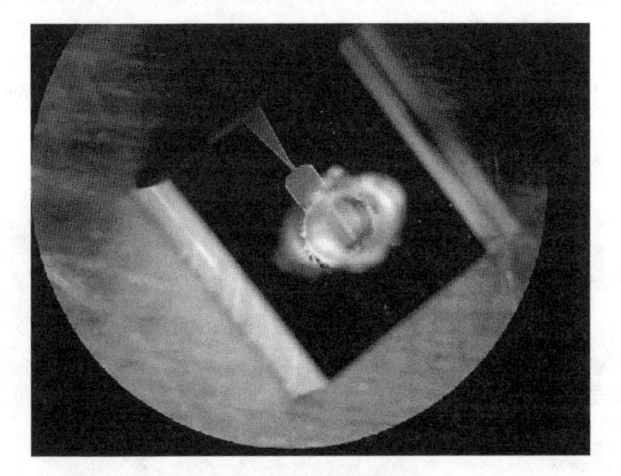

Fig. 9.23. Microthruster chip

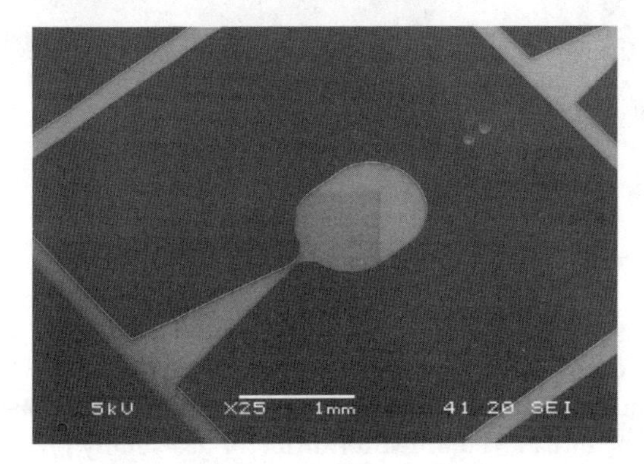

Fig. 9.24. AIST-Microspace Micronozzle

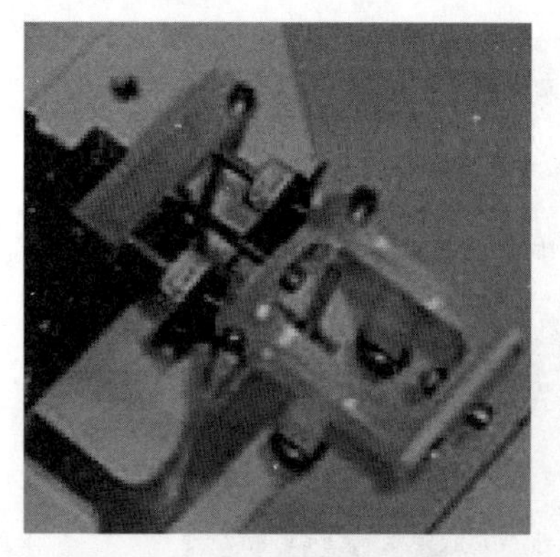

Fig. 9.25. Microthruster assembly

With this kind of assembly, a group of two thrusters occupies a volume of about 50 cc for a mass of about 100 g. The use of the commercial microvalves from the company Lee and the Intersema pressure sensor allows the conception of a much more compact thrusters group. In a volume of 5 cc and with a mass of less than 20 g it has been possible to assembly a group of 3 thrusters for 3 different thrust vectors. Such compact assembly allows faster thrust dynamic. The more compact design requires the use of 3 machined aluminium parts with grooves for the gas channels as shown in the Fig. 9.26 (Manzoni 2003).

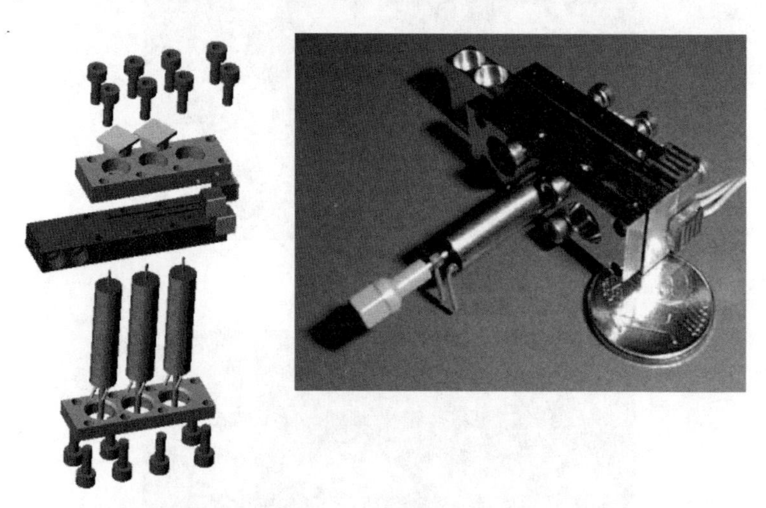

Fig. 9.26. High dynamic microthruster assembly

The micronozzles have been mounted on a microbalance realised by the author, to measure μN thrust in vacuum. A typical measurement of thrust is based on the detection of the acquisition. Averaging the balance oscillations carries out the calculation of the thrust. The thrust for different gas supply pressures and vacuum conditions is presented in the Fig. 9.27.

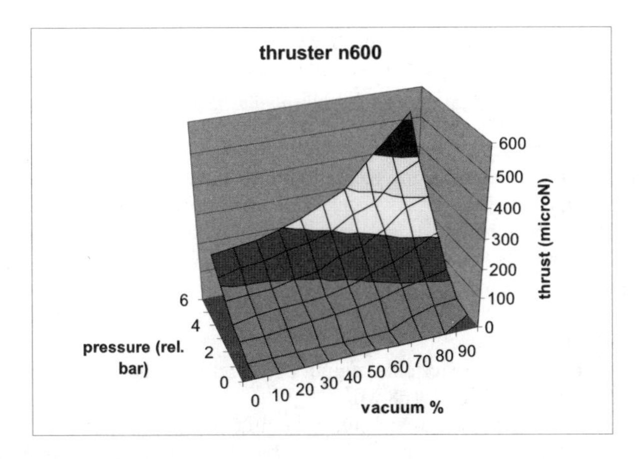

Fig. 9.27. Thrust experimental results

9.5 Conclusions

The initial investigation we performed on the state of the art of microsystems for aerospace application has given a clear indication of the opportunity to develop a micropropulsion system due to the complete absence of such subsystem available and ready for use for micro and nanosatellites (Manzoni 2003). Such indication is confirmed by the several similar activities ongoing worldwide under funding of the main space agencies. We can conclude that most of the devices necessary to realise a micropropulsion system for attitude control of miniaturised spacecrafts are under study and are available for experimental mission. Such missions are in the planning programs of some space agency and will show results in the immediate future that will help to select the more efficient and reliable technologies which will enable a future commercial and strategic use of miniaturised spacecrafts. Nevertheless, the complete development of micropropulsion system requires additional research work. Such additional directions have been collected in a strategic view in the Road-Map of Micropropulsion which has been endorsed by the European Space Agency and constitute the backbone of the Harmonisation and Funding Program of ESA (Lang 2002).

9.6 References

Ellinghaus F (2004) Solar thrusters sailor. @http://solar-thruster-sailor.info/sts/sts.html

Epstein AH, Senturia SD (1977) Microheat engines, gas turbines and rocket engines, Technical Report of MIT microengine project, AIAA 1977

Gad-el-Hak M (1999) The fluid mechanics of microdevices, J of Fluids Engineering 121:5-33

Gilmore D (1997) Nanotechnology for spacecraft thermal control. In: Helvajian H (ed) Micro and nanotechnology for space systems. The Aerospace press, pp 97-101

Graziani F, Ferrante M, Palmerini GB, Santoni F, Tortora P, Muscinelli M, Pirillo F (2000) Mechanical tests for low-cost microsatellite programs. Proc. of the 51st Int. Astronautical Congress, October 2-6, Rio de Janeiro, Brazil

Helvajan H (1995) Microengineering technology for space systems. Aerospace Corporation report no. ATR-95(8168)-2

Janson S (1995) Spaceraft as an assembly of ASIMs. In: Helvajian H (ed) Microengineering technology for space systems. The Aerospace press, pp 97-02

Jigatch, Leipunsky I, Kuskov M, Pchechenkov P, Larichev M, Gogulya M, Krasovsky V (2002) A technique to prepare aluminized nanosized energetic composition. Nanotech2002, Houston

Lang M (2002) Technical dossier on mapping chemical propulsion – micropopulsion. ESA documentation, TOS-MPC/2168/ML

Liu L (2003) Asian efforts on ano and micro satellite R&D and opportunities for EU-Asia Cooperation. Reports, 4th Round Table on Micro/NanoTechnology for Space. ESTEC

Liu L (2003) Nanotechnology opportunity report. 2nd edn., CMP Scientifica

Manzoni G (2000) Design of a highly integrated micropropulsion system for microsatellites attitude control, JPC, Hunstville (USA)

Manzoni G, et al. (2000) Miniaturized pyrotechnic initiator. ES-MPI final report, ESA contract 14679/00/NL/CK

Manzoni G, Vaccari L, Fabrizio ED (2001) Design and prototyping of a highly integrated micropropulsion system for microsatellites attitude control. MST 2001, Düsseldorf (D)

Manzoni G (2003) Nanotechnology & nanosatellites: a visionary picture with 15 orders of magnitude field of view. Asia Pacific Nanoelectronics Exhibition and Technology Forum, Taiwan

Manzoni G, Miotti P, DeGrandis F, Vaccari L, Marmiroli B, Perennes F, DiFabrizio E, Tortora P (2003) Components for chemical micropropulsion systems, an overview on strategies, concepts, studies, prototypes and concrete possibilities. ESA 4th Round table on micro-nanotechnologies for space, ESTEC

Manzoni G, Tortora P, Matsumoto S, Maeda R (2004) Microthrusters and microturbines technologies and prototypes: A roadmap for nanosatellites propulsion and power ISTS2004, Japan

Micci MM, Ketsdever D (2004) Micropropulsion for small spacecraft, Progress in Astronautic and Aeronautics 187:10-16

Nakasuka S, Tsuda Y, Sako N, Eishima T (2003) Cubesat: Future direction beyond education. Proc. of the first international CubeSat symposium, Tokio, pp 20-25

Oates GO (2004) Aerothermodynamics of gas turbine and rocket propulsion. AIAA Press, Education Series, ISBN 1-56347-241-4

Rossi C (1999) Design, fabrication and thrust prediction of solid propellant microthrusters for space application. Proc. of the Design, test and microfabrication of MEMS and MOEMS, Paris

Santoni F, Tortora P, Graziani F, Manzoni G, DiFabrizio E, Vaccari L (2002) Micropropulsion experiment on UNISAT-2 – 0-7803-7321 IEEE Press

Sedky S (1997) IR microbolometer made of polycristalline silicon germanium alloys. 2nd Round Table on Micro/Nano Technologies for Space, 15-17/10/1997, ESTEC, The Netherlands

Stenmark (2000) Feasibility demonstration of the micropropulsion cold gas thrusters system. 3rd Round table on Micro/Nano technologies for Space, ESTEC, The Netherlands

10 Carbon Nanotube Production and Applications: Basis of Nanotechnology

S. E. Iyuke[1] and N. P. Mahalik[2]

[1]School of Process and Materials Engineering, University of Witwatersrand, South Africa
[2]Department of Mechatronics, Gwangju Institute of Science and Technology, South Korea

10.1 Introduction

Micromanufacturing and nanotechnology possess the same basic goal of promise to change the present and future of manufacturing from the conventional top down to a new bottom up, atom-by-atom technique to achieve precise and specific objects and processes. Carbon nanotube or cylinder of carbon atoms is one of such materials produced atom by atom whose mechanism is yet to be clearly understood. Carbon nanotubes have diameters ranging from less than 1 to 100 nm, and exhibit unique mechanical, electronic and magnetic properties, which perhaps have made their research the most active and interesting field in carbon related science in the recent years. This chapter therefore discusses carbon nanotube technology, production and applications since there is presently tremendous interest in their commercialisation for both near and future applications.

10.2 Nanotechnology and Carbon Nanotube Promises

Nanotechnology is a technology that owes its basis on the prefix nano, a Greek word, which means (dwarf) billionth (1×10^{-9}) meter dimensions. It is a relatively new technology that deals, studies or possesses the ability to systematically organise and manipulate properties and behaviour of matter and even builds matter at the atomic and molecular levels. It may be seen as the next technological rush since the Silicon Valley outburst in electronic miniaturisation, and the emerging cutting-edge technology. It is certainly setting the pace for the creation of functional devices, manufacturing and fabrication, materials and systems at the molecular level, atom by atom, to create large structures with fundamentally precise and specific molecular organisation. Two interrelated specialities in the manufacturing industry that point at nanotechnology as a common focus in their trend of developments are miniaturisation and ultra-precision processing, because both are moving towards dimensions that lie in several nanometer range. According to Rocco, three basic attributes of nanotechnology are: (1) at least one dimension of 1-100 nm is present, (2) there are processes that exhibit fundamental control over the physical and chemical characteristics of structures at molecular scales, and (3) the molecular structures otherwise known as nanostructured materials can be combined into larger structures, that are appreciable to the ordinary human vision. A promising group of nanostructured materials is the nanotubes, which are currently fabricated from various materials such as boron nitride, molybdenum, carbon (carbon nanotube), etc. However, at the moment, carbon nanotubes seem to be superior and most important due to their unique nanostructures with interesting properties, which suit them to a tremendously diverse range of applications in micro- or nanoscale electronics, biomedical devices, nanocomposites, gas storage media, scanning probe tips, etc.

Carbon nanotubes can be described as graphene sheets rolled into seamless cylindrical shapes as presented in Fig. 10.1 (Ebbesen 1994; Poole and Owens 2003) with such a small diameter that the aspect ratio (length/diameter) is large enough to become a one-dimensional structure in terms of electronic transmission. Since its discovery in 1991 by Iijima while experimenting on fullerene and looking into soot residues, two types of nanotubes have been made, which are single walled carbon nanotubes (SWNTs) and multi walled carbon nanotubes (MWNTs). SWNT consists only of a single graphene sheet with one atomic layer in thickness, while MWNT is formed from 2 to several tens of graphene sheets arranged concentrically into tube structures. The SWNTs have three basic geometries, armchair, zigzag and chiral forms as presented in Fig. 10.1.

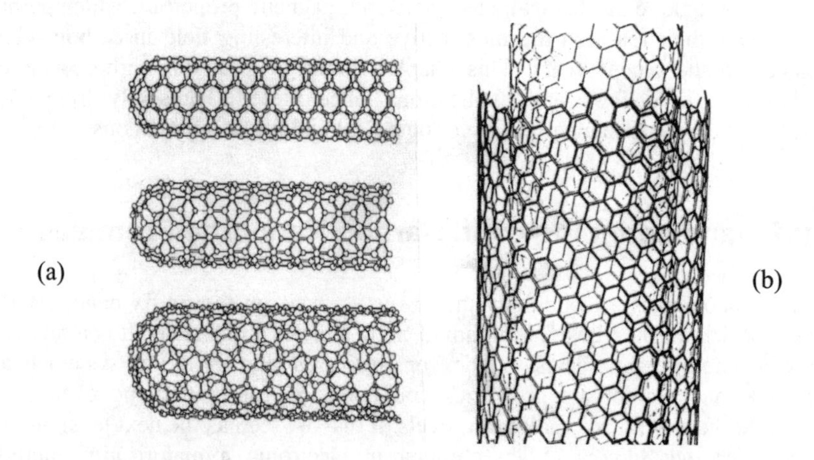

Fig. 10.1. Graphics showing types of carbon nanotubes (a) SWNT: from top is the armchair, zigzag and bottom, chiral form; (b) double wall helicoidal form typifying MWNT (Ebbesen 1997; Poole 2003)

They are promising one-dimensional (1D) periodic structure along the axis of the tube, where dispersion relations predict interesting electronic characteristics. Electronic transport confinement in radial direction is maintained by the monolayer thickness of the nanotubes, while circumferentially, periodic boundary conditions are imposed leading into 1-D dispersion relations for electrons and photons in SWNTs. Various experiments and simulations have estimated that about one-third of the nanotubes are metallic and two-third semiconducting as a result of the tube diameter and the chiral angle between its axis and the zigzag direction. Quite a number of resistivity measurements on MWNTs had revealed that some are metallic as much as graphite, and others are semiconducting having resistivity value in the order of 5 greater than the former. Experiments had affirmed that armchair carbon nanotubes are metallic (Ebbesen 1997). Multi walled nanotubes on the other hand, not only have shown metallic and semiconducting properties, they are also being used for fuel storage such as hydrogen and methane (Iyuke 2001, 2004). It is therefore interesting to infer that carbon nanotube is emerging as a building block for nanotechnology, nanoelectronics, nanomanufacturing and nanofabrication, etc.

10.3 Growing Interest in Carbon Nanotube

The report of Kroto et al. in 1985 on the discovery of "bulkyball", fullerene C_{60} molecule led into a new beginning in carbon materials and a huge interest in their sciences. Sumio Iijima of NEC Laboratory in Tsukuba observed MWNTs with high-resolution transmission electron microscopy (TEM) in 1991, and two years later he observed SWNTs. While for the first time, researchers at the William Rice University synthesised bundles of SWNTs in 1996, and most recently, in 2003 Fakhru'l-Razi, Iyuke and co-workers synthesised SWNTs intertwined bundles that are similar to the ship anchor rope shown in Fig. 10.2. This trend had originated opportunities for quantitative experimentations on carbon nanotube science. As soon as 2000, the growing interest in nanotubes has resulted in overwhelming out pour of publications and patents as compared to other nanomaterials. A relatively comprehensive survey by Gupta and Dwivedy has presented internationally patenting activities from 1987 to 2001 with 226 inventions drawn from the United States (56%), Japan (28%), South Korea (9%) and the remainder from other countries to reveal the current trend of interest in developing carbon nanotube technology towards commercialisation. Table 10.1 presents selected inventions of carbon nanotube science and technology. The growing interest in carbon nanotubes has currently attracted a vast volume of literature, while the coverage in R&D and discussions in general indicate great promises into the future. This is due to the fact that carbon nanotubes possess tremendous potentials for applications as a result of their unique properties in thermal and electrical conductivities, high strengths and stiffness, etc. These properties are directly connected to the carbon atom as the building block, which in turn is so unique among the elements in its ability to exist in a wide variety of structures and forms.

Table 10.1 Selected inventions in carbon nanotubes research in chronological order (Gupta and Dwivedy 2004)

Invention	Qty.	Reference
Carbon fibrils, method for producing same and compositions containing same	34	Tennet et al. 1987
Carbon fibrils and method for producing same	23	Tennet et al. 1992
Uncapped and thinned carbon nanotubes and process	23	Green and Tsang 1994
Carbon nanotubule enclosing a foreign material	21	Ajayan et al. 1993
Process of isolation of carbon nanotubes from a mixture containing carbon nanotubes and graphite particles	14	Uchida et al. 1994
Method and device for the production of carbon nanotubes	13	Ohshima et. al. 1996
Process for the separation of carbon nanotubes from graphite	11	Ikazaki et al. 1997
Storage of hydrogen in layered nanostructures	23	Rodriguez et al. 1997
Process for separating components from gaseous streams	24	Rodriguez et al. 1997
Method of purifying carbon nanotubes	13	Ebbesen et al. 1997
Process for purifying uncapping and chemically modifying carbon nanotubes	19	Hiura and Ebbesen 1997
Method for producing encapsulated nanoparticles and carbon nanotubes using catalytic disproportionation of carbon monoxide	12	Nolan et al. 1998
Method of forming carbon nanotubes on a carbonacous body, composite material obtained thereby and electron beam source element using same	16	Yamamoto et. al. 1998
Method for making carbon nanotubes	12	Olk, 1998
Field emission device having nanostructured emitters	39	Debe, 1998
Graphitic nanotubes in luminescence assays	8	Massey et al. 1999
Nanometer-scale microscopy probes	9	Lieber et. al. 2000
Field emission cold-cathode device	27	Nakamoto, 2000
Field emission electron source	20	Zettl and Cohen 2000
Method for manufacturing carbon nanotubes as functional elements of MEMS devices	17	Mancevski, 2000
Method of making ropes of single walled carbon nanotubes	9	Smalley et al. 2001
Electron-emitting source and method of manufacturing the same	13	Uemura et al. 2001
Process for fabricating article comprising aligned truncated carbon nanotubes	9	Jin et al., 1999
Article comprising enhanced nanotube emitter structure and process for fabricating article	13	Jin et al. 2001

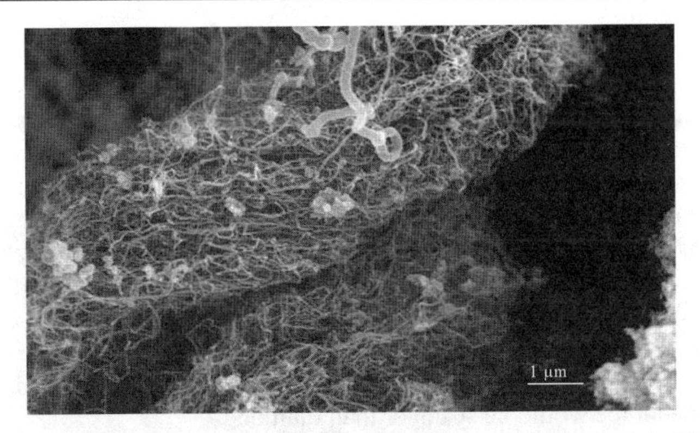

Fig. 10.2. Typical SWNTs intertwined into bundles of ropes (Fakhru'l et al. 2003)

10.4 Structure and Properties of Carbon Nanotubes

Carbon is unique among the column IV elements of the periodic table with sp^2 hybridised bonding, hexagonal ground state graphite in the condensed phase, close to being two-dimensional (2-D) semi-metal, and anisotropic. Si, Ge, Sn and Pb are sp^3 hybridised in cubic solid ground states. Diamond, another form of carbon is a three-dimensional material and isotropic, while the fullerene and carbon nanotubes discovered more than a decade ago have zero- and one-dimensional forms of carbon, respectively. The nanotube may be capped by the hemisphere (½-section) of the fullerene, C_{60} structure during production. Since the discovery of fullerenes a great deal of investigations both theoretically and experimentally, have been focused into these interesting and unique nanostructures (Fig. 10.3), and tremendous applications have been identified and others proposed for the future of nanotechnology.

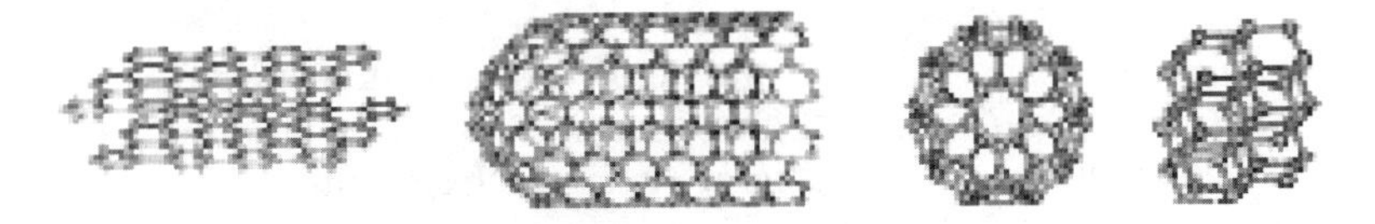

Fig. 10.3. Schematics of typical grapheme layers and other carbon nanostructured carbon materials (Graphite, Carbon nanotube, Fullerence, Diamond)

Carbon being the number 6 element in the periodic table has its electronic configurations arranged in the quantised atomic orbitals as $1s^2$ $2s^2$ $2p^2$ at the ground state. Graphite crystal structure forms strong interplanar bonds with its nearest neighbours using the occupied 2_s, $2p_x$ and $2p_y$ orbitals normally denoted as sp^2 hybridisation, while the remaining $2p_z$ orbital forms weaker interplanar bonding used to hold the planes above and below, that enable the planes to slide, thereby results into the semimetallic characteristics of graphite. Diamond structure on the other hand possesses carbon atom that is bonded to four nearest neighbours with 2_s, $2p_x$, $2p_y$ and $2p_z$ orbitals in tetrahegdral sp^3 configuration that makes diamond hard and strong. Hence while diamond is isotropic, graphite is anisotropic. Thus in similar vein, the structure of carbon nanotube is analogous to that of graphite, but the only difference is when the graphene sheets are rolled to produce tubes. As mentioned earlier, carbon nanotube consisting of one cylindrical graphite sheet is the single walled nanotube, but if several cylinders are nested, it is called multi walled nanotube, with interlayer space range of 0.34-0.36 nm, which is almost similar to the typical atomic spacing in graphite. C-C bonds length in carbon nanotube have been observed by Bonard et al. (Bonard et al. 2001) to be 0.14 nm, which is shorter than the C-C bonds in diamond, implying that carbon nanotube would be stronger than diamond.

Carbon nanotube structures are obtained by rolling up the carbon lattice in one of the symmetry axes or along any other direction that is different from the symmetric axis to produce a zigzag, armchair or chiral tubes, respectively, as shown in Fig. 10.4(b) (http://www.nas.nasa.gov/Groups/SciTech/nano/images/images.html).

As depicted in Fig. 10.4(a), the nanotube can be seen as a two-dimensional honeycomb lattice of a single layer of graphene, which was likened by Iijima in 1991 to the Bravais lattice vectors of the graphite sheet to map the circumference around a cylinder. A lattice vector, $\vec{B}$ may be defined with two primitive lattice vectors $\vec{R}_1$ and $\vec{R}_2$ (Fig. 10.4(a)) and a pair of indices (*m, n*), which are integers that denote the position of carbon atom on the two-dimensional hexagonal lattice, and the chiral vector of the nanotube (Mintmire 1997; Saito 2003) that is normal to the tube axis as,

$$\vec{B} = m\vec{R}_1 + n\vec{R}_2 \tag{10.1}$$

Thus rolling up of the sheet as the placement of the atom at $(m=0, n=0)$ on the atom at (m,n) would give nanotubes of different structures as,

$$(m,n) = \begin{cases} (m=n) & \text{Armchair or serpentine structure} \\ (m, n \neq 0 \text{ and } n \neq m) & \text{Chiral structure} \\ (m, n=0) & \text{Zigzag or saw tooth structure} \end{cases} \tag{10.2}$$

Due to space limitation, this work would not be able to discuss details of the subject, thus the reader is advised to consult more subject specific literatures (Saito et al. 2001; Mintmire and White 1997) for a better coverage. As a result of the C-C bond strength, extremely small diameter of the carbon atom and the π-electrons available from the sp^2 configuration of graphite, carbon nanotubes provide quite a number of remarkable mechanical and electronic properties as summarised by Hoenlein et al. presented in Table 10.2. The diameter, *d* of the nanotube can be determined from the indices *(m,n)* from Dresselhaus & Avouris (2003),

$$d = \frac{\sqrt{3}\ell_{C-C}(n^2 + nm + m^2)^{1/2}}{\pi} = \frac{B}{\pi} \tag{10.3}$$

Where, l_{C-C} is the C-C bond length (1.42Å), and B the length of the chiral vector $\vec{B}$ (Yamg, 2003). It is obvious therefore that the study on the helicity of nanotube has indicated a great deal of its application in microelectronics. Coupled with other potential applications, such as in gas storage, biomedicine etc, have spurred divergent researches, looking into scaleable and efficient production techniques as discussed in the next sections.

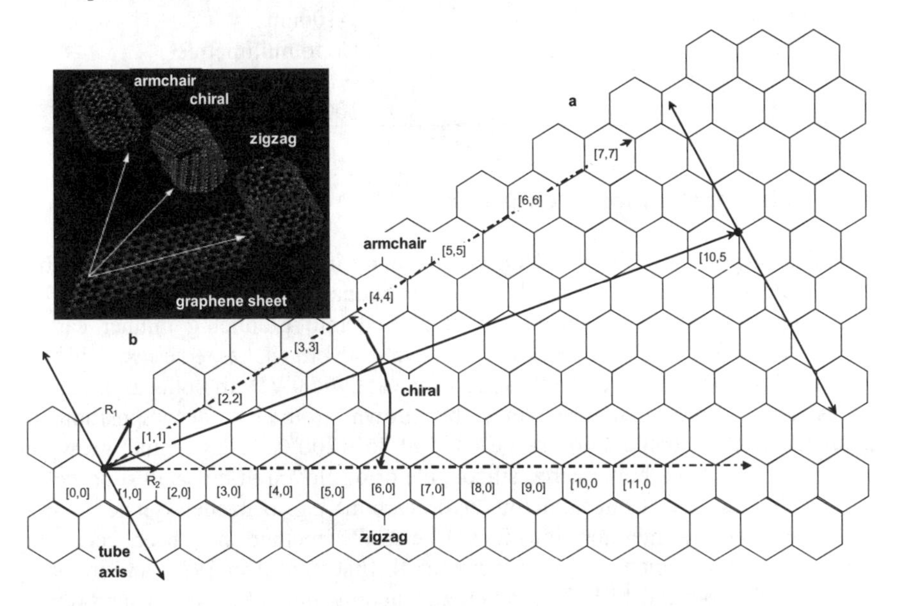

Fig. 10.4. (a) Carbon nanotube structures are obtained by rolling up carbon lattice to form a zigzag, armchair or chiral tube; atom at position (10,5) is projected on (0,0), while others are shown with dotted lines to project armchair and zigzag tubes respectively. (b) graphical representations of morphologies projected from the graphene sheet (Bonard 2001).

10.5 Production of Carbon Nanotube

The growth of carbon nanotubes during synthesis and production is believed to commence from the recombination of carbon atoms split by heat from its precursor. Due to the overwhelming interest, enormous progress is being made in the synthesis of carbon nanotubes. Although a number of newer production techniques are being invented, three main methods are the laser ablation, electric arc discharge and the chemical vapour deposition. The last is becoming very popular because of its potential for scale up production, hence it is discussed first in this section.

Table 10.2. Selected electrical and mechanical properties of carbon nanotubes (Hoenlein et al. 2003)

Characteristics	Measure
Electrical conductivity	Metallic or semiconducting
Electrical transport	Ballistic, no scattering
Energy gap (semiconducting)	E_g(eV) ≈*1/d* (nm)
Maximum current density	~10^{10} A/cm^2
Maximum strain	0.11% at 1 V
Thermal conductivity	6 kW/Km
Diameter	1-100nm
Length	Up to millimetres
Gravimetric surface	> 1500 m^2/g
E-modulus	1000 GPa, harder than steel

10.5.1 Chemical Vapour Deposition

Chemical vapour deposition (CVD) methodology in producing carbon nanotubes bears little or no difference with the conventional vapour grown carbon fibre technology. As the latter technology improved into producing thinner carbon fibres from several micrometers to less than 100 nm, it has emerged with the advent of carbon nanotubes synthesised from catalytic CVD. In both cases, carbon fibres and carbon nanotubes may be grown from the decomposition of hydrocarbons at temperature range of 500 to 1200^{0}C. They could grow on substrates such as carbon, quartz, silicon, etc or on floating fine catalyst particles, e.g. Fe, Ni, Co, etc. from numerous hydrocarbons e.g. benzene, xylene, natural gas, acetylene, to mention but few. The CVD method has been receiving continuous improvement since Yacaman et al. first used it in 1993, and in 1994, Ivanov et al. produced MWNTs. Patterned silicon wafers of porous n^+ and p type plain types were used to grow regular arrays of MWNTs as shown in Fig. 10.5(a) & (b) (Fan et al. 2000), and SMNTs were grown as low and high whiskers population density fastened to the filament fibrils for carbon fibre surface treatment shown in Fig. 10.5(c & d) (Iyuke et al. 2004). SWNTs also produced from floating catalyst CVD was presented earlier (Fakhru'l et al. 2003).

The schematic of typical catalytic chemical vapour deposition system, shown in Fig. 10.6 was equipped with a horizontal tubular furnace as the reactor. The tube was made of quartz tube, 30 mm in diameter and 1000 mm in length. Ferrocene and Benzene vapours acting as the catalyst (Fe) and carbon atom precursors respectively were transported either by argon, hydrogen or mixture of both into the reaction chamber, and decomposed into the respective ions of Fe and carbon atoms, resulting into carbon nanostructures. The growth of the nanostructures occurred in either the heating zone, before or after the heating zone, which is normally operated between 500^{0}C and 1150^{0}C for about 30 min. The flow of hydrogen gas was 200ml/min, while Argon gas was used to cool the reactor as reported elsewhere (Dana 2004).

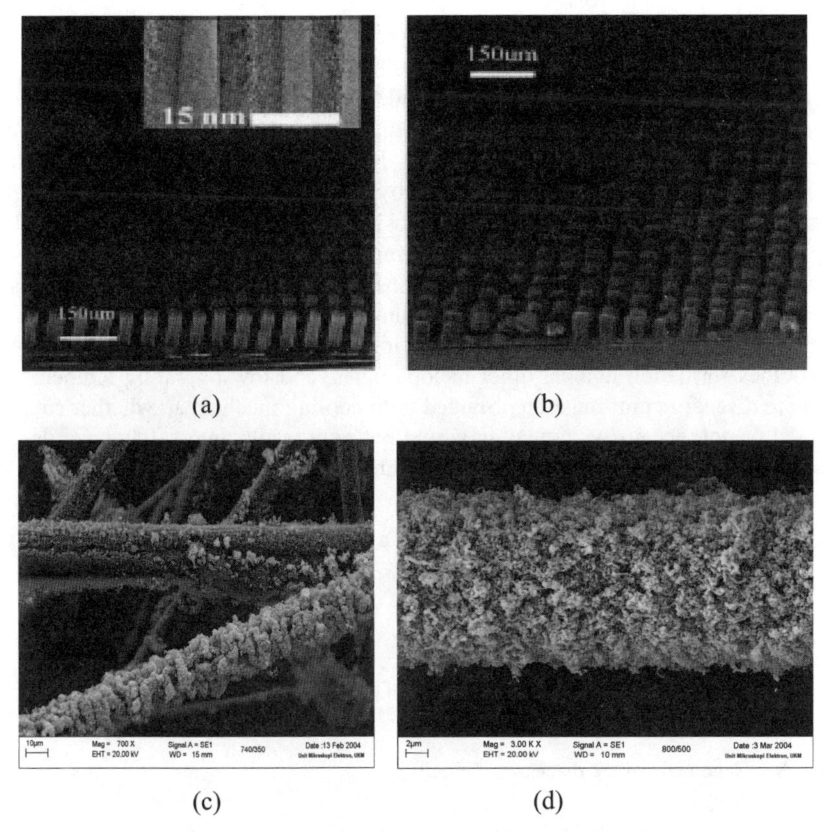

Fig. 10.5. MWNTs arrays grown on (a) n^+ type porous silicon and (b) p type plain silicon substrates (Fan et al., 2004); while whiskers of SWNTs were grown on the surfaceof carbon fibrils (C) sparsely growing whiskers, (d) highly populated whiskers (Iyuke et al. 2004)

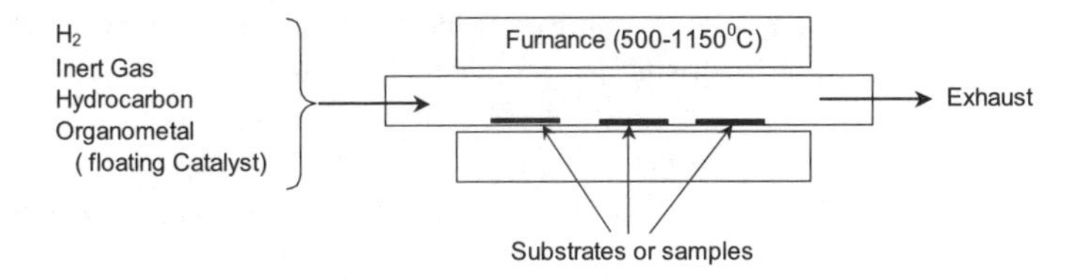

Fig. 10.6. Schematic of catalytic CVD operated either as floating catalyst or substrate catalyst

10.5.2 Arc Discharge

The arc discharge method produces a number of carbon nanostructures such as fullerenes, whiskers, soot and highly graphitised carbon nanotubes from high temperature-plasma that approaches 3700°C (Ebbesen 1997). The first ever produced nanotube was fabricated with the DC arc discharge method between two carbon electrodes, anode and the cathode in a noble gas (helium or argon) environment (Kroto et al. 1985). Schematic representation of a typical arc discharge unit is presented in Fig. 10.7. Relatively large scale yield of carbon nanotubes of about 75% was produced by Ebbesen and Ajayan with diameter between 2 to 30 nm and length 1 μm deposited on the cathode at 100 to 500 Torr He and about 18 V DC. It has conveniently been used to produce both SWNTs and MWNTs as revealed by Transmission Electron Microscope (TEM) analysis. Typical nanotubes deposition rate is around 1mm/min and the incorporation of transition metals such as Co, Ni or Fe into the electrodes as catalyst favours nanotubes formation against other nanoparticles, and low operating temperature. The arc discharge unit must be provided with cooling mechanism whether catalyst is used or not, because overheating would not only results into safety hazards, but also into coalescence of the nanotube structure.

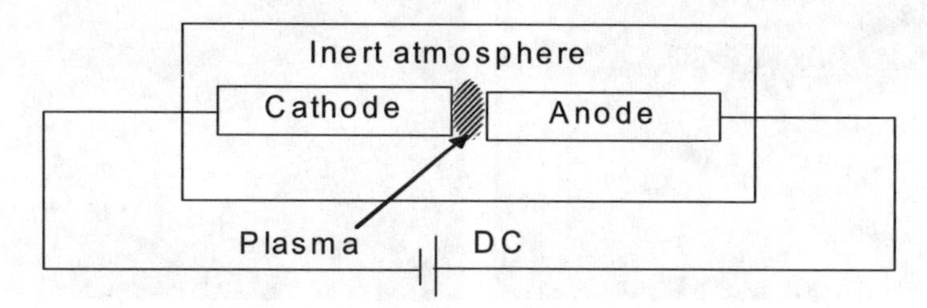

Fig. 10.7. Schematic of arc discharge method

10.5.3 Laser Ablation

Laser ablation technique involves the use of laser beam to vaporise a target of a mixture of graphite and metal catalyst, such as Co or Ni at temperature approximately 1200°C in a flow of controlled inert gas (argon) and pressure (Fig. 10.8), where the nanotube deposits are recovered at a water-cooled collector at much lower and convenient temperature. The method was used in early report (Thess et al 1996) to produce ropes of SWNTs with remarkably uniform narrow diameters ranging from 5-20 nm, and high yield with graphite conversion greater than 70-90%.

The bundles entangled into a 2-D triangular lattice via the van der Waals bonding to achieve lattice constant equal to 1.7 nm. The metal atom (catalyst) due to its high electronegetivity, deprived the growth of fullerenes and thus a selective growths of carbon nanotubes with open ends were obtained. Changing the reaction temperature can control the tubes diameters, while the growth conditions may be maintained over a higher volume and time, when two laser pulses are employed. However, by the virtue of relative operational complexity, the laser ablation method appears to be economically disadvantageous, which in effect hampers its scale up potentials as compared to the CVD method.

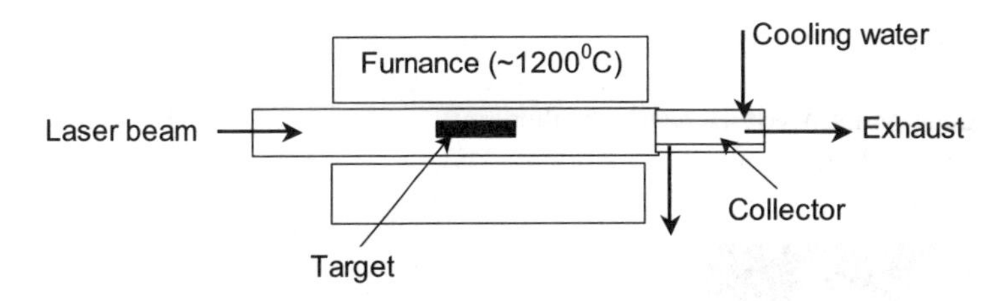

Fig. 10.8. Schematic of laser ablation method

10.5.4 Mechanisms of Growth

Several mechanistic pathways are being proposed for carbon nanotube growth depending on the production techniques. However, a huge ambiguity exists in an effort to adopt a generalised mechanism for products of the continuously changing synthetic methods, talk less of the type of nanotubes (SWNTs or MWNTs). Saito, Dresselhaus and co-worker referred to this scenario as different school of thoughts, where one assumes that the growth mechanism involves C_2 dimers absorption to close the tube with cap that is assisted by its pentagonal defects. Another assumption is that the nanotubes get opened up during the growth while carbon atoms are added at the open ends for length propagation. In their discussion (Saito et al. 2001) specifically on arc discharge method, the proposition is that the nanotube grows axially at the open ends, and if chiral, the additions of one hexagon continuously to the open ends occur as C_2 dimers are absorbed at the active dangling bond edge sites. But if the addition of carbon atoms is not according to the discussed order, the C_2 dimer absorption may result into capping the tube with pentagon, while if the carbon addition involves C_3 trimer, hexagon could merely be added. In SWNTs growth, armchair type would result from the absorption of a single C_2 dimer to add a hexagon, while the zigzag edge requires one C_3 trimer to initiate the growth. The thickening mechanism of the nanotube has been related to the corresponding vapour grown carbon fibre mechanism (Ebbesen 1997).

Moreover, in the study of the growth mechanism of carbon nanotubes synthesised by hot-filament CVD by Chen et al (2004) it was reported that after formation of metallic island from the catalyst, a stretching force elongated the metallic nanoparticles until it was broken into two parts, where one part stays at the base of the nanotube while the other remains at the tip. A relatively similar phenomenon can be induced from the work of Danna 2004 that the SWNT was nucleated on the metal catalyst nanoparticle, which broke into two parts, while one remains at the base, the other continue to propagates the elongation of the nanotube until it drops from the tip to terminate the growth process (Fig. 10.9). Fig. 10.9(a) presents with arrow 1, the non-growing open tube end, and arrow 2 points at the still growing tube tip holding the catalyst for growth propagation While Fig. 10.9(b) presents the tube growing phase and the open terminal end when the metal catalyst drops off. The maximum length of carbon nanotube obtained was in the average of 12 μm. As proposed earlier (Chen et al. 2004), during the growth period the portion of metal catalyst at the base of the tube keeps liquid state, and thus it is easily broken into parts, which are located in spots within the length of the tubes as seen with TEM images (Dana 2004).

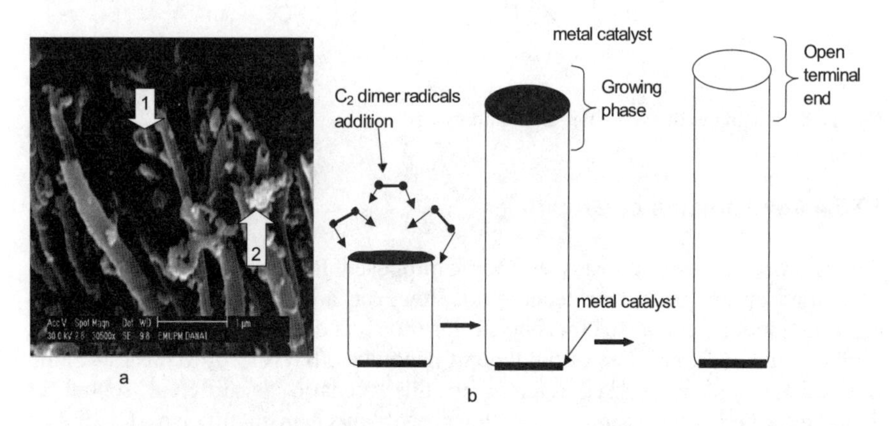

Fig. 10.9. Depiction of SWNTs growth mechanism (a) arrow 1 shows open-ended terminal end, while arrow 2 points at a growing end of a nanotube carrying the metal catalyst (Danna 2004) (b) presents the addition of C_2 dimers to the growing end containing the metal catalyst preceding the growth termination with an open end of the nanotube.

10.5.5 Purification of Carbon Nanotube

Most often than none, carbon nanotubes are found in the midst of other carbon materials, such as soot, other amorphous carbon, carbon nanoparticles, etc during production. In order to obtain good quality nanotubes for specific applications therefore, isolation, separation or post production purification process must be employed. Thus a post pre-treatment technique must be mastered to accompany any promising large scale production method. The basic purification methods that have been used with appreciative success are gas phase, liquid phase and intercalation techniques (Ebbessen 1997), while others that have been tried and still being developed are filtration, chromatography and centrifugation processes. The contaminants with large particle sizes or the ones agglomerated to the nanotubes can easily be treated in an ultrasonic bath to disengage the tubes from the particles (Ebbesen 1992). They could also be separated easily due to their weight differences from the nanotubes by dispersing the powder in a solvent, such as alcohol or ether and subsequently centrifuged. The gas phase treatment, commonly known as oxidation, involves treating the samples in air or oxygen at temperatures 650 to 750^{0}C. This has been successful because it is believed that the nanotubes survive the oxidation much longer than other carbon nanoparticles, because they are longer and are consumed from the tips inwards. The liquid phase oxidation has been used, dispersing the samples in standards oxidants e.g. H_2NO_3, H_2SO_4 or $KMnO_4$. The acidified solution of potassium permanganate has been found to be most successful in purifying and opening the end cap of the tubes (Ebbesen 1997). Intercalation post-production treatment of carbon nanotubes applies the difference in oxidation rates of graphite and intercalated graphite to remove impurities from open graphite (Ebbesen 1997; Ikazaki et al. 1994). Thus close ended nanotubes and other nanoparticles, which are not easily intercalated can be separated from graphitic flakes or other open graphitic structures. Typical treatment steps (Ikazaki et al. 1994) for an arc discharge nanotubes involved immersion of the sample in a $CuCl_2$-KCl / molten salt mixture at 400^0C for 168 hours. Excess salt was removed with deionised water after cooling and the cupper salt reduced to the metal in He/H_2 medium at 500^0C, heating at 10^0/min for an hour. Other purification techniques involve refluxing the sample in acid, centrifuged and followed with cross flow membrane filtration to remove catalyst particles and amorphous carbon. Colloidal suspension has been used as a function of particle size to purify the tube without physical damage and size exclusion chromatography was used for purification and size selection for MWNTs. Annealing at about 3000^0C has been employed to remove defects from carbon nanotube and improve graphitisation; and to distinguish the chiral types of SWNTs. Current can be passed between metal electrodes, whereby conducting nanotubes are burnt while the semiconducting ones remain purified (Mamalis et al. 2004).

10.6 Applications of Carbon Nanotubes

Carbon nanotube after its discovery has attracted overwhelming number of applications and newer applications are being sought due to the unusual properties it displays in the field of engineering materials. As a result of the newness of carbon nanotube technology, most of the potential applications discussed in this section are closely related to its obvious physical properties. Few of such selected potential applications include electrical transport, hydrogen adsorption, catalysis, artificial muscles, mechanical reinforcement, fuel cells, field emission and computers. It is expedient to mention however that this list does not in any capacity exhaust the uniqueness of carbon nanotubes.

10.6.1 Electrical Transport of Carbon Nanotubes for FET

Transport characteristics of carbon nanotubes have shown a tremendous impact in the microelectronic technology, especially in nanoscale electronic devices. The transport measurements of nanotubes follow different patterns depending on the nanotubes package as a sample such as a SWNT, a single rope or bundle of SWNTs, a single bundle of MWNTs or a single MWNT. The extremely thin nanotube is characterised as almost perfect one-dimensional conductor that behaves as single electron charging, resonant tunnelling at specific energy level, as well as proximity induced superconductivity at low temperatures. While the one-dimensional Luttinger liquid phenomenon is attained at relatively high temperatures, whereby tunnelling conductance into these tubes obey the power law suppression as a function of temperature and bias voltage. In an electrical resistance experiment involving a single MWNT at temperature as low as 20mK and zero magnetic field, a logarithmic conductance decrease was observed with decreasing temperature and saturation less than 0.3 K (Langer et al 1996). But two-dimensional weak localisation was observed when a positive magnetoresistance was obtained in the presence of a magnetic field that is perpendicular to the axis of the nanotube.

In the comprehensive review on carbon nanotube by Popov a number of electrical transports of carbon nanotubes results were presented (Fig. 10.10). Such as the one with bundles of SWNTs with bridging contacts separated by 200-500 nm where the I/V curve exhibits strong suppression of conductance near zero voltage for temperature less than 10K (Fig. 10.10(a)). The linear response conductance, G of the bundle as a function of the gate V_g showed peaks separated by regions (Fig. 10.9(b)) of very low conductance (Bockrath et al 1997). Conductance of a single SWNTs on a two-probe measurement showed metallic with resistance of tens of kiloohms, while a tunnel barrier of high resistance about 1 MΩ was observed (Tans et al. 1997). Here, the I/V curves showed a clear gap of about zero bias voltage and at higher voltages, the current increases stepwise as presented in Fig. 10.9(c). The gap change with V_g around zero bias voltage signifies Coulomb charging of the nanotube, which is a peculiar metallic characteristic of carbon nanotubes.

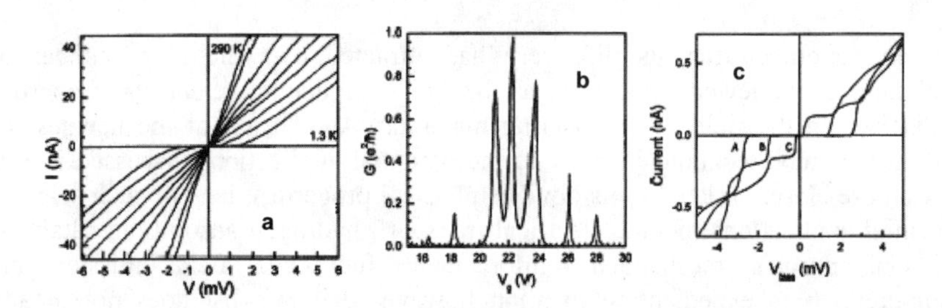

Fig. 10.10. Electrical transport characteristics (a) I/V curves of a 12 nm diameter bundle of 60 SWNTs (Bockrath et al 1997), (b) conductance, G against gate voltage V_g at temperature of 1.3 K with peak spacing of about 1.5 V, (c) I/V curves of a single SWNTs at varying gate voltage V_g (Tans et al. 1997).

In the carbon nanotube transport measurements by Tans et al (1998), a single semiconducting nanotube was contacted by two Pt electrodes on a SiO_2 layer over a Si substrate (Fig. 10.11(a)). It exhibited field - effect transistor characteristics, when voltage was applied to the gate electrode, and the nanotube was switched from a conducting to non-conducting acting as an insulator, which was also tested at room temperature. This observation implies great application potentials in the microelectronics of transistors. Fig. 10.11(b) (Tans et al. 1998) presents typical I/V curves of the nanotube FET, where varying the gate voltage from positive to negative, the curves changed from large gap of insulating non-linear to strongly metallic characteristics. The phenomenon has been explained with the current conventional semi-classical band-bending models (Popov 2004).

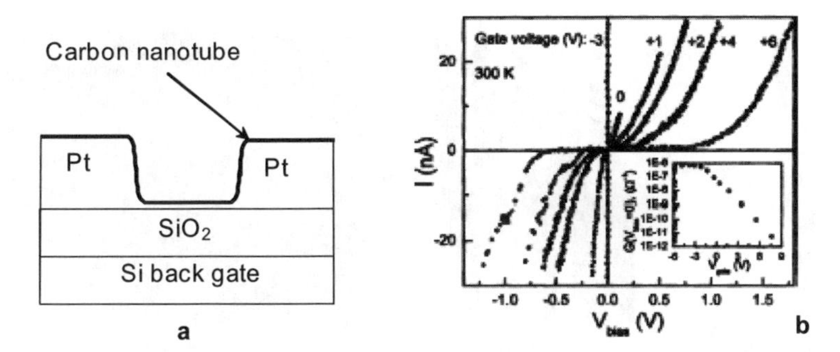

Fig. 10.11. Carbon nanotube base fields effect transistor (FET) (a) schematic of the device (b) I/V curves of the FET for different values of the gate voltage V_g (Tans et al. 1998)

10.6.2 Computers

An obvious objective in modern computer development is the increase of number of switches while reducing size and increasing storage capacity. Thus the extremely thin, light weight nanowire of carbon nanotube, which has been displayed to exhibit metallic, semiconducting and insulating characteristics, becomes excellent materials for smaller switches and smarter computer chips. Its superiority over the conventional metal wire such as copper is reflected in its low heat resistance even at diameter of 2 nm, whereby large currents can be carried at different interconnections without heating or melting. Carbon nanotubes possess very high thermal conductivity, which make them good heat sinks, making it possible to transfer heat rapidly away from the chips. Similar to Tans et al (1998) work discussed above, it has been observed in a carbon nanotube based gold electrodes gate FET device (Poole and Owens 2003) that a small voltage applied to the gate changed the conductivity of the nanotube up to a factor greater than 10^6. This magnitude is comparable to the conventional silicon FETs, and the switching time of the device was estimated to be very fast at the clock speeds of a tetrahertz, almost 10^4 times faster than the present processors. One of the several models proposed for the physical description of current transport in carbon nanotube (CNT) based FET (CNTFET) is the vertical transistors concept that allows higher packing densities because the source and drain areas are arranged on top of one another. Thus a three-dimensional structure where the active devices are not bound to the surface of the usual mono-crystalline Si wafer was simulated from a 1 nm diameter, 10 nm long SWNT with a coaxial gate and a gate dielectric with 1 nm oxide are presented in Fig. 10.12 (Hoenlein et al. 2003). In their report, recently published works of two p-type CNTFETs were compared with three silicon MOS-field-effect transistors (Si-MOSFET) as shown in Table 10.3. The parameters of the CNTFETs were much better than the best ones for the MOSFET devices. This is thus a clear indication of the wonderful performance of the future CNTs based computers and other microelectronics.

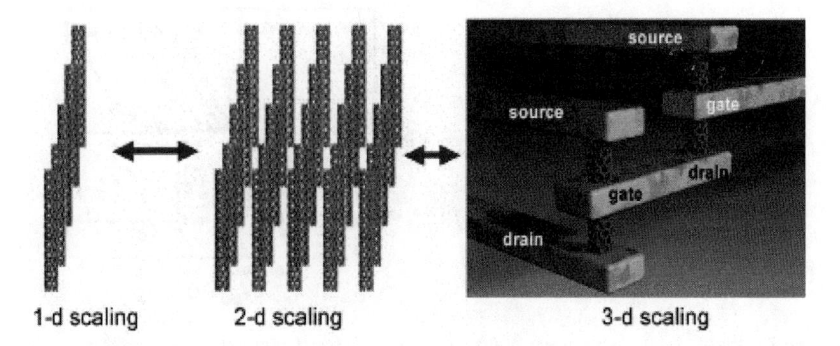

Fig. 10.12. Three-dimensional modelling of vertical CNTFETs (Hoenlein et al. 2003) for better performed chips and switches

Table 10.3. Performance comparison of selected CNTFETs with Si-MOSFETs

Parameters	p-CNTFET 1.4×10^{-3} nm (1V)	p-CNTFET 3×10^{-3} nm (1.2 V)	MOSFET 1×10^{-4} nm (1.5 V)	MOSFET 10 nm (1.2 V)	MOSFET 14 nm (0.9 V)
Drive current Ids (mA/μm)	2.99	3.5	1.04 nFET 0.46 pFET	0.450 nFET 0.360 pFET	0.215 pFET
Transconductance (μS/Am)	6666	6000	1000 nFET 460 pFET	500 nFET 450 pFET	360 pFET
S (mV/dec)	80	70	90	125 101	71
On-resistance (Ω/Am)	360	342	1442 nFET 3260 pFET	2653 nFET 3333 pFET	4186 pFET
Gate-length (nm)	1400	2000	130	10	14
Normalised gate-oxide (1/nm)	80/1 = 80	25/8 = 3.12	4/2 = 2	4/1.7 = 2.35	4/1.2 = 3.33
Mobility($cm^2/(Vs)$)	1500	3000	-	-	-
Ioff (nA/μm)	-	1	3	10	100

10.6.3 CNT Nanodevices for Biomedical Application

The numerous properties of carbon nanotubes (CNTs) as metals, semiconductors, electron field emitters and electromechanical actuators (often known as artificial muscles) have made them extremely valuable as well as provide a great future for biomedical applications such as microsurgical and diagnostic devices, artificial limbs, implants like artificial ocular muscles, hearts, etc. This section therefore discusses briefly future applications of CNTs in artificial muscles and X-ray generation by CNT-based field emission in the biomedical laboratory.

10.6.4 X-Ray Equipment

X-ray machines generate exceedingly high frequency, short wavelength, high-energy electromagnetic waves that penetrate the body during medical diagnostic and therapeutic practices. X-ray equipment has been in use for several years to obtain tissue photographs of tumour, skeletal fractures, deformations, etc. Although they have been part of quality and modern medicine, however, they could be extremely dangerous whether in a short or long term exposures thereby causing nausea, vomiting, dizziness, sterility, burns, genetic mutations, cancers, and death if used incorrectly or in excess (Carr and Brown 1998). The conventional thermionic cathodes x-ray tube has a metal filament that is resistively heated to temperature over 1000^0C to emit electrons, which is in turn are targeted and bombarded on a metal anode to emit x-rays. The high temperature requirement in thermionic cathode is a limitation that is not experienced with the field emission mechanism because electrons are emitted at room temperature and controllable voltage. Nevertheless, field emission faces another inherent bottlenecks since most diagnostic applications require tube current in the order of 10-100 mA and 30-150 KV operating voltage, which are almost impossible to attain for the field emission x-ray device. In an effort to circumvent these problems, Cheng, Zhou and co-workers (Yue et al. 2002) developed a CNT-based x-ray tube that could emit sufficient x-ray flux for diagnostic imaging and photography (Fig. 10.13). The device, presented in Fig. 10.13(a) consisted of a field emission metal cathode coated with SWNTs, and the gate electrode was a tungsten mesh that was 50-200μm distance from the cathode.

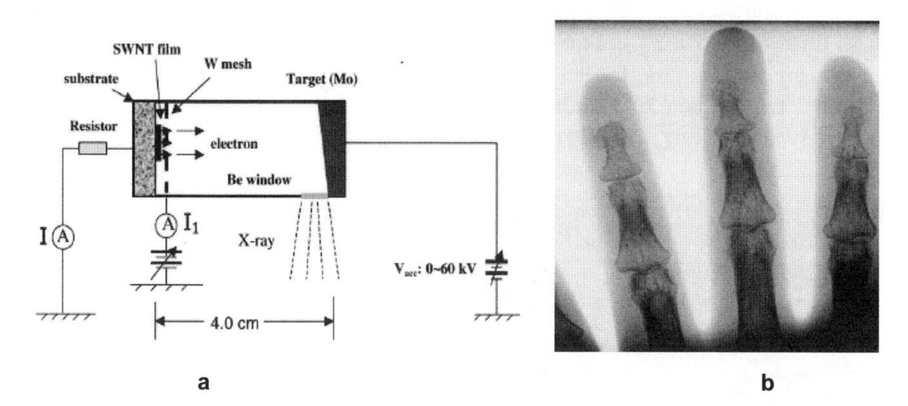

Fig. 10.13. X-ray device by SMNTs-based field emission experiments at University of North Carolina (a) schematic of the CNT-based field emission x-ray emitter (Yue et al. 2002), (b) X-ray image of humanoid fingers (Cheng Zhou 2003)

Applying a relatively low voltage between the gate and the cathode produced electrons from the cathode, which were accelerated and bombarded on a copper target to produce x-ray beam through the Be window in a high voltage applied between the gate and the anode. They observed that 30mA emission current from a relatively small carbon nanotube cathode, and release x-ray wave forms that could be programmable pulse and repetitive in rate. Interestingly, the x-ray flux produced was sufficient to image humanoid fingers shown in Fig. 10.13(b) (Yue et al. 2002). Some obvious advantages of the CNT-based x-ray device proposed over the usual thermionic x-ray tube could include:

- Prolonged life span of the x-ray tube,
- Significant size reduction of the x-ray device for industrial or medical applications,
- Focus electron beam with programmable pulse width and repetition rate,
- Low temperature cathode tube.

10.6.5 CNTs for Nanomechanic Actuator and Artificial Muscles

Single walled carbon nanotubes have been known to deform when electrically charged. This implies that CNTs possess the characteristics of an actuator, which convert electrical energy to mechanical energy, or vice versa. In 1999, Baughman et al (1999) used "bucky" paper made as sheets of bundles of SWNTs, otherwise referred to as "artificial" muscles to demonstrate CNTs characteristics as actuators (Vohrer et al. 2004). The explanation given for the actuation properties was a double layer charge injection that produces dimensional changes in covalently bonded direction for conjugated CNTs materials (Sun et al. 2002). The nanostructure of the artificial muscles observed with scanning electron microscope (SEM) is shown in Fig. 10.14(a), where the CNTs are tightly packed into bundles with most of their axes in horizontal planes, and consisting 3 by 20 mm strips of 25-50 μm in thickness. In demonstrating the actuation properties of the artificial muscles, two of the CNTs strips were bonded together with a double-stick Scotch tape in to sheets (Fig. 10.14(b)).

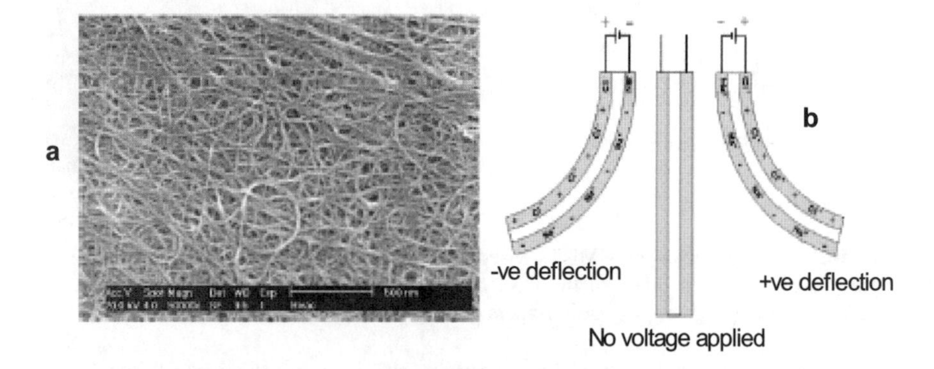

Fig. 10.14. Actuation properties of artificial muscles (bucky paper) (a) SEM image showing the nanostructure, and (b) actuation characteristics of the artificial muscles displaying positive or negative deflection when voltage was applied (Baughman et al. 1999)

The sheets were placed in a 1M sodium-chloride electrolyte solution. Upon applying few volts as shown in Fig. 10.13(b), about 1 cm deflection was obtained and changing the polarity of the voltage could reverse the phenomenon. Because the response of the device is dependent on the expansion of opposite electrodes and oscillation of cantilever could be produced when an AC current is applied, making it a bimorph cantilever actuator (Poole and Owens 2003). Moreover, since individual bundle of CNTs would behave in the same pattern, then employing only three filaments would create a truly miniaturised or nanosized actuators that could revolutionise the microelectromechanical systems (MEMSs) and nanoelectromechanical systems (NEMSs) technologies, especially in biomedical applications where nanotechnology would promise a great future. A number of CNT-based nanodevices have been proposed in the literature, with actuation effects such as nanotube tweezers as CNT scanning tunnelling microscope (STM) tip, attached to the cantilever arm of an atomic force microscope (AFM), simulated nanogears, etc.

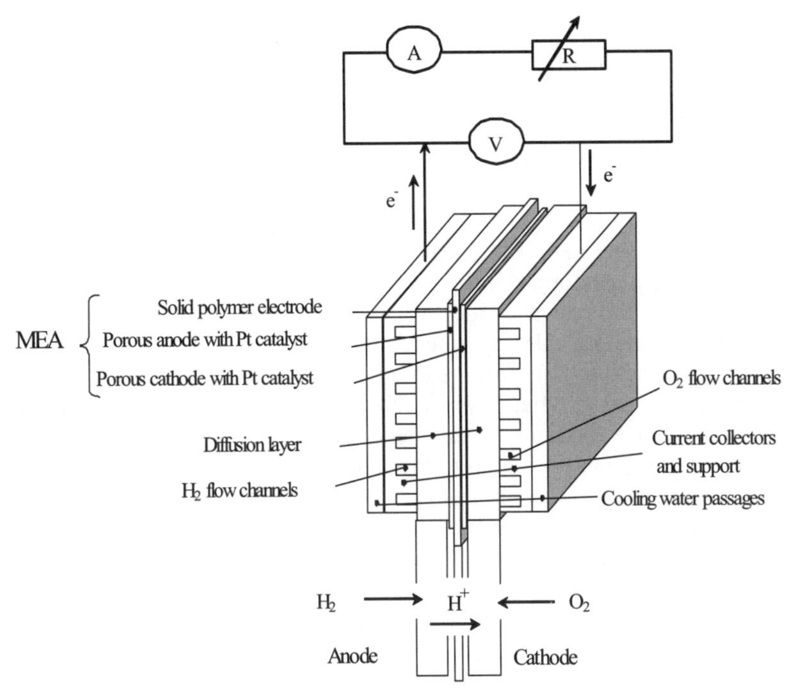

Fig. 10.15. Schematic diagram of MEA and single cell testing apparatus (Iyuke et al. 2003)

10.6.6 Fuel Cells

In the year 2004, especially at the last quarter of the year, motorists and other forms of energy generators from fossil fuel around the world whimpered as prices of petroleum products especially petrol soared to unprecedented levels. Consumers at homes stared in disbelief as gas and electricity prices increased with no glimpse of immediate end. In spite of these problems coupled with environmental pollution problems such as global warming caused by the exhaust releases from the internal combustion engines, the world has no option other than to look into a safe and environmental friendly hydrogen fuel cell technology for a lasting solution. Thus the overwhelming advantages of proton exchange membrane (PEM) fuel cell amongst its four competing potential commercialisation counterparts (alkaline, phosphoric acid, molten carbonate and solid oxide fuel cells) have aroused concerns in research and development in energy and environment. PEM fuel cell (Fig. 10.15) is one of the currently known five types of fuel cell that convert chemical energy of fuels (e.g. H_2 and O_2) directly into electricity, heat and water. PEM fuel cell is commonly claimed as the most promising fuel cell due to its portable power and residential applications. This section therefore discusses the roles of carbon nanotubes in acquiring better performed and cost saving in size reduction of the membrane electrode assembly (MEA) and hydrogen storage, which have remained teething problems in fuel cells commercialisation and production efforts.

10.6.7 Membrane Electrode Assembly

One major component in the PEM fuel cell is the MEA often referred to as the heart of the PEM fuel cell. The MEA consists of a sheet of proton conducting polymer electrolyte membrane with two Pt/C electrodes, which are the anode and cathode bonded to the opposite sides of the membrane sheet. The arrangement is then compressed on both sides by grooved bipolar plates or grooved end plates in the case of single cell, to transport the H_2 and O_2/air respectively to the electrodes, as shown in Fig. 10.15 (Iyuke et al. 2003). Perfluorinated sulfonated cation exchanger membrane (Nafion^{TM}) has been commonly used in MEA for PEM fuel cell to provide mechanical support, insulation and as solid electrolyte for H^+ transport. These qualities have been improved further in the study of Marken and co-workers where the electrochemical behaviour of MWNTs/Nafion^{TM} modified electrodes in 5 mM $Fe(CN)_6^{4-}$ solution and 0.1 MKCl. The cyclic voltammograms obtained for Nafion^{TM} alone and MWNTs/Nafion^{TM} blends are presented in Fig. 10.16(a)&(b), respectively showing the oxidation and reduction peaks observed are at +0.25 and +0.19 V having 63mV inter peaks separation at scanning rate of 0.1 V/s with the modified MWNT/Nafion^{TM} electrode. In comparing Fig. 10.16(a & b), higher peak current responses for oxidation and reduction and inter peaks separation of 290 mV for the Nafion^{TM} coated glassy carbon electrode due to the incorporated MWNTs were observed (Tsai et al. 2004). It was thus explained that the higher activity of the MWNTs/Nafion^{TM} modified electrode was due to the higher surface electroactivity provided by the MWNTs as compared to the electrode coated with only Nafion^{TM}. These observed improved qualities therefore justify better workability, life span, chemical, pressure and heat stability of the MWNTs/Nafion^{TM} coated electrode as membrane for PEM fuel cell application. On the other hand, Pt/graphite electrode has been the usual electrocatalytic electrode in PEM fuel cell (Iyuke et al 2003). However, in the electrochemical and electrocatalytic studies in a PEM fuel cell (Tang et al. 2004) of a finely and highly dispersed Pt on carbon nanotubes (Pt/CNT) electrode as compared to Pt/graphite electrode showed a remarkable superiority over the latter. The insert of Fig. 10.16(c) presents the TEM image of Pt dispersed on the CNTs. Line I represents typical cyclic voltammogram of CNT electrode, which is the background current with characterised capacity current, and in this case larger than the graphite electrode, Line II. Thus these results are considered breakthrough in obtaining elegant MEA, which would result into miniaturised PEM fuel cell stack.

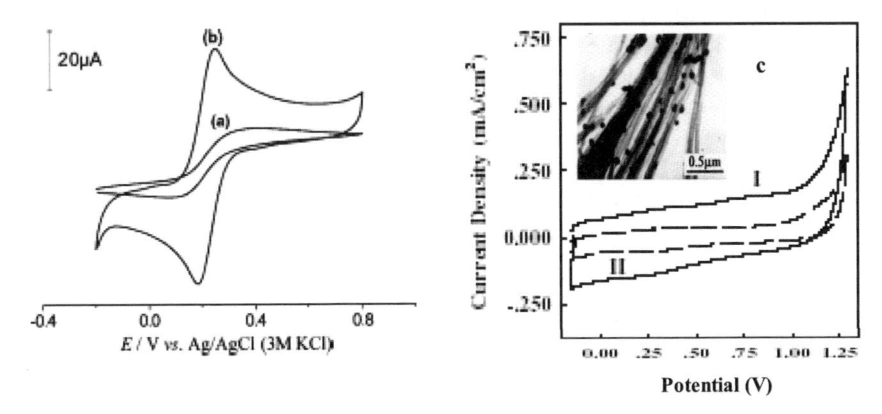

Fig. 10.16. Membrane electrode assembly (MEA) components and performance comparison (a) cyclic voltammetric response for modified Nafion™, (b) cyclic voltammetric response for MWNTs/ Nafion™ modified glassy carbon electrode, and (Tsai et al. 2004), (c) linear voltammograms of Pt/CNT and Pt/graphite electrodes in 50 cm^3/min air flow bubbled H_2SO_4 aqueous solution: Line I, CNT electrode; Line II, graphite electrode; Insert is the TEM image showing Pt catalyst dispersion on well aligned CNTs (Tang et al. 2004)

10.6.8 Mechanical and Electrical Reinforcement of Bipolar Plates with CNTs

As mentioned earlier, cost is one of the key issues hampering or delaying full bloom of PEM fuel cell application in the automobile. The high cost is not only from the MEA but also the bipolar plates. Hitherto, these plates are machined from Poco™ graphite or carbon/polymer composites, which are brittle during machining of the flow channels (Fig. 10.15). However, efforts to research into other potential materials such as metal-based bipolar plates, carbon-filled polymers or carbon-carbon composites have not yielded adequate success because cost remains high and performance is still unsatisfactory. Obviously the metal-based plates would succumb to corrosion in fuel cell stack environment, and the cations that will be released will not only enhance membrane resistance but also lead to the electrode catalyst poisoning. As for polymer filled with carbon, the carbon load would normally be greater than 50% by volume to attain good electrical conductivity for the bipolar plate (Barbir et al. 1999). Unfortunately, injection moulding proposed for mass and economic manufacturing of bipolar plates will be constrained by the high carbon concentrations due to difficulty in processing. Meanwhile, improvising with compression moulding will trade off the benefits expected because it is slow, as it requires cooling and products discharging (Barbir et al. 1999). Furthermore, high carbon concentration in the polymer would substantially decrease the strength and ductility and also impose adverse effects on the tensile strengths of the composites. However, Wu and Shaw (2004a) indicated some rays of hope from the problems when they demonstrated triple continuous carbon-filled polymer blends in injection moulding of polyethylene terephthalate (PET)/polyvinylidine fluoride (PVDF) blended with CNTs. They observed that the blend exhibited 2500%, 36% and 320% improvement in electrical conductivity, tensile strength and elongation, respectively over PET blended with CNTs at the same carbon concentration. In an increased effort to affirm the earlier claims with similar blends, Wu and Shaw (2004b)obtained improved mechanical and electrical properties for CNT-filled polymers over ordinary polymer blends as presented in Table 10.4.

Table 10.4. Tensile strength, electrical conductivity and resistivity of polymer and CNT-polymer blends (Wu and Shaw 2004b) for PEM fuel cell bipolar plates

Parameter	PET	PVDF	CNT-filled PET/PVDF (6 vol.% CNT)	CNT-filled PET (6 vol.% CNT)	% improvement in CNT-filled PET/PVDF over CNT-filled PET
Elongation at the rate of break 9%)	2.2	1400	5.1	1.2	325
Tensile stress at the rate of yield (MPa)	34	32	-	25	-
Tensile stress at the rate of break (MPa)	34	54	34	25	36
Conductivity (S/cm)			0.059	0.0023	
Resistivity, ρ (cm)			16.95	430	

10.6.9 Hydrogen Storage in CNTs

The world today is in search for a novel material to store hydrogen as fuel for a clean, renewable and environmental friendly technology, such as fuel cell, automotive, stationary power generation etc. Quite a number of publications have proposed materials ranging from metals and alloys in their hydride forms, carbons to cryogenic methods for hydrogen adsorption. Amongst these prospective candidates, carbon materials such as carbon nanotube and graphite nanofibre (GNF) are receiving the most attention due to their large capacity to adsorb hydrogen, though still controversial. The United States department of Energy has set targets of 6.5 wt% and 62 kg H_2/m^3 for on-board hydrogen storage at ambient temperature for fuel cell powered vehicles. It has been envisaged that a compact passenger vehicle powered by fuel cell would require 4 kg H_2 for a 400 km driving range (Yang 2003). Since the first report on hydrogen storage in SWNTs by Dillon et al (1997) quite a number of works have been reported in the literature. Thus in conjunction with the review by Poole Jr and Owens (2003) subsequent experimental results on H_2 storage in SWNTS, MWNTs, GNFs are presented in chronological order in Table 10.5. In general the vast evidences available affirm the progressive developments of hydrogen storage for PEM fuel cell that will alleviate the petroleum reserves and price problems as well as the threatening global environmental issues.

Table 10.5. Chronology of hydrogen storage reports in carbon nanotubes (SWNT and MWNT) and graphite nanofibres (GNF)

Material	H_2 wt % storage	Conditions T(K), P(MPa)	Author & Reference
SWNT	~5-10	133, 0.04	Dillon et al. 1997
MWNT	11.26	300, 9	Chambers et al. 1998
GNF	67.55	300, 12	Chambers et al. 1998
SWNT	~8	80, 11.2	Ye et al. 1999
SWNT	4.2	300, 10.1	Liu et al. 1999
GNF	35	300, 11	Park et al. 1999
Li/MWNT	20	473-673. 0.1	Chen et al. 1999
K/MWNT	14	300, 0.1	Chen et al. 1999
Li/MWNT	2.5	473-673, 0.1	Yang et al 2000
K/MWNT	1.8	300, 0.1	Yang et al 2000
K/MWNT	1.3	300, 0.1	Pinkerton 2000
MWNT	3.98	300, 10	Li et al. 2001
GNF	>0.7	296, 11	Tibbetts et al. 2001
SWNT	0.05	296, 3.59	Tibbetts et al 2001
GNF	0.7	295, 10.5	Poirier et al. 2001
SWNT	1.5	300, 0.08	Hirscher et al. 2001
GNF	~15	300, 12	Gupta & Srivastava 2001
GNF	6.5	300, 12	Browning et al 2002
MWNT	0.65	300-373, 0.1	Lueking & Yang 2002
MWNT	3.6	298, 7	Lueking & Yang 2002
SWNT	1.0	253, 6	Luxembourg et al 2004

10.7 References

Ajayan et al. (1993) US Patent no. 5457343

Barbir F, Braun J, Neutzler J (1999) Properties of molded graphite bipolar plates for PEM fuel cell stacks, J. New Mater. Electrochem. Systems 2:197-201

Baughman RH, Cui CX, Zakhidov AA, Iqbal Z, Spinks GM, Wallace GG, Mazzoldi A, Rossi DD, Rinzler AG, Jaschinski O, Roth S, Kertesz M (1999) Carbon nanotube actuator, Science 284:1340-1346

Bockrath M, Cobden DH, McEuen PL, Chopra NG, Zettle A, Thess A, Smalley RE (1997) Single-electron transport in ropes of carbon nanotubes, Science 275:1922-1927
Bonard MJ, Kind H, Stockli T, Nilsson OL (2001) Field emission from carbon nanotubes: The first five years, Solid-State Electronics 45:893-896
Browning DJ, Gerrard ML, Lakeman JB, Mellor IM, Mortimer RJ, Turpin MC (2002) Nano Lett. 2:201-207
Carr JJ, Brown JM (1998) Introduction to biomedical equipment technology. 3rd ed. Prentice Hall, pp 605
Chambers A, Park C, Terry R, Baker K, Rodriguez MN (1998) Hydrogen storage in graphite nanofibres, J. Phy Chem. B 102:4253
Chen P, Wu X, Lin J, Tan KL (1999) Science 285: 91-95
Chen X, Xu RWJ, Yu D (2004) TEM investigation of the growth mechanism of carbon nanotubes synthesized by hot-filament chemical vapor deposition, Micron 35:455
Cheng Y, Zhou O (2003) Electron field emission from carbon nanotubes, C. R. Physique 4:1021
Danna ABM (2004) Characterization and adsorption studies of carbon nanotubes. MS thesis, Universiti Putra Malaysia
Debe KM (1998) US Patent 5726524
Dillon AC, Jones KM, Bekkedahl TA, Kiang CH, Bethune DS, Heben MJ (1997) Storage of hydrogen in single-walled carbon anotubes. Nature 386:377
Doris B, Ieong M, Kanarsky T, Zhang Y, Roy RA, Dokumaci O, Ren Z, Jamin F, Shi L, Natzle W, Huang H, Mezzapelle J, Mocuta A, Womack S, Gribelyuk M, Jones EC, Miller R, Wong HP, Haensch W (2002) IEDM Tech. Dig. 1:267
Dresselhaus G, Avouris P (2003) Carbon Nanotubes. In: Dresselhaus MS, Dresselhaus G, Avouris P (eds.) CNT, Springer, pp 1-9
Ebbesen TW, Ajayan PM (1992) Large-scale synthesis of carbon nanotubes, Natur 358:220
Ebbesen T, Ajayan PM, Hiura H (1997) US Patent 5641466
Ebbesen TW (1997) Production and Purification of Carbon Nanotube. In: Ebbesen W. T. (ed) Carbon Nanotubes: Preparation and Properties. CRC Press, New York, pp 139
Ebbesen TW (1994) Carbon Nanotubes, Annu. Rev. Mater. Sci. 24:235
Fakhru'l A, Iyuke SE Ali MA, Danna ABM Al-Khatib MF (2003) Carbon Nanoporous Balls Adsorb Methane better than Activated Carbon at Room Temperature and Pressure, Asia Pacific Nanotechnology Forum News Journal 2:1-6
Fan S, Liang W, Dang H, Franklin N, Tombler T, Chapline M (2000) Carbon Nanotube arrays on Silicon Substrates and their Possible Application. Phys E: Low-Dimensional Syst Nanostructures 8:179
Ghani T, Ahmed S, Aminzadeh P, Bielefeld J, Charvat P, Chu C, Harper M, Jacob P, Jan C, Kavalieros J, Kenyon C, Nagisetty R, Packan P, Sebastian J, Taylor M, Tsai J, Tyagi S, Yang S, Bohr M, (1999) IEDM Tech. Dig. 415
Green MLH, Tsang SC (1994) US Patent 5346683
Gupta BK, Srivastava ON (2001) Int. J. Hydrogen Energy 26: 857
Gupta RK, Dwivedy I (2004) International Patenting Activity in the Field of Carbon Nanotubes, Current Applied Physics, in press
Han J, Globus A, Jaffe R, Deardorff G (1997) Molecular dynamics simulations of carbon nanotube-based gears, Nanotechnology 8:95
Hirscher M, Mecher M, Haluska M, Weglikowska UD, Quintel A, Duesberg GS, Choi YM, Downes P, Hulman M, Roth S, Stepanek I, Bernier P (2001) Appl. Phys. A 72:129
Hiura H, Ebbesen T (1997) US Patent 5698175
Hoenlein W, Kreupl F, Duesberg GS, Graham AP, Liebau M, Seidel R, Unger E (2003) Carbon Nanotubes for Microelectronics: status and furure prospects, Materials Science & Engineering C 23:663

http://www.nas.nasa.gov/Groups/SciTech/nano/images/images.html.
Iijima S (1991) Helical Microtubules of Graphite Carbon. Nature 354:56
Ikazaki et al. (1997) US Patent 5695734
Ikazaki F, Oshima S, Uchida K, Kuriki Y, Hayakawa H, Yumura M, Takahashi K, Tojima K (1994) Chemical purification of carbon nanotubes by use of graphite intercalation compounds, Carbon 32:1539
Ivanov V, Nagy JB, Lambin P, Lucas A, Zhang XB, Zhang XF, Bernaerts D, Tendeloo GV, Amelinckx S, Landuyt JV (1994) The study of Carbon Nanotubules Produced by Catalytic Method, Chem. Phys. Lett. 223:329
Iyuke SE, Yasin FM, Fakhru'l A, Shamsudin S, Mohamad AB, Daud WRW (2000) FCCVD-coating of multifilament carbon Fibres by whiskerization with Fe catalyst. Proc. of the ICCE 11, South Carolina, p 297
Iyuke SE (2001) Hydrogen storage in the interplanar layers of graphite nanofibres synthessed from palm kernel shell. Proc. of the Chemistry and Technology Conference, PIPOC2001, Malaysia, pp 203-209
Iyuke SE, Mohamad AB, Kadhum AAH, Daud WRW, Chebbi R (2003) Improved membrane and electrode assemblies for proton exchange membrane fuel cells, J of Power Sources 114:195-200
Iyuke SE, Ahmadun FR, Chuah TG, Danna ABM (2004) Methane storage as solid- and liquid-like forms in and on carbon nanotubes, 4th Int. Mesostructured Materials Symposim, IMMS, pp 291-292
Javay A, Kim H, Brink M, Wang O, Ural A, Guo J, McIntyre P, McEuen P, Lundstrom M, Dai H (2000) Nat. Mater 1:241
Jin S, Kochanski PG, Zhu W (1999, 2001) US Patent 6283812; US Patent 6250984
Kroto HW, Heath JR, O'Brien SE, Curl RF, Smalley RE (1985) C_{60}: buckminsterfullerene, Nature 318:162
Langer L, Bayot V, Grivei E, Issi JP, Heremans JP, Olk CH, Stockman L, Haesendonck CV, Bruynseraede Y (1996) Quantum transport in a multi-walled carbon nanotube, Phys. Rev. Lett. 76:479
Li X, Zhu HW, Ci L, Xu C, Mao ZQ, Wei BQ, Liang J, Wu D (2001) Effects of structure and surface properties on carbon nanotubes' hydrogen storage characteristic, Carbon 39:2077
Lieber MC, Charles M, Wong SS, Woolley TA, Joselevich E (2000) US Patent 6159742
Lueking A, Yang RT (2002) Hydrogen spillover from a metal oxide catalyst onto carbon nanotubes: Implications for hydrogen storage, J. Catal. 206:165
Liu C, Fan YY, Liu M, Cong TH, Cheng MH, Dresselhaus MS (1999) Hydrogen storage in single-walled carbon nanotubes at room temperature, Science 286:1127
Luxembourg D, Flamant G, Guillot A, Laplaze D (2004) Hydrogen storage in solar produced single-walled carbon nanotubes, Materials Science and Engineering B 108:114
Mamalis AG, Vogtlander AOG, Markopoulos A (2004) Nanotechnology and nanostructured materials: trends in carbon nanotubes, Precision Engineering 28:16-20
Mancevski V (2000) US Patent 6146227
Massey JR, Martin TM, Dong L, Lu M, Fischer A, Jameison F, Liang P, Hoch R, Leland KJ (1999) US Patent 5866434
Mintmire JW, White CT (1997) Properties: Theoretical Predictions, In: Ebbesen WT (ed) Carbon nanotubes: Preparation and properties. CRC Press, New York, pp 191
Nakamoto M (2000) US Patent 6097138
Nakayama Y (2001) Scanning probe microscopy installed with nanotube probes and nanotube tweezers, Ultramicroscopy 91:49-52
Nolan EP, Cutler AH, Lynch DG (1998) US Patent 5780101

Ohshima S, Yumura M, Kuriki Y, Uchida K, Ikazaki F (1996) US Patent 5482601
Olk HC (1998) US Patent 5753088
Park C, Anderson PE, Chambers A, Tan CD, Hidalgo R, Rodriguez NM (1999) J. Phys. Chem. B. 103:10572
Park JH, Kim JN, Cho SH, Kim JD, Yang RT (1998) Chem. Eng. Sci. 3:3951
Pinkerton FE, Wicke BG, Olk CH, Tibbetts GG, Meisner GP, Meyer MS, Herbst JF (2000) J. Phys. Chem. B 104:9460
Poirier E, Chahine R, Bosc TK (2001) Int. J. Hydrogen Energy, 26; 831
Poole PC, Owens JF (2003) Introduction to nanotechnology. John Wiley & Sons, pp 114-132
Popov VN (2004) Carbon nanotubes: properties and application, Materials Science and Engineering R 43:61
Rocco CM (2001) National Science Foundation, Official who oversees the nanotechnology initiative. Scientific American 285:3
Rodriguez et al. (1997) US Patent 5653951
Rodriguez MN, Nelly M, Baker RTK (1997) US Patent 5626650
Rosenblatt S, Yaish Y, Park J, Gore J, Sazonova V, McEuen PL (1999) Nano Lett. 2:869
Saito R, Dresselhaus G, Dresselhaus MS (2001) Physical properties of carbon nanotubes. Imperial College Press, London
Smalley RE, Colbert TD, Guo T, Rinzler AG, Nikolaev P (2001) US Patent 6183714
Sun G, Kurti J, Kertesz M, Baughman RH (2002) Dimensional changes as a function of charge injection in single-walled carbon nanotubes, J. Am. Chem. Soc. 124:15076
Tang H, Chen JH, Huang ZP, Wang DZ, Ren ZF, Nie LH, Kuang YF, Yao SZ (2004) High dispersion and electrocatalytic properties of platinum on well-aligned carbon nanotube arrays, Carbon 42:191-203
Tans SJ, Devoret MH, Dai H, Thess A, Smalley R, Geerlings LJ, Dekker C (1997) Individual single-wall carbon nanotubes as quantum wires, Nature 386:474
Tans (SJ), Verschueren (ARM), Dekker C (1998) Room-temperature transistor based on a single carbon nanotube, Nature 393:49
Tennet et al. (1987) US Patent 4663230
Tennet et al. (1992) US Patent 5909615
Thess A, Lee R, Nikolaev P, Dai H, Petit P, Robert J, Xu C, Lee YH, Kim SG, Rinzler AG, Colbert DT, Scuseria GE, Tomane´k D, Fischer JE, Smalley RE (1996) Crystalline ropes of metallic carbon nanotubes, Science 273:483
Tibbetts GG, Meisner GP, Olk CH (2001) Hydrogen storage capacity of carbon nanotubes, Filaments, and vapor-grown fibers. Carbon, 39:2291
Tsai YC, Chen JM, Li SC, Marken F (2004) Electroanalytical thin film electrodes based on a NafionTM-multi-walled carbon nanotube composite, Electrochemistry Communications 6:917
Uemura S, Nagasako S, Takeshi (Mie, JP), Yotani, Junko (Mie, JP), Morikawa, Mitsuaki (Mie, JP), Saito, Yahachi (Mie, JP) (2001) US Patent 6239547
Vohrer U, Kolaric I, Haque MH, Roth S, Weglikowska UD (2004) Carbon nanotube sheets for the use as artificial muscles, Carbon 42:1159
Wu M, Shaw L (2004a) A novel concept of carbon-filled polymer blends for applications of PEM fuel cell bipolar plates, Int. J. Hydrogen Energy, in press
Wu M, Shaw L (2004b) On the improved properties of injection-molded, carbon nanotube-filled PET/PVDF blends, J. Power Sources 136:37
Yacaman MJ, Yoshida MM, Rendon L, Santiesteban JG (1993) Catalytic growth of carbon microtubules with fullerene structure, Appl. Phys. Lett. 62:202
Yamamoto K, Koga Y, Fujiwara S (1998) US Patent 5773834

Yang RT (2000) Hydrogen storage by alkali-doped carbon nanotubes: Revisited. Carbon 38:623

Yang RT (2003) Adsorbents: Fundamentals and applications. John Wiley & Sons, Inc. Canada, pp 231

Ye Y, Ahn CC, Witham C, Fultz B, Liu J, Rinzler AG, Colbert D, Smith KA, Smalley RE (1999) Hydrogen adsorption and cohesive energy of single-walled carbon nanotubes, Appl. Phys. Lett. 74:2307

Yu B, Chang L, Ahmed S, Wang H, Bell S, Yang C, Tabery C, Ho C, Xiang Q, King T, Bokor J, Hu C, Lin M, Kyser D (2002) IEDM Tech. Dig. 251:1-5

Yue GZ, Qiu Q, Gao B, Cheng Y, Zhang J, Shimoda H, Chang S, Lu JP, Zhou O (2002) Generation of continuous and pulsed diagnostic imaging x-ray radiation using a carbon-nanotube-based field-emission cathode, Appl. Phys. Lett. 81:355

Zettl AK, Cohen ML (2000) US Patent 6057637

11 Carbon based Nanostructures

Qianwang Chen and Zhao Huang

University of Science and Technology of China, Hefei 230026, PR of China

11.1 Introduction

Carbon is quite an important element in our lives. Existing abundantly in the earth and the universe, it can be widely used in the future technology. Pure carbon exists in four different crystalline forms: Diamond, Graphite, Fullerenes and Nanotubes. Other common carbon includes amorphous carbon, charcoals, soot and glassy carbon (microcrystalline forms of graphite). Before C_{60} was discovered in 1985, Diamond and graphite are the two well-known forms of crystalline carbon. Then, in 1991, Iijima discovered carbon nanotubes (CNTs). Scientists and engineers are interested in this exotic carbon structure and have discovered series of theory and method for the synthesis of CNTs. There are still other special carbon nanostructures, such as nanoballs, nanofibers. Each structure may have its own theory and can be applied in special fields. This chapter provides an overview of current science and technology of carbon nanostructures, with a special focus on the synthesis of CNTs and fullerenes and their applications.

11.2 History of Fullerenes

Harold Kroto, James Heath, Sean O'Brien, Robert Curl, and Richard Smalley first discovered the C_{60} molecules in 1985. C_{60} is a molecule that consists of 60 carbon atoms, arranged as 12 pentagons and 20 hexagons. The shape is the same as that of a soccer ball (Fig. 11.1). Based on a theorem of the mathematician Leonhard Euler, one can show that a spherical surface entirely built up from pentagons and hexagons must have exactly 12 pentagons. Depending on the number of hexagons, molecules of different sizes are obtained, such as C_{36}, C_{70}. However, C_{60} is the object to which scientists show most interest. It was named fullerene because of its structural similarity to the geodesic domes of the architect R. Buckminster Fuller. Initially, C_{60} could only be produced in tiny amounts. So there were only a few kinds of experiments that could be performed on the material. Things changed dramatically in 1990, when Wolfgang Krätschmer, Lowell Lamb, Konstantinos Fostiropoulos, and Donald Huffman discovered how to produce pure C_{60} in much larger quantities. This opened up completely new possibilities for experimental investigations and started a period of very intensive research. Nowadays it is relatively straightforward to mass-produce C_{60}.

Fig. 11.1: C_{60}'s shape

The discovery of C_{60} has stimulated a large activity in chemistry. It opened up the new branch of fullerene-chemistry that studies the new families of molecules based on Fullerenes. By 1997 about 9000 fullerene compounds were known. In 1991 the alkali-intercalated material K_3C_{60} was found to become superconducting at 18 K, the record for organic superconductors (Iwasa 1994). During the next month the critical temperature was increased to Tc = 35 K in $RbCs_2C_{60}$. These phases exhibit a polymer structure, which is air-stable (Winter et al. 1992). In the same year TDAE C_{60}-Fullerene was supposed to be an itinerant ferromagnet (Stephens 1992). However, its optical conductivity does not display a clear metallic component, which indicates that TDAE-C_{60} is not an itinerant ferromagnet. In 2002, C_{60}-$CHBr_3$ was prepared as a 117-Kelvin superconducting system (Schon et al. 2001). In the last few years there has been a growing interest in the synthesis of fullerene oligomers, and many researchers have attempted to polymerize fullerene C_{60} by various methods, such as photoirradiation, high-pressure compression at high temperatures.

11.3 Structure of Carbon Nanotubes (CNTs)

CNTs are surely the most important nanostructures both from a fundamental point of view and for future applications. Initially, CNTs were simply be sorted by single-walled carbon nanotube (SWCNT) and multi-walled carbon nanotube (MWCNT) (Fig. 11.2). Paid most attention by scientists and engineers, different types of CNTs have been synthesised via various processes.

11.3.1 Y-shaped

Y-shaped junctions and rings composed of single wall CNTs are proposed as possible nanosized electronic devices exhibiting quantum interference mechanisms. The defect consists of six heptagons, according to the Crespi's rule, localized exactly at the bifurcation area. Fig. 11.3 shows a clear Y-shaped nanotube (Lee et al. 2002), while Fig. 11.4 further shows two Y-shaped junctions form a Y-shaped ring (Grimm et al. 2003).

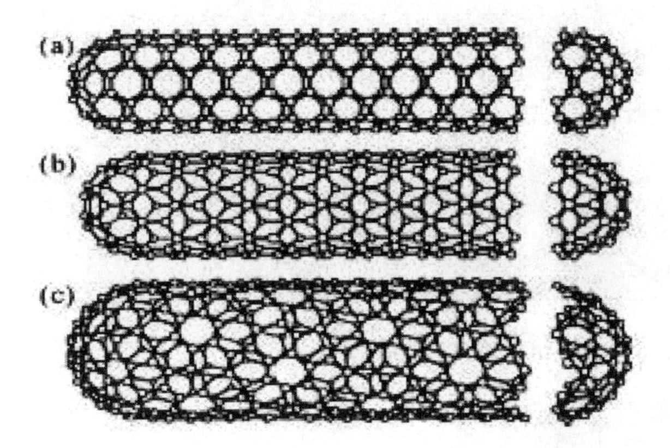

Fig. 11.2: Left: three types of SWCNT with different chiralities; the difference in structure is shown at the open end of the tubes. a) armchair structure; b) zigzag structure; c) chiral structure (Ajayan et al. 2003)

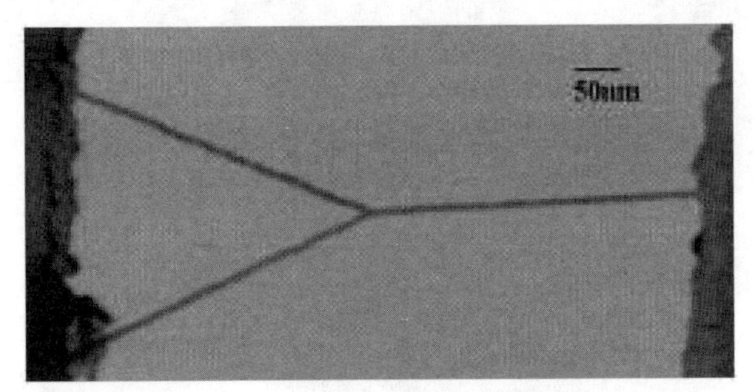

Fig. 11.3: Y-shape nanotube

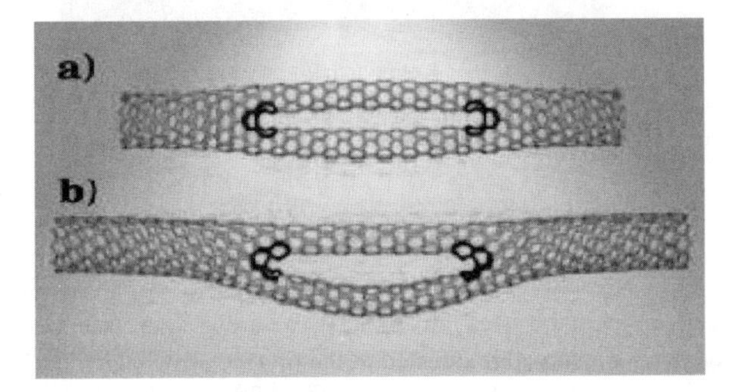

Fig. 11.4: Y-shape ring with a) equal, b) different ring-arm length

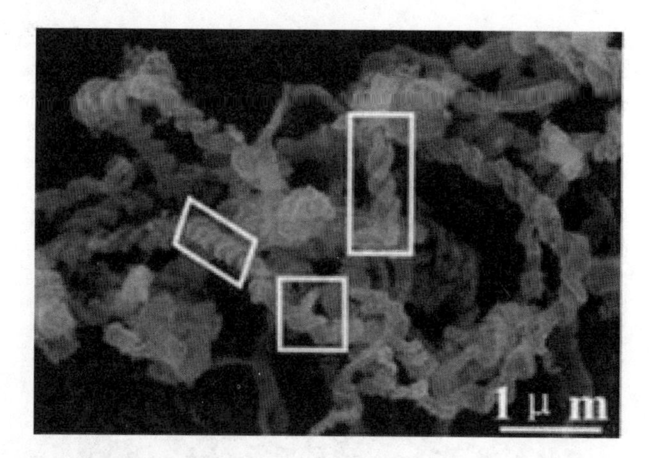

Fig. 11.5: FESEM image of double helical CNTs

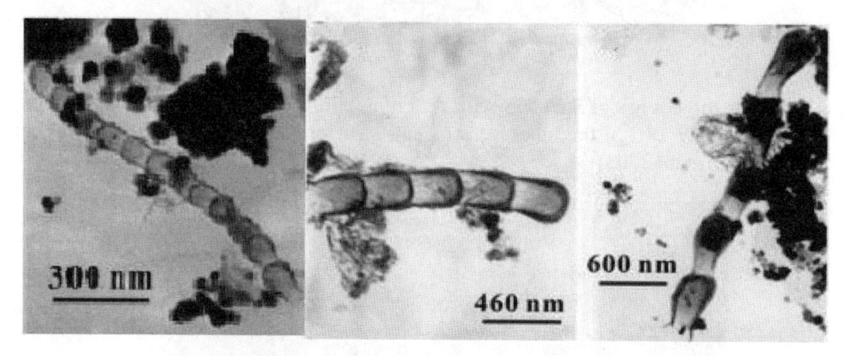

Fig. 11.6: TEM images of CNTs with a bamboo-like structure

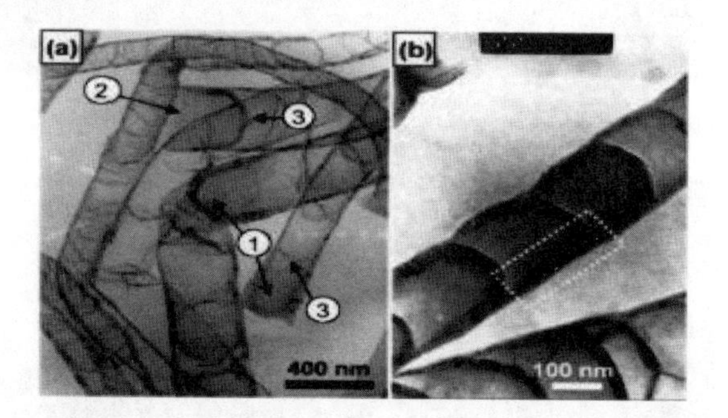

Fig. 11.7: (a) All CNTs have a bamboo structure. There are closed tips with no encapsulated catalytic particles (see arrows ①), an open root (see arrow ②), and compartment layers with a curvature directed to the tip (see arrows ③). (b) The CNT has the compartment layers regularly at a distance of about 200 nm. The thickness of the wall increases at the joint of the compartment layers (see the dotted box)

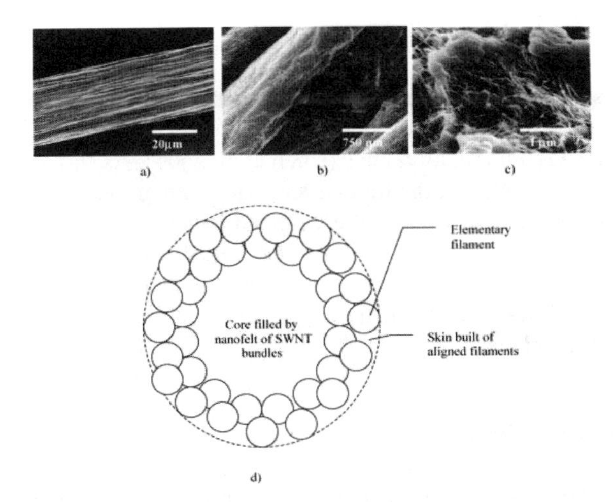

Fig. 11.8: Hierarchical morphology of SWCNT fibers. (a) outer surface of fiber composed of aligned elementary filaments; (b) elementary filaments of 0.2-2 μ m diameter built of packed SWCNT bundles; (c) nanofelt of SWCNT bundles of 10-30 nm diameter; and (d) skin-core model of a SWCNT fiber

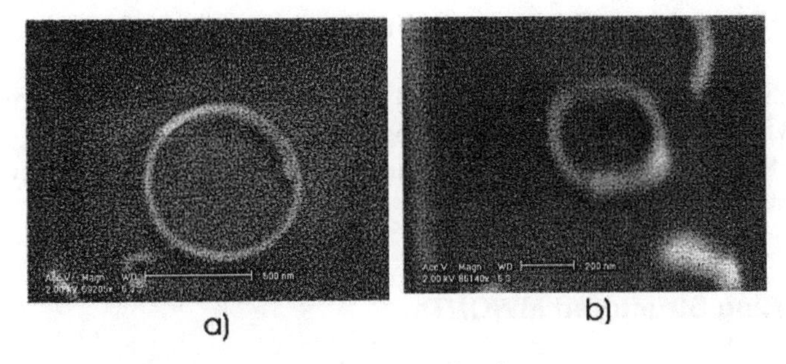

Fig. 11.9: SEM images of multiwalled carbon nanotube rings

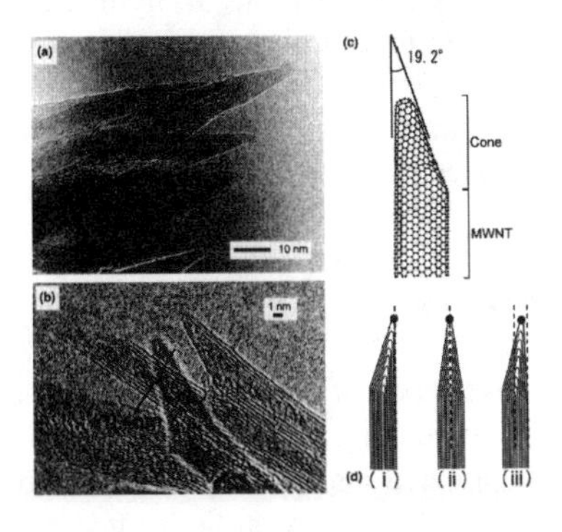

Fig. 11.10: Cone shape end caps of MWNTs

Fig. 11.11: Cylindrical structures

11.3.2 Double Helical

Researchers at Hefei National Laboratory for Physical Sciences at Microscale recently synthesized perfect double helical CNTs by chemical reduction of supercritical CO_2 with alkali metals (Chen et al. 2005). As shown in Fig. 11.5, one tube wrapped another, and the tubular structure with an opened cap can be clearly observed. The outer diameter of the wrapped pair is about 220 nm, and diameter of each single CNT is around 110 nm. The helical structure of CNTs is similar to that in protein. It is observed that the double helical CNTs are imaged to have nearly parallel lines; the interlayer separation is about 0.34 nm, consistent with graphitic atomic planes.

11.3.3 Bamboo-like Structure

In a similar experiment described above, bamboo-like structured CNTs were also synthesized. From Fig. 11.6, it is noted that no encapsulated solid particle in the closed compartment is observed. Same structure of CNTs has ever been prepared by thermal chemical vapor deposition (Fig. 11.7) (Lee and Park 2000).

11.3.4 Hierarchical Morphology Structure

The PCS method provides a promising way to produce highly porous SWCNT fibers with exceptional mechanical properties (Neimark et al. 2003). As shown in Fig. 11.8, SWCNT fibers possess a hierarchical structure with several levels of structural organization.

11.3.5 Ring Structured MWCNTs

Rings of typically 0.5 μm in diameter have been observed in carbon nanotube deposits produced catalytically by thermal decomposition of hydrocarbon gas (Fig. 11.9) (Ahlskog and Seynaeve 1999) with an atomic force microscope and a scanning electron microscope. The ring formation was interpreted as single turn coils with a short overlap between the beginning and end of the coiled nanotube, but the toroidal interpretation cannot be ruled out.

11.3.6 Cone Shape End Caps of MWCNTs

Metal-free method some times produce these types of CNTs. Three types of cone shape end caps of MWCNTs have been observed (Fig. 11.10) (Koshio et al. 2002). Image (i) shows the vertex of the cone located on a line extended from the outermost wall of the nanotube. Image (ii) shows a vertex on the center axis of the nanotube, just like a pencil point. Image (iii) shows a vertex located between the center axis and the line extended from the outermost wall of the nanotube. Most compartments have cone-like structures, but several long relatively cylindrical structures (Fig. 11.11) (Wang et al.2004).

11.4 Structure of Fullerenes

Fullerenes are another exotic nanostructure of carbon. Initially, considering C_{60}, fullerenes were usually thought to have a shape of regular sphere. Along with the in-depth research on fullerenes, other types have been discovered, such as cone shape and tube shape. Here the structures of some new fullerenes will be introduced.

11.4.1 Structure of C_{48} Fullerenes

A recent research has constructed 27 isomers for the C_{48} cage and these are shown in Fig. 11.12 (Wu et al. 2004). It is found that the most stable C_{48} structure has the smallest number of shared pentagonal bond (N_{55}) and no squares and deltas, and all the reported structures are higher in energy. In comparison with the available isomers from literature, C_{48} isomer (1) with the least number of shared pentagonal C-C bonds (N_{55} = 6) is the most stable structure and has the largest vertical ionization potential. However, the low-lying isomers 2-4 are very close in energy, and the most stable triplet state is structure 2.

11.4.2 Toroidal Fullerenes

In finding structural conditions for preferable fullerenes, scientists investigated a variety of covering patterns, *e.g.*, combinations of C_5, C_6, C_7, and polygons, arranged in different ways, in view of lowering the total energy of possible molecular structures. Toroidal structure was first identified (Diudea et al. 2003). A toroidal surface can be covered with hexagons by cutting out a parallelogram from a graphite sheet, rolling it up to form a tube and finally gluing its two ends to form a torus. A 4-valent square-tiled toroid is transformed into 3-valent hexagonal (and other polygonal) lattices either by simple cutting procedures or by some more elaborated operations, such as leapfrog and related transformations (Fig. 11.13).

11.4.3 Structure of C_{60}, C_{59}, C_{58}, C_{57}

Recently, Hu and Ruckenstein studied the structures and stabilities of several defect fullerene clusters—C_{59}, C_{58}, and C_{57}—formed by removing one, two, and three adjacent carbon atoms from a C_{60} cluster (Hu et al 2003). A more advanced structures and stabilities for all possible isomers of the defect fullerene clusters of

C_{60}, C_{59}, C_{58}, and C_{57} have been investigated by Lee SU and Han YK. When one atom is removed from C_{60} cluster, one 5-membered ring and two 6-membered rings are destroyed and replaced with 5- and 8-membered rings, denoted as the C_{59}_5-8 isomer, or 4- and 9-membered rings, the C_{59}_4-9 isomer (Fig. 11.14). It is noteworthy that there are obvious relationships between structure and stability of the defect fullerene clusters. First, the unsaturated carbon atom favors being located at a 6-membered ring rather than a 5-membered ring. Second, the most stable isomers prefer to have newly-formed 5-membered rings, rather than newly-formed 4-membered rings.

11.4.4 The Smaller Fullerene C_{50}

The smaller non-IPR fullerenes, predicted to have unusual properties because of their adjacent pentagons and high curvature, are so labile that their properties and reactivity have only been studied in gas phases. Fullerenes smaller than C_{60} are predicted to have unusual electronic, magnetic, and mechanical properties that arise mainly from the high curvature of their molecular surface (Xie et al. 2004). Fig. 11.15 describes the typical structure of $C_{50}Cl_{10}$ capturing C_{50}. Similar element may be applied to produce $C_{54}Cl_8$ and $C_{56}Cl_{10}$ (Chen 2004).

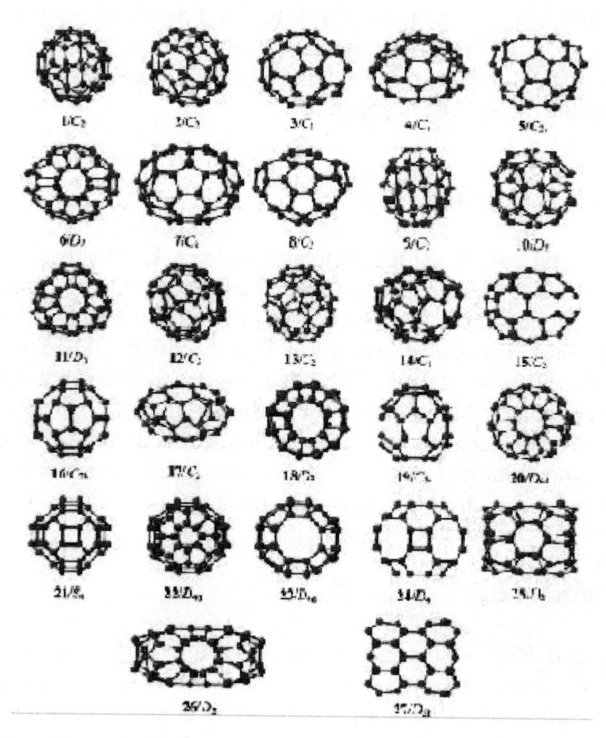

Fig. 11.12: B3LYP/6-31G* C_{48} isomers

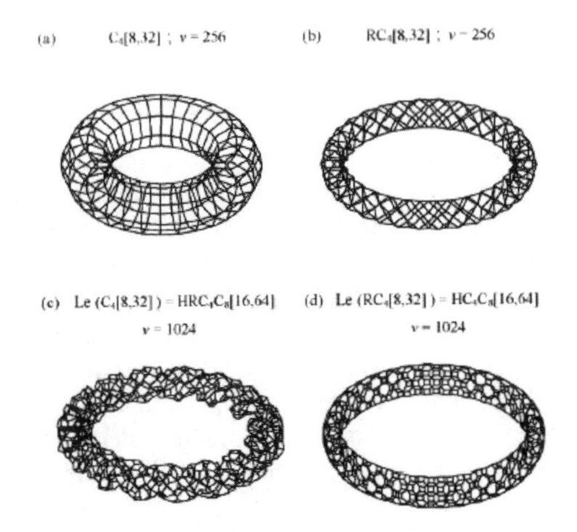

Fig. 11.13: Leapfrog transforms of two 4-valent tori: C_4 (8, 32) andRC_4 (8, 32)

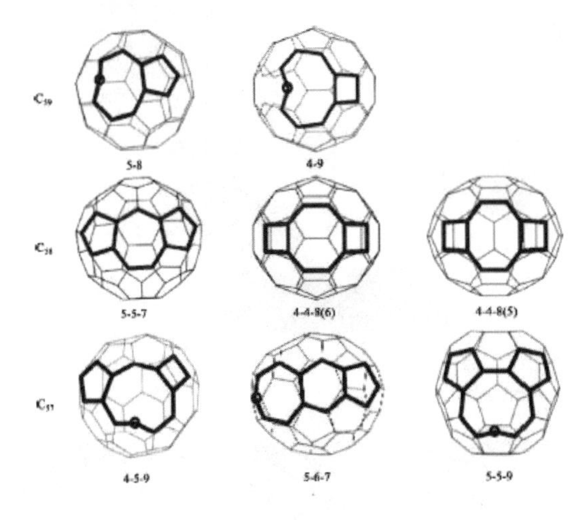

Fig. 11.14: Optimized C_{59}, C_{58}, and C_{57} clusters

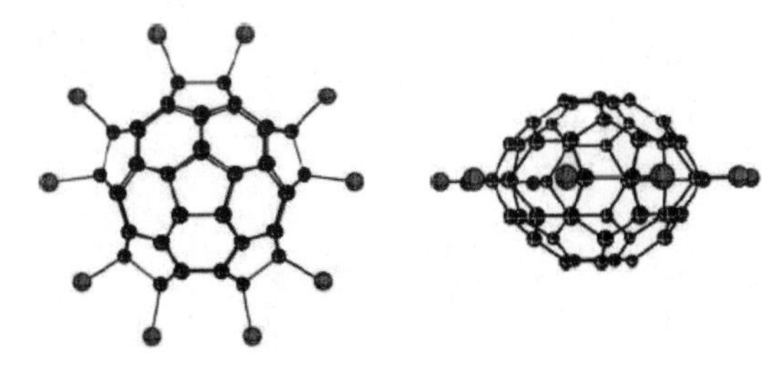

Fig. 11.15: The B3LYP/6-31G* optimized structure of $C_{50}Cl_{10}$ (D_{5h})

11.5 Structure of Carbon Nanoballs (CNBs)

Carbon nanoballs belong to the fullerenes family, but they are a little different in structure. Other fullerenes as C_{60} are single walled, but nanoballs are multi walled. It is void in the center of carbon nanoballs but, in the wall of nanoballs, there is no hollow structure or channel as in the case of MWCNTs. Because their structures are similar to the onions, scientists call them bucky onions. Nanoballs in Fig. 11.16 have diameters of between 500 nm and 1μm (Liu 2002). Recently, our group produced hollow carbon nanoballs by chemical reduction of carbon dioxide with metallic Li (Fig. 11.17). By studying the defect of bucky onions it is found that the self-interstitial concentration remains low and is only slightly enriched near the onion centre. The loss of atoms from the outer surface of the bucky onions by sputtering acts as a vacancy source for the whole onion (Sigle et al. 1997).

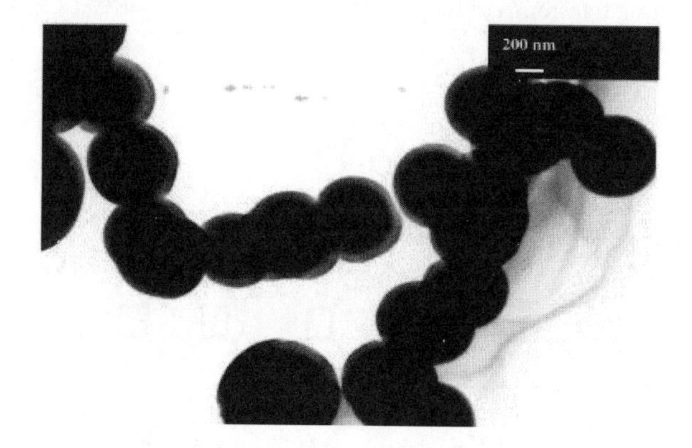

Fig. 11.16: TEM view of nanoballs

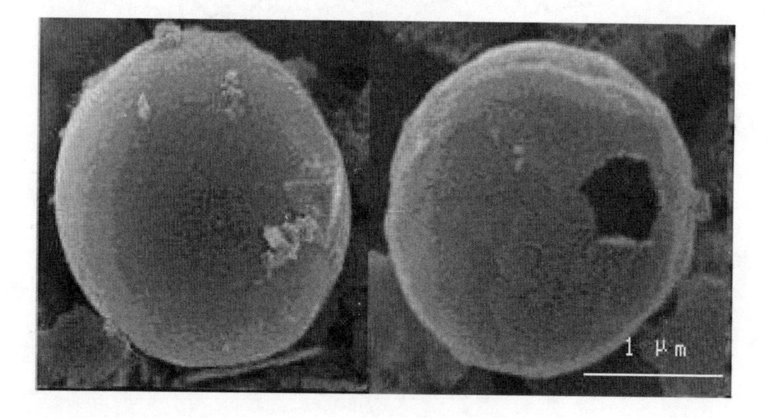

Fig. 11.17: SEM images of hollow carbon nanoballs

11.6 Structure of Carbon Nanofibers (CNFs)

Carbon nanofibers have recently received large attention for industrial application. Various shapes of CNFs have been discovered in the last several years, such as hexagonal, corn-shaped and helical ones.

11.6.1 Hexagonal CNFs

A newly research pointed that a transverse section of platelet and tubular CNFs had a hexagonal shape, not a round shape, very different from the traditional ones (Fig. 11.18) (Tanaka et al 2004). It is observed in the platelet CNF, the hexagonal planes were stacked perpendicular to the fiber axis.

11.6.2 Corn-shaped CNFs

Corn-shaped CNFs contain a corn angle θ. θ can be deduced from the projected dimensions of the tilted cones. Normally, the θ can be decreased with the increase ofsubstrate temperature (Hayashi et al. 2004). Fig. 11.19 gives out several CNFs with different angle θ.

11.6.3 Helical CNFs

There exist two helical fibers that had opposite helical senses, but had identical cycle number, coil diameter, coil length, coil pitch, cross section, and fiber diameter (Fig. 11.20) (Qin et al. 2004). As shown in Fig. 11.20, we can observe two helical fibers with irregular tips (A) in different morphologies and (B) in identical morphologies. This was a novel mirror-symmetric growth mode.

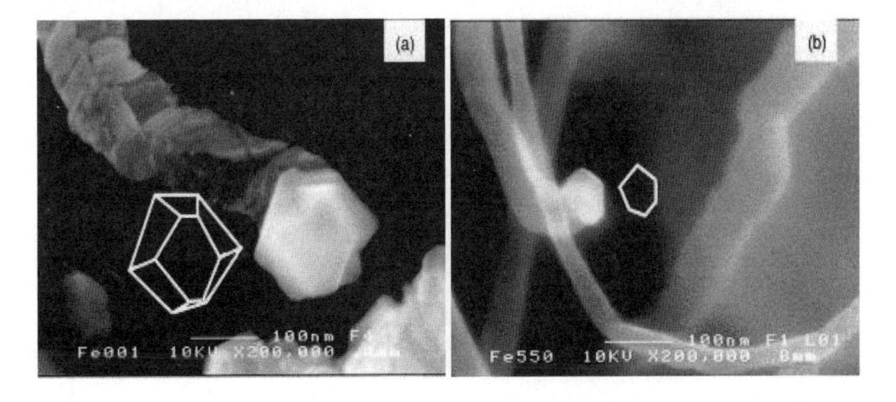

Fig. 11.18: Magnified HRSEM photographs of (a) platelet CNF, and (b) tubular CNF

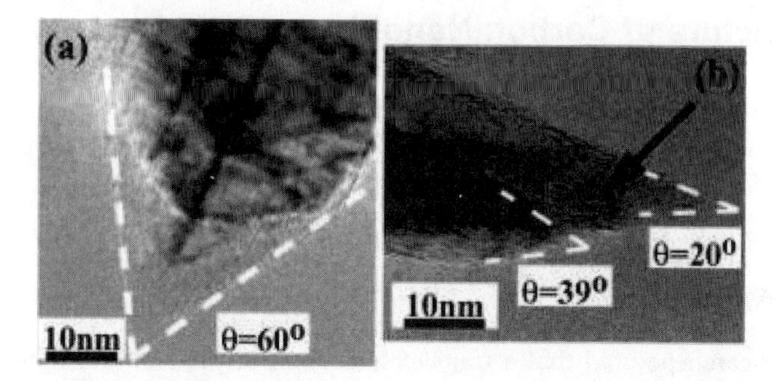

Fig. 11.19: HRTEM images of three types CNFs with corn angles of 20°, 39° (b), and 60° (a), respectively

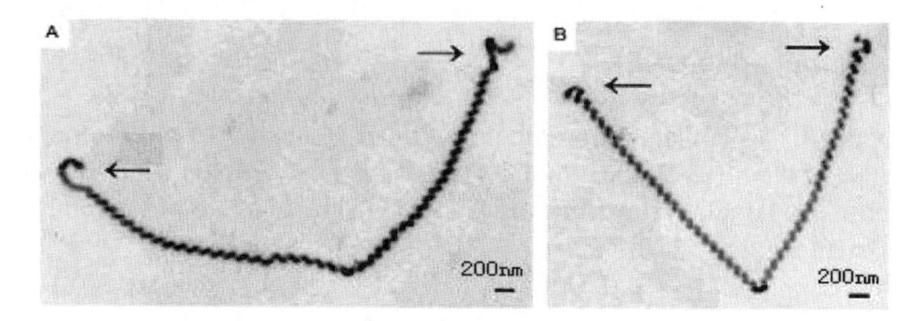

Fig. 11.20: TEM views of two opposite helical fibers

11.7 Porous Carbon

Porous carbon materials, especially those containing nanopores, micropores or mespores, are being used in various applications such as super capacitors, specific adsorbents and catalyst supports. More recently a kind of glassy carbon has been used as a preform for the reaction forming of structural ceramics. Several structures prepared by different method are shown in Fig. 11.21 (Wang et al. 2003).

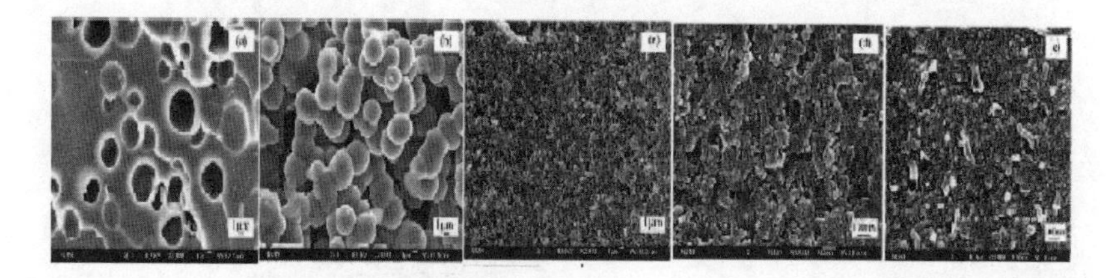

Fig. 11.21: Morphological changes of the porous carbon materials prepared with different glycols (a) PC–EG; (b, d and e) PC–DEG, PC–DPG and PC–TEAG, respectively; (c) TEG

11.8 Properties of Carbon Nanostructures

Carbon nanostructures have attract considerable attention due to their special properties. The most important properties of carbon nanotubes and fullerenes are stated below. Carbon nanotubes have excellent molecular, electronic, optical and mechanical properties due to their nearly one-dimensional structure.

11.8.1 Molecular Properties

A perfect SWCNT has no functional groups. Therefore these quasi-1D cylindrical aromatic macromolecules are chemically inert. However, curvature-induced pyramidalization and misalignment of the π-orbitals of the carbon atoms induces a local strain, and carbon nanotubes are expected to be more reactive than a flat graphene sheet (Niyogi et al. 2002). From the standpoint of chemistry, it is conceptually useful to divide the carbon nanotubes into two regions: the end caps and the sidewall. The end caps will always be quite reactive, irrespective of the diameter of the carbon nanotubes.

11.8.2 Electronic Properties

CNTs are quasi-one-dimensional materials made of sp^2-hybridized carbon networks and have been the subject of extensive research and discussion. In particular, the electronic structure of a single CNT has been studied theoretically, which predicts that CNT becomes either metallic or semiconducting depending on its chiral vector, i.e., boundary conditions in the circumference direction. An electron in a nanotube is a massless neutrino on a cylinder surface with a fictitious Aharonov- Bohm flux determined by its structure. A nanotube becomes a metal or a semiconductor, depending on whether the amount of the flux vanishes or not.

The electron accepting ability of C_{60} has been explored extensively. Apparently, oxidative electrochemistry of fullerenes is not as rich as reductive electrochemistry. Scientists have also discovered that the electronic properties are not the same among different types of fullerenes. For example, chirality in fullerene molecules could alter significantly the electronic properties of a cage-like system. Regarding quasi-spherical giant fullerenes, it is demonstrated that heptagons and additional pentagons produce dramatic changes in the charge distribution.

11.8.3 Optical Properties

A newly EMA method indicates that the effect of the larger internal radii of the nanotubes cannot be neglected (Lu et al. 2001). The optical activity of chiral nanotubes disappears if the nanotubes become larger (Damnjanovic et al. 1999). Pristine C_{60} is a molecule of high symmetry, with p-electrons delocalized along the whole 3-D structure. These properties make C_{60} an interesting material in the nonlinear optical field (Brusatin et al. 2002). It also shows a wide variety of uncommon physical properties ranging from optical limiting to superconductivity and photoconductivity. It has been clearly observed that the final linear and nonlinear optical properties of fullerenes are strictly related to the matrix used for inclusion, to the possibility of functionalizing the pristine C_{60}, and to the processing protocol of the solid nanocomposite. Another work got a similar result that the lowering of symmetry and the presence of a crystal field in fullerenes affect the selection rules and the energies of the intermolecular excitations (shift and splitting of degenerate electronic levels). The optical properties of fullerenes depend equally on intra- and intermolecular processes (Makarova 2001).

11.8.4 Mechanical Properties

The mechanical properties of CNTs have attracted much attention since they were discovered in 1991. CNT exhibits extraordinary mechanical properties: The Young's modulus is over 1 Tera Pascal. It is stiff as diamond. The estimated tensile strength is about 200 Giga Pascal. Weak region of mechanical strength of CNT was the center of the tube, not the connection region, which meant that the connection region was stronger than the tube itself (Abe et al. 2004). The mechanical strength at the connection region must be significantly improved for the real application and one plausible candidate might be the particle method.

11.8.5 Periodic Properties

Despite their complexity in structure, fullerenes still have their own periodic properties (Torrens 2004). First, the properties of the fullerenes are not repeated; only, and perhaps, their chemical character. Second, the relationships that any fullerene p has with its neighbor p + 1 are approximately repeated for each period (p stands for the number of edges common to two pentagons). Nanoballs are electric materials. Due to easily adsorptions of amorphous carbon and impurities to the surface of nanoballs, their electric capability may decrease (Liu et al. 2001). Another specific property is the polymerization. Two bucky onions in similar dimension may open the chemical bond of the atoms in out wall and connect together to form a stable dimer. Nanofibers have many different types including exotic properties. For example, vapor-grown carbon nanofibers (VGNFs) have electric properties (Wei et al. 2004). Their electric resistance of composite prepared from the polyester-grafted carbon nanofiber and poly(ethylene glycol) suddenly increased in methanol vapor over 1000 times, and returned to initial resistance when it was transferred into dry air. Graphic carbon nanofibers (GCNFs) have high dispersion and electrocatalytic properties. Researches still underway are exploring the novel properties of nanofibers.

In summary, some interesting properties of several types of carbon nanostructures have been discussed. Properties of nanotubes and fullerenes attract scientists so deeply that their characters are discovered and concluded systematically. The study of nanoballs and nanofibers is still in its infancy and requires substantial, prolonged development, many new exotic properties are waiting for exploiting.

11.9 Synthesis

In this section, different techniques for the synthesis of various carbon nanostructures are summarized. Firstly, growth mechanism is explained briefly. Then the detailed introduction to each synthesis method of nanotubes, fullerenes, nanoballs and nanofibers will be shown. The most results summarized are almost appeared in the literature in the last 5 years, standing for the update proceeding of the synthesis of carbon nanostructures. They are many known ways to prepare carbon nanotubes. Generally speaking, carbon nanotubes are produced by three main techniques: arc discharge, laser ablation and chemical vapor deposition. Fullerenes are produced earlier by arc discharge, methods of benzene flame, sputtering and electron beam evaporation were developed a little later. Nanoballs are usually produced as outgrowth in the processes producing carbon nanotubes, so chemical vapor deposition can be described as the technique of production. Nanofibers can be produced by traditional techniques, such as chemical vapor deposition et al, but we will mainly discuss new techniques such as "nanofiber seeding", and ethanol flame. In the sequel the growth mechanism will be described.

11.9.1 Carbon Nanotubes

The growth mechanism of nanotubes is not well understood; different models exist, but some of them cannot unambiguously explain the mechanism. The metal or carbide particles seem to be necessary for the growth because they are often found at the tip inside the nanotube or also somewhere in the middle of the tube. In 1972 Baker et al. proposed a model of the growth of carbon fibers, which is shown in Fig. 11.22(a). It is suggested that acetylene decomposes at 600 °C on the top of a nickel cluster on the support. The dissolved carbon diffuses in the cluster, precipitates on the rear side and forms a fiber. The carbon diffuses through the cluster due to a thermal gradient formed by the heat release of the exothermic decomposition of acetylene. The activation energies for filament growth were in agreement to those for diffusion of carbon through the corresponding metal (Fe, Co, Cr). Oberlin et al. proposed a variation of this model. The fiber is formed by a catalytic process involving the surface diffusion of carbon around the metal particle, rather than by bulk diffusion of carbon through the catalytic cluster. In this model the cluster corresponds to a seed for the fiber nucleation (Fig. 11.22(b)). The metal cluster can have two roles: 1) acting as a catalyst for the dissociation of the carbon-bearing gas species; 2) carbon diffuses on the surface of the metal cluster or through the metal to form a nanotube. The most active metals are Fe, Co and Ni, which are good solvents for carbon (Kim et al. 2001). For SWCNT it is supposed that the nanoparticles have to be smaller than those for MWCNT (Dai et al. 1996), but this is in contradiction to the arc-discharge in which SWCNT grows radial from one bigger metal cluster (Saito 1995).

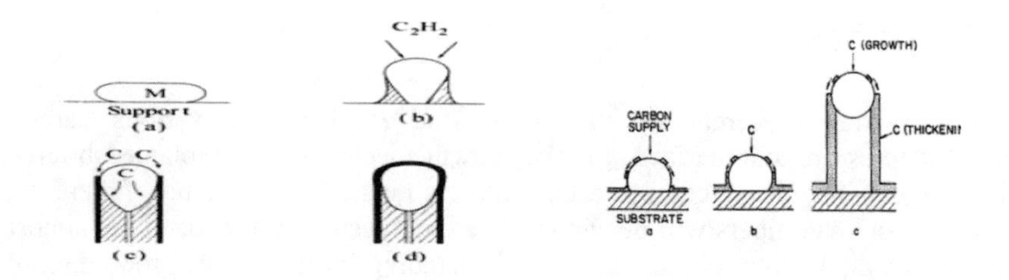

Fig. 11.22: Growth model of vapor grown carbon fibres; the metal cluster acts as a seed for the growth

11.9.2 Fullerenes

In recent years, attempts have been made both theoretically and experimentally in order to unveil the mystery of the growth mechanism of fullerene. Experiments on resistive evaporation of two carbon rods with different degrees of enrichment in ^{13}C have revealed that the carbon vapor consists of smaller clusters before the process of fullerene formation starts (Ebbesen 1992). Further, experimental gas ion chromatography studies of the structure of carbon clusters revealed the possible paths of fullerene formation through the coagulation of monocycles and gradual transformations of polycyclic structures into fullerene cages. These elementary carbon clusters such as monocyclic rings or polycyclic rings are termed as precursors from which the fullerenes are supposed to be formed by the successive stacking with the different carbon belts. It is quite natural to presume that the fullerene cages must have been generated from appropriate precursors. Wakabayashi and Achiba have suggested a kinetic-ring stacking model where fullerenes can be constructed by stacking proper-sized carbon rings. Dias has proposed a qualitative circumscribe algorithm based on the graph theoretical footings, which is really akin to the kinetic-ring stacking model. This algorithm can be stated as a method of encircling a monocyclic/polycyclic precursor with appropriate carbon belts in order to generate the fullerene cage structures. It has been further mentioned that successive circumscribing of qualified conjugated hydrocarbons with a combination of pentagonal and hexagonal rings terminates at fullerenes when the number of pentagonal rings reaches 12 or terminates at a system that can be capped to give fullerenes with 12 pentagonal rings. A new research on small fullerenes revealed there is a chance of bond cleavage of the polycyclic precursors as the growth process proceeds towards the cage formation.

On the other hand, the monocyclic precursors are found to have significantly lower deformation energies than the polycyclic precursors.

11.9.3 Nanoballs

Carbon nanoballs are usually produced with carbon nanotubes. Especially in metal catalyzed reaction, nanoballs can be found easily. Growth mechanism of carbon nanoballs is stated below. The nanoballs are usually formed with relatively small particles, and carbon saturation can occur due to its large surface-to-volume ratio (Park et al. 2002). Therefore the small nanoparticles will become rapidly saturated with carbon at the initiation of the CNT growth, and the subsequently occurring diffusion out to the surface forms the graphite rind. The graphite rind at the surface of the nanoballs immediately deactivates their catalytic role by inhibiting incorporation of the carbon species from the ambient gas and stops nucleation and growth from the particle. The observation indicates that the carbon species arriving subsequent to the nanoball formation cannot form crystalline carbon layers without the catalytic role of the small nanoparticles. The mechanism can be also applied in the reaction of $Fe(CO)_5$ catalyzed pyrolysis of pentane (Liu et al. 2002). Carbon nanoballs were found in the rear end of the high temperature region of the quartz tube. When a lower carrier gas flow rate was used, nanoballs could be found over an extended area and this suggested that the lower carrier gas flow rate favored the formation of carbon nanoballs.

11.9.4 Nanofibers

Baker proposed the root growth mechanism of carbon nanofibers (CNFs) in 1989. Observation of many growth sequences has shown that both the addition of a second metal to the catalyst and also the strength of the metal-support interaction can play an important role in modifying the filament growth characteristics. The growth of so-called vapor-grown carbon nanofibers (VGCNFs) can be classified into the 'hollow-cored mechanism' (Fig. 11.21(b)) (Pan et al. 2004), because VGCNFs were obtained by the following processes: A hollow and highly crystalline graphite carbon nanotubes were formed as the inner region and then, a low crystalline graphitic pyrocarbon layer was deposited on the nanotubes as an outer layer. The mechanism of the catalytic growth of CNF can be concluded to different growth steps. A schematic representation of the mechanism of steady-state growth is given in Fig. 11.23, which is based on the review of de Jong and Geus. The above mentioned steps show that the steady state growth process is a delicate balance between the dissociation of the carbon-containing gases, carbon diffusion through the particle, and the rate of nucleation and formation of graphitic layers. The growth mechanisms and synthesis techniques of carbon nanotubes, fullerenes, carbon nanoballs and carbon nanofibers has been summarized. Details of growth mechanisms of them show that carbon nanostructures are basically produced from various shapes of graphite curliness. And so many synthetic methods have been developed. Especially, considerable progress has been made in preparation of carbon nanostructures by simple chemical processes.

Table 11.1. A brief summary of the major synthesis methods

Sstructures	Synthesis method	Characteristic and conditions
CNT	Arc discharge	SWNT synthesis: use metal catalyst as anode, inert gas. MWNT synthesis: arc-plasma method in water, different helium pressures.
	Laser ablation	Costly technique, Primarily pure SWCNTs with good diameter control and few defects.
	Chemical vapor deposition	Easiest to scale up to industrial production, simple process, long SWCNT, diameter controllable, pure, but usually produced MWCNT with defects.
	Chemical reaction method	Simple process, inexpensive technique, products can be controlled.
Fullerenes	radiofrequency (RF) thermal plasma method	Lower gas velocities, more voluminous plasma flames, longer residence time of reactive species, feed rate of powders can be independently changed.
	Laser vaporization	Synthesis of multishell fullerenes, gram amounts production, can be controlled so as to selectively synthesize certain types of multishell fullerenes and other multishell carbon clusters.
	RF-inductive coupled plasma discharge	Synthesis of both fullerenes and endohedral metallofullerenes, allows longer reaction times, the temperature of the RFICP reactor can be adjusted by changing its RF field frequency.
	Flame combustion	Can form fullerenes, fullerenic nanostructures and fullerenic soot, conditions of fuel/oxygen ratio, chamber pressure, and inert gas dilution are sensitive, formed directly from curved structures in the soot.
CNB	Fe-catalysed grown by thermal chemical vapor deposition	Formed only from pentane, lower carrier gas flow rate favored the formation of CNBs, high temperature gives longer contact times.
	Cobalt catalyst grown production	Acid treatment removes the encapsulated cobalt, increasing the content of cobalt can make high content of products.
CNF	Microwave plasma-enhanced chemical vapor deposition	A new type of corn-shape carbon nanofibers (CCNFs) with metal-free tips was fabricated.
	nanofiber seeding	A rapidly convenient method to obtain thin, substrate-supported, transparent films of nanofibers, requiring nanometer scale control of surface architecture.
	Ethanol flame	Synthesis of CNFs with solid-core, much simpler, more economic, easier to be controlled.

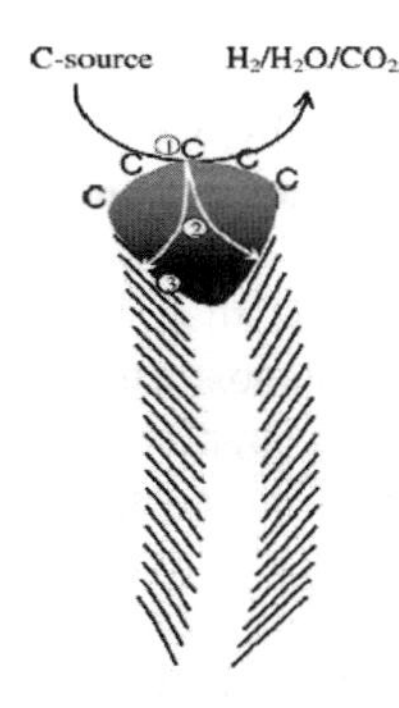

Step 1: decomposition of carbon-containing gases on the metal surface. Step 2: carbon atoms dissolve in and diffuse through the bulk of the metal. Step 3: precipitation of carbon in the form of a CNF consisting of graphite

Fig. 11.23: Schematic representation of the catalytic growth of a CNF using a gaseous carbon-containing gas

11.10 Potential Applications of Nanostructures

11.10.1 Energy Storage

Graphite, carbonaceous materials and carbon fiber electrodes are commonly used in fuel cells, batteries and other electrochemical applications (Daenen et al. 2003). CNTs can be considered as excellent energy storage material because of their small dimensions, smooth surface topology and perfect surface specificity. In 2004 a novel molecular ferrocene-zinc porphyrin-zinc porphyrin-fullerene (Fc-ZnP-ZnP-C-60) tetrad was reported to have energy storage properties (Guldi et al. 2004). Hydrogen storage, lithium intercalation, and electrochemical supercapacitors in some carbon-based materials will be discussed in the following section.

11.10.2 Hydrogen Storage

Because of their large surface area, low density, and hollow structure, CNTs have been considered to be potential materials for hydrogen storage (Bacsa et al. 2004). Ye et al. measured hydrogen adsorption at low temperatures (70 K) on single-walled CNT samples and concluded that the adsorption was proportional to the BET specific surface area (Ye et al. 1999). Other researchers, however, reported that hydrogen storage capacity was higher in CNT bundles wherein the adsorption sites were created under pressure (Shiraishi et al. 2002). Purification of the CNTs by oxidative acid treatments or by heating in inert gas decreases the hydrogen storage. Also, increasing the specific surface area does not necessarily increase the hydrogen storage capacity. There seems to be a correlation between the pore volume at low pore diameters (<3 nm) and hydrogen storage capacity.

Hydrogen uptakes in different carbon nanostructures at 10 MPa and 293 K using grand canonical Monte–Carlo numerical simulations have been calculated recently (Guay et al. 2004). Simulations indicate that pure carbon nanostructures could not reach a hydrogen uptake of 6.0 wt.%. The amount of adsorbed hydrogen in SWCNTs, MWCNTs and graphic nanofibers is lower than 1.4 wt.% for the optimum porosity around 0.70 nm. For standard carbon nanostuctures in which the porosity is similar to the bulk interplanar distance in graphite (0.34 nm), the amount of adsorbed hydrogen is lower than 0.6 wt.%. The material properties of hydrogenated fullerenes such as $C_{60}H_{36}$ are also under scrutiny and some of the features revealed so far may indicate a certain potential in the context of hydrogen storage (Vasilev et al. 2004). In contrast to the attractive forces operative in connection with the storage of hydrogen on nanotubes, which are essentially weak intermolecular interactions, hydrofullerenes feature comparatively strong s C–H bonds. This in turn raises the issue of the means whereby the stored hydrogen may be released.

11.10.3 Lithium Intercalation

Molecular dynamic simulation predicted that the alkali metal storage capacity of SWNT bundles is substantially higher than that of the intercalated graphite and disordered carbon, the current anode materials in rechargeable Li-ion batteries (Zhao et al. 2000). Experimentally, the reversible capacity was found to depend on the quality and morphology of the SWCNT containing materials. The reported values vary in the range of $Li_{1-1.6}C_6$ (Gao et al. 1999). The reversible Li storage capacity increased from LiC_6 in close-end SWCNTs to LiC_3 after etching, which is twice the value observed in intercalated graphite. The enhanced capacity is attributed to Li diffusion into the interior of the SWCNTs through the opened ends and sidewall defects (Fig. 11.24) (Shimoda et al. 2002).

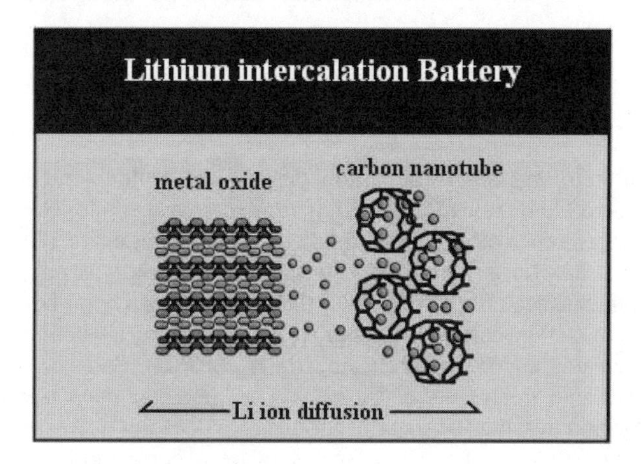

Fig. 11.24: Li intercalation battery model

In a previous study, it is reported the electrochemical intercalation of lithium into multiwall carbon nanotubes where it has been shown that lithiated MWCNTs present a 'neck-lace' structure and that nanotubes could be reversibly intercalated by lithium (Maurin et al. 2000). It is also revealed electrolyte composition can affect the lithium intercalation into carbon nanotubes by introducing side reactions like exfoliation of graphite layers caused by co-intercalation of solvent molecules. The low specific capacity observed may be due to the low proportion of MWCNTs in the raw material. Fullerenes and filamentous carbon structures can also be inserted lithium by the catalytic graphitization (dissolution-precipitation) method in melts of cast iron (Lee et al. 2003). As a result, the graphitic product could reversibly intercalate more than 300 mAh/g equivalent of lithium. The first-cycle irreversible capacity was a mere 14%. The coulombic efficiency seemed to stabilize at values > 99% from the fifth cycle.

11.10.4 Electrochemical Supercapacitors

Supercapacitors have a high capacitance and potentially applicable in electronic devices. Typically, they are comprised two electrodes separated by an insulating material that is ionically conducting in electrochemical devices. The capacity of an electrochemical supercap inversely depends on the separation between the charge on the electrode and the counter charge in the electrolyte. Carbon nanotubes have been considered as the ideal material for supercapacitors due to their high utilization of specific surface area, good conductivity, chemical stability and other advantages (Chen et al. 2004). To increase the specific capacitance of supercapacitors based on carbon nanotubes, activation and surface modification can be carried out by using KOH and concentrated nitric acid respectively (Deng et al. 2004). The results showed that activation enlarged the BET specific surface area of the CNTs and hence increased the specific capacitance of the supercapacitors. A new form of carbon nanofiber web was prepared that is considered to be a suitable material for the electrode of a supercapacitor exhibiting high capacitance (Fig. 11.25) (Kim et al. 2004). The capacitance of the electrical double-layer capacitor was strongly dependent on the specific surface area, micropore volume, and resistivity of the samples. A previous research indicated a controlled rate of fiber formation as well as proper selection of the precursor might play key roles in determining the capacitive behavior (Adhyapak et al. 2002).

Recently researches discovered bimodal porous carbons with both micropores and meso- or macropores selectively synthesized by a SiO_2 colloidal crystal-templating process could make a high performance electrical double-layer capacitor (Moriguchi et al. 2004). The electrical double-layer capacitance per surface area of the templated porous carbons was much larger than those of commercially available activated carbons with high surface areas. The surface of meso- and macropores generated in the porous carbons shows a highly efficient electrical double-layer capacitive property; the specific capacitance per surface area originating from meso- and/or macropores was estimated to be 20 +/- 2 muF /cm^2.

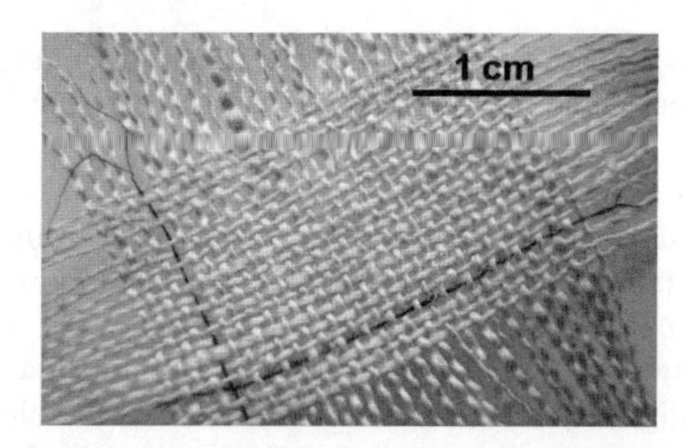

Fig. 11.25 Fabric web as materials of supercapacitors produced by nano technique graduate school in Dallas, USA

11.10.5 Molecular Electronics with CNTs

11.10.5.1 Field Emitting Devices

Carbon nanotubes are promising electron emitters because of their sharp geometries that lead to significant external field enhancement, as well as their mechanical strength. However, distinguishing the emission due to an individual SWCNT from that due to surrounding structures is a challenge (Nojeh et al. 2004). Previous results show that adsorbates on CNTs have great influence on the field emission properties of CNTs. It is observed that the electric field initiates the significant step-like jumps of field emission current, which is believed to be due to the formation of absorbate enhanced-tunneling configurations between absorbates and field emitting CNTs (Yeong et al. 2004). Examples of potential applications for nanotubes as field emitting devices are flat panel displays, gas-discharge tubes in telecom networks, and electron guns for electron microscopes, AFM tips and microwave amplifiers. Recently, an electron industry in Japan explored colorful FED monitor using a CNT based field-emission cathode (Fig. 11.26). IBM Company has made the same progress too. Compared with former silicon FED, the index of electric current capability has been improved twice.

Fig. 11.26: 5.6-inch colorful FED monitor and 12.1 inch homochromatic FED

11.10.5.2 Transistors

CNTs show promise in overcoming limitations as low charge-carrier mobilities because their carrier mobility exceeds even common semiconductor materials (Zinn et al. 2003). Not only do CNTs exhibit very high strength, but also their flexibility makes them a promising material for the development of large-scale flexible electronics. The research team said that the low switching voltage is due to the high carrier mobility of the CNT network. Other advantages such as flexible and could be made inexpensively also make CNTs good materials for transistors. An example of application in audions of CNTs is shown in Fig. 11.27.

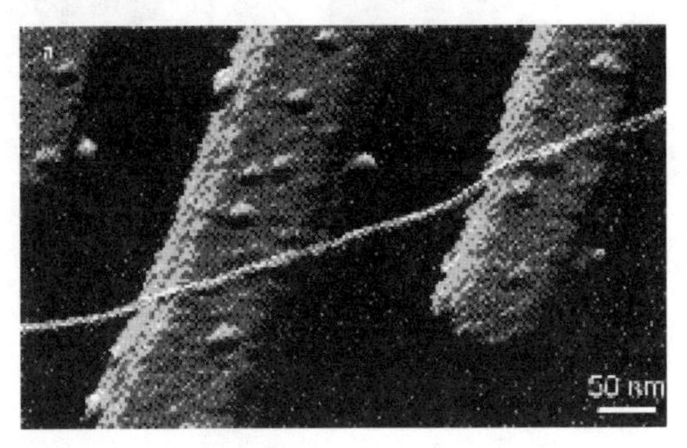

Fig. 11.27 CNT applied in electronic audion

11.10.5.3 Nanoprobes and Sensors

CNTs might constitute well-defined tips for scanning probe microscopy (Dai et al. 1996). Because of their flexibility, the tips are resistant to damage from tip crashes, while their slenderness permits imaging of sharp recesses in surface topography. Another important application of CNTs is their use as tips of AFM or STM (Fig. 11.28) (Zhao et al. 2002); CNTs are used as nanoprobes. The CNT tips offer several advantages: a) have intrinsically small diameters, which in the case of single-walled nanotubes can be as small as 0.5 nm; b) have high aspect ratios that allow them to probe deep crevices and trench structures; c) can buckle elastically, that limits the force applied by the AFM probe and reduce deformation and damage to biological and organic samples; d) can be modified at their ends to create functional probes. As for sensing applications, carbon nanotubes have some advantages: small size with larger surface; high sensitivity, fast response and good reversibility at room temperature as a gas molecule sensor; enhanced electron transfer when used as electrodes in electrochemical reactions; and easy protein immobilization with retention of activity as potential biosensors (Zhao et al. 2002). Because of the very rich electrochemical properties of fullerenes, their exploitation for the development of chemical sensors for sensitive and selective detection of different analytes, commenced a few years ago (Sherigara et al. 2003). Due to a sequential reversible oxidative and reductive electron transfers, fullerenes act as electron mediators and, hence, operate as electron relays for activation of oxidations or reductions of target substances.

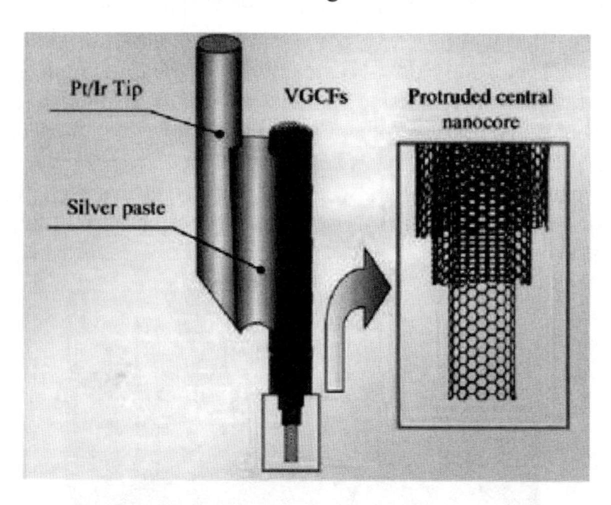

Fig. 11.28: Use of MWCNT as AFM tip

11.11 Composite Materials

Carbon nanostructures and C-60 fullerenes show unique properties, which make these structures ideally suited for the fabrication of advanced composite materials (Dai et al. 2001). These carbon nanostructures provide an important means for making advanced composite materials with polymers, which have been demonstrated to show properties characteristic of both constituent components with interesting synergetic effects. For example, photovoltaic effects, arising from the photoinduced charge transfer at the interface between conjugated polymers as donors and a C_{60} film as acceptor, suggests interesting opportunities for improving energy-conversion efficiencies of photovoltaic cells based on conjugated polymers (Sariciftci et al. 1992). Indeed, increased quantum yields have been obtained by addition of C_{60} to form heterojunctions with conjugated polymers, such as poly(p-phenylenevinylene), poly (2-methoxy-5-(2¢-ethyl-hexyloxy)-pphenylenevinylene) (MEH-PPV), poly (3-alkylthiophenes) (P(3AT)s), and platinum-poly-yne. Another example: Due to its delocalized p-electron orbital fullerene C_{60} has been shown to exhibit third-order optical responses comparable to those of certain conjugated polymers (Innocecenzi et al. 2000). Dissimilar CNTs may be joined together allowing them to form molecular wires with interesting electrical, magnetic, NLO, and mechanical properties attractive for a variety of potential applications. Indeed, CNTs have been proposed as new materials for electron field emitters in panel displays, single-molecular transistors, scanning probe microscope tips, gas and electrochemical energy storage, catalyst and protein/deoxyribonucleic acid (DNA) supports, molecular-filtration membranes, and artificial muscles. Composite materials from carbon nanofibers were already discovered. Vapor-grown carbon nanofibers (VGCFs), used as reninforcements for thermoplastic matrices, have potential applications as conducting polymers, enhancing both stiffness and thermal stability (Lozano et al. 2000).

11.12 Summary

Carbon-based nanomaterials and nanostructures including CNTs and fullerenes play an important role in nanoscale science and technology. This charpter has given an overview of the current status of fullerenes, CNTs and related carbon nanostructures. This field shows several important directions in basic research, including electronic transport, mechanical, field emission properties and chemistry. The perspective for applications also appears very bright, for example, the main avenues of potential applications of carbon nanotubes are: conducting nanowires; field emitters; nanotips for Scanning Tunneling Microscope and ultimate reinforcement fibers for composites. Their realization depends on good preparation methods and precise characterization. It is believed exciting results are born to the world.

11.13 References

Abe H, Shimizu T, Ando A, Tokumoto H (2004) Simple thermal chemical vapor deposition synthesis and electrical property of multi-walled carbon nanotubes, Phys. E-low-dime. & nanostru. 24:42-43

Adhyapak PV, Maddanimath T, Pethkar S (2002) Application of electrochemically prepared carbon nanofibers in supercapacitors, J. Pow. Sour. 109:105

Ahlskog M, Seynaeve E (1999) Ring formations from catalytically synthesized carbon nanotubes, Chem. Phys. Lett. 300:202

Ajayan PM, Ebbesen TW (1997) Nanometre-size tubes of carbon, Rep. Prog. Phys. 60:1025

Bacsa R, Laurent C, Morishima R (2004) Hydrogen storage in high surface area carbon nanotubes produced by catalytic chemical vapor deposition, J. Phys. Chem. B 108:12718

Baker RTK (1989) Catalytic growth of carbon filaments, Carbon 27:315

Brusatin G, Signorini R (2002) Linear and nonlinear optical properties of fullerenes in solid state materials, J. Mater. Chem. 12:1964

Chen QL, Xue KH, Shen W (2004) Fabrication and electrochemical properties of carbon nanotube array electrode for supercapacitors, Elec. Acta 49:4157

Chen ZF (2004) The smaller fullerene C_{50}, isolated as $C_{50}Cl_{10}$, Chem. Int. ed., 43:4690

Cyvin SJ (1991) Enumeration and classification of benzenoid hydrocarbons further developments for constant-isomer series, J. Chem. Inf. Comput. Sci. 31:413

Daenen M (2003) The wondrous world of carbon nanotube, http://students.chem.tue.nl/ifp03/default.htm

Dai HJ, Rinzler AG, Nikolaev P (1996) Single-wall nanotubes produced by metal-catalyzed disproportionation of carbon monoxide, Chem. Phys. Lett. 260:471

Dai HJ, Hafner JH, Rinzler AG (1996) Nanotubes as nanoprobes in scanning probe microscopy (1996) Nature 384:147

Dai LM, Mau AWH (2001) Controlled synthesis and modification of carbon nanotubes and C-60: Carbon nanostructures for advanced polymer composite materials, Adv. Mater. 13:899

Damnjanovic M, Milosevic I, Vukovic T, Sredanovic R (1999) Full symmetry, optical activity, and potentials of single-wall and multiwall nanotubes, Phys. Rev. B 60:2728

De Jong KP, Geus JW (2000) Carbon nanofibers: Catalytic synthesis and applications, Catal. Rev.-Sci. Eng. 42:481

Deng MG, Zhang, Hu YD (2004) Effect of activation and surface modification on the properties of carbon nanotubes supercapacitors, Phys.-Chim. Sinica 20:432

Dias JR (1993) Fullernens to benzenoids and the leapfrog algorithm, Chem. Phys. Lett. 204:486

Diudea MV, John PE, Graovac A (2003) Leapfrog and related operations on toroidal fullerenes, Crot. Chem. Acta 76:153

Ebbesen TW, Tabuchi J, Tanigaki K (1992) The mechanistics of fullerene formation, Chem. Phys. Lett. 191:336

Gao B, Kleinhammes A, Tang XP (1999) Electrochemical intercalation of single-walled carbon nanotubes with lithium (1999) Chem. Phys. Lett. 307:153

Grimm D, Muniz RB Latge A (2003) From straight carbon nanotubes to Y-shaped junctions and rings, Phys. Rev. B 68:193407

Guay P, Stansfield BL, Rochefort A (2004) On the control of carbon nanostructures for hydrogen storage applications, Carbon 42:2187

Guldi DM, Imahori H, Tamaki K (2004) A molecular tetrad allowing efficient energy storage for 1.6 s at 163 K, J. Phys. Chem. A 108:541

Hayashi Y, Tokunaga T, Soga T (2004) Corn-shape carbon nanofibers with dense graphite synthesized by microwave plasma-enhanced chemical vapor deposition, Appl. Phys. Lett., 84:2886

Hebard AF, Rosseinskky MJ (1992) Superconductivity at 18 K in potassium-doped C60 (1992) Nature 350:600

Hu YH, Ruckenstein E (2003) Ab initio quantum chemical calculations for fullerene cages with large holes, J. Chem. Phys. 119:10073

Innocenzi P, Brusatin G, Guglielmi M (2000) Optical limiting devices based on C-60 derivatives in sol-gel hybrid organic-inorganic materials, J. Sol-Gel Sci. Technol. 19:263

Katayama T, Araki H, Yoshino K (2002) Multiwalled carbon nanotubes with bamboo-like structure and effects of heat treatment, J. Appl. Phys. 91:6675

Kim C, Kim JS, Kim SJ (2004) Supercapacitors prepared from carbon nanofibers electrospun from polybenzimidazol, J. Elec. Soci. A 151:769

Kim MS, Rodriguez NM, Baker RTK (1991) The interaction of hydrocarbons with copper nickel and nickel in the formation of carbon filaments, J. Catal. 131:60

Koshio A, Yudasaka M, Iijima S (2002) Metal-free production of high-quality multi-wall carbon nanotubes, in which the innermost nanotubes have a diameter of 0.4 nm, Chem. Phys. Lett. 356:595

Lee CJ, Park J (2000) Growth model of bamboo-shaped carbon nanotubes by thermal chemical vapor deposition, Appl. Phys. Lett. 77:3397

Lee SU, Han YK (2004) Structure and stability of the defect fullerene clusters of C-60: C-59, C-58, and C-57, J. Chem. Phys. 121:3941

Lee YH, Jang YT, Choi CH (2002) Direct nano-wiring carbon nanotube using growth barrier: A possible mechanism of selective lateral growth, J. Appl. Phys. 91:6044
Lee YH, Pan KC, Lin YY (2003) Graphite with fullerene and filamentous carbon structures formed from iron melt as a lithium-intercalating anode, Mater. Lett. 57:1113
Lin WH, Mishra RK, Lin YT, Chi J (2003) Computational studies of the growth mechanism of small fullerenes: A ring-stacking model, Chem. Soc. 50:575
Liu HW, Hou SM, Liu SJ (2001) Investigation on the structure and electric properties of Bucky onions, Acta Phys.-Chim. Sini. 17:427
Liu XY, Huang BC, Coville NJ (2002) The $Fe(CO)_5$ catalyzed pyrolysis of pentane: carbon nanotube and carbon nanoball formation, Carbon 40:2791
Lou ZS, Chen CL, Chen QW (2005) Formation of variously shaped carbon nanotubes in carbon dioxide–alkali metal (Li, Na) system, Carbon 43:1104
Lozano K (2000) Vapor-crown carbon-fiber composites: Processing and electrostatic dissipative applications, Metal & Mater. Soc. A 52:34
Lu WG, Dong JM, Li ZY (2001) Optical properties of aligned carbon nanotube systems studied by the effective-medium approximation method, Phys. Rev. B 63:33401
Makarova TL (2001) Electrical and optical properties of pristine and polymerized fullerenes, Semiconductors 35:243
Maurin G, Bousquet C, Henn F (2000) Electrochemical lithium intercalation into multiwall carbon nanotubes: a micro-Raman study, Sol. State Ionics 136:1295
Moriguchi I, Nakahara F (2004) Colloidal crystal-templated porous carbon as a high performance electrical double-layer capacitor material, Ele. Sol. Sta. Lett. 7:A221
Neimark AV, Ruetsch S, Kornev KG (2003) Hierarchical pore structure and wetting properties of single-wall carbon nanotube fibers, Nano letters 3:419
Niyogi S, Hamon MA, Hu H (2002) Chemistry of single-walled carbon nanotubes, Acco. Chem. Rese. 35:1105
Nojeh A, Wong WK, Baum AW (2004) Scanning electron microscopy of field-emitting individual single-walled carbon nanotubes, Appl. Phys. Lett. 85:112
Oberlin A, Endo M, Koyama T (1976) Filamentous growth of carbon through benzene decomposition, J. Crystal Growth 32:335
Pan CX, Liu YL, Cao F (2004) Synthesis and growth mechanism of carbon nanotubes and nanofibers from ethanol flames, Micron 35:461
Park JB, Choi GS (2002) Characterization of Fe-catalyzed carbon nanotubes grown by thermal chemical vapor deposition, J. Crystal Growth 244:211
Qin Y, Zhang ZK, Cui ZL (2004) Helical carbon nanofibers with a symmetric growth mode, Carbon 42:1917
Rodriguez NM (1993) A review of catalytically grown carbon nanofibers, J. Mat. Res. 8:3233
Saito Y (1995) Nanoparticles and filled nanocapsules, Carbon 33:979
Sariciftci NS, Braun D, Zhang C (1992) Photoinduced electron-transfer from a conducting polymer to buckminsterfullerene, Science 258:1474
Schon JH, Kloc C, Batlogg B (2001) High-temperature superconductivity in lattice-expanded C-60, Science 293:2432
Shimoda H, Gao B, Tang XP (2002) Lithium intercalation into opened single-wall carbon nanotubes: Storage capacity and electronic properties, Phys. Rev. Lett. 88:15502
Shiraishi M, Takenobu T (2002) Hydrogen storage in single-walled carbon nanotube bundles and peapods, Chem. Phys. Lett. 358:213
Sigle W, Redlich P (1997) Point defect concentration development in electron-irradiated bucky onions, Phil. Maga. Lett. 76:125
Stephens PW, Cox D (1992) Lattice structure of the fullerene ferromagnet TDAE-C-60, Nature 355:331

Tanaka A, Yoon SH, Mochida I (2004) Preparation of highly crystalline nanofibers on Fe and Fe-Ni catalysts with a variety of graphene plane alignments, Carbon 42:591

Toebes ML, Bitter JH, VanDillen AJ, De Jong KP (2002) Impact of the structure and reactivity of nickel particles on the catalytic growth of carbon nanofibers, Cata. Today 76:33

Vasil'ev YV, Hirsch A, Taylor R, Drewello T (2004) Hydrogen storage on fullerenes: hydrogenation of $C_{59}N$ center dot using $C_{60}H_{36}$ as the source of hydrogen, Chem. Commun. p 1752

Wang YX, Tan SH, Jiang DL (2003) Preparation of porous carbon derived from mixtures of furfuryl resin and glycol with controlled pore size distribution, Carbon 41:2065

Wang YY, Tang GY, Koeck FM (2004) Experimental studies of the formation process and morphologies of carbon nanotubes with bamboo mode structures, Diamond and Related Materials 13:1287

Wei G, Fujiki K, Saitoh H, Shirai K, Tsubokawa N (2004) Surface grafting of polyesters onto carbon nanofibers and electric properties of conductive composites prepared from polyester-grafted carbon nanofibers, Poly. J. 36:316

Winter J, Kuzmany H (1992) Potassium-doped fullerene KxC-60 with x=0, 1, 2, 3, 4 and 6, Solid State Comm. 84:935

Wu HS, Xu XH, Jiao HJ (2004) Structure and stability of C-48 fullerenes, J. Phys. Chem. A 108:3813

Xie SY, Gao F, Zheng LS (2004) Capturing the labile fullerene[50] as $C_{50}Cl_{10}$, Science 304:699

Ye Y, Ahn CC, Witham C (1999) Hydrogen adsorption and cohesive energy of single-walled carbon nanotubes, Appl. Phys. Lett. 74:2307

Yeong KS, Thong JTL (2004) Effects of adsorbates on the field emission current from carbon nanotubes, Appl. Surf. Sci. 233:20

Yildirim T, Barbedette L (1996) Synthesis and properties of mixed alkali-metal-alkaline-earth fullerides, Phy. Rev. B 54:11981

Zhao J, Buldum A, Han J, Lu JP (2000) First-principles study of Li-intercalated carbon nanotube ropes, Phys. Rev. Lett., 85:1706

Zhao Q, Gan ZH, Zhuang QK (2002) Electrochemical sensors based on carbon nanotubes, Electroanalysis 14:1609

Zinn AA (2003) Flexible transistors with high carrier mobilities made from carbon nanotubes, MRS Bulletin 28:789

12 Molecular Logic Gates

Chandana Karnati and Hai-Feng Ji

Louisiana Tech University, Ruston, LA 71272

12.1 Introduction

It was predicted that scale-down development of silicon microchips would reach its limit in 2012. This chapter introduces molecular logic gates in contrast to established traditional form of gates based on silicon microchips. The roadblocks arise from fundamental physical constraints as well as monetary restrictions. The scientific barriers include very thin oxide layers i.e at the four-atom-thick level resulting in inadequate insulation thereby, causing charge leakage. Moreover, silicon looses its fundamental band structure when restricted to very small sizes. The oxide thickness limit and cannot simply be overcome by technological improvements. However, this may be overcome by using a different solution than further thinning of the oxide (Schulz 1999). Molecules, due to their discrete orbital levels, possess large energy level separations at room temperature and at the nanometer-size level make them independent of broadband properties. Construction of new fabrication lines for each generation of chips is required to maintain the chip manufacturing process, by contrast, molecular construction is a bottom-up technology that uses atoms to build nanometer-sized molecules that could further self-assemble into a desired computational circuitry. The bottom-up approach gives rise to the prospect of manufacturing electronic circuits in rapid, cost-efficient and flow-through processes. These processes can be analogous to the production of photographic film, with overall enormous cost savings over traditional microchip fabrication. Moreover, molecular density is high compared to solid-state devices (Tour 2000).

12.2 Logic Gates

Logic gates are the fundamental components of the digital circuits, which process binary data encoded, in electrical signals. According to the predefined logic functions they transduce electrical inputs into electrical outputs (Mitchell 1995).

Logic gates can be classified as single-input and multiple-input logic gates, based on the 4 possible output patterns for a single-input i.e if the input is 0 the output can be 0 or 1 (two choices) and if the input is 1 the output is 0 or 1 (two choices). Each one of these four output bit patterns corresponds to a logic type: PASS0 which always outputs 0, PASS1 which always outputs 1, YES always obeys the input, NOT always opposes the input whatever the input is. OR, AND, INHIBIT, etc fall under multiple-input logic gates. The symbol and the truth table of three basic logic gates, NOT, OR, AND, are shown in Fig. 12.1.

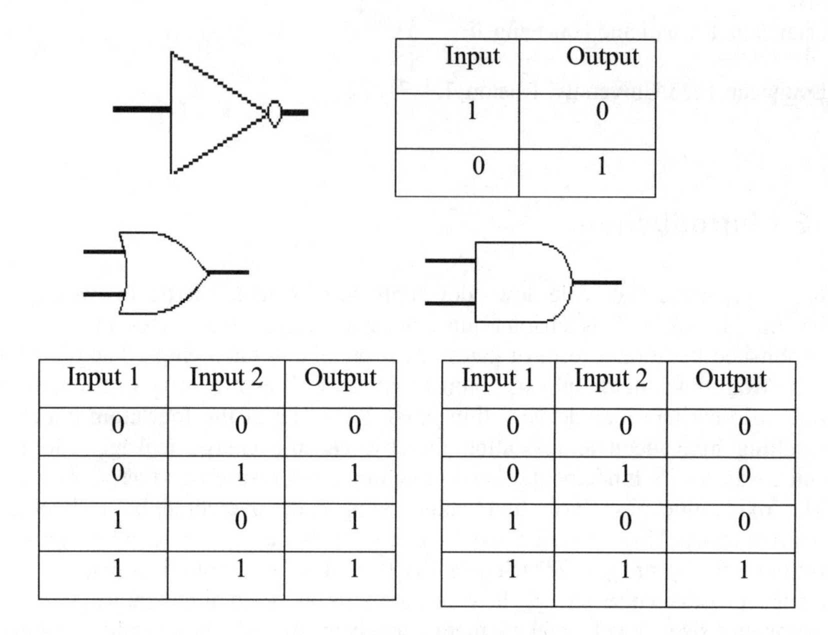

Input	Output
1	0
0	1

Input 1	Input 2	Output
0	0	0
0	1	1
1	0	1
1	1	1

Input 1	Input 2	Output
0	0	0
0	1	0
1	0	0
1	1	1

Fig. 12.1. Three basic logic gates and their corresponding truth tables

The AND and OR gates convert two inputs (I1and I2) into a single output (O). In the AND gate the output is 1 when both the inputs are 1, the output is 0 in all the other three cases. Logic gates can be interconnected and the logic functions of the resulting arrays are the combination of the operations performed by individual gates.

Combinational logic circuits can be assembled by connecting the input and output terminals of the three basic logic gates AND, NOT and OR. Simple examples of the combinational logic circuits able to convert two binary inputs (I1 and I2) into a single output (O) according to the corresponding truth tables are summarised in Fig. 12.2.

Simple organic molecules in which binary operations could be performed are called as molecular switches. They adjust their structural and electronic properties when stimulated with chemical, electrical, or optical inputs. Chemical systems require proper anology between the molecular switches and the logic gates in order to execute AND, NOT, OR and more complex logic functions.

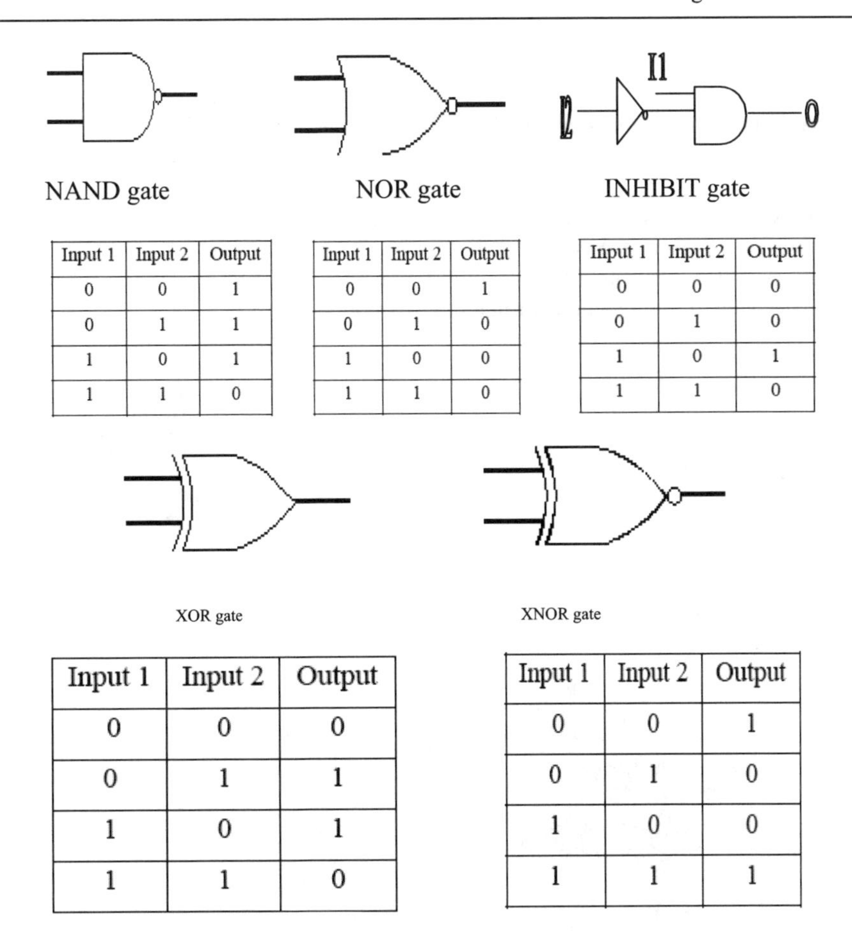

NAND gate

Input 1	Input 2	Output
0	0	1
0	1	1
1	0	1
1	1	0

NOR gate

Input 1	Input 2	Output
0	0	1
0	1	0
1	0	0
1	1	0

INHIBIT gate

Input 1	Input 2	Output
0	0	0
0	1	0
1	0	1
1	1	0

XOR gate

Input 1	Input 2	Output
0	0	0
0	1	1
1	0	1
1	1	0

XNOR gate

Input 1	Input 2	Output
0	0	1
0	1	0
1	0	0
1	1	1

Fig. 12.2. Combinational logic circuits with two inputs and one output

12.3 Fluorescence based Molecular Logic Gates

The ultimate goal of miniaturisation is to use molecular assemblies as memory, processing and mechanical devices. To acieve these goals, it is essential to transform molecular structures between two or more states in response to external signals such as photonic, chemical, electrochemical, or magnetic stimuli, and to tailor readable output(s) such as electronic or optical signals that reflect the molecular state. Electronically transduced photochemical switching is possible in organic monolayers and thin films, enzyme monolayers, redox-enzymes tethered with photoisomerisable groups, enzymes reconstituted onto photoisomerisable FAD-cofactors etc. Photoinduced electron transfer process (PET) in many organic compounds, leads to OFF-ON and ON-OFF mechanism. Numerous organic molecules can follow this simple signal transduction protocol, and execute simple

NOT and YES operations. For example, pyrazole derivative **1** emits only when the concentration of H^+ is low. Photoinduced electron transfer from the central pyrazoline unit to the pendant benzoic acid quenches the fluorescence of the protonated form. Thus, the emission intensity switches from a high to low value when the concentration of the H^+ changes from low to high value. This inverse relation between the chemical input (concentration of H^+) and the optical output (fluorescence intensity) corresponds to a NOT operation (de Silva 1993). On the contrary, PET process in molecule 2 executes a YES operation, where the fluorescence of anthracene is turned ON with presence of H^+.

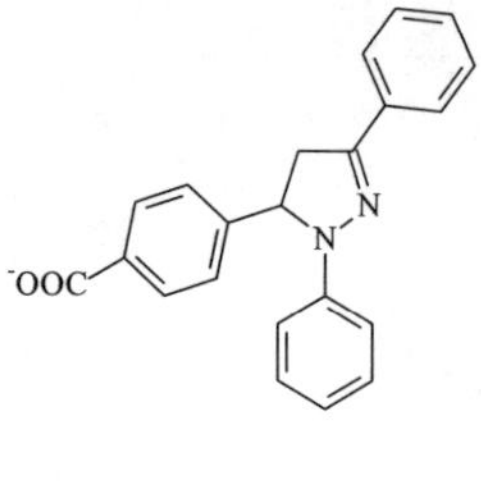

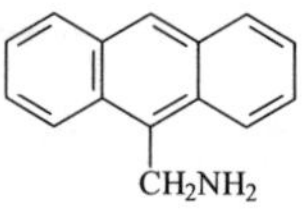

In OR gates the output is 0 only when both the inputs are 0 and the output will be 1 in all the other three cases. To execute this operation compounds should be able to respond to two input signals i.e., they have to produce a detectable signal when one or both the inputs are applied. This can be readily realised by many photophysical processes, which have been an active area of current photochemistry process study. For example, compound **3** generates an optical output (fluorescence) in response to one or two chemical inputs (K^+ and Rb^+ metal cations). In the absence of metal cations, PET from the tertiary amino groups to the anthracene appendage quenches the fluorescence (de Silva 1993).

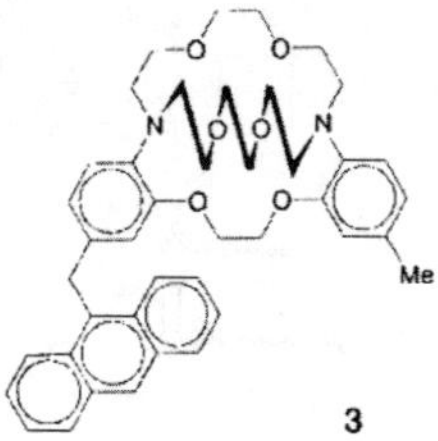

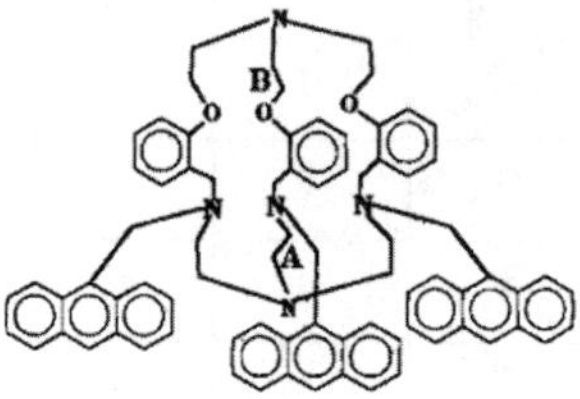

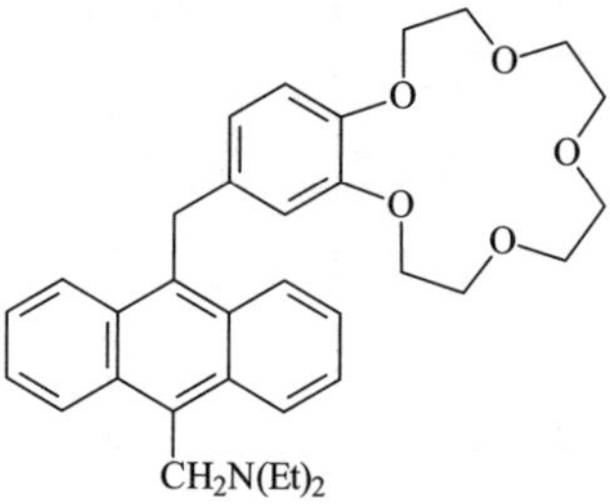

In another example, the cryptand compound **4** containing three anthryl moieties is capable of functioning as a very efficient multi-input OR logic gate. The fluorophores in these systems do not show any fluorescence due to an efficient PET from nitrogen lone pairs to the anthracene moiety. However, the fluorescence is recovered to different extents in the presence of transition metal ions like Cu(II) and Ni(II), inner-transition metal ions like Eu(III) and Tb(III) ions along with H^+

and Zn(II) (Ghosh 1996). The first fluorescence based molecular AND gate was introduced by de Silva in his work on Na^+ detection (de Silva **1997**). Molecule **5** fluoresces when amine receptor is bound to H^+ and benzo-15-crown-5 ether to Na^+. This is the ON state with high fluorescence intensity emitted from the anthracene moiety. When the amine group is proton-free, it serves as an efficient PET donor ($\Delta G_{PET} \sim -0.1$ eV) to the fluorophore, which is separated by only a methylene group. In the presence of protons in sufficient concentration and in the absence of Na^+, the protonated amino methyl moiety behaves as an electron-withdrawing group on the anthracene fluorophore permitting rapid PET from the benzocrown ether moiety ($\Delta G_{PET} \sim -0.1$ eV) a short distance away and resulted in the quenching of the fluorescence of **5**. The ion-induced fluorescence emission spectral behavior of these systems is illustrated in Fig. 12.3, giving an excellent example for a digital AND logic gate (de Silva 1997). Compound **6** also generates an optical output (fluorescence) only when both two chemicals exist (H^+ and K^+ or Na^+ metal cation) (AND gate) (de Silva 1993).

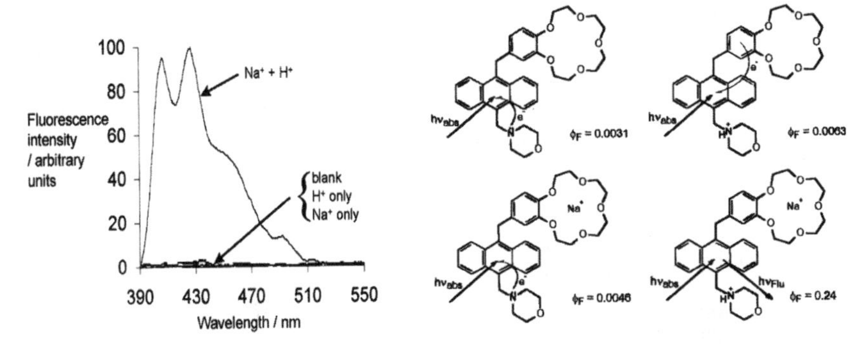

Fig. 12.3. Fluorescence emission spectra of gate **5** in MeOH excited at 377 nm under the four ionic conditions necessary to test its digital AND logic gate function. The PET processes are also shown

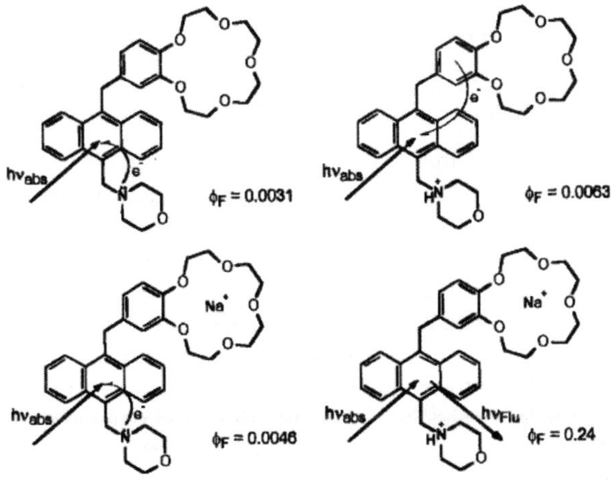

6

7

This molecule exhibited an unique logic gate system that combines an AND gate and an INHIBIT gate as described in Fig. 12.4 by two representations. For example, **2** combines an AND gate and an INHIBIT gate with the input notations IN_1, IN_2, and IN_3 signifying presence (1) or absence (0) of potassium ion (K^+), proton (H^+), and sodium ion (Na^+) in solution.

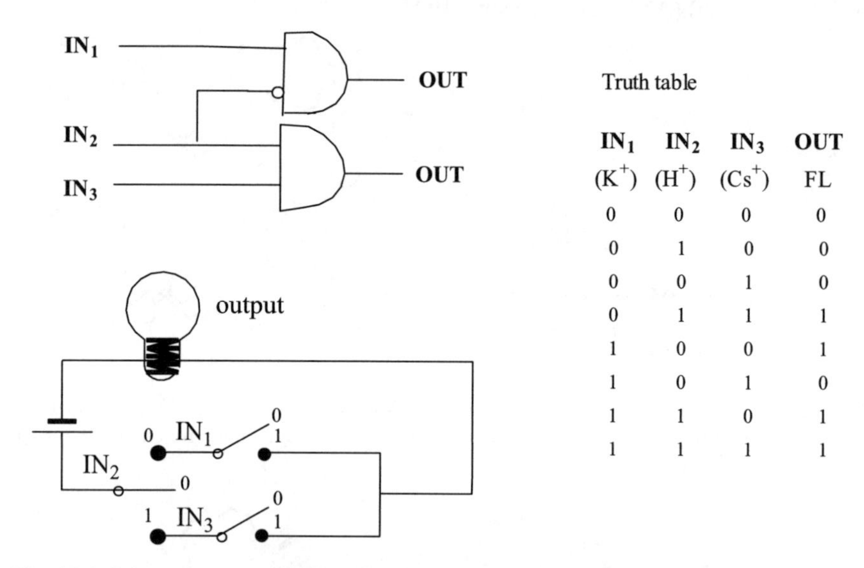

IN_1 (K^+)	IN_2 (H^+)	IN_3 (Cs^+)	OUT FL
0	0	0	0
0	1	0	0
0	0	1	0
0	1	1	1
1	0	0	1
1	0	1	0
1	1	0	1
1	1	1	1

Fig. 12.4. Schematic representation of **7** showing an AND gate and an INHIBIT gate logics switched by H^+. Inset shows the truth table for this logical gating performance

Monoaza-18-crown-6 ether and boronic acid receptor linked compound **8** shows selective fluorescent enhancement with D-glucosamine hydrochloride in aqueous solution at pH 7.18. This system also behaves like an AND logic gate where in the fluorescence recovery is observed only when ammonium cation and a diol compound are supplied (Cooper 1997).

Flavylium derivative **9** can exist in more than two forms (*multistate*) that can be interconverted by more than one type of external stimulus (*multifunctional*). This can be taken as the basis for an optical memory system with multiple storage and non-destructive readout capacity. For the 4'hydroxyflavylium compound, light excitation and pH jumps can be taken as inputs, and fluorescence as outputs that exhibits an AND logic function (Pina 1998, 1999).

8

9

(Structural transformations of 4'hydroxyflavylium compound)

Single molecules with several functional groups can be used for mimicking complicated multi logical functions. For example, compound **10** converts two chemical inputs into an optical output mimicking a NAND function. At neutral pH, the anthracene fluorophore emits fluorescence. After the addition of either H^+ or adenosine triphosphate (ATP), the fluorescence remains but the optical output diminishes in intensity when both chemical inputs (H^+ and ATP) are applied. Protonation of the aligoamino arm leads to the association of **10** with ATP resulting in the quenching phenomenon (Albelda 1999).

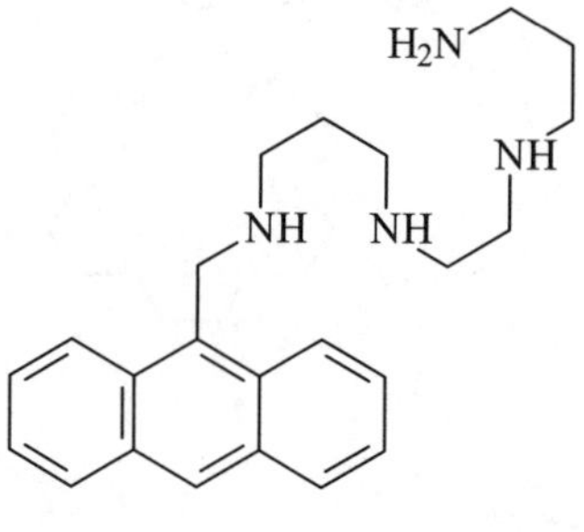

10

Similar to OR gate, NOR gate can readily be realised. The photophysical properties of metal complexes using lanthanide ions as emitting moieties can also be used to develop luminescent molecular level devices. Compound **11** converts two chemical inputs into an optical output according to a NOR operation. For Terbium complex **11**, H^+ and O_2 are the chemical inputs and lanthanide luminescence is the output. In the presence of H^+, excitation at 304 nm is followed by the fluorescence of the phenanthridine appendage. In the absence of H^+, the excitation of the phenanthridine fragment is followed by intersystem crossing, electron transfer to the Tb center and delayed emission from the lanthanide. In the presence of O_2, the tripled state of the phenanthridine appendage is quenched and the Tb emission is not observed. Thus, the optical output (Tb emission) is detected only when the two chemical inputs (H^+ and O_2) are not applied (Parker 1998).

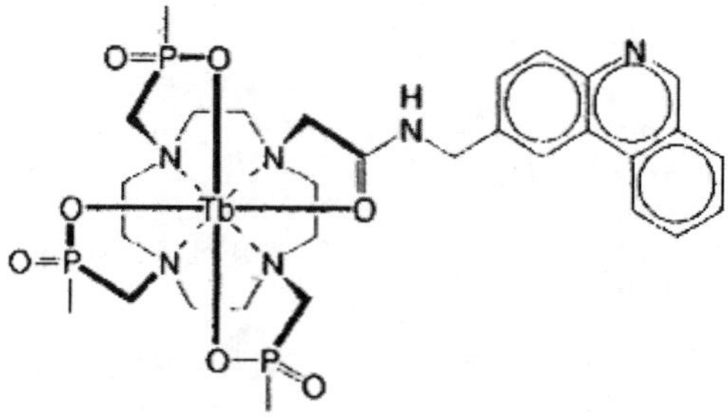

11

In another example, the anthracene fluorescence of compound **12** under neutral pH conditions can be observed only in the absence of Cu^{2+} and Ni^{2+} (chemical inputs). The binding of one of the cations to the aligoamino arm efficiently quenches the anthracene fluorescence. The behavior of the compound **12** corresponds to molecular NOR gate (Alves 2001).

12

Another two NOR gate systems are shown in Fig. 12.5. Complexation or input of either H^+ or Zn^{2+} cation with **13** is expected to quench fluorescence output. Molecule **14** is a spatially overlapping “fluorophore-receptor” system. Its receptor, which is reminiscent of 2,2’; 6’,2’’-terpyridyl, complexes H^+ or Hg^{2+}. The fluorescence output of **14** is seriously quenched by either cation and the NOR truth table is satisfied. The mechanism of fluorescence quenching with Hg^{2+} involves a nonemissive ligand-to-metal charge transfer (LMCT) excitation (de Silva 1999a).

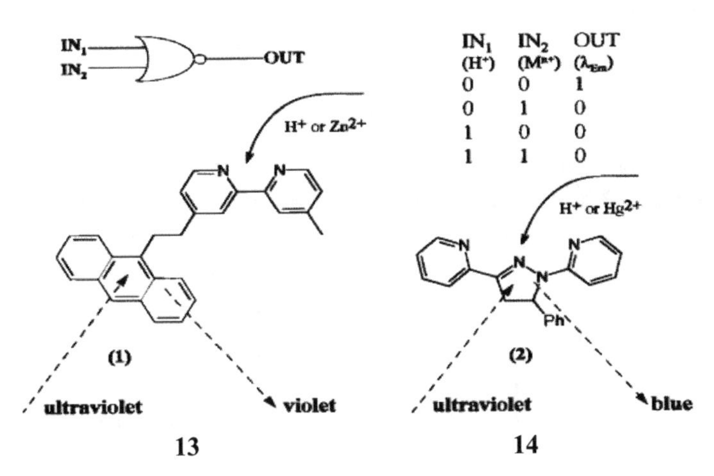

Fig. 12.5. Molecular-scale implementation of **NOR** logic gates

Tb(III) based quinolyl derived macrocyclic 1,4,7,10-tetraazacyclododecane (cyclen) conjugate **15,** executes the INHIBIT function with H^+ and O_2 as the chemical inputs and Tb luminescence being the optical output. Under basic conditions, the complex **32** has a weak emission band at 548 nm in either aerated or degassed solutions. After the addition of H^+, the emission intensity dramatically increases in the absence of O_2. The protonation of the quinoline fragment facilitates the energy transfer from the quinoline excited state to the lanthanide excited state, enhancing the emission intensity. The presence of O_2 supresses the energy transfer process. Thus the emission intensity is high only when the concentration of H^+ is high and that of O_2 is low. If a positive logic convention is applied to the system, the signal transduction of the **32** translates into the truth table of the INHIBIT circuit (Gunnlaugsson 2000, 2001).

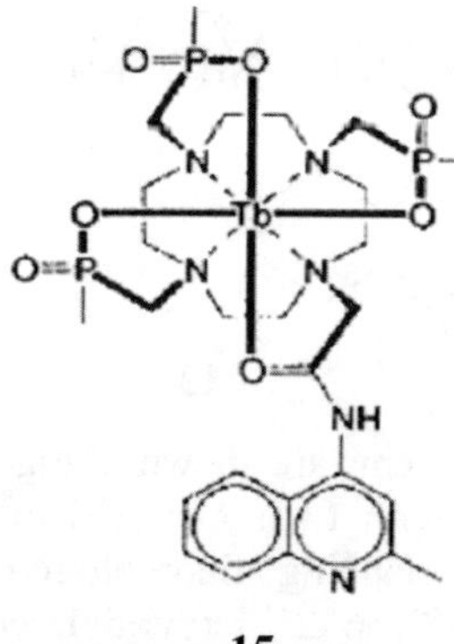

15

Functional integration also succeeded in achieving a more complex INHIBIT logic operation with a rather simple supramolecule **16** (Fig. 12.6) (de Silva 1999).

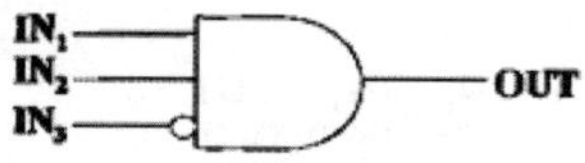

IN_1 (Ca^{2+})	IN_2 (β-CD)	IN_3 (O_2)	OUT (λ_{Em})
0	0	0	0
0	1	0	0
1	0	0	0
1	1	0	1
0	0	1	0
0	1	1	0
1	0	1	0
1	1	1	0

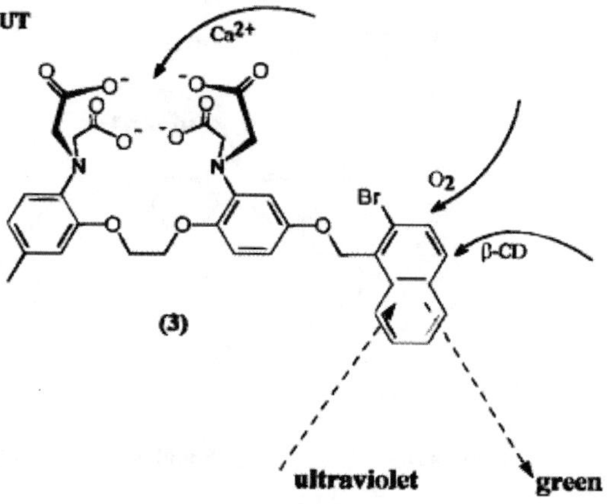

16

Fig. 12.6. Molecular-scale implementation of INHIBIT logic gates

Photochromic spiropyrans (SP) via proton transfer or energy transfer is another logical gate system (Fig. 12.7). Oxidation potential of Fe^{2+} is significantly reduced after coordination with the open form of spiropyran, thus, the electron transfer between the tetrathiafulvalene (TTF) unit and Fe^{3+} can be photocontrolled in the presence of SP. Radical cation of fluorescein formed by oxidation with ferric ion can be reduced to the corresponding neutral species upon UV light irradiation in the presence of SP (Scheme 1), just like the TTF^{+}. Radical cation of fluorescein shows strong emission at 480 nm when excited at 410 nm compared to its neutral form. Hence, a promising redox fluorescence switch based on fluorescein can be constructed. A redox fluorescence switch based on fluorescein and a photochromic molecular switch based on spiropyran through the Fe(II)/Fe(III) redox couple is realised. The concatenation of these two molecular switches can mediate the transduction of two kinds of external inputs into one kind of optical output, which corresponds to the function of an INHIBIT logic gate (Guo 2004).

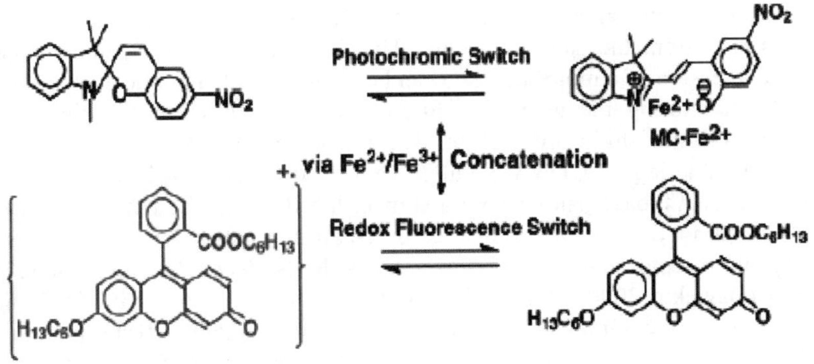

Fig. 12.7. Photochromic switch of spiropyran

12.4 Combinational Logic Circuits

A molecule with two different receptors that possess four distinct, spectroscopically distinguishable states that is a molecule that indicates whether only one, the other, both, or none of the binding sites are occupied by substrates was designed. The molecule shown below shows a multifold signal expression in the presence or absence of one or two different types of metal ions i.e., Ca^{2+} and Ag^{+} (Rurack 2001).

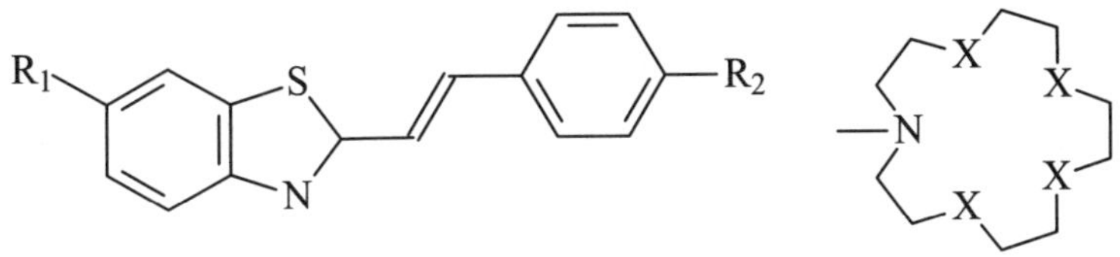

17: R_1=A15C5, R_2=$AT_4$15C5

18: R_1=A15C5, R_2=$N(CH_3)_2$

19: R_1=$N(CH_3)_2$, R_2=$AT_4$15C5

A15C5: X=O

$AT_4$15C5: X=S

In terms of digital action, **17** can be employed as a bimodal logic gate. A strongly decreased absorbance at 440 nm is only found in the presence of both ions (transducing a NAND event, with digital action = 1 for Ca^{2+}/Ag^{+} = 0/0, 0/1, and 1/0) and a strong fluorescence signal, when excited at 440 nm, is only found for Ag^{+}. The latter case with a digital signal = 1 only for Ca^{2+}/Ag^{+} = 0/1 represents a noncommutative INHIBIT logic function.

12.5 Reconfigurable Molecular Logic

In all the above cases each species was attributed with one specific logic label, but in certain circumstances one can change the coding to define different logic expressions from the same molecule. This is called as reconfigurable logic. A logic expression of a molecular gate can be changed with an optical expression by simply monitoring the gate at a different wavelength when significant input-induced spectral shifts are present. The phenomenon is known as wavelength reconfigurable logic. A DNA-binding dye 4',6'-diamidino-2-phenylindole(DAPI) (**20**) signals AT base pairing with a shift in the fluorescence emission spectrum. The signaling follows Watson-Crick base-pairing rules. The dye with its two phosphate receptor sites functions as a molecular NAND gate with nucleotides as inputs (Baytekin 2000). TRANSFER logic is equivalent to a YES (referring to $input_1$), whereby $input_2$ is not involved with the output. Reconfiguration of molecular logic gates can also be achieved by considering different types of outputs. For example, molecule **21** has four states that are distinguishable by means of absorption and emission maxima as well as emission quantum yield and lifetime. So there is no classification until operating conditions are restricted. For this compound, if the absorbance at 440 nm is taken as the output a NAND logic function can be attributed while a 2-input INHIBIT gate, if emission is taken as the output (Rurack 2001). Extending the wavelength-reconfigurable logic leads to the idea of superposition of logic gates. This function inherits the multichannel nature of polychromatic light to obtain different, simultaneous logic expressions from the same species. All the 4 single-input functions YES, NOT, PASS0, PASS1 can be demonstrated with a non-modular *chromophore-receptor* system **22** (Fig. 12.8), a single H^+ chemical input., and light transmittance output at carefully chosen wavelengths, because of large H+ induced displacement of the absorption band. All the 4 wavelengths can be observed simultaneously (de Silva 2002).

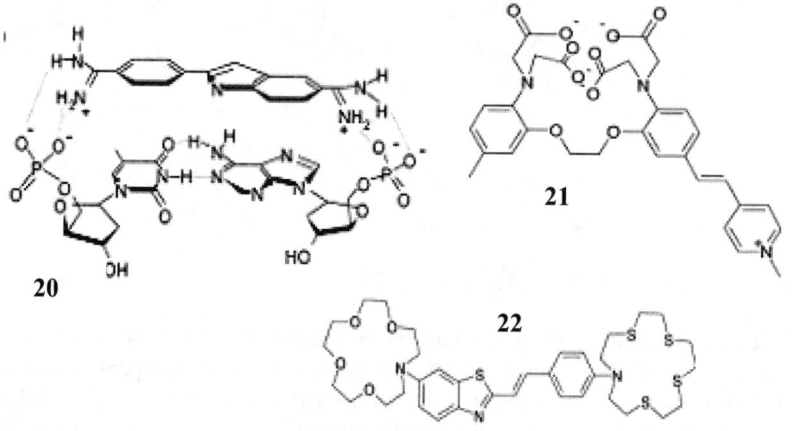

Fig. 12.8. Molecules exhibiting reconfigurable switching (**20**), TRANSFER logic (**21**), and SUPERPOSED logic respectively (**22**)

The complexity in the logic operations is also achieved using supramolecular systems with *fluorophore-spacer-receptor* configuration. Different logic operations such as NOT, OR etc can be integrated with this supramolecule (de Silva 1999).

12.6 Absorption based Molecular Logic Gates

The absorption spectra of crown spirobenzopyrans compound **23** that is responsive to the combination of ionic and photonic stimuli was developed to mimic an AND gate (Inouye 1997).

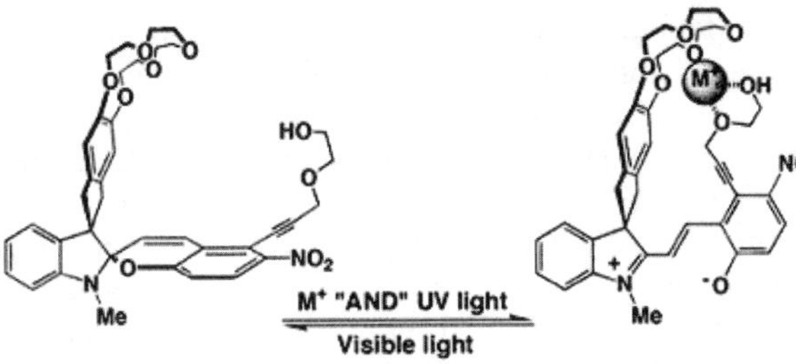

23

Dual-mode "co-conformational" switching is observed in catenanes incorporating bipyridinium and dialkylammonium recognition sites. These catenanes undergo co-conformational switching upon one-electron reduction of the two bipyridinium units. An AND logic behavior could be exhibited using acid/base stimulations by one of the catananes in its reduced form (Ashton 2001).

Biscrowned malachite green leuconitrile **24** showed clear-cut switching from the powerful cation binding to the perfect cation release, so-called all-or-none type switching in the cation binding upon photoionisation. It is deduced that similar type malachite green derivatives carrying bis(monoazacrown ether) moiety with different sizes undergo photochemical control of cation complexation as efficient as that of compound **24**. The clear-cut photoinduced switching system of cation binding with **24** is promising for device applications (Kimura 1997).

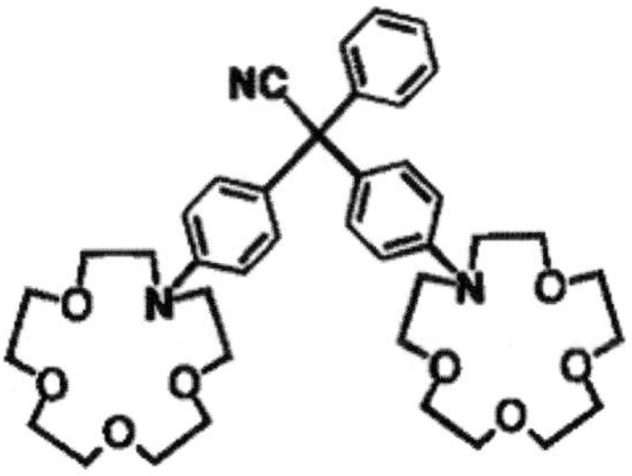

24

Redox switches: Molecular redox switches based on helical metal complexes in which an iron can occupy one of the two distinct cavities was synthesised. Iron has two distinct oxidation states Fe(II) and Fe(III) with distinct spectral properties which helps in the quantitative monitoring of the switching process. Treatment of the chiral complex 5-Fe(III) with ascorbic acid results in rapid reduction, with the appearance of the purple-red absorption of the 5-Fe(II)-bipyridil complex and translocation of iron from the internal hydroxamate binding sites to the external bipyridil sites. Subsequent oxidation reaction of 5-Fe(II) for few minutes with ammoniumpersulfate at 70°C causes reversal of this process manifested by the disappearance of the purple-red color and regeneration of the light brown color of the 5-Fe(III) (Zelikovich 1995).

Supramolecular systems can be also used to implement the combinational logic functions (Raymo 2004). In particular, certain host-guest complexes can execute XOR and XNOR functions. Individual molecular components could be used to reproduce complex logic circuits rather than networks of communicating molecular gates. Relatively complex combinational logic functions able to transduce two or three inputs into one or two outputs can also be reproduced at the molecular level. They are assembled connecting AND, NOT, and OR gates.

With the careful selection of the output parameter, both XOR and XNOR functions can be implemented on single compound, compound **25**. Compound **25** shows an absorption band at 390 nm. The complexation of Ca^{2+} in the tetracarboxylate cleft shifts this band to shorter wavelengths while protonation of the quinoline fragment pushes the same band to longer wavelengths. However, when both the inputs are added simultaneously, the opposing effects cancel each other resulting in un-alteration of the position of the absorption band. The absorbance probed at 390 nm decreases only when one of the two chemical inputs is applied. In contrast, the transmittance at the very same wavelength increases when only one of the two inputs is applied. Thus if a positive logic convention is applied to inputs and output, this compound reproduces a XNOR function when the absorbance is taken as the output, and a XOR function, when the transmittance is considered as the output (de Silva 2000).

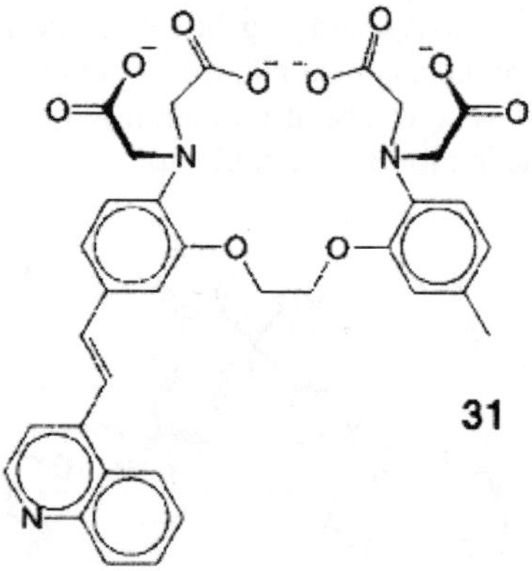

31

25

Switching action is possible between two independent molecular components, such as merocyanine and spiropyran, or 4,4'-pyridylpyridinium and 4,4'-bipyridinium, switched by visible light. Under the irradiation condition the spiropyran SP (Fig. 12.9) switches to the protonated merocyanine MEH upon acidification. The transformation of the colorless SP into the yellow-green MEH is accompanied by the appearance of an absorption band in the visible region. Upon irradiation with visible light, MEH releases a proton, switching back to the colorless form SP, and its characteristic absorption band disappears. This reversible process can be exploited to control the interconversion between the base BI and its conjugated acid BIH. These processes can be monitored following the photoinduced enhancement and thermal decay, respectively, of the current for the monolectronic reduction of the 4,4'-bipyridinium dication (Raymo 2003a-b, 2001). Spyropyran can be switched to two merocyanine forms ME and MEH when input with chemical and light stimulations as shown in Fig. 12.8. The differences in the absorption and emission properties of the three states can be exploited to follow anticlockwise and clockwise switching cycles starting and ending with SP.

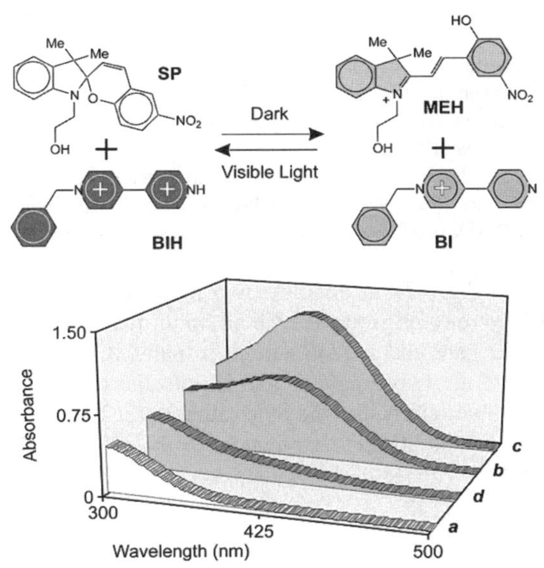

Fig. 12.9. Absorption spectra (1×10^{-4} M, MeCN, 25 °C) of an equimolar solution of SP and BIH at initial time (a), after 1 day (b) and 5 days (c) in the dark, and after the subsequent visible irradiation for 15 min (d)

This three-state molecular switch detects three input signals (***I*1**) ultraviolet light, (***I*2**) visible light and (***I*3**) H^+ and responds to these stimulations generating two output signals (***O*1**) the absorption band at 401 nm of MEH and (***O*2**) that at 563 nm of ME. Each signal can be either ON or OFF and can be represented by a binary digit, i.e., it can only take two values 1 or 0. Thus, the molecular switch *reads* a string of three binary inputs and *writes* a specific combination of two binary outputs. The logic circuit involved and the truth table are shown in the Fig. 12.10 (Raymo 2001a, 2002).

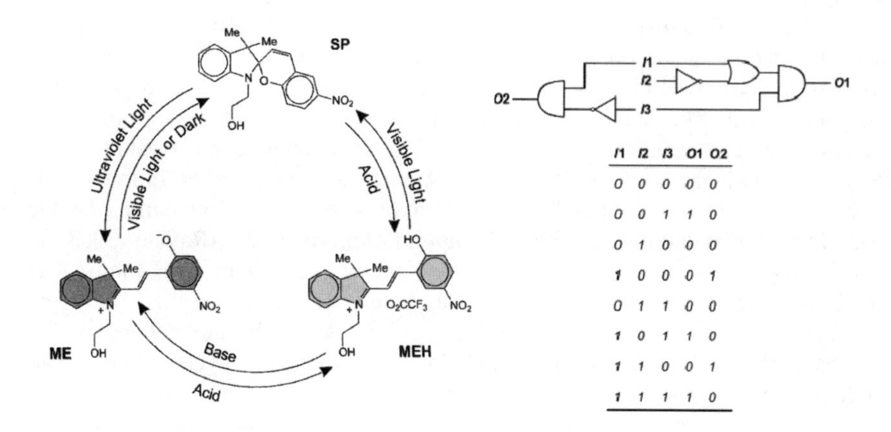

I1	I2	I3	O1	O2
0	0	0	0	0
0	0	1	1	0
0	1	0	0	0
1	0	0	0	1
0	1	1	0	0
1	0	1	1	0
1	1	0	0	1
1	1	1	1	0

Fig. 12.10. Ultraviolet light (I1), visible light (I2), and H+ (I3) inputs induce the interconversion between the three states SP, ME, and MEH. The colorless state SP does not absorb in the visible region. The yellow±green state MEH absorbs at 401 nm (O1). The purple state ME absorbs at 563 nm (O2). The truth table illustrates the conversion of input strings of three binary digits (I1, I2, and I3) into output strings of two binary digits (O1 and O2) operated by this three-state molecular switch. A combinational logic circuit incorporating nine AND, NOT, and OR operators correspond to this particular truth table.

The half-adder (Fig. 12.11), requires two molecular switches. This particular combinational logic circuit processes the addition of two input digits (I1 and I2) operating an AND gate and a XOR circuit in parallel. The XOR portion of the half-adder converts the two inputs I1 and I2 into the output O1. The AND gate converts the same two inputs into the other output O2. The role of the two outputs O1 and O2 in binary additions is equivalent to that of the unit and ten digits in decimal arithmetic. The tetracarboxylate receptors **26** and **27** satisfy these conditions.

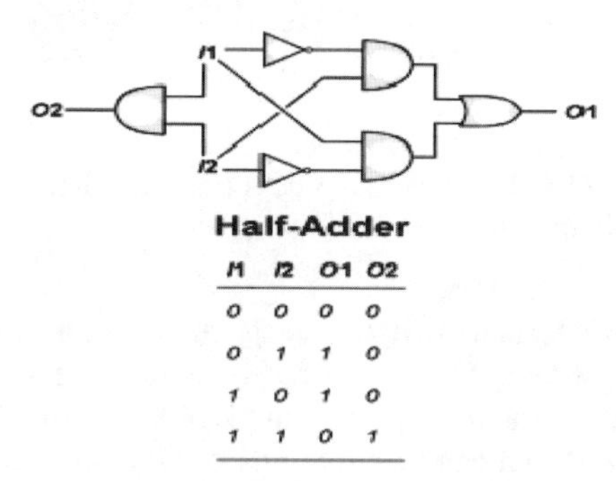

I1	I2	O1	O2
0	0	0	0
0	1	1	0
1	0	1	0
1	1	0	1

Fig. 12.11. Combinational logic circuits and their corresponding truth tables

The quinoline derivative **26** has an absorption band at 390 nm with a transmittance of 5% in H_2O. Increase in the concentrations of H^+ and Ca^{2+} increases the transmittance. But the application of both the inputs simultaneously keeps the transmittance low. Thus, the transmittance at 390 nm (O1) is high when the concentration of Ca^{2+} (I1) is high and that of H^+ (I2) is low and vice versa. The molecule **27** works similar to **26**. Hence, when the two molecular switches **26** and **27** are co-dissolved in H_2O, they can be operated in parallel with the two inputs I1 and I2. Of course, individual molecular switches cannot share the very same Ca^{2+} and H^+ ions. They actually share the same type of chemical inputs producing in response the two outputs O1 and O_2. If a positive logic convention (low = 0, high = 1) is applied to all signals, the signal transduction behavior of the two molecular switches translates into the truth table of the half-adder (Fig. 12.12) (Raymo 2002).

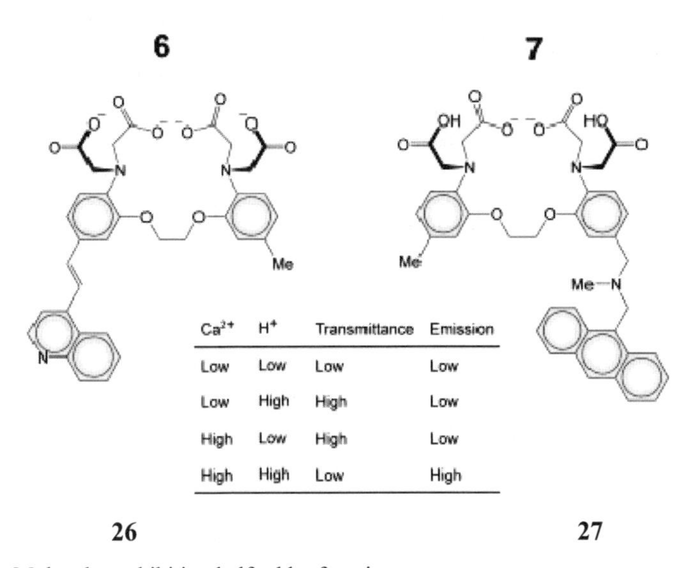

Ca^{2+}	H^+	Transmittance	Emission
Low	Low	Low	Low
Low	High	High	Low
High	Low	High	Low
High	High	Low	High

Fig. 12.12. Molecules exhibiting half-adder functions

The fullerene C_{60} molecule has many unique properties such as large non-linear susceptibility, fast response time, extremely high quantum efficiency, strong broadband reverse saturable absorption, low fluorescence quantum yield, high rigidity, high degree of symmetry, solubility, stability, capability to form thin films and crystals, and flexibility to tune its kinetic and spectral properties by the addition of chemicals and hence, emerged as an excellent material for molecular photonic applications. The transmission of a CW probe laser beam at 885 nm corresponding to the peak absorption of the S1 state through C_{60} in toluene is switched by a pulsed pump laser beam at 532 nm that excites molecules from the ground state. The transmission of the probe beam was completely switched off (100% modulation) by the pump beam at a pump intensity of 100 MW/cm^2 and a pulse width 1.5 ns, with switch on and off times of 2.5 and 7.5 ns, respectively. Hence it was able to design all-optical NOT and the universal NOR and NAND logic gates with multiple pump laser pulses (Ami 2001; Singh 2004). Energy transfer in higher electronic state serves as a new direction for molecular logic gates. These systems involve polyatomic molecules and, the number of available states participating in these energy transfers is an important consideration. A complicated logic circuit involving AND, OR and XOR operations can be executed based on sequential forward S_2-S_2 energy transfer and back S_1-S_1 energy transfer (cyclic energy transfer) for a system comprising of an azulene and zinc porphyrin (Yeow 2003). ICT chromophore-receptor systems also constitute complicated integrated systems. Different dyes for example **28-29** (Fig. 12.13) when integrated with Tsien's calcium receptors, allows to perform two- or three-input OR or NOR operations when input with multiple ions. *receptor1-chromophore-receptor2* systems employing the different chromophore dyes like **30-32** (Fig. 12.13) selectively target two ions in to the receptor terminals. The wavelength of observation can be used to configure the system to demonstrate various logic operations most important being the XOR operation with a transmittance output. Two ions exert opposite effects on the chromophore excited state and are responsible for the behaviour needed for XOR operation. Additional logic operations like "integrated XOR with a preceding OR operation" can be performed with **31** in the fluorescence mode. This system requires three input signals. Simple measurements with UV/V spectrophotometer unearth various logic functions within chromophores (Fig. 12.13) integrated with one or two receptors (de Silva 2002)

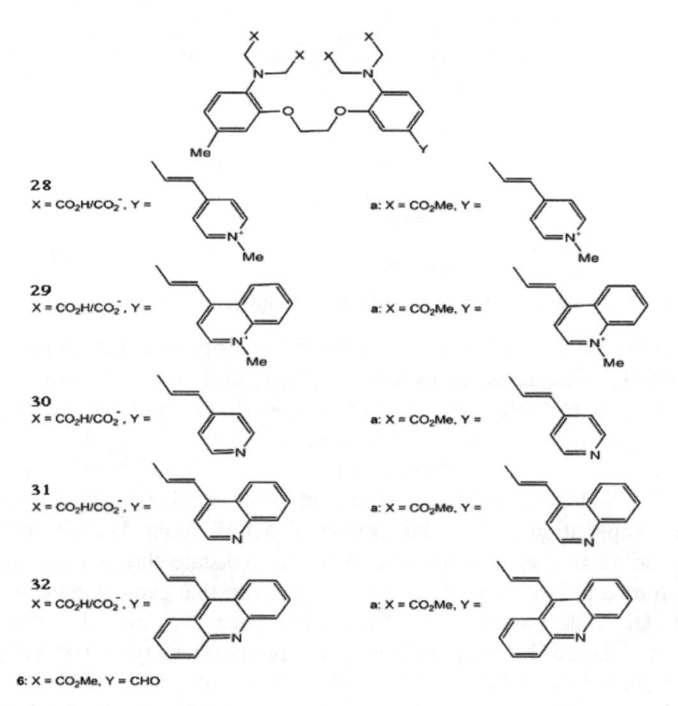

Fig. 12.13. Molecular chromophores

12.7 Molecular Logic Gates: Electronic Conductance

The molecular logic gates listed so far were investigated in solutions. The major challenge is the identification of methods to incorporate these functional molecules into solid-state devices while maintaining their signal transduction abilities. SAMs are the most promising for the ordering of wires and devices on the Au surfaces, with rigid rod systems standing nearly perpendicular to the surface, the thiol groups dominating the adsorption sites on the Au (Tour 2000).

Electronically configurable molecular-based logic gates have been fabricated from an array of configurable switches, each consisting of a monolayer of redox-active rotaxanes sandwiched between metal electrodes. The redox properties of these molecular compounds, as measured in solution, do translate well into solid-state device properties. A single monolayer of the catenane anchored with amphiphilic phospholipids counterions and sandwiched between an n-type polycrystalline Si (poly-Si) electrode and a Ti/Al top electrode can act as a switching device. This device exhibits hysteretic (bistable) current/voltage characteristics. The switch can be opened at 12 V, closed at 22 V, and read between 0.1 and 0.3 V and may be recycled many times. The switches were read by monitoring current flow at reducing voltages in air and at room temperature. The switches were irreversibly opened by applying an oxidizing voltage across the device. Once the conditions for addressing and configuring the individual devices were determined, linear arrays of devices into AND and OR wired logic gates were configured. The truth table of any AND gate is such that a high response is only recorded when both inputs are high. The high and low current levels of those gates were separated by factors of 15 and 30, respectively, which is a significant enhancement over that expected for wired logic gates (Collier 1999).

The switching mechanism is illustrated in Fig. 12.14. The ground state "co-conformer" [A^o] of this catenane has the TTF unit located inside the cyclophane. Upon oxidation, the TTF unit becomes positively charged, and the Coulombic repulsion between TTF1 and the tetracationic cyclophane causes the crown ether to circumrotate to give co-conformer [B^1], which will reduce back to [B^o] when the bias is returned to 0 V. This bistability is the basis of this device. The energy gap between the highest occupied and lowest unoccupied molecular orbitals for co-conformer [B^o] must be narrower in energy than the corresponding gap in co-conformer [A^o], implying that, in a solid state device, tunneling current through the junction containing [B^o] will be greater.

Thus, this co-conformer represents the "switch closed" state, and the ground state co-conformer [A^o] represents the "switch open" state (Collier 2000). A molecule in two-dimensional molecular electronics circuits is shown in the Fig. 12.15. Addressing an array of bistable [2]rotaxanes through a two dimensional crossbar arrangement provides the device element of a current-driven molecular electronic circuit. The development of the [2]rotaxane switches through an iterative, evolutionary process is described in (Luo 2002). The arrangement reported here allows both memory and logic functions to use the same elements (Luo 2002).

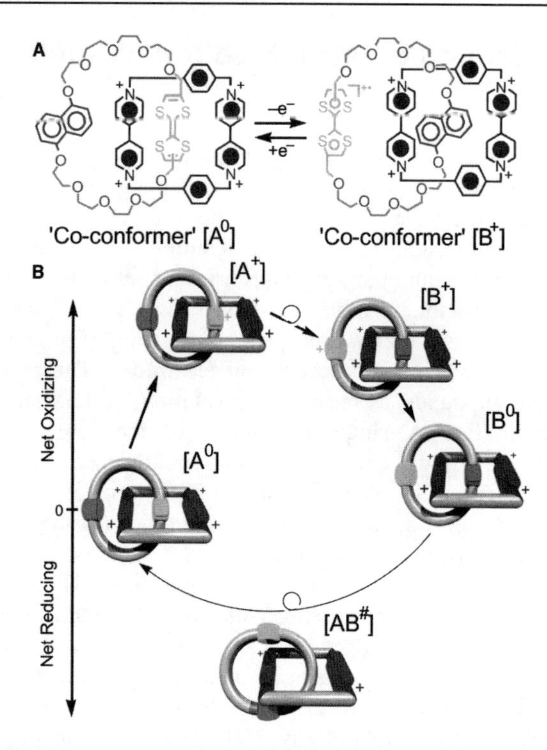

Fig. 12.14. Switching mechanism of a rotaxane molecule. A system that was made even more complicated by integrating the switchable molecules in two-dimensional molecular electronics circuits shown in the Fig. 12.15

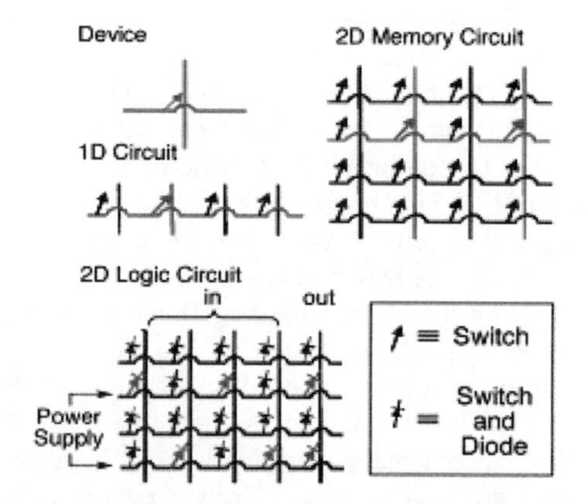

Fig. 12.15. A simple yet versatile circuit architecture, known as a crossbar, is shown at increasing levels of complexity in terms of both fabrication and function. Each junction is a switching device, with black arrows corresponding to open switches and red arrows to closed switches. Wires that are utilised to address or modify the switches are highlighted in red, except for the case of the 2D Logic Circuit in which the configuration of the circuit (where the junctions are both switches and diodes) has been indicated.

Another electron-conductance based logical gates are based on polyacetylene that consists of a chain of carbon atoms held together by alternating double and single bonds through which electrons are thought to travel in small packets called solitons. Hence, different structures of polyacetylene can be used as soliton switches. Soliton switches can be used much like relays and combined to perform the simple logic functions required to form Boolean gates. (Groves 1995).

12.8 Conclusions

Since the emergence of the concept, molecular logic gates have evolved to one of most active areas in molecular devices. Small size is the main advantage of molecular devices. In addition, the range of response times for molecular devices is in the order of femtoseconds while the fastest present devices operate in the nanosecond regime resulting in a 10^6 increase in speed. (Tour 2000). Also molecular-based systems offer distinct advantages in uniformity and potential fabrication costs. However, challenges still remain.

- Molecules are easier to synthesise but are difficult to arrange on a surface or a 3-D array and cannot be ensured that they would stay in a place.
- Due to higher density, heat dissipation has a million-fold increase in circuit density and an enormous cooling fan would be needed to prevent the ensuing meltdown or decomposition of the molecules. As part of the new design scenarios molecular devices could function by electrostatic interactions produced by small reshapes of the electron density due to the input signals. Electrostatic potential interactions between molecules would transport the information throughout the CPU and there is no need for electron current or electron transfer as in the present devices.
- Another challenge is the switching frequency of the molecular devices. It is normally smaller than 1000 Hz, which is substantially smaller than that of MOSFET of 10^{11} Hz or 10^{12} of superconductor device.

These concerns provide a real challenge for molecular electronic computing research. Although a few architectural constructions of molecular wires and devices have been achieved, extensive R&D work still need to be done to tackle these challenges satisfactorily.

12.9 References

Ami S, Joachim C (2001) Logic gates and memory cells based on single C60 electromechanical transistors, Nanotechnology 12:44-50

Ashton PR, Baldoni V, Balzani V, Credi A, Andreas Hoffmann HAD, Maria-Victoria Martinez-Diaz, Raymo FM, Fraser Stoddart J, Venturi M (2001) Dual-mode co-

conformational switching in catenanes incorporating bipyridinium dialkylammonium recoznition sites, Chem. Eur. J. 7:3482-3485

Collier CP, Mattersteig G, Wong EW, Luo Y, Beverly K, Sampaio J, Raymo FM, Stoddart JF, Heath JR (2000) A catenane-based solid state electronically reconfigurable switch, Science 289:1172

Collier CP, Wong EW, Belohradsky M, Raymo FM, Stoddart JF, Kuekes PJ, Williams RS, Heath JR (1999) Electronically configurable olecular-based logic gates, Science 285:391

Cooper CR, James TD (1997) Selective d-glucosamine hydrochloride fluorescence signalling based on ammonium cation and diol recognition, Chem. Commun. 15:1419

de Silva AP, Dixon IM, Nimal Gunaratne HQ, Gunnlaugsson T, Maxwell PRS, Rice TE (1999a) Integration of logic functions and sequential operation of gates at the molecular-scale, J. Am. Chem. Soc 121:1393

de Silva AP, Dixon IM, Gunaratne HQN, Gunnlaugsson T, Maxwell PRS, Rice TE (1999b) Integration of logic functions and sequential operation of gates at the molecular-scale, J. Am. Chem. Soc. 121:1393

de Silva AP, McClenaghan ND (2004) Molecular-scale logic gates, Chem. Eur. J. 10:574

de Silva AP, McClenaghan ND (2002) Simultaneously multiply-configurable or superposed molecular logic systems composed of ICT chromophores and fluorophores integrated with one- or two-ion receptors, Chem. Eur. J. 8:4935

de Silva AP, Nimal Gunaratne HQ, McCoy CP (1997) Molecular photonic AND logic gates with bright fluorescence and "Off-On" digital ction, J. Am. Chem. Soc. 119:7891

Ghosh P, Bharadwaj PK (1996) Ni(II), Cu(II), and Zn(II) cryptate-enhanced fluorescence of a trianthrylcryptand: A potential molecular photonic OR operator, J. Am. Chem. Soc. 118:1553

Groves MP, Carvalho CF, Prager RH (1995) Switching the polyacetylene soliton, Mater Sci. Eng. C3:181

Gunnlaugsson T, Mac Donail DA, Parker D (2001) Lanthanide macrocyclic ouinoyl conjugates as luminiscent molecular-level devices, J. Am. Chem. Soc. 123:12866

Gunnlaugsson T, Mac Dónail DA, Parker D (2000) Luminescent molecular logic gates: The two-input inhibit (INH) function, Chem. Commun. 1:93

Guo X, Zhang D, Tao H, Zhu D (2004) Concatenation of two molecular switches via a Fe(II)/Fe(III) couple, Org. Lett 6:2491

Inouye M, Akamatsu K, Nakazumi H (1997) New crown spirobenzopyrans as light-and ion-responsive dual-mode signal transducers, J. Am. Chem. Soc. 119:9160

Ji HF, Dabestani R, Brown GM, (2000) A supramolecular fluorescent probe, activated by protons to detect cesium and potassium ions, mimics the function of a logic gate, J. Am. Chem. Soc. 122:9306

Kimura K, Mizutani R, Yokoyama M, Okamoto R, Doe H (1997) All-or-none type photochemical switching of cation binding with malachite green carrying a bis(Monoazacrown ether) moiety, J. Am. Chem. Soc. 119:2062

Luo Y, Patrick Collier C, Jeppesen JO, Nielsen KA, DeIonno E, Ho G, Perkins J, Tseng HR, Yamamoto T, Fraser Stoddart J, Heath JR (2002) Two-dimensional molecular electronics circuits, Chemphyschem 3:519

Mitchell RJ (1995) Microprocessor systems: An introduction. Macmillan, London

Pina F, Maestri M, Balzani V, (1999) Photochromic flavilium compounds as multistate/multifunction molecular-level system, Chem. Comm. 2:107

Pina F, Roque A, Melo MJ, Maestri M, Belladelli L, Balzani V (1998) Multi state/multifunctional molecular-level systems: Light and pH switching between the various forms of a synthetic flavylium salt, Chem, Eur. J. 4:1184

Raymo FM (2002) Digital processing and communication with molecular switches, Adv. Mater 14:401
Raymo FM, Alvarado RJ, Giordani S, Cejas MA (2003) Memory effects based on intermolecular photoinduced proton transfer, J. Am. Chem. Soc. 25:2361
Raymo FM, Giordani S (2001) Signal processing at the molecular level, J. Am. Chem. Soc. 123:4651
Raymo FM, Giordani S (2004) Molecular logic gates. Encyclopedia of Nanoscience and Nanotechnology 5:677
Raymo FM, Giordani S, White AJP, Williams DJ (2003) Digital processing with a three-state molecular switch, J. Org. Chem. 68:4158
Rurack K, Koval'chuck A, Bricks JL, Slominiskii JL (2001) A simple bifunctional fluoroionophore signalling different metal ions either independently or cooperatively, J. Am. Chem. Soc. 123:6205
Schulz M (1999) The end of the road for the silicon, Nature 399:729
Singh CP, Roy S (2004) Dynamics of all-optical switching in C60 and its application to optical logic gates. Opt. Eng. 43:426
Tour JM (2000) Molecular electronics: Synthesis and testing of components, Acc. Chem. Res. 33:791
Yeow EKL, Steer RP (2003) Energy transfer involving higher electronic states: A new direction for molecular logic gates, Chem. Phys. Lett. 377:391
Zelikovich L, Libman J, Avraham A (1995) Molecular redox switches based on chemical triggering of iron translocation in triple-stranded helical complexes, Nature 374:790

13 Nanomechanical Cantilever Devices for Biological Sensors

T. S. Kim, D. S. Yoon and J. H. Lee

Microsystem Research Center, Korean Institute of Science and Technology, Republic of South Korea

13.1 Introduction

In the last 20 years, technology has been divided into many subtechnologies. For example, mechatronics has been divided into microelectromechanical systems (MEMS) or nanoelectromechanical systems (NEMS). Furthermore, various subtechnologies such as MEMS, NEMS and biotechnology have been combined to form new fields such as nanobiotechnology. These dramatic technological changes are due to the efforts of researchers to overcome the technological hurdles faced by conventional technology. With respect to nanobiotechnology, there is great interest in micromechanical biosensors and nanomechanical biosensors that use the nanomechanical technology of MEMS and NEMS. Micromechanical and nanomechanical biosensors are devices that measure physical quantities by utilizing variations in the physical properties of specifically fabricated microstructures that originate from biological interactions. In microcantilever biosensors, the cantilever transduces the recognition event from its receptor immobilised surface (for example, a DNA probe and an antigen or antibody) into a mechanical response (for example, static displacement and resonance frequency). The mechanical response can then be detected with different methods. Advances in MEMS and NEMS technology have facilitated the development of microcantilever biosensors and they have given the biosensors many advantages. For example, a greatly reduced size, high sensitivity, increased minimum detectable sensitivity, and greater reliability. Furthermore, these advances have enabled existing devices to be scaled down to a micrometer regime or even a nanometer regime. As a result, these devices have better sensitivity. As devices are fabricated on a microscale or even a nanoscale, they offer us an opportunity to explore new basic scientific phenomena that occur only when the dimensions are small (for example, the binding force measurements of protein folding and DNA hydrogen bonding). Although biosensing tools are currently undergoing a further stage of development, a remarkable breakthrough is needed to obtain a biosensor system that is practical and portable. Arntz et al. suggested the following primary requirements for a future generation of biosensors (Arntz et al. 2003).

- The combination of a biologically sensitive part with a physical transducer for specific and quantitative detection of analytes
- The ability of label-free detection of the biological interaction
- The scalability of the sensors to allow massive parallelisation
- Sensitivity of the detection range applicable for in vivo problems

A microcantilever biosensor is compatible with these requirements. Compared with other mechanical biosensors based on quartz crystal microbalance and surface acoustic waves, the microcantilever biosensor has strong potential for parallelisation and scalability. Another advantage of the microcantilever biosensor is that it is basically a label-free and highly sensitive device. Further, they have peculiar properties of inducing static deformation or resonance frequency shifts that are related to biological interactions such as antigen-antibody binding and DNA hybridisation. The biological application of cantilevers is a promising field in academic and industrial research because of their usefulness in various biological tests such as bioassays and the binding force measurement of biomolecule-biomolecule interactions. Another trend in microcantilevers is that the dimensions of a mechanical cantilever are still being reduced, and this trend indicates the advent of a new generation of NEMS. Scaling down the dimensions of microcantilevers improves sensitivity, spatial resolution, energy efficiency and the speed of response. Furthermore, the scale-down is expected to achieve a mass sensitivity of around 10^{-19} g (Davis et al. 2000). As a result, nanomechanical cantilevers will be able to detect specific molecular interactions, cell adhesion and chemical gases with miniscule quantities. The combination of a scaled-down cantilever and the nanoscale phenomena that occur on the surface of the cantilever give the sensor an ability that is up to or beyond the femto and atto regime.

In this chapter, we introduce two types of micromechanical biosensors: a microcantilever and a nanocantilever. The schematics to explain the basic principles of a cantilever's mechanics and detection scheme are presented. The summary of various methods of microcantilever fabrication is given. The discussion on measurement and readout techniques is included. Many bioassay examples with reference to categories such as DNA, protein, and cell detection are also given. Finally, we analyse the current status of microcantilever biosensors and discuss various breakthroughs that are needed for the popularisation and commercialisation of microcantilevers.

13.2 Principles

A cantilever's response to stimuli such as mechanical stress, temperature change, and chemical and biochemical interactions can be measured in a variety ways.

Some responses that are of special interest are cantilever deformation, the change in resonance frequency, and the current between the cantilever and a surface. Microcantilever-based sensors are categorised into two main measurement regimes: static deformation (static mode; DC mode) and resonance response variation (dynamic mode; AC mode). The theoretical background for these modes of operation is provided. Static cantilever deflection can be caused by external forces exerted on the cantilever, as in the case of an AFM, or by intrinsic stresses generated on the cantilever surface or within the cantilever. However, the resonance response variation can be caused by a mass loading effect or by viscous damping from various media.

13.3 Static Deformation Approach

If assume that no external gravitational, magnetic, and electrostatic forces exist, cantilever deformation may be related to a gradient of the mechanical stress generated in the devices. It is well known that uniform surface stress on isotropic material tends to either increase the surface area (compressive stress) or decrease the surface area (tensile stress). If one fails to compensate for this effect with an equal stress on the opposite side of the thin plate or beam, the entire structure will be permanently bent. The different stresses (σ) that act on both sides of a cantilever, which may or may not be a composite, cause permanent bending. This phenomenon is well described by Stoney's formula (Stoney 1909), in which the radius of curvature of the bent cantilever is expressed as follows:

$$\frac{1}{R}=\frac{6(1-\nu)}{Et^2}\Delta\sigma \tag{13.1}$$

Where, R is the cantilever's radius of curvature, ν is Poisson's ratio, E is Young's modulus for the substrate, t is the thickness of the cantilever, and $\Delta\sigma$ is the differential surface stress. The differential surface stress means the difference between the surface stress on the top and bottom surfaces of the microcantilever. Initially, the surface stress of each side, $\Delta\sigma_{s1}$ and $\Delta\sigma_{s2}$, are in equilibrium. When they become unequal, the cantilever bends. From the radius of curvature, R, the tip displacement can be expressed as

$$\Delta z=\frac{3l^2(1-\nu)}{Et^2}\Delta\sigma \tag{13.2}$$

Where l is the length and Δz is tip deflection and the surface stress is expressed as a function of the deflection of the cantilever Δz. This equation shows a linear relation between the cantilever bending and the differential surface stress. It can be used to detect the adsorption of an extra layer on the surface of a cantilever.

The adsorption must occur on one side of the cantilever in order to maximise the relative stress between the two sides ($\Delta\sigma$) and, as a result, to maximise the output signal of the sensor. An exemplar of this kind of measurement is Berger's work on the self-assembled monolayers of alkanethiols. They monitored the phase transition of a layer of alkanethiols that was added to the cantilever: these molecules are alkane chains with a thiol group (-SH) grafted at one end. They then used an Au-coated cantilever with one side used for the formation of self-assembled monolayers, which are covalently immobilised through the thiol group. When we assume that the surface differential stress is proportional to the number of molecules absorbed, the mass added $\Delta\sigma m_a$ to the cantilever is expressed as follows:

$$\Delta s = C\,\Delta m_a \tag{13.3}$$

Where, C is a proportionality constant that depends on the sticking coefficient of absorbed molecules (Datskos et al 2001). The maximum deflection for the cantilever length l is given by

$$\Delta z_{\max} = \frac{l^2}{2R} \tag{13.4}$$

$$\text{and}\quad \Delta z_{\max} = C\frac{3l^2(1-\nu)}{Et^2}\Delta m_a \tag{13.5}$$

In this case, Δm_a results from the modulation of the Gibbs surface-free energy of the absorbed side. For a noble metal coated with an organic modifying substrate, Δz_{max} can be expressed as follows:

$$\Delta z_{\max} = \frac{3l^2(1-\nu)}{Et^2}\left(\sigma_{sub} - s_1 - E_{binding}\frac{m_s}{M} - W_{steric}\right) \tag{13.6}$$

Where, σ_{sub} is the surface tension of the nonfunctionalised cantilever, s_1 is the surface stress of chemical interaction side of cantilever, $E_{binding}$ is the binding energy per mole, m_s is the surface mass, M is the molar mass of the absorbed molecules, and W_{steric} is the energy of the steric repulsive forces between the molecules in the substrate. As the concentration of the molecules or analytes increases, the absorbed analyte increases, causing the cantilever to bend.

13.4 Resonance Mode Approach

The resonant frequency of an oscillating cantilever can be written as:

$$f = \frac{1}{2\pi}\sqrt{\frac{k}{m^*}} \tag{13.7}$$

Where, k is the spring constant and m^* is the effective mass of the cantilever. For a rectangular cantilever, the following equation expresses the spring constant for the vertical deflection as derived for a load at the end:

$$K = \frac{F}{h} = \frac{Ewd^3}{4L^3} \tag{13.8}$$

Where, F is the load applied at the end of the cantilever, h is the resulting end deflection, E is the modulus of elasticity for the cantilever material, and w, d, and L are the width, thickness, and length of the microcantilever beam, respectively. The fundamental frequency can be written as

$$f = \frac{1}{2\pi}\sqrt{\frac{k}{m^*}} = \frac{d}{2\pi(0.98)L^2}\sqrt{\frac{E}{\rho}} \tag{13.9}$$

Where, ρ is the density of the cantilever material and $m^*$is the effective mass of the cantilever. Note that $m^*=nm_b$, where m_b is the mass of the cantilever beam and the value of n is either 0.14 for a 0.06 N/m V-shaped silicon nitride cantilever or 0.18 for a 0.03 N/m V-shaped silicon nitride cantilever or 0.24 for a rectangular cantilever. From Eq.13.7, it is obvious that changes in the mass and in the spring constant cause the resonance frequency to change. Therefore, the resonant frequency shift can be written as follows (Brath et al. 2002):

$$df(m^*, K) = (\frac{\partial f}{\partial m^*})dm^* + (\frac{\partial f}{\partial K})dK = \frac{f}{2}(\frac{dK}{K} - \frac{dm^*}{m^*}) \tag{13.10}$$

By designing cantilevers with localised adsorption areas at the terminal end of the cantilever (end loading), we can minimise the contribution of the differential surface stress [the dK/K terms in Eq.]. In that case, the changes in the resonance frequency can be attributed entirely to the mass loading. However, the changes in the spring constant may be due to a number of causes, such as changes in the elastic constant of the surface film and changes in the dimensions of the cantilever or coatings.

A group from the IBM Zurich research division reported an example of measuring mass loading (Berger et al. 1996). They used some zeolite crystals attached to a cantilever to monitor the water adsorption in the pores of the material, which has a high surface-to-mass ratio, as a function of the humidity. They detected an adsorbed water mass in the picogram range. In the case of a multilayer cantilever, such as a piezoelectric unimorph cantilever, the resonant frequency is expressed as follows:

$$f = \frac{\upsilon_n^2}{2\pi}\frac{1}{L^2}\sqrt{\frac{(\overline{EI})}{(\rho_{np}h_{np} + \rho_p h_p)}} \quad (13.11)$$

Where $\overline{EI}$ is

$$\frac{1}{(\overline{EI})} = \frac{12}{w}\frac{E_{np}\,h_{np} + E_p\,h_{np}}{E_{np}^2\,h_{np}^4 + E_p^2\,h_p^4 + E_{np}\,E_p\,h_{np}h_p(4h_{np}^2 + 6h_{np}\,h_p + 4h_p^2)} \quad (13.12)$$

E is Young's modulus, I is the moment of inertia, h is the thickness, w is the width, L is the length, σ_n^2s the dimensionless nth-mode eigenvalue, and the subscripts np and p denote the elastic nonpiezoelectric layer (Lee et al. 2004). Because the nonpiezoelectric layer is composed of multilayers of different materials, we used the rule-of-mixtures approach to combine each individual modulus (Kaw et al. 1997). For example, when a PZT nanomechanical cantilever, which comprises multilayers of SiO_2, SiN_x, Ta, Pt, PZT, Pt, and SiO_2, has a thickness of 2.0 μm, we can calculate the resonant frequency of the piezoelectric unimorph cantilever structure that is clamped to one end. Theoretically, the first resonant frequency is 16,470 Hz for a structure with the dimensions of 100μm × 300μm and 61,384 Hz for a structure with the dimensions of 50μm × 150μm. Note that because the dimensions of the cantilever decreased from 100μm × 300μm to 50μm ×150μm, the spring constant of the cantilever increased from 1.48 N/m to 5.92 N/m. If molecular adsorption (induced mass) does not affect the spring constant of the cantilever, the beam spring constant, k, remains constant during the mass loading. The adsorbed mass can then be calculated by using the variation in the resonance frequency of a cantilever, as in the following equation:

$$\frac{(f_1^2 - f_2^2)}{f_1^2} = \frac{\Delta m}{m} \quad (13.13)$$

Where, f_2 is the resonance frequency after the mass loading and f_1 is the resonance frequency before the mass loading. However, surface adsorption can cause a change in the spring constant, dk, due to changes in the surface stress. When the adsorption-induced changes in the spring constant and mass are small, the resonance frequency after adsorption can be approximated by the following equation:

$$f_2 = f_1[1 + \frac{1}{2}(\frac{\delta k}{k} - \frac{\delta m^*}{m^*})] \quad (13.14)$$

Where, k is the initial spring constant, m* = 0.14 × (the mass of the cantilever), δk is the change in the spring constant, δm is the change in the mass, and f_1 and f_2 are the resonance frequencies before and after adsorption (Cherian and Thundat 2002). The parameters f_1 and f_2 are experimentally measured values; δm^* is obtained from the surface excess values and the surface area of one side of the cantilever; and m* is calculated from the dimensions and the density of the cantilever.

In general, theoretical estimations based on commercially available microcantilevers (AC method) show a minimum detectable mass density of 0.67 ng/cm^2. When the active area of the structure is taken into account, a minimum detectable mass of 10^{-15} g can be achieved (Martin et al. 1987). Furthermore, by using NEMS to reduce the dimensions of the mechanical transducer, it is possible to achieve a mass sensitivity of around 10^{-19} g. However, when a sensor is operated in a liquid with the aid of the resonance mode, the resonance peak and its quality factor, Q, both shift towards lower values because of damping (Butt et al. 1993). This damping considerably reduces the achievable resolution in terms of changes in the minimum detectable mass. To solve this problem, Mehta et al. and Tamayo et al. proposed methods that enhance the Q factor of oscillating cantilevers in liquids and, hence, the resolution.

13.5 Heat Detection Approach

A composite cantilever behaves bimetallically because it is made out of several layers of different materials with different mechanical and thermal properties (Young's modulus, E; and the expansion coefficient, α). This behavior means that every change in the state of a cantilever's layer can induce the entire cantilever to bend. The simplest use of this property is heat detection and, for uniform heating of a two-layered cantilever, the deflection of the cantilever, z, is described as follows (Marie 2002):

$$\Delta z = \frac{5}{4}(\alpha_1 - \alpha_2)\frac{t_1 + t_2}{t_2^2 \kappa}\frac{l^3}{(\lambda_1 t_1 + \lambda_2 t_2)w}P \qquad (13.15)$$

Where, κ is a device parameter, α_i is the thermal expansion coefficient, t_i is the thickness of the two layers, λ_i is the thermal conductivity of the two layers, l and ω are the length and width of the cantilever, and P is the total power on the sensor. A thin silicon cantilever coated with a 0.4 μm aluminium layer has a sensitivity in the picojoule range. Thus, the cantilever can be used as a very small and sensitive power meter. This bimetallic effect can also be observed if there is a reaction (such as the phase transition of a deposited layer of material or a catalytic reaction on the surface of a cantilever with gas physical adsorption) on only one side of the cantilever. These reactions will then involve a miniscule amount of solid material (picoliter).

13.6 Microfabrication

Most commercial microcantilevers and nanocantilevers focus on silicon or silicon nitride cantilevers due to their availability and easy integration with silicon-based technology (Lavrik et al. 2004).

13.6.1 Si-based Cantilever

Typically, fabrication of suspended microstructures, such as a cantilever transducer, consists of deposition, patterning, and etching steps that define, respectively, thickness, lateral sizes, and the surrounding of the cantilever. One of the frequently used approaches in microcantilever fabrication involves deposition of a sacrificial layer on a prepatterned substrate followed by deposition of a structural material layer (such as a silicon nitride layer or a polysilicon layer) using LPCVD or PECVD processes. By varying the conditions of these deposition processes, the stress and stress gradient in the deposited layers can be minimised so that suspended structures do not exhibit significant deformation after they are released by etching of the sacrificial layer. The cantilever shapes can be defined by patterning the silicon nitride film on the top surface using photolithography followed by reactive ion etching (RIE). Photolithographic patterning of the structural material (silicon nitride or polysilicon) on the bottom surface is used to define the mask for an anisotropic bulk etch of Si. The silicon substrate is then etched away to produce free-standing cantilevers.

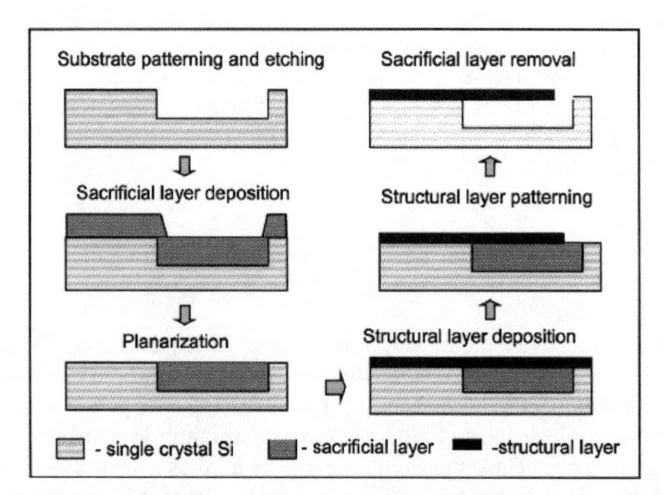

Fig. 13.1. Illustration of the steps in a process flow used for fabrication of silicon nitride cantilevers. The process involves a deposition of a structural silicon nitride layer on a silicon wafer with a prepatterned sacrificial layer. The cantilever shapes can be defined by patterning the silicon nitride film on the top surface using photolithography followed by reactive RIE (Lavrik et al. 2004)

Using a similar sequence of processes, single-crystal silicon cantilevers can be created with the difference that doping of silicon or epitaxy of a doped silicon layer substitutes deposition of a silicon nitride layer.In order to avoid any bulk micromachining, such as through-etch of silicon in KOH, various cantilever fabrication processes based on the use of a sacrificial layer were developed. These processes frequently rely on silicon oxide as a material for the sacrificial layer. The use of a sacrificial layer for fabrication of silicon nitride microcantilever is illustrated in Fig. 13.1. While the use of a sacrificial layer introduces additional restrictions on the material choice, it enables process flows that are fully compatible with standard complementary-metal-oxidesemiconductor (CMOS) chip technology. These silicon-based nanomechanical cantilevers mainly rely on the cantilever bending that comes from surface stress when the target DNA or protein molecules are specifically bound to their receptors that are immobilised on the surface of the cantilever. However, silicon-based nanomechanical cantilevers require an expensive optical apparatus because they cannot acquire a direct electrical signal. In addition, there are generally optical limitations such as a narrow dynamic range and parasitic deflection in optical measurements.

13.6.2 Piezoresistive Integrated Cantilever

A full Wheatstone bridge was symmetrically placed on a chip, with two resistors on the cantilevers and two resistors on the substrate. One cantilever can be used as the reference cantilever and the other can be used for the measurement. The materials used to fabricate cantilevers are single-crystal silicon, amorphous silicon, and microcrystalline silicon. The starting materials of the single-crystal silicon cantilevers is silicon-on-insulator wafers with a 220 nm silicon membrane after thinning down the top silicon layer, and a 400 nm silicon dioxide intermediate layer. Fig. 13.2 shows SEM photographs of piezoresistive sensor arrays and a schematic cross section. Amorphous silicon (580°C) and microcrystalline silicon (610°C) cantilevers were fabricated from 55 nm of silicon nitride with LPCVD on silicon wafers. Next, 150 nm of amorphous silicon and microcrystalline silicon layers were separately deposited with LPCVD. The wafers were divided into two groups for boron-ion implantation at 30 keV with a dose of 53×10^{13} cm^{-2} or 53×10^{14} cm^{-2} for single-crystal silicon and 53×10^{14} cm^{-2} or 53×10^{15} cm^{-2} for amorphous and microcrystalline silicon layers. The piezoresistors were patterned and then defined using SF_6 RIE. The contact pad and the cross beam of all the wafers were implanted with 53×10^{15} cm^{-2} boron ion, and a mask was used to preclude the resistance of the cross beam and to form a good electrical contact with metal. Subsequently, 280 nm of silicon nitride was deposited as a protective layer and as an etch mask for the later wet etching. The doped boron was activated at 950°C for 10 min or 1050°C for 30 min. The cantilevers were defined from the front. The silicon nitride was etched with RIE, and then the cantilevers were released by KOH front-side etching. The channel thickness had been controlled to be about 50 mm. Contact holes were opened with the help of LPCVD SiO_2. Finally, 20/500 nm Ti/Al metal wiring film for the piezoresistor was made by e-beam evaporation.

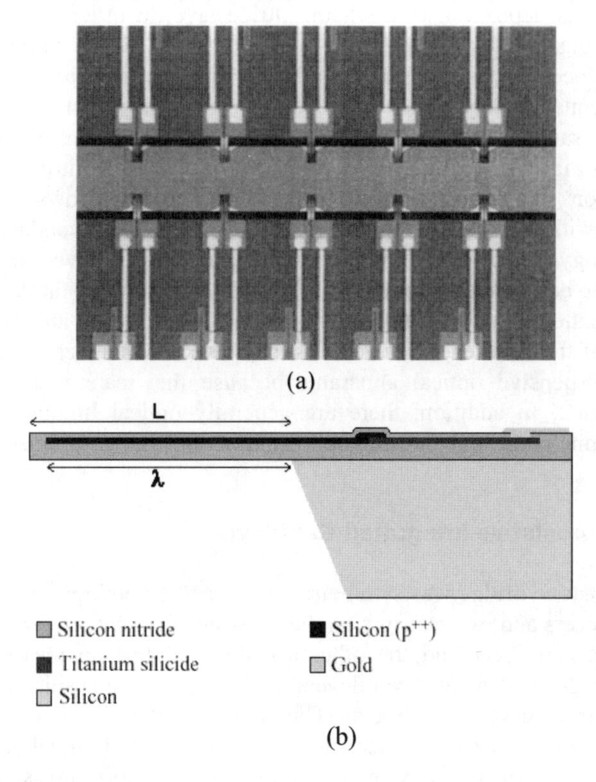

Fig. 13.2. (a) SEM photgraph of a piezoresistive cantilevers. Ten cantilevers were designed on the two sides of a channel, and the cantilever size is 120㎛× 338㎛, and (b) schematic cross section of cantilever and substrate (Yu et al. 2002)

13.6.3 Piezoelectric Integrated Cantilever

Piezoelectric material is considered to be smart because it enables the driving and sensing of mechanical resonance to be done electrically. Because of this quality, we can use a $Pb(Zr_{0.52}Ti_{0.48})O_3$ (PZT) monolithic microcantilever to electrically measure the direct mechanical resonance. Fig. 13.3 shows some fabricated nanomechanical PZT cantilevers. The monolithic PZT thin film microcantilevers are fabricated in eight main steps. In the first step, the substrates that form the PZT capacitors were 4 inch Si (100) wafers covered with 1.5 μm thick low-stress silicon nitride (SiN_x) deposited by LPCVD. The PECVD silicone oxide was then deposited with a thickness of 2000 μm. The bottom electrode was prepared by sputtering a thin Ta adhesion layer followed by a Pt layer with a thickness of 1500 μm. The PZT films were deposited by a diol-based sol-gel route. For a metal-ferroelectric-metal capacitor structure, a Pt layer was deposited for the top electrode by DC sputtering. The Pt for the top electrode and the PZT for the piezoelectric material were etched by RIE and ICP, respectively. The patterning of the Pt bottom electrode was carried out with RIE etching. SiO_2 thin film was then deposited and patterned for an electrical passivation layer via hole patterning. Cr/Au with a 30/100 nm for electrical pads was formed by evaporation and a lift-off process. The cantilever window and the back SiN_x window were patterned with RIE. The bulk silicon was then wet etched using a TMAH silicon etchant. Finally, SiN_x etching was carried out with RIE for forming cantilever.

13.7 Measurement and Readout Technique

The deflection of the cantilever is measured with various systems. The most common sensors that are used to sense cantilever bending include optical levers, interferometry and a piezoresistive element. Resonance frequencies are mainly measured with optical components that use an external oscillator. Alternatively charge coupled devices (CCDs) that record through an optical microscope are used to detect a self-oscillating cantilever (Davis et al. 2000). Recently, we have demonstrated self-exciting and sensing piezoelectric microcantilevers for protein detection without an external oscillator and optical component.

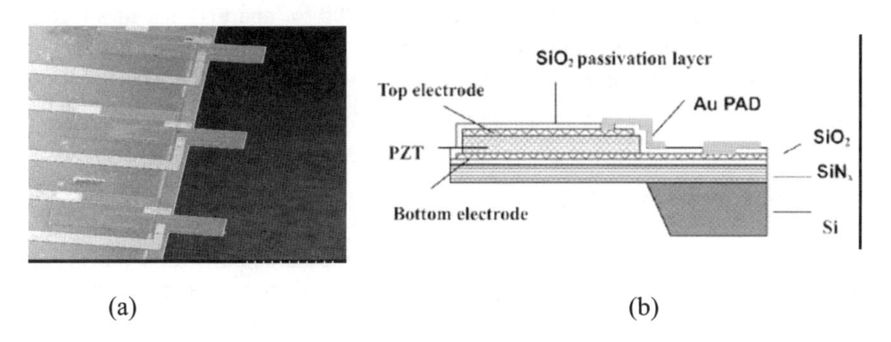

Fig. 13.3. (a) SEM photographs of the micromachined PZT cantilever arrays designed for the dual purpose of electrical self-excitation and sensing, and (b) schematic side view of unimorph PZT monolithic microcantilever (Lee et al. 2004).

13.7.1 Optical Method

One of the most common techniques used to measure the deflection of a cantilever is the so-called beam bounce or optical lever technique. In this system, a laser beam is reflected off the surface of the cantilever and directed into a photosensitive detector (PSD). The position of the reflected beam, and hence the cantilever deflection, is determined by subtracting the PSD output. When the cantilever bends, the reflected laser light moves on the PSD surface. However, an optical lever technique does not measure the displacement but the bending curvature at the point where the laser hits the cantilever. Most commercial PSDs are based on a quadrant photodiode that consists of four cells. The merits of the optical lever method are its linear response, simplicity, and reliability. The optical lever does not require the positioning of components directly above the cantilever. Meyer et al. reported that they could measure 10^{-14} m of displacement by using a micro-mirror on the cantilever surface (Meyer et al. 1988). Nevertheless, the resolution of the optical lever method is typically limited more than interferometry.

13.7.2 Interferometry

An interferometer provides a more sensitive technique for measuring the deflection of a cantilever. This technique is based on the interference of a reference laser beam with one reflected by the cantilever. Interferometry is highly sensitive and can provide a direct and absolute displacement. Rugar et al. developed a deflection sensor based on the interference of light between the cleaved end of an optical fiber and the back of a cantilever; they achieved subnanometer deflection and were able to use single-spin magnetic resonance microscopy (Ruger et al. 1989,1992,1998). Bonaccurso et al. proposed an interferometric system based on two optical fibers for the parallel surface stress monitoring of two cantilevers that were only 200 μm apart (Bonaccueso et al. 2000). Nevertheless, although interferometry is highly sensitive, it has several disadvantages; for example, it requires tedious positioning of the fiber, it works poorly in liquids, and it works well for small displacements.

13.7.3 Piezoresistive Method

The piezoresistive sensing method requires stress measurement with a Wheatstone bridge when a piezoresistive material such as doped silicon is strained. Piezoresistive materials change their resistance when strained. Most materials change the cross section and length under a load and thus change their resistance. However, the resistance change of piezoresistive material is larger than that which can be accounted for by geometrical effects. When voltage is applied to a Wheatstone bridge with resistors of identical initial resistance R, the differential output voltage can be measured by the following equation:

$$\Delta V_{out} = \frac{1}{4}\frac{\Delta R}{R}V_{in} \tag{13.16}$$

In general, a chip that uses a piezoresistive method has four resistors on the chip. Two of the resistors are placed on the cantilever: one resistor is an active cantilever that reacts with the analyte in the sample; the other resistor is a passive cantilever that filters out signals that are identical for both piezoresistors. The deformation of a cantilever from variations in the surface stress can generally be measured from changes in the resistance of the piezoresistive materials, such as doped single-crystal silicon and doped poly silicon.

The piezoresistive technique has several advantages. For example, its readout system is a part of the chip, it needs no optical components, it can take measurements in non-opaque liquids, and it is comparable to CMOS fabrication. Its main advantage, however, is that it needs no alignment. In the case of an optical lever, there are typically two alignment steps that require physical positioning: first, a laser must be aligned to the end of the cantilever; and second, a split-photodiode must be aligned to the laser beam that reflects off the cantilever. When using the piezoresistive technique, the resistor bridge can be balanced simply by changing the resistance of one of the elements. For applications with low temperature or an ultrahigh vacuum, physical alignment is difficult; hence, the piezoresistor is a simple alternative. However, the disadvantage is that the deflection resolution for the piezoresistive readout system is only one nanometer whereas the deflection resolution for an optical readout is one Angstrom. Another disadvantage is that heat from the working current causes a temperature fluctuation in the cantilever, which leads to parasitic cantilever deflection and piezoresistive changes.

13.7.4 Capacitance Method

The capacitance method is based on measuring the capacitance from changes in displacement between a conductor on the cantilever and a fixed conductor on the substrate. Because the cantilever is separated from the fixed conductor by a small gap, the cantilever deformation can generate a capacitance variation that is inversely proportional to the separation gap. Developed by several authors, this method provides high sensitivity and absolute displacement and comparable to integrated MEMS . However, the capacitance method has trouble working in an electrolyte solution due to the Faradic current. Furthermore, it affects from the variation in the dielectric constant of the medium (Baxter 1997). As an alternative to NEMS, a quantum bell system is promising as a branch of the capacitance method (Erbe et al. 2001).

13.7.5 Piezoelectric Method

Instead of relying on the cantilever bending that comes from surface stress, the adsorbed mass can be detected by monitoring changes in the mechanical resonant frequency of a piezoelectric cantilever. A detection mechanism that uses a shift in the resonance frequency has the following advantage: as long as the bio-agent stays on the surface of the cantilever, the postadsorption resonance frequency remains unchanged. Accordingly, the adsorbed molecule can still be detected long after the adsorption takes place (Yi et al. 2002). However, nowadays most researchers rely solely on an external oscillator to measure the shift in the resonant frequency. They also require complex optical components to detect mechanical resonance because silicon-based cantilevers cannot acquire a direct electrical signal from the cantilever. Piezoelectric material is considered to be smart because it enables the driving and sensing of mechanical resonance to be done electrically. Due to a piezoelectric effect, charges are induced in the piezoelectric capacitor layer when a cantilever is deformed. For a unimorph cantilever, the tip deflection from a cantilever deformation can be expressed as follows in terms of the piezoelectric coefficient (d_{31}), the cantilever length, and the applied voltage (Smits et al. 1991):

$$\delta(V)=\frac{-3d_{31}\,s_{11}^{Si}\,s_{11}^{p}\,h_{Si}\,(h_{Si}+h_p)L^2}{K}\Delta V \tag{13.17}$$

Where,

$$K=4s_{11}^{Si}s_{11}^{p}h_{Si}(h_p)^3+4s_{11}^{Si}s_{11}^{p}(h_{Si})^3h_p+(s_{11}^{p})^2(h_{Si})^2+(s_{11}^{Si})^2(h_p)^4+6s_{11}^{Si}s_{11}^{p}(h_{Si})^2(h_p)^2 \tag{13.18}$$

In this equation, Si and p denote the elastic poly-silicon layer and the piezoelectric thin film, while d_{31}, s_{11}, h and L are the piezoelectric coefficient, the compliance, the layer thickness, and the cantilever length, respectively. From the above equation, we can use the direct piezoelectric effect (the displacement converted to an electrical signal) to achieve a readout system, and we can use the reverse piezoelectric effect (the electrical signal converted to displacement) to vibrate the cantilever.

Minne et al., for example, integrated a piezoresistive cantilever with a piezoelectric ZnO ceramic layer to perform z-axis actuation while the force sensing was executed by a piezoresistive cantilever (Minne et al. 1995). They also pointed out that the electromechanical coupling effects between a piezoresistive sensor and a piezoelectric actuator strongly influence the sensor output signals. In the meantime, a similar idea for actuating the z-axis was realised by integrating a silicon cantilever with a piezoelectric PZT ceramic layer while the force sensing was done by optical sensors (Fujii et al. 1996). We demonstrated a smart self-excited PZT microcantilever with force sensing and actuation abilities (Lee et al. 1999). For a small mass detection system, we showed that a PZT micromachined cantilever has a gravimetric sensitivity of 300 cm^2/g, which enables detection of 5 ng mass using Au metal deposition (Lee et al. 2002). We demonstrated piezoelectric actuating and optical sensing. More recently, we fabricated with the composition of $SiO_2/Ta/Pt/PZT/Pt/SiO_2$ on an SiN_x supporting layer for the dual purpose of electrical self-excitation and for protein detection below a picogram regime (Lee et al. 2004). We conducted dual sensing and actuating by changing the sweeping frequency on the cantilever's actuator while simultaneously using a monolithic sensor on the cantilever to measure the sensing signal. The charge induced by the resonance of the actuator was measured by the sensor as a variation in the DSP signal-processed voltage, which was controlled by a PC. The main advantage of the piezoelectric method is that it can convert a mechanical signal into a relatively clear electrical signal. This ability can also provide a novel tool for qualifying and quantifying biomolecules without any sample labeling or bulky optical apparatus.

13.8 Biological Sensing

Cantilever sensors have great potential in genomics research. DNA hybrisation is a prominent example of molecular recognition, which is fundamental to the biological processes of replication, transcription, and translation. In cantilever, surface stress arises from the Watson-Crick base pairing between unlabeled oligonucleotide and their surface-immobilised binding partners (Fritz et al. 2000).

13.8.1 DNA Detection

The most common technique is to measure the deflection of a cantilever by using an optical lever when only one side of the cantilever is coated with receptor molecules. By changing the surface stress only on the functionalised cantilever side for molecular recognition with respect to the other side, we can bend the cantilever. Furthermore, surface stress changes can be measured in situ during the self-assembly of alkanthiol on Au by means of a micromechanical sensor (Berger et al. 1997). One hot topic is the detection of single base-pair variations in DNA (such as single nucleotide polymorphisms), which are a source of biological diversity and several diseases. By measuring changes in surface stress with a technique of optical beam deflection, Fritz et al. demonstrated the sensitive and specific transduction of DNA hybridisation and receptor-regand binding into the direct nanomechanical response of a microfabricated cantilever. Using cantilever arrays with a 1μm thick and 500μm long cantilever, they used a thin Au layer to immobilise thiol-modified oligonucleotide. They monitored ssDNA hybridisation with two parallel microcantilevers when the differential deflection of the microcantilevers enabled discrimination of two identical 12 mer oligonucleotides with a single base mismatch (Fig. 13.4). Defined on the basis of noise, the method's limit of detection was estimated to be 10 nM. Hansen et al. obtained a similar result using 10 mer oligonucleotides. Baselt et al. proposed the use of a magnetic bead on a microcantilever sensor in order detect biological species such as cells, proteins, toxins, and DNA at concentrations as low as 10^{-18} M . A force-amplified biological sensor uses a sandwich immunoassay technique to take advantage of the high sensitivity of force microscope cantilevers for the purpose of detecting the presence of as little as one superparamagnetic particle bound to a cantilever (Fig. 13.5). By applying a magnetic field and measuring cantilever deflection, one can detect the presence of single micrometer-sized magnetic bead on a cantilever.

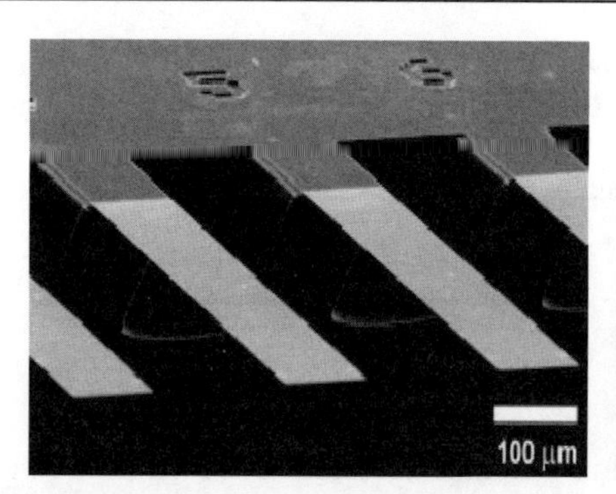

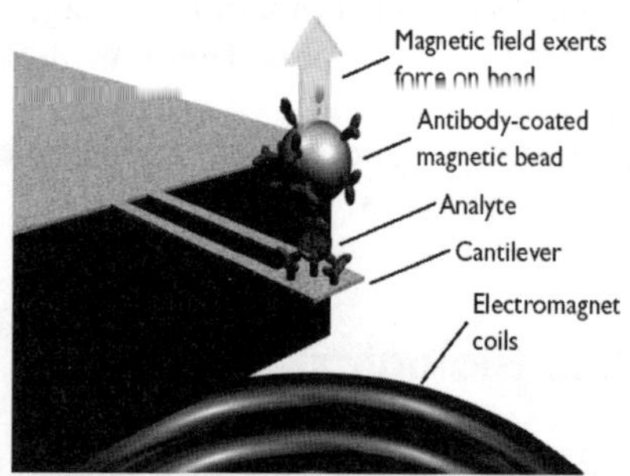

Fig. 13.4. SEM (left image) of a section of a microfabricated silicon cantilever array and scheme illustrating the hybridisation experiment (Right images) at IBM Zurich

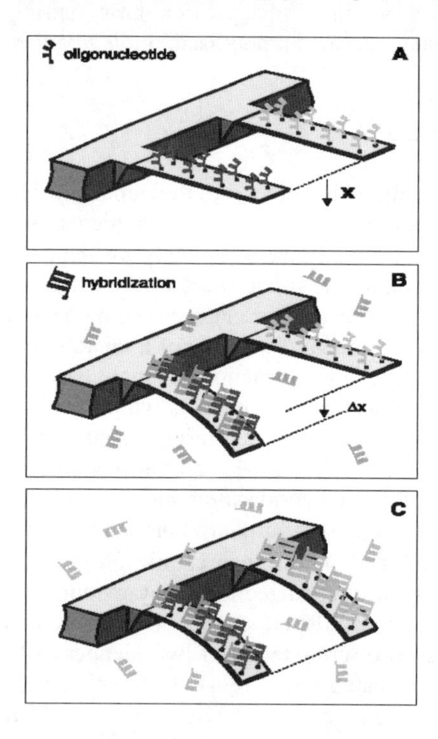

Fig. 13.5. Force amplified biological sensor (FABS) concept. A cantilever-beam force transducer senses the presence of magnetic beads, the number of which is proportional to the concentration of analyte in the sample (Baselt et al. 1996). Each cantilever is functionalised on one side with a different oligonucleotide base sequence. Oligonucleotide base sequences differ by one base only. They are synthesised with a spacer and a thiol group at one end to covalently bind to the gold coated cantilever surface: (a) differential deflection signal is set to zero; (b) after injection of the first complementary oligonucleotide (green), hybridisation occurs on the cantilever providing the matching sequence (red), increasing the differential signal $\Delta x = 9$ nm; (c) injection of with the second oligonucleotide (blue) to bend (Fritz et al.2000)

13.8.2 Protein Detection

To accurately diagnose and closely monitor complex diseases such as cancer, we need to be able to quantitatively detect multiple proteins. A series of proteins that are directly relevant to diseases has recently become available, and the series makes it possible to precisely diagnose the disease by detecting the right proteins. In general, the application for protein detection can be realised if we use an optical lever for interactions that are protein antigen-antibody specific. Arntz et al. demonstrated continuous label-free detection of two cardiac biomarker proteins (creatin kinase and myoglobin) by using an array of microfabricated cantilevers that were functionalised with covalently anchored anti-creatin kinase and anti-myoglobin antibodies . Their method of detection enables biomarker proteins to be detected by using the optical lever method to measure the surface stress generated by antigen-antibody molecular recognition. Furthermore, by calculating the differential deflection signals of sensor cantilevers, they can use reference cantilevers to eliminate thermal drifts, undesired chemical reactions and the turbulence from injections of liquids. The sensitivity achieved for myoglobin detection is below 20 µg/ml. By measuring surface stress, Moulin et al. monitored the surface stress changes associated with the nonspecific adsorption of immunoglobin G and bovine serum albumin (BSA) on an Au surface. They also showed the ability to differentiate between the adsorption of low-density lipoproteins and the oxidised form on heparin. Moreover, they observed that compressive surface stress arose from the adsorption of immunoglobin G, and that tensile surface stress arose from the adsorption of BSA. These phenomena were attributed to the different packing and deformation of each protein on the Au surface. By using optical lever detection to measure the surface stress that operates in a static mode, Butt et al. showed unspecific BSA adsorption on a hydrophobic microcantilever surface. Raiteri et al. showed that the specific binding of a herbicide to a cantilever coated with the herbicide's antibody causes a permanent deflection (Raiteri et al. 1999). They also reported that myoglobin can be detected with a differential pair of cantilevers, one of which is functionalised with monoclonal antimyoglobin antibodies (Raiteri et al. 2001). Albarez et al. showed how the specific binding of the antibodies on a cantilever's sensitised side can be measured with nanomolar sensitivity for the detection of the pesticide DDT.

Wu et al. reported a cantilever-based biosensor that is sensitive enough to be used in a diagnostic assay for the protein markers of prostate cancer. Using an optical lever method, they detected a low concentration of proteins with prostate-specific antigen (PSA). Furthermore, they measured the calibration curve for the PSA at concentrations of 10^{-2} to 10^{6} ng/mL by using a series of similar cantilever transducers and they plotted the curve in terms of the changes generated in the surface stress. Wee et al. reported an electromechanical biosensor that used self-sensing piezoresistive microcantilevers for the electrical detection of PSA and C-reactive protein (CRP) as specific markers of prostate cancer and cardiac disease . The electrical detection of antigen-antibody (Ag-Ab) specific binding was accomplished via changes in the surface stress changes with the aid of a direct nanomechanical response of microfabricated self-sensing microcantilevers. A piezoresistive sensor integrated with the microcantilever measures how the film resistance varies in relation to the surface stress caused by biomolecule-specific binding. When specific binding occurs on a functionalised Au surface, surface stress is induced throughout the cantilever, resulting in cantilever bending and changes in the resistance of the piezoresistive layer. Wee et al. observed that the sensor output voltage is proportional to the injected antigen concentration (without antigen and at 10 ng/ml, 100 ng/ml, and 1 μg/ml). Moreover, PSA and CRP antibodies were found to be very specific for their respective antigens. This phenomenon indicates that the self-sensing microcantilever approach is beneficial for detecting disease markers, and that the piezoresistive microcantilever sensor system is applicable to miniaturised biosensor systems. The piezoelectric detection method can be another measurement example for the observation of protein-protein interaction. Lee et al. present the resonant frequency change of piezoelectric microcantilevers due to the combination of mass loading and the spring constant variation that arises from the antigen-antibody interaction of CRP. The experimentally measured shift in the resonant frequency is larger than the theoretically calculated shift by two orders of magnitude due to the compressive stress that arises from the CRP antigen-antibody interaction. Lee et al. also demonstrated a novel electrical measurement under a controlled ambient temperature and humidity for label-free detection of a PSA as low as 10 pg/mL . Fig. 13.6 shows the fluorescence scanner images as a function of the interacted PSA Ag concentrations with (a) 100 ng/ml, (b) 10 ng/ml, (c) 1 ng/ml, and (d) 100 pg/ml, and, as a negative control, with (e) 1 mg/ml of BSA. Also, figure. Fig. 13.6 shows the experimental results of the first changes in the resonant frequency as a function of the PSA Ag concentration. The results indicate that the dimensions of the cantilever affect its detection sensitivity. This method allows PSA proteins to be detected via an electrical measurement of the changes in the resonant frequency that are generated by the molecular interaction of the antigen and the antibody.

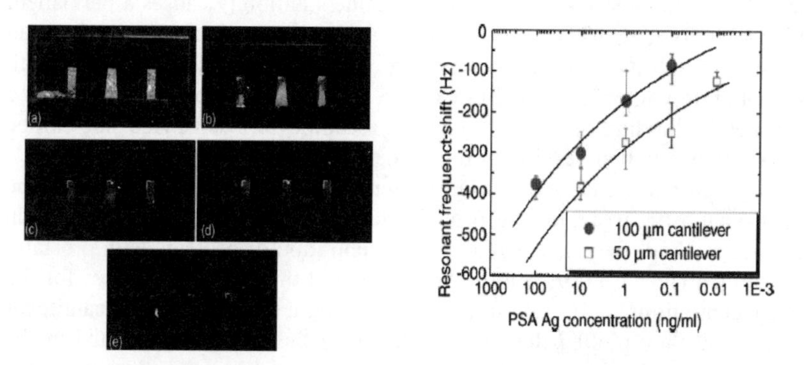

Fig. 13.6. The fluorescence scanner images (Left) as a function of the interacted PSA Ag concentrations with (a) 100 ng/ml, (b) 10 ng/ml, (c) 1 ng/ml, and (d) 100 pg/ml, and, as a negative control, with (e) 1 mg/ml of BSA. Also, experimental first resonant frequency changes as a function of PSA Ag concentration (Right). The results clearly indicate that the dimensions of the cantilever affect its detection sensitivity (Lee et al. 2004)

Most biological assays reportedly use thin film cantilevers because thin film cantilevers can be fabricated more easily than thick film cantilevers. However, thick film cantilevers have an advantage of being able to induce a very large driving force, which can be beneficial for acquiring the desirable sensitivity of a cantilever sensor in viscous fluid. Kim et al. reported the biosensing of a PSA biomarker with the aid of a thick film cantilever biosensor in a liquid environment. To fabricate the thick film cantilever, they used screen printing of PZT paste and standard MEMS processing, and they conducted a real-time bioassay by using the thick film cantilever in a liquid cell made of PDMS. They observed that when injected with a liquid sample of 10 ng/ml PSA, the resonant frequency of the cantilever decreases and there is a resonant frequency change of 200 Hz after stabilisation.

13.8.3 Cell Detection

Cell detection with a microcantilever is a promising approach for rapid, ultrasensitive, and economical detection. As an alternative approach for detecting microbial pathogens and viruses, resonant mechanical sensors are more effective than static deflection. In this approach, additional mass loading of the adsorbed microbial cells causes a change in the resonant frequency of the oscillator. Moreover, the resonant frequency of nanomechanical devices with prefabricated immunospecific areas, yields a high sensitivity that enables detection in a femptogram regime (Ilic 2001). Antonik et al. proposed sensing the mechanical responses to external chemical stimuli, of living cells cultured directly on the surface of a cantilever. With a microcantilever operating in an oscillating mode, Ilic et al. weighed and counted, in air, the number of bacteria adsorbed onto an antibody-coated cantilever by monitoring shifts in its resonance frequency. It has been shown that (Fig. 13.7 (left)) an electron micrograph of an Escherichia coli O157:H7 bacterium on a silicon nitride cantilever. Because the cantilever is coated with a layer of antibodies to this bacteria, the binding is specific to this strain of pathogen. The measured resonant frequency shift as a function of the additional cell loading was observed and correlated to the mass of the specifically bound Escherichia coli O157:H7 cells, as shown in Fig. 13.7.

To detect the mass of individual virus particles, Gupta et al. present the microfabrication and application of arrays of silicon cantilever beams with a nanoscale thickness. The dimensions of the fabricated cantilever beams were in the range of 4μm to 5μm in length, 1μm to 2μm in width and 20 nm to 30 nm in thickness. They demonstrated the detection of a single vaccinia virus particle with an average mass of 9.5 fg. This particle is a member of the Poxviridae family and forms the basis of the smallpox vaccine. Using the frequency spectra of the cantilever beams, they were able to detect the airborne virus particles because of the thermal and ambient noise and the optical detection method.

The size of the cell is comparable to the size of the fabricated cantilever. The resonant frequency shift of this single cell can be detected with a cantilever vibrating in air. By creating even smaller vibrating devices of the type described above and by operating in a reduced pressure environment, we should be able to detect attograms of mass change. This ability is a clear advantage of smaller mechanical devices.

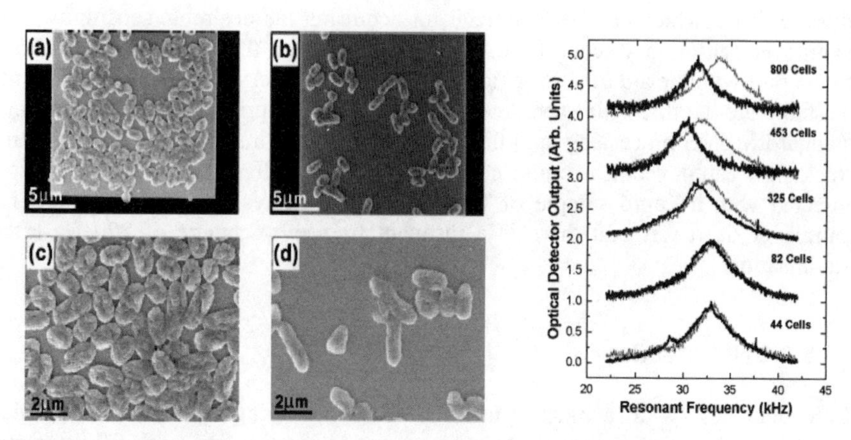

Fig. 13.7. Scanning electron micrographs (left image) of cells bound to the immobilised antibody layer on the surface of various cantilevers and the measured thermal noise spectra (right image), due to the transverse vibrations of the cantilevers before and after cell attachment. The observed small signal-to-noise ratio is caused by the low reflectivity of the silicon nitride. (Ilic 2001, 2000))

13.9 Conclusions

Microfabricated cantilevers are a versatile platform for real-time in situ measurements of the physical, chemical, and biochemical properties of physiological fluids such as gas, liquid, and biological samples. The primary advantages of these microcantilevers are that they have a very high sensitivity because of their ability to detect cantilever motion with subnanomolar resolution. As a result of cantilever design engineering and functionalisation, they also have the capability of physical, chemical, and biological sensing. Another unique advantage of cantilever sensors is that the deformation and shift in resonance frequency that occur simultaneously during the analysis provide complementary information about the interactions between the cantilever and the environment.

Although cantilever sensors have numerous advantages in practical use, they have several problems that should be solved before they can be popularised and commercialised. Functionalisation of the cantilevers is very important in the development of the technique. In particular, the markers that recognise specific analytes should be separately immobilised on individual cantilevers. This immobilisation can enable multiple sensing of numerous chemicals and biomarkers related to environmental pollutants, as well as infectious or genetic diseases. Another requirement for the commercial use of cantilever sensors is that they should have a platform for a multiarray system in which individual cantilever transducers are arranged into a chip integrated with on-chip electronic circuitry. Moreover, for point-of-care application, the microfabricated microfluidic chip has to be combined with the multicantilever array chip for sample preparation of whole blood or serum. Accordingly, the integration of individual cantilevers into a biochip and the microfluidic interface are a promising research area, and further development of the cantilever array biosensor will provide real-time monitoring of multiple biomarkers. According to the literature, the interest of researchers in cantilever sensors is currently shifting from the individual cantilever sensor to the multiple cantilever array chip integrated with a microfluidic chip, cantilevers, and electronic circuitry.

13.10 References

Arntz Y, Seelig JD, Lang HP, Zhang J, Hunziker P, Ramseyer JP, Meyer E, Hegner M, Gerber C (2003) Label-free protein assay based on a nanomechanical cantilever array, Nanotechnology 14:86

Abedinov N, Grabiec P, Gotszalk T, Ivanov T, Voigt J, and Rangelow IW (2001) Micromachined piezoresistive cantilever array with integrated resistive microheater for calorimetry and mass detection, J. Vac. Sci. Technol. A 19:2884

Alexander S, Hellemas L, Marti O, Schneir S, Elings V, Hansma PK, Longmire M, Gurley J (1989) An atomic-resolution atomic-force microscope implemented using an optical lever. J. Appl. Phys. 65:164

Alvarez M, Calle A, Tamayo J, Lechuga L, Montoya A (2003) Development of nanomechanical biosensors for detection of the pesticide DDT. Biosens. Bioelectron. 18:649

Amantea R, Goodman LA, Pantuso F, Sauer DJ, Varhese M, Villianni TS, White LK (1998) Progress toward an uncooled IR imager with 5-mK NETD, Infrared Technology and Applications, vol. XXIV 3436, pp 647

Amantea R, Knoedler CM, Pantuso FP, Patel VK, Sauer DJ, Tower JR (1997). In: Orlando, FL (ed) Proc. of the SPIE, vol. 3061, pp 210-215

Antonik,MD, D'Costa NP, Hoh JH (1997) A biosensor based on micromechanical interrogation of living cells, IEEE Eng. Med. Biol. 16:66

Arntz Y, Seelig JD, Lang HP, Zhang J, Hunziker P, Ramseyer JP, Meyer E, Hegner M and Gerber CH (2003) Label-free protein assay based on a nanomechanical cantilever array, Nanotechnology 14:86-89

Baselt DR, Gil U Lee, Richard J. Colton, J (1996) Biosensor based on force microscope technology, J. Vac. Sci. Technol. B 14 :789-793

Baxter LK (1997) Capacitive Sensors, Design and Applications, IEEE Press, New York.

Berger R, Delamarche E, Lang HP, Gerber C, Gimzewski JK, Meyer E, Guntherodt HJ (1997) Surface stress in the self-assembly of alkanethiols on gold, Science 276:2021

Berger R, Gerber CH and Gimzewski JK (1996) Analytical methods & instrumentation. Special Issue μ TAS'96, pp 74-77

Blanc N, Brugger J, Rooij NF de, Duerig U, (1996) Scanning force microscopy in the dynamic mode using microfabricated capacitive sensors, J. Vac. Sci. Technol. B14:901

Bonaccurso E, Butt H-J, Franz V, Stepputat M, Raiteri R (2000) A new microcantilever-based surface stress sensor for operation in liquids. Cantilever sensors and nanostructures, Heidelberg

Brath FG, Humphery JAC (2002) Sensors and sensing in biology and engineering. Springer, New York, pp 337-345

Britton CL, Jones RL, Oden PI, Hu Z, Warmack RJ, Smith SF, Bryan WL, Rochelle JM (2000) Multiple-input microcantilever sensors, Ultramicroscopy 82:17-20

Butt HJ, Siedle P, Seifert K, Fendler K, Seeger T,. Bamberg E, Weisenhorn AL, Goldie K, Engel A (1993) Scan speed limit in atomic force microscopy, J. Microsc. 169:75-80

Butt HJ, (1996) A Sensitive method to measure changes in the surface stress of solids, J. Colloid Interf. Sci. 180:251

Chen GY, Thundat T, Wachter EA, Warmack RJ (1995) Adsorption-induced surface stress and its effects on resonance frequency of microcantilever, J. Appl. Phys. 66:1695

Cherian S, Thundat T (2002) Determination of adsorption-induced variation in the spring constant of a microcantilever, Appl. Phys. Lett. 80:2219

Datskos PG, Sepaniak MJ, Tripple CA, Lavrik N (2001) Photomechanical chemical microsensors, Sensors and Actuators B 76:393

Davis ZJ, Abadal G, Kuhn o, Hansen O, Grey F, Boisen A (2000) Fabrication and characterisation of nanoresonating devices for mass detection, J. Vac. Sci. Technol. B 18:162

Erbe A, Weiss C, Zwerger W, Blick RH (2001) Nanomechanical resonator shuttling single electrons at radio frequencies, Phys Rev. Lett. 87:96106

Fritz J, Baller MK, Lang HP, Rothuizen H, Vettiger P, Meyer E, Guntherodt HJ, Gerber C, Gimzewski JK (2000) Translating biomolecular recognition into nanomechanics, Science 288:316

Fujii T, Watanabe S (1996) Feedback positioning cantilever using lead zirconate titanate thin film for force microscopy observation of micropattern, Appl. Phys. Lett. 68:467

Gupta A, Akin D, Bashir R (2004) Single virus particle mass detection using microresonators with nanoscale thickness, Appl. Phys. Lett. 84:1976

Hansen KM, Ji HF, Wu G, Datar R, Cote R, Majumdar A, Thundat T (2001) Cantilever-based optical deflection assay for discrimination of DNA single-nucleotide mismatches, Anal. Chem. 73:1567

Ilic B, Czaplewski C, Craighead HG, Neuzil P, Campagnolo C, Batt C (2000) Mechanical resonant immunospecific biological detector, Appl. Phys. Lett. 77:450

Ilic B, Czaplewski D, Zalalutdinov M, Craighead HG, Neuzil P, Campagnolo C, Batt C (2001) Single cell detection with micromechanical oscillators, J. Vac. Sci. Technol. B 19:2825

Kaw AK (1997) Mechanics of composite materials. CRC press

Kim HJ, Kim YB, Park J, Kim TS (2003) Biological element detection sensor application of micromachined PZT thick film cantilever. IEEE Sensors, Toronto, Canada

Lavrik NV, Sepaniak MJ, Datskos PG (2004) Cantilever transducers as a platform for chemical and biological sensors, Rev. Sci. Instrum. 75:2229

Lee C, Itoh T, Suga T (1999) Self-excited piezoelectric PZT microcantilevers for dynamic SFM with inherent sensing and actuating capabilities, Sensors and Actuators A72:179

Lee JH, , Yoon KH, Hwang KS, Park J, Ahn S, Kim TS (2004) Label free novel electrical detection using micromachined PZT monolithic thin film cantilever for the detection of C-reactive protein. Biosens, Bioelectron. 20:269

Lee JH, Kim TS, Yoon KH, (2004) Effect of mass and stress on resonant frequency shift of functionalised $Pb(Zr_{0.52}Ti_{0.48})O_3$ thin film microcantilever for the detection of C-reactive protein, Appl. Phys. Lett. 84:3187

Lee JH, Yoon KH, Kim TS (2002) Characterisation of resonant behavior and sensitivity micromachined PZT cantilever, Integr. Ferroelectr. 50:43

Lee JH, Hwang KS, Park J, Yoon KH, Yoon DS, Kim TS (2005) Immunoassay of prostate0specific antigen (PSA) using resonant frequency shift of piezoelectric nanomechanical cantilever, Biosens. Bioelectron. 20:2157

Martin Y, Williams CC, Wickramasinghe HK (1987) Atomic force microscope - force mapping and profiling on a sub 100-A scale, J. Appl. Phys. 61:4723

Mehta A (2001) Manipulation and controlled amplification of Brow-nian motion of microcantilever sensors, Appl. Phys. Lett. 78:1637

Mertens J, Finota E, Thundat T, Fabre A, Bourillot E (2003) Effects of temperature and pressure on microcantilever resonance response, Ultramicroscopy 97:119

Meyer G and Amer NM (1988) Novel optical approach to atomic force microscopy, Appl. Phys. Lett. 53:1045

Minne SC, Manalis SR, Quate CF (1995) Parallel atomic force microscopy using cantilevers with integrated piezoresistive sensors and integrated piezoelectric Actuators, Appl. Phys. Lett. 67:3918

Moulin AM, O'Shea SJ, Badley RA, Doyle P, Welland ME (1999) Measuring surface-induced conformational changes in protein, Langmuir 15:8776

Raiteri R et al.(1999) Sensing of biological substances based on the bending of microfabricated cantilevers, Sensor Actuat. B-Chem 61:213

Raiteri R, Grattarola M, Butt HJ, Skladal P (2001) Micromechanical cantilever-based biosensors, Sens. Actuators B 79:115

Rodolphe M (2002), DNA Hybridisation investigated by microcantilever-based sensor. M. Sc. thesis, MIC Technical University of Denmark

Rugar D, Mamin HJ, and Guethner P (1989) Improved fiber-optic interferometer for atomic force microscopy, Appl. Phys. Lett. 55:2588

Rugar D, Mamin HJ, and Terris BD (1998) Force microscope using a fiber-optic displacement sensor, Rev. Sci. Instrum. 59:2337

Rugar D, Yannoni CS, and Sidles JA (1992) Mechanical detection of magnetic resonance, Nature 360:563

Smits JG, Choi W-S (1991) The Constituent equations of piezoelectric heterogeneous bimorphs, IEEE Trans. Ultrason. Ferroelectr. Freq. Control 38:256

Tamayo J (2001) Chemical sensors and biosensors in liquid environment based on microcantilevers with amplified quality factor, Ultramicroscopy 86:167

Thaysen J, Boisen A, Hansen O, and Bouwstra S (2000) Atomic force microscopy probe with piezoresistive read-out and a highly symmetrical wheatstone bridge arrangement, Sens. Actuators, A 83:47

Thundat T, Wachter EA, Sharp SL, Warmack RJ (1995) Detection of Mercury Vapor Using Resonating Cantilevers, Appl. Phys. Lett. 66:1695

Tortonese M, Barrett RC, and Quate CF (1993) Piezoresistive silicon cantilever Method, Appl. Phys. Lett. 62:834

Wee KW, Kang GY, Park J, Kang JY, Yoon DS, Park JH, Kim TS (2005) Novel electrical detection of label-free disease marker proteins using piezoresistive self-sensing micro-cantilever, Biosens. Bioelectr. 20:1932

Wu GH, Data RH, Hansen KM, Thundat T, Cote RJ, Majumdar A (2001) Bioassay of prostate-specific antigen (PSA) using microcantilevers, Nat. Biotechnol. 19:856

Yi JW, Shih WY, Shih W-H (2002) Effect of length, width, and mode on the mass detection sensitivity of piezoelectric unimorph cantilevers, J. Appl. Phys 91:1680

Yu X, Thaysen J, Hansen O, Boisen A (2002) Optimisation of sensitivity and noise in piezoresistive cantilevers, J. Appl. Phys. 92:6296

14 Micro Energy and Chemical Systems (MECS) and Multiscale Fabrication

B. K. Paul

Nano/Micro Fabrication Facility, Oregon State University

14.1 Introduction

Over the past 40 years, there has been a large and growing emphasis on the fabrication of devices with ever-decreasing dimensions spread over multiple length scales. These so-called multiscale systems are here defined as integrated systems with sub-millimeter features spanning more than three orders of magnitude. One example of a multiscale system is the integrated circuit (IC) where feature sizes may range four or more orders of magnitude from 100 nanometer gates to millimeter-scale bonding pads. The drive toward multiscale systems is obvious. On one end, smaller gate sizes permit higher component densities improving both speed and cost. At the other end, macro-scale pads are required for electrical interconnection with the macro-world. As shown in Table 14.1, other examples of multiscale systems involve mechanical miniaturisation including microelectromechanical systems (MEMS), micro total analysis systems (μTAS) and micro energy and chemical systems (MECS).

Over three decades ago, mechanical miniaturisation was introduced through pioneering biomedical research at Stanford (Samaun et al. 1973; Middlehoek et al. 1980). Since then, many applications have been developed such as automotive accelerometers, inkjet printers, microelectronics cooling, point-of-use chemical synthesis, and man-portable power generation to name a few. MEMS were the first of these devices to develop. MEMS are highly miniaturised electromechanical devices used for microscale energy conversion (transduction). Typical applications are as miniature sensors and actuators such as accelerometers for automotive air bags, thermal inkjet printheads, micromirror arrays for computer projection and read/write magnetic memory heads. Typical feature sizes are on the order of one to ten μm. MEMS technology enables highly complex assemblies without the need for mechanical assembly. The basis of MEMS fabrication is integrated circuit (IC) processing and silicon micromachining techniques. The traditional manufacturing engineering community did not contribute much to the process development necessary to enable ICs and MEMS. This was in part due to the fact that the science base needed to manipulate matter for the processing of electrons is fundamentally different than the manufacturing science for producing structural systems. Then to, the precision in ICs and MEMS were at levels previously unattainable in traditional manufacturing. Feature sizes in these systems were on or below the order of traditional manufacturing tolerances. Whole new fields of dimensional and compositional analysis were required. Finally, in traditional manufacturing, systems integration across multiple length scales was provided mainly through mechanical assembly, which could be managed by unskilled laborers. Consider the mechanical assembly of airplanes or automobiles, which could involve hundreds or thousands of workers in contrast with automated IC integration. Consequently, manufacturing engineers did not perceive their skill set to be applicable at the small scale.

Table 14.1. Different types of miniaturisation in use today

Parameter	ICs	MEMS	μTAS	MECS
Function	Signal processing	Signal acquisition	Lab-on-chip	Process intensificati on
Primary materials	Semiconductor	Silicon	Silicon, glass and polymer	Metal, ceramics and polymers
Key elements	Transistors	Transducers	μFluidic pumps and valves	μChannel arrays
Feature size	100 nm	μm	Tens of μm	25 to 250 μm
System size	mm to cm	mm to cm	mm to cm	Mm to meters

However, new microfluidic technologies are beginning to emerge providing the manufacturing engineer an opportunity to participate in the multiscale revolution. μTAS (also known as BioMEMS) are microfluidic systems used for chemical, biological or biochemical manipulation and analysis. These microfluidic devices may incorporate many of the transduction concepts from ICs and MEMS into microchannels or microwells to collect data and information based on assays performed on nanoliters or picoliters of fluid. A popular application of μTAS recently has been DNA analysis-on-a-chip which has contributed significantly to the decoding of the human genome (Weigl et al. 1999). Due to their ability to replace long, arduous chemical assaying procedures, μTAS systems are sometimes referred to as "lab-on-a-chip" technology. In constrast, MECS are microfluidic devices, which rely on highly-paralleled, embedded microchannels for the bulk processing of mass and energy. There are numerous advantages of microfluidic systems that are common to both μTAS and MECS. First of all, microfluidic devices all have high surface area to volume ratios, which shortens the diffusion distances and provides high rates of heat and mass transfer. Secondly, certain conditions such as high pressures can generally be sustained in these systems, which are hard to sustain in macro-scale systems since, in the case of high pressures, the surface areas are reasonably small and the resultant forces are also small. Lastly, the functional parts of the systems all operate with small volumes of fluid. Consequently, these systems permit excellent temperature control with the ability to rapidly mix and quench materials. The major difference in μTAS and MECS technology is the volume of fluid processed. MECS devices are typically based on arrays of microchannels in order to handle much larger volumes of fluid and also typically have more demanding chemical and thermal property requirements. Consequently, the overall size of MECS devices ranges between a few millimeters for portable power systems to over one meter for distributed fuel reforming.

Fig. 14.1 shows a small slice of the current and future microfluidic applications that will benefit from nano and microtechnology integration. Already much of the biological revolution has benefited from the accelerated heat and mass transfer available within microchannels. Continued advances in lab-on-a-chip technology are fueling radical innovations in medicine. Similar microtechnology is beginning to show promise as a means for economically producing nanomaterials. Other multiscale trends are toward the decentralised processing of mass and energy. In the future, residential air conditioning will be made "ductless" through the application of many distributed micro heat pumps resulting in large energy savings. Microchannel reactors with nanostructured catalysts will make on-site waste cleanup a reality lending to the realisation of "green" manufacturing. Portable kidney dialysis based on multiscale technology will make life more manageable (and cost effective) for a whole new generation of kidney patients.

New fabrication approaches are being developed to advance these technologies. While material processing requirements for MEMS revolve around electromechanical integration leading to IC processing and silicon micromachining, material processing for μTAS generally involves polymer processing due to near ambient device operating conditions. However, many of the devices shown in Fig. 14.1 are MECS devices. Unlike MEMS and μTAS devices, MECS devices require highly-paralleled arrays of microchannels made from more traditional engineering materials. MECS devices are produced by a fabrication approach known as microlamination (Paul et al. 1999). Microlamination consists of the patterning and bonding of thin layers of material, called laminae, to generate monolithic devices with embedded microchannel features.

Within this context, a new role for the manufacturing engineer emerges as new methods for economical multiscale fabrication are investigated. Future challenges lie in the breadth of material and dimensional integration required. Compare the mechanical assembly of an airplane with wingspans on the order of tens of meters and electrical subassemblies on the order of millimeters (four orders of magnitude) with the integration of thermal electric generation (TEG) superlattices with resolution on the sub-nanometer scale into a portable power generator with overall system sizes approaching one meter (over nine orders of magnitude). New multiscale fabrication methods emphasise fabrication, assembly and characterisation of more traditional engineering materials including metals, ceramics and selected polymers due to a need for different thermal and chemical properties. Advances in material science are needed to refine grain sizes and provide better compositional homogeneity at the small scale. Further, many of these monolithic, integrated systems are produced by shaping and joining discrete pieces of material – the domain of the manufacturing engineer.

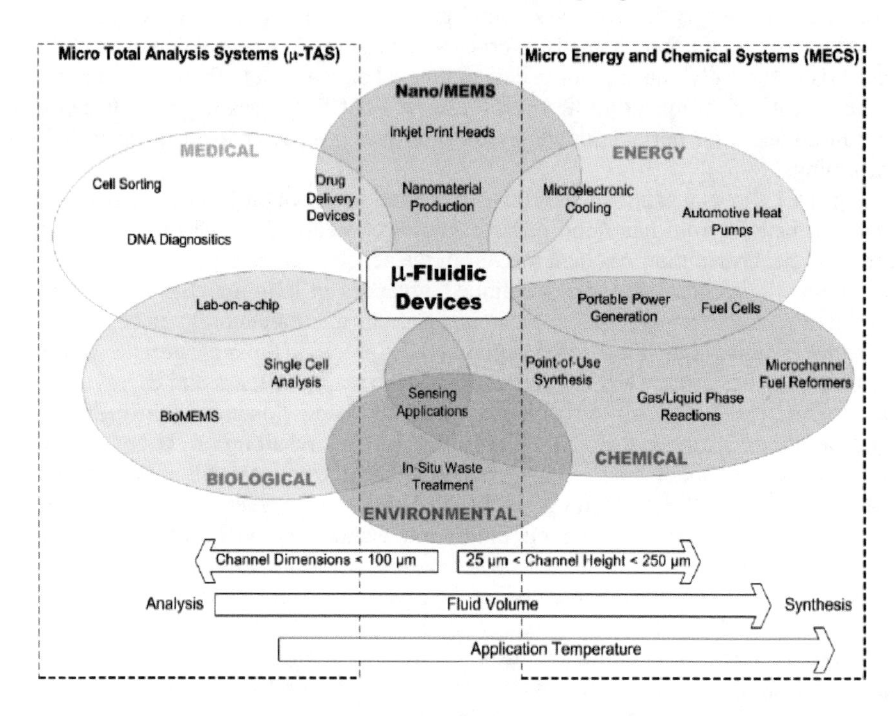

Fig. 14.1. Microfluidic applications-Multiscale fabrication (Courtesy: Pluess)

This chapter includes the findings from seven year's of study of microlamination processes conducted at the Oregon State University (OSU) Nano/Micro Fabrication (NMF) Facility. In sum, significant progress has been made toward understanding the source and effect of shape variation within microlaminated structures. Shape variation is defined as any change from the specification of the product's design such as rough surfaces, improper alignment or warpage. Other studies have been conducted to understand the economic drivers within microlamination processes. Also, new microlamination methods have been developed based on new material and geometry requirements. Future efforts are needed to overcome the economic and technological challenges of microlamination processes in order for MECS technology to become of substantial industrial and societal benefit. This chapter is organised into three sections. The first section provides a justification and foundation for microlamination and the current research being conducted in multiscale fabrication. The second section consists of findings from specific investigations into technological issues within specific microlamination architectures. The final section explores the economics of MECS device fabrication via microlamination and provides implications for future research and development.

14.2 Micro Energy and Chemical Systems

The objective of this section is to lay the groundwork for a deeper discussion of MECS fabrication techniques. Specifically an overview of MECS is provided including a discussion of important MECS applications. An overview is given of some of the fabrication challenges associated with the large-scale production of MECS devices. The section concludes with an introduction to microlamination.

Micro Energy and Chemical Systems (MECS) are bulk microfluidic devices, which rely on embedded microstructures for their function. The overall size of MECS devices range between microscale systems, such as microelectromechanical systems (MEMS), and macroscale systems such as automobile engines and vacuum pumps. These mesoscopic systems are expected to provide a number of important functions where a premium is placed on either mobility, compactness, or point-of-use synthesis. The benefits of decentralised and portable energy and chemical systems are realised by the enhancement in heat and mass transfer performance within high surface-to-volume ratio microchannels. Other benefits of microchannels include rapid temperature changes, excellent temperature control, fast mixing, and the opportunity of operating at elevated pressures (Ameel et al. 1997, 2000; Peterson 2001). Implementations of MECS in heat transfer and chemical applications include microelectronic cooling systems (Kawano et al. 1998; Little 1990), chemical reactors (Martin et al. 1999; Matson et al. 1999), fuel processing (Daymo et al. 2000; Tonkovich et al. 1998), and heat pumps (Drost and Friedrich 1997; Drost et al. 1999, 2000) among others.

MECS devices are primarily focused on taking advantage of the extremely high rates of heat and mass transfer available in microscale geometries to radically reduce the size of a wide range of energy, chemical and biological systems. Typically a MECS device will have dimensions on the order of 1 to 10 cm but it will include embedded microscale geometries such as arrays of parallel microchannels where individual channel dimensions are on the order of 100 microns in width by several mm in height. By taking advantage of the enhanced rates of heat and mass transfer in the embedded microscale geometries, researchers have typically been able to reduce the size of a variety of energy and chemical systems by a factor of 5 to 10 (Brooks et al. 1999; Tonkovich et al. 1998; Warren et al. 1999). Typically MECS devices are fabricated in metals, ceramics or polymers and therefore cannot depend on conventional microfabrication techniques. Microlamination has proved to be an economical and flexible method for fabricating MECS devices (Porter et al. 2002). MECS devices are occasionally referred to as mesomachines and in Europe, research in this area is identified as process intensification.

14.2.1 Heat and Mass Transfer in MECS Devices

Typical MECS devices include individual microscale geometries with dimensions on the order of 100 microns. Single-phase flow in a 100-micron wide channel will in almost all cases of interest be laminar, with Reynolds numbers typically between 1 and 100. In laminar flow, the residence time required to have the fluid reach thermal equilibrium with the walls of the channel decreases as D^2 where D is the width of the channel. If we decrease the channel width by a factor of 10, we will decrease the required residence time by a factor of 100. A heat exchanger with 100 micron wide channels would be $1/100^{th}$ the size of a heat exchanger with 1 mm channels for the same heat transfer rate and it can be shown that for laminar flow, the increased pressure drop associated with the smaller channels is exactly offset by the decreased length of the channel required to deliver the same heat transfer rates. In effect we get a theoretical factor of 100 reduction in size with no increase in pressure drop. In practice microchannel heat exchangers have demonstrated heat fluxes 3 to 5 times higher than conventional heat exchangers. Ultimately, by going to smaller channels we will reduce the diffusion barrier to the point where other processes such as axial conduction along the heat exchanger determine residence time for thermal equilibrium. At that point, further reductions in channel dimensions will not result in reduced residence time. While single-phase flow in microchannels is well understood, processes that involve wall effects such as phase change and surface tension are significantly different in microscale geometries. Boiling is an example. The limited experimental investigations of boiling in microchannels suggest that the process of phase change in a microchannel is different from macro scale devices. However, published performance for microchannel evaporators and condensers show a significant improvement in performance when compared to single-phase microchannel devices, suggesting that as our understanding of phase change in microchannels improves we can expect a further improvement in the performance of microchannel heat exchangers. Mass transfer is similar to heat transfer in microchannels. The residence time that reactants need to reach complete conversion in a catalytic gas phase reactor also decrease by D^2 if the residence time is dominated by diffusion rather then reaction kinetics. Many important reactions such as steam reforming are catalytic gas phase reactions where the primary barrier to short residence time is the time it takes for the reactant to diffuse to the catalyst site and for the products to diffuse back into the bulk flow. Reductions in residence time and reactor size on the order of a factor of 20 have been reported in the literature. Ultimately, by going to smaller channels we will reduce the diffusion barrier to the point where reaction kinetics will determine residence time for complete chemical conversion.

14.2.2 Applications of MECS Technology

The combination of size reductions and improved process control associated with MECS and the promise of high volume mass production (using microlamination) means that small modular energy and chemical systems will be available at a reasonable cost. The availability of compact and modular energy and chemical systems means that MECS can be applied to both portable and distributed applications.

For heat transfer applications, example technologies include compact, high performance heat exchangers for thermal management and waste heat recuperation. Microscale combustion systems have been developed which take advantage of microchannel heat transfer to produce extremely compact high flux combustion systems. This technology can be used within compact heating and cooling schemes or for compact power generation or propulsion. Mechanically-constrained ultra thin film gas absorption and desorption has been used to reduce the size of absorption-cycle heat pumps and gas absorbers by a factor of 5 to 10. The application of compact heat pumps to microelectronics cooling could significantly reduce heat dissipation needs (Kawano et al. 1998; Little 1990). It is estimated that distributed, compact industrial heat pumps could improve building heating and cooling costs by 25% by eliminating ducting and heat pump cycling losses. Compact heat-actuated heat pumps could provide cooling for an automobile while using exhaust heat as the main energy source. Efforts are currently being made to adapt these systems to man-portable cooling for chemical and biological warfare suits. Targets are to eliminate the need for batteries weighing less than a third of a conventional system.

For chemical reactions where reaction rates are limited by diffusion, microreaction technology has been demonstrated within hydrocarbon fuel processing systems for fuel cells and proposed for distributed environmental clean-up systems and distributed point-of-use chemical synthesis. Hydrogen production systems for automotive fuel cells can reduce the size of automotive fuel reformers by a factor of 10. Integrated fuel cells and fuel reformers will lead to high energy density battery replacements (Benson and Ponton 1993). It is estimated that these types of systems could provide a person with 10 We for one week in a system weighing less than 1 kg. It is also expected that microscale fuel cells and thermoelectric generators capable of generating 50 to 100 milliwatts of power for up to one year are possible with a weight that is between one third and one tenth of the weight of an electrochemical battery system.

MECS microreactor technologies provide opportunities for distributed environmental restoration, in-situ resource processing and CO_2 remediation and sequestration. Examples of environmental restoration applications include catalytic reactor systems for the distributed destruction of PCBs and the in-situ cleanup or neutralisation of contaminated aquifers (Koeneman et al. 1997). MECS-based high flux carbon dioxide absorption units and chemical processing systems could be used for the production of rocket fuel on the moon and on Mars greatly reducing the payload and cost of manned space flight. High flux carbon dioxide absorption units could be used for the removal and sequestration of carbon dioxide from power plant exhaust. Finally, the application of MECS microrection technology to the distributed production of nanoparticles and designer molecules could avert health risks and safety concerns surrounding the distribution of nanotechnology for consumption. Microchannel technology can also facilitate the miniaturisation of biologically-based processing systems by accelerating the diffusion of cell media, oxygen and other nutrients and by providing excellent temperature control. This technology might be important for high throughput enzymatic reactors and life support systems for tissue-based sensors.

14.3 MECS Fabrication

MECS is a revolutionary technology that has the potential to impact many aspects of how we live. The availability of small, inexpensive but high capacity MECS devices will increasingly make the distributed processing of mass and energy a reality. However, to realise this vision, we must be able to economically and reliably fabricate large numbers of MECS devices.

14.3.1 Challenges

While MECS devices do include geometries with microscale dimensions, the development of the technology has emphasised materials and fabrication techniques that are significantly different from conventional silicon-based micromachining (Paul and Peterson 1999; Martin et al. 1999). Key differences include materials, operating conditions, system sizes and feature sizes. Silicon is not a good material for most MECS devices. Silicon is brittle and has a high thermal conductivity, which can be a problem for many MECS applications (Peterson 1999). Most MECS devices are fabricated out of metals or polymers with some work being done on the fabrication of ceramic MECS devices. One challenge is to develop fabrication approaches that are appropriate for the many different materials and operating conditions encountered in MECS devices.

Many MECS devices operate at high temperatures with chemically reactive products. Often microreactors will operate at 500 to 600°C and in some applications, designers would like to operate the microreactors at temperatures approaching 1000°C. Suitable fabrication techniques exist for materials that can operate at temperatures up to 600°C but above 600°C difficulties in microlamination of refractory metals and ceramics is a significant challenge. MECS technology will require better methods for fabricating high temperature and chemically inert microscale geometries. Regarding system size, MECS applications will often require arrays of microscale geometries that cover between 1 and 400 cm^2 and this must be at a low cost. Many conventional microfabrication techniques cannot economically process large surface areas. This requires the development of new approaches or a significant improvement in the capabilities of existing microfabrication techniques.

Much of the promise of MECS depends on the availability of a large number of inexpensive distributed components that can compete with the centralised processing of mass and energy. To be competitive, MECS devices must be extremely inexpensive. This will require fabrication techniques that are appropriate for high volume low cost production with dimensional features across the nano, micro and meso-scales. Traditional prototyping techniques such as diffusion bonding will need to give way to continuous processing techniques such as solder paste bonding and other low cost bonding methods.

14.3.2 Feature Sizes

The most critical dimension in most MECS devices is the height (or width whichever is smaller) of the microchannel though the integration of nano-scale features within microchannels has begun to challenge this assumption. Microchannel dimensions can range from 25 to 250 μm high and 25 μm to tens of mm wide. The sizes of microchannels are determined from a variety of perspectives. From a product design perspective, microchannel heights and lengths impact the residence time of fluid molecules within microchannel reactors or heat exchangers as well as the pressure drop across these devices. In addition, particulate matter within bulk fluids sometimes can present challenges in channel blockage and clogging. For example, consider the separation of hydrogen from gasoline for fuel cell applications. Gasoline contains random particles that can make bulk processing difficult in microchannels with heights below about 100 μm. Further, consider the effect of microchannel size on the overall device shape. MECS devices normally consist of a sequence of unit operations designed either for heat transfer or chemical processing. In order to get bulk fluid flows into the microchannels, a series of headers or plenums are needed each successively decreasing in size. Another way of looking at it is that the headers connect the microchannels to the outside world. Since the major value-added function of MECS devices is carried out in microchannels, generally it is desirable to reduce the volume of header with respect to the volume of microchannel. Due to scaling laws, as the height of microchannels decrease in size, the length of those channels also decrease. Given that the header volume stays relatively the same size, at some point the overall header volume begins to dominate the overall device volume leading to awkward device configurations (e.g. automotive radiators). From a fabrication perspective, as the size of microchannel features decrease, tolerances have more effect on device performance. In other words, a 5 μm error will have more impact on fluid dynamics and heat transfer (due to flow maldistribution) in a 25 μm-high channel than in a 250 μm-high channel. If processing could be performed in a silicon platform, tolerances would permit going well below 5 μm. However, most MECS devices are large compared with IC and MEMS devices and, therefore, bulk material costs are significant. Therefore, monocrystalline silicon is not practical. Further, many MECS devices require thermal or chemical properties not available in silicon architectures. Therefore, many MECS applications need to make use of engineering materials.

The tolerances of many conventional manufacturing processes have been found suitable for microchannels making the economical production of MECS devices more feasible. For example, it has been found that reasonable precision of microchannel heights can be achieved at micro and meso-scale dimensions through the lamination of shim stock. The advantage of this approach is that the precision tolerances are controlled by an economical, high-volume process such as rolling. In this case, the scale of channel height (the critical feature) is dictated by the thickness of the shim stock, which can normally be sourced down to 25-μm thick. Further, early results showed that the *warpage* produced during the diffusion bonding of copper foils limited microchannel dimensions to no smaller than 80 μm high (Krause et al. 1994) though that constraint has since been lifted.

14.3.3 Microlamination

Current microelectronic integrated circuits (IC) are predominately silicon-based. MECS, on the other hand, require the mechanical and thermal properties provided by other materials. Many thermal applications require low thermal conductivity materials to reduce axial heat transfer (Peterson 1998, 1999). Other requirements for subcomponents could be for highly fatigue resistant material for springs or magnetic steels for generator and motor cores. Many of the prevailing MEMS microfabrication technologies are based on silicon, polymers, or electroplated pure metals (having high thermal conductivity). Adapting these MEMS fabrication techniques for the construction of some MECS devices would be difficult to achieve.

A second requirement for MECS construction is the need for high-aspect-ratio features. Microchannel arrays with >30:1 aspect ratios are commonly needed for heat exchangers and regenerators. Other MECS designs may call for a small gap between adjacent sub-components where the gap is maintained for the entire length of the structure. Other MECS requirements call for heterogeneity in fabrication materials where electrical and magnetic sections may require a metal and dielectric sections may require a polymer or ceramic. Furthermore, electronic circuits may be needed in the overall design of MECS to process information. While MEMS fabrication platforms provide excellent electromechanical integration, these capabilities are limited to planar structures as opposed to three-dimensional, bulk structures.

Finally, MECS must be able to offer all of these capabilities at low cost in order to compete with conventional macro-scale energy and chemical conversion devices. The most notable high-aspect ratio MEMS fabrication technology is LIGA (Becker et al 1986; Ehrfeld et al. 1997). In addition to being primarily a polymer forming method, LIGA is dependent upon highly capital-intensive synchrotron X-ray generation. LIGA and its lower cost derivatives all use lithographic techniques for mold making and electrofoming for mold filling. Weaknesses of this approach include limited material selection, limited geometric complexity (two dimensional structures), and inconsistent pattern-transferring methods (Walsh et al. 1996). Other net-shape microfabrication techniques have been exploited including laser-beam, electron-beam, ion-beam, electrochemical, electrodischarge, and mechanical methods for material removal or deposition. However, all of these approaches are either serial in nature and, therefore, lack the capability of economical mass production, involve single layer thin film forming and, therefore, provide limited aspect ratios, or are incapable of producing embedded microchannel geometry. No well-established micromechanical fabrication method currently exists for addressing MECS device fabrication requirements in a low-cost, high-volume manner.

For over ten years, Oregon State University and the Pacific Northwest National Laboratory (Richland, Washington) have been developing microlamination techniques for producing high aspect ratio MECS devices. The fabrication methods being pursued by these groups rely on the patterning and bonding of thin layers of material (laminae) to make monolithic assemblies with embedded microchannel circuits (Fig. 14.2). Microlamination is not a process rather, it is a processing architecture. All microlamination techniques involve three steps: 1) lamina patterning; 2) lamina-to-lamina registration to form a stack; and 3) bonding of the stack into a monolith.

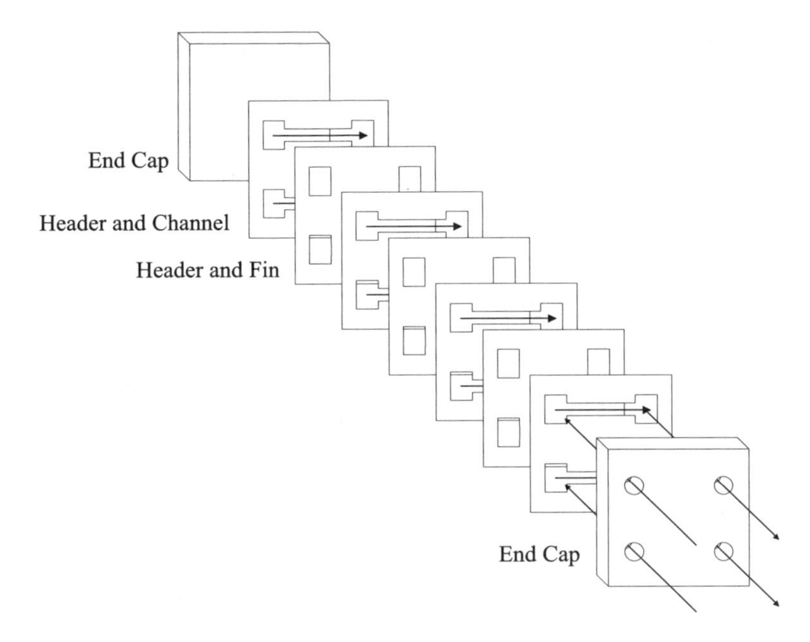

Fig. 14.2. Microlamination scheme used to fabricate a dual micro-channel array. Arrow shows direction of flow

Typical lamina patterning techniques used for metals include photochemical machining, wire electrodischarge machining, electrochemical machining, punching and laser machining (Wegeng and Drost 1994). Photochemical and electrochemical machining have the potential to produce 'blind' features which can be important for reducing the number of laminae per stack or for creating free-standing features such as posts needed to prevent microchannel warpage. Patterning processes for polymers include hot embossing, laser machining, injection molding and punching. In addition to being computer-controlled and, therefore, not requiring expensive, inflexible photomasks, laser machining in polymers has the added advantage of patterning laminae without leaving a burr. Laser burrs in the laser machining of metals can cause difficulties in bonding laminae. Hot embossing and injection molding both provide the ability to produce blind features. Hot embossing offers the ability to produce thinner laminae, which can be important for reducing device volumes. These molding and forming techniques all require tooling which can be produced using more conventional microfabrication techniques. Variants of these polymer-patterning processes have been studied for patterning green tapes for producing patterned ceramic substrates.

Several registration techniques have been employed in microlamination methods including pin, edge, and self-alignment. The technique used depends on the bonding process chosen. The Thermally-Enhanced Edge Registration (TEER) technique has been demonstrated to be an effective technique for registering laminae during thermal bonding processes (Thomas and Paul 2002, 2003). The TEER technique employs the difference in CTE between the bonding fixture and the laminated material to produce a registration force on the laminae at the bonding temperature. By making the bonding fixture from a material that has a lower CTE than the laminae, a clearance will exist between the fixture and the laminae at room temperature, making the loading of laminae simple compared with other mechanical alignment methods. At the same time, TEER has been shown to achieve a layer-to-layer registration accuracy as small as 2 μm for small microlaminated structures (Thomas and Paul 2002).

Bonding methods that have been explored for metals include diffusion bonding, diffusion brazing, diffusion soldering and solder paste bonding (Kanlayasiri and Paul 2004; Gabriel et al. 2001; Paul et al. 2004). Advantages of diffusion bonding include the ability to work with a homogeneous material. Diffusion brazing and soldering can result in greatly reduced pressure and temperature conditions yielding more robust production methods (less dimensional error). Solder paste-bonding offers the promise of improved economics and the ability to integrate electrical circuits with fluidic circuits. Polymer bonding methods have included solvent welding, thermal bonding, adhesive bonding and ultrasonic welding (Paul et al. 2003). Ultrasonic welding can be easily automated but, along with adhesive bonding, can yield poor channel geometries including particulate matter within microchannels. Solvent welding can leave behind residual solvent which may leach out in subsequent device operation. Thermal bonding can require longer cycle times.

14.4 Dimensional Control in Microlamination

Microlamination has the capacity to fabricate metal, polymer and ceramic devices with high aspect ratios in large production volumes. This has been demonstrated in industry where microlamination has been used to mass produce ink-jet print heads. Development of microlamination methods for production of MECS devices has required research into new lamina patterning processes, bonding techniques, and registration methods.

14.4.1 Effects of Patterning on Microchannel Array Performance

As suggested, microchannel devices are well known for their superior performance in heat and mass transport applications. Much of this is due to the large surface area to volume ratios made possible by the high-aspect-ratio microchannels (HARM) within these devices. In order for these HARM devices to be efficient, pressure drop through these devices must be minimised. Past research has shown that as the channel size decreases within microchannels, the fluid flow behavior within the laminar flow regime of the channel begins to deviate from the traditional Navier-Stokes theory (Pfahler et al. 1990; Rahman and Gui 1993; Zemel et al. 1984). Wu and Little (1983) and Peng et al. (1994) both found that the wall effect in microchannels had a significant impact on the laminar flow behavior within the microchannels. Parameters given for influencing the wall effect included the method used to form microchannel walls. Typically, in metal microlamination, the device is formed from cold-rolled shim stock, which possesses a very smooth surface finish. Therefore, sidewall geometries in metal microlaminated structures are not expected to play a large role in the microfluidic wall effect. A microchannel, however, (Fig. 14.3) formed by microlamination may have significant endwall surface roughness depending upon the method of machining used during the lamina-patterning step. Many techniques have been used for lamina patterning including laser micromachining, photochemical etching, and electrochemical machining. Laser micromachining has been used to form microchannels for microlamination because of its ability to quickly adapt to complex geometry. The endwalls formed by the laser micromachining process are known to be rough. Photochemical etching has been used in microlamination due to its relatively low cost when compared with laser machining.

Matson et al. (1997) suggested the use of photochemical etching to pattern highly complex stainless steel shims to fabricate laminated microchannel chemical reactors with a relatively low cost per shim (less than US$1). However, in photochemical etching processes, the costs of waste treatment and disposal can be greater than the processing costs (Datta and Harris 1997). An alternative process with fewer waste disposal problems is electrochemical micromachining (EMM). EMM is a well-established technique that has been used in the electronics industry to machine thick and thin films. This method is known to provide a very smooth endwall surface. It is expected that for microlaminated channels there exists an aspect ratio limit at which the method of lamina patterning no longer significantly influences the microchannel pressure drop. Knowledge of this limit would be beneficial for device designers of cross-flow and counter-flow microchannel devices where maximum aspect ratio limits are known to exist. Efforts to quantify the end wall effect in microlamination have been reported in Paul et al. (1999). In this study, stainless steel microchannels with various aspect ratios were fabricated with machining techniques resulting in both good (EMM) and poor (deep UV laser micromachining) surface finish. Pressure drop was then tested across the microchannels. The results were compared to see whether there is a difference in the pressure drop performance of the microchannels due to these two machining methods and at what aspect ratio the difference becomes insignificant.

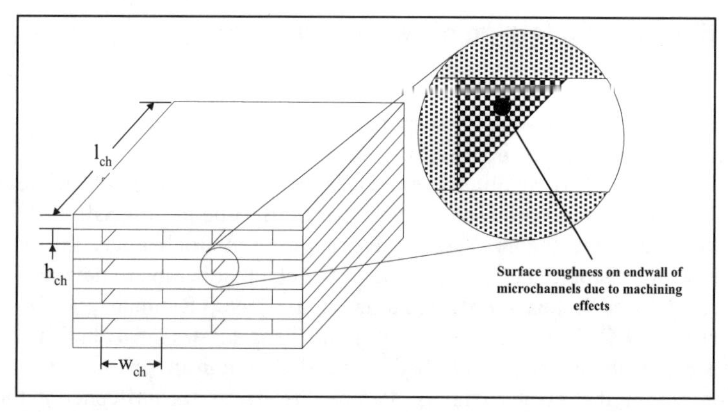

Fig. 14.3. Schematic defining terms associated with a microlaminated geometry

14.4.2 Theory

For fluid flow in a channel, it is well known that the friction factor is not only a function of Reynolds number, but it also depends on the shape of the channel cross section. With the same flow area, the friction factors are different if the channel cross sections are different. Surface roughness of the channel is another important factor for flow in the microchannels because of large relative roughness although the effect is negligible in macro scale fluid (Wu and Little 1983; White 1994). In this experiment, two different shapes of the channel cross-section were considered--rectangular and trapezoid. The difference in the shapes of cross section is caused by different machining methods used in the experiment. The EMM process provides a trapezoidal cross section with smooth end walls whereas laser machining gives a rectangular cross section with rougher end walls. In this study, the range of the volumetric flow rate Q for the microchannel test was set and the pressure drop Δp across the test module was measured. The test module was assumed to have a specific cross-sectional area A and a channel length L. The results are presented in form as a friction factor f and hydraulic diameter D_h. The friction factor is defined as:

$$f = \left(\frac{2 \Delta p A^2}{\rho L Q^2} \right) D_h \tag{14.1}$$

Where, ρ is the density of liquid and the hydraulic diameter D_h can be calculated from:

$$D_h = \frac{4 \cdot A}{WettedPerimeter} \tag{14.2}$$

For laser micromachining, the cross-section area is rectangular, so the hydraulic diameter is equal to $D_h=2WH/(W+H)$, where W and H are the channel width and channel height, respectively. On the other hand, the EMM process produces a trapezoidal cross-section. We can calculate the hydraulic diameter in Eq.14.2 by substituting the cross-sectional area and the wetted perimeter of this trapezoidal structure. The cross-sectional area of this structure is equal to the average of the top width *a* and bottom width *b* of the channel multiplied by the channel height *H*, or $A=(a+b)H/2$, where the wetted perimeter will equal to $a+b+2\{H^2+(1/4(a-b)^2\}^{-1/2}$. The Reynolds number, *Re*, can be calculated from

$$Re=\frac{QD_h}{A\nu} \tag{14.3}$$

Where, ν is the dynamic viscosity of the liquid. As stated in the Moody chart for the laminar flow regime, the relationship of the friction factor in laminar flow, and the Reynolds number is linear, $f=C_f/Re$, where C_f is the friction coefficient for the system. In this study, a simple analysis of the effect of the surface roughness generated by different fabrication techniques on friction coefficients within single channel devices was investigated.

14.4.3 Microchannel Fabrication

Simple single channel devices (Fig. 14.4) were laminated in 304 stainless steel for performing the analysis. The bottom end plate and top interconnect layer were machined from 3 mm thick sheet stock. One batch of the middle stainless steel laminae was machined by one-sided through-mask EMM. The important parts of the patterns were 3 mm diameter sumps and the microchannel. Channel widths and depths explored are shown in Table 14.2. These dimensions were verified by metallography after fabrication and testing of the microchannels. The EMM laminae were set to be machined in the transpassive regime of the electrochemical system to provide a smooth finish. The laser micromached laminae were patterned using a Nd:YAG laser (266 nm) with an average power of 300 mW, an effective spot size of 35 μm and a frequency of 4.5 kHz. Each machined lamina was then diffusion bonded between the two thick end caps with mirror-like finishes (R_a = 0.2 μm). Both end caps were polished to provide extremely smooth surfaces for the bonded microchannel to minimise the effect of sidewall friction. Details of the fabrication procedures are given elsewhere (Wattanutchariya et al. 2003).

The bonded microchannels were pressure drop tested at various flow rates in the laminar flow regime. According to Wu and Little (1983) and Peng et al. (1994) the transition zone from laminar flow to turbulent flow in microchannels occurs much earlier than conventional values. They reported that for different test devices with hydraulic diameters ranging from 0.1 to 0.3 mm, the transition zone varies from Reynolds numbers of 200 to 900. In this study, the maximum Reynolds number was kept to no more than 200, which was still in the laminar flow regime.

Table 14.2. Dimensions and theoretical friction coefficients of experimental channels. Theoretical friction coefficients are from Shah and London (Shah and London 1978)

Channel No.	Machining Method	Top Width (μm)	Bottom Width (μm)	Depth (μm)	Length (μm)	Hydraulic Diameter (μm)	Theoretical C_f
1	Laser	225.4	225.4	78.2	12000	116.1	67.7
2	Laser	319.3	319.3	77.2	12000	124.3	73.5
3	Laser	474.1	474.1	72.3	12000	125.5	80.0
4	Laser	563.8	563.8	74.3	12000	131.2	81.7
5	Laser	678.4	678.4	74.3	12000	133.9	83.8
6	Laser	1037	1037	77.2	12000	143.4	87.3
7	EMM	277.2	162.8	76.2	12000	106.3	62.8
8	EMM	372.2	306.8	74.3	12000	119.9	72.0
9	EMM	507.7	454.2	74.3	12000	127.6	80.0
10	EMM	591.3	510.5	75.3	12000	130.3	76.0
11	EMM	656.2	656.2	78.2	12000	139.7	82.9
12	EMM	1296.3	1111.1	73.4	12000	133.7	84.0

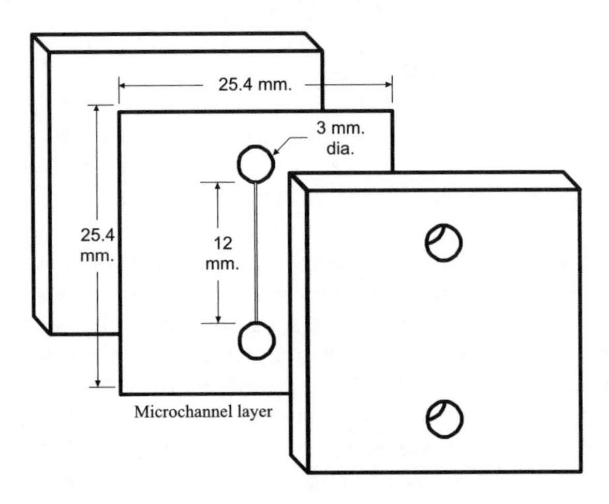

Fig. 14.4. Schematic of a microchannel layer and end caps

14.4.4 Results

To obtain the microchannel dimensions, metallography was performed on the test devices. Fig. 14.5 and Fig. 14.6 represents the cross-section views of the diffusion bonded test devices of channel number 3 and 9, which were fabricated by laser micromachining and by electrochemical micromachining respectively. Fig. 14.7 shows a comparison of the theoretical and experimental friction coefficients for the set of microchannels from which the channels in Fig. 14.6 came. A summary comparison of the theoretical and experimental friction coefficients are shown in Fig. 14.8 as a function of aspect ratio.

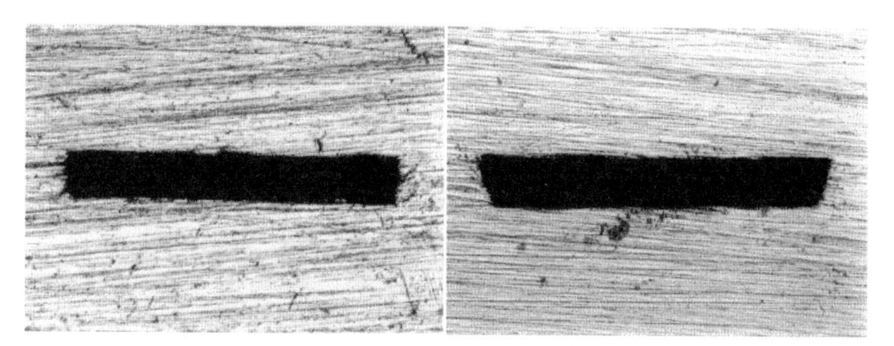

Fig. 14.5. Laser Micromachining; **Fig. 14.6.** EMM

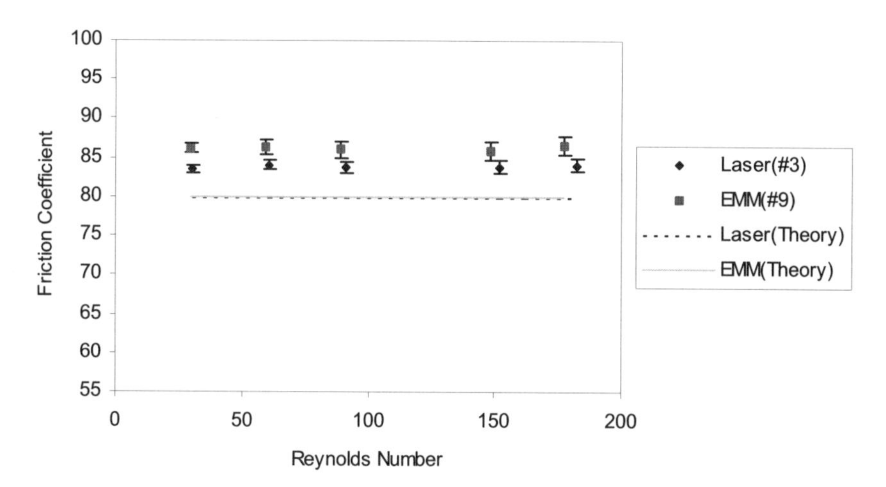

Fig. 14.7. Friction coefficient as a function of Reynold's number for channels 3 (aspect ratio 7:1) and 9 (aspect ratio 6:1). The theoretical friction coefficients happen to coincide for these two channels

From the results, the friction coefficients of both laser-machined channels and electrochemically-machined channels were found greater than the corresponding theoretical values and the differences from theory are significantly greater for laser-machined channels. In Fig. 14.8, at 2:1 aspect ratio, the friction coefficient of the laser-machined microchannel differs by about 40% from that of the electrochemically-machined microchannel. As the aspect ratio of the microchannels increases, the difference in friction coefficient decreases generally and when the aspect ratio reaches 9:1, the value of friction coefficient becomes almost the same. This provides some evidence that the machining method does play a significant role in MECS device performance at lower aspect ratios. Lower aspect ratio features produced by methods yielding higher surface roughnesses will result in increased flow friction causing the need for greater pumping powers. Based on this analysis, it would seem that this impact on fluid flow behavior in microchannels is directly related to the end wall surface roughness of the microlaminated microchannels. Further, this particular study indicates that the effect of the machining method in microlamination is no longer significant if the aspect ratio of the microchannel is greater than about 9:1.

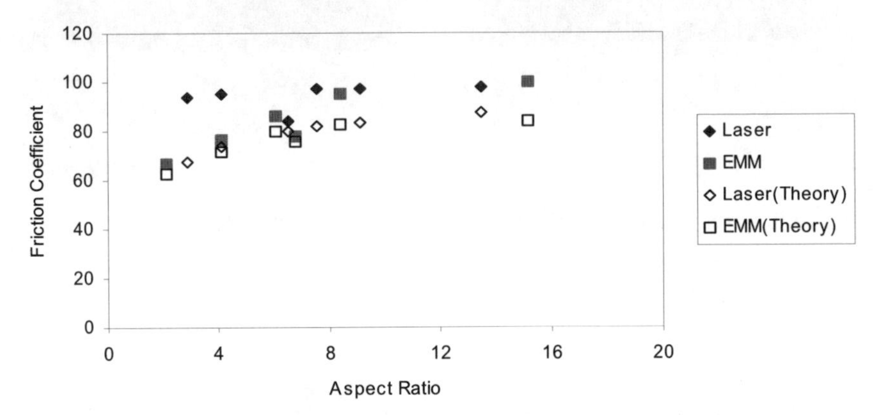

Fig. 14.8. Average friction coefficient and aspect ratios of the microchannels

It is also interesting to note that the difference between the theoretical and experimental values for the friction coefficient in the electrochemically-machined channels is minimal below an aspect ratio of about 8:1 but then increases significantly beyond this point. This phenomenon can be explained by noting that at low aspect ratios, where the end walls have strong influence, the surface roughness of end walls dominates pressure drop across the channel. Since EMM is known to produce perhaps the best surface finish of all non-traditional machining processes, the good agreement between the experiment and theory in those channels at low aspect ratio is not surprising. The deviation from theoretical friction coefficient at higher aspect ratio is explained in Fig. 14.9. Investigation of the higher aspect ratio channels showed a slight deflection in the sidewalls of these channels resulting in a decrease in the cross-section of the channels. This suggests that at aspect ratios above 10:1, sidewall deflection appears a more dominant constraint to aspect ratio than machining method. It is interesting that this type of shape variation appeared even though the thickness of the sidewall end plates was increased significantly to circumvent this problem. It is anticipated that this type of fin warpage behavior will become more pronounced at higher aspect ratios especially inside of spatially-intensified MECS devices where fin thicknesses will be more on the order of microchannel heights. Therefore, the effect of patterning methods on end wall surface roughness should be considered in the design and fabrication of microlaminated MECS devices with low aspect ratio features.

Fig. 14.9. Cross section of channel number 12

14.5 Sources of Warpage in Microchannel Arrays

Several sources have been identified for causing warpage within microchannel arrays. As shown above, one source of microchannel warpage can be the application of bonding pressure on regions of external laminae adjacent to microchannel regions. Another noted source of warpage can come directly from warpage in the raw shimstock, foil or film. Out of the box, even annealed, cold-rolled shimstock can have flatnesses greater than 50 μm on a 50 μm thick foil (Wattanutchariya and Paul 2004). Consequently, flattening procedures are generally required for precision microchannel array fabrication prior to the patterning step. Unannealed metal foils are particularly susceptible to warpage during the bonding cycle due to the relaxation of residual stresses built up in the foils during cold rolling or electrodeposition (in the case of electroplated foils). Perhaps the greatest source of warpage can come from sideloads on the foils either due to friction between laminae, friction between external laminae and the fixture or due to the registration technique. Slower temperature ramps in the bonding cycles can help to ameliorate some of these causes of warpage. Also, application of bonding pressure at the bonding temperature is a must to avoid sideloads caused by friction. Pin registration is particularly troublesome in thermal bonding processes if the coefficient of thermal expansion between the fixture and the laminae are different. As an example, the following analysis explains the effect that the TEER technique can have on fin buckling and registration behavior. Fig. 14.10 illustrates a simple TEER setup. Several laminae are aligned to one another with the use of a TEER fixture. The distance between the posts in the TEER fixture is referred to as the fixture slot width (W_s). At any temperature, the allowance (A) between the width (W) of the laminae and W_s is:

$$A(T) = W(T) - W_s(T) = W \cdot (1 + \alpha_{ss}\Delta T) - W_s \cdot (1 + \alpha_g \Delta T) \qquad (14.4)$$

Where, A is the allowance between the fixture and the laminae which are both a function of temperature (T), α_{ss} is the CTE of stainless steel, α_g is the CTE of graphite and ΔT is the change in temperature from room temperature.

When applying the TEER technique during a diffusion bonding cycle, the metal laminae and the bonding fixture are heated up to a specific temperature, normally in the range of about 60% to 75% of the melting temperature of the base material. For diffusion bonding of metallic structures, this temperature can be 1000 °C or higher. Within a TEER procedure, the fixture and laminae will expand differently as the temperature is raised resulting in a registration force on the laminae at the bonding temperature. However, if this registration force is higher than the critical buckling load of the microchannel fin, buckling of the fin will occur. This is illustrated in Fig. 14.11. When buckling develops, flow channels that are separated by the buckled fin are no longer uniform resulting in flow maldistribution and a drop in the effectiveness of the microchannel array. Fig. 14.12 shows nonuniform microchannels produced as a result of fin buckling. Fig. 14.13 illustrates the result from a previous investigation conducted by Thomas of the TEER technique for microlamination. The graph shows the relationship between the allowance of the bonding fixture and laminae versus the amount of misalignment. This picture suggests that there is an interval of allowance between the laminae and the bonding fixture, where the misalignment is somewhat constant. From this graph, it is observed that once interference is established the accuracy of registration is not improved further. On the other hand, if this interference extends beyond a certain point, the fin will buckle. This indicates that interference should remain within a particular range. This range should constrain the allowable tolerance of the fixture, governed by the desired registration accuracy and the buckling of the laminae.

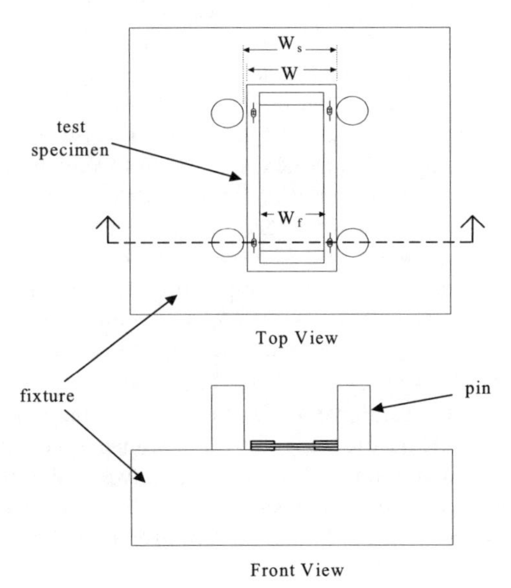

Fig. 14.10. Loading of laminae into a TEER fixture

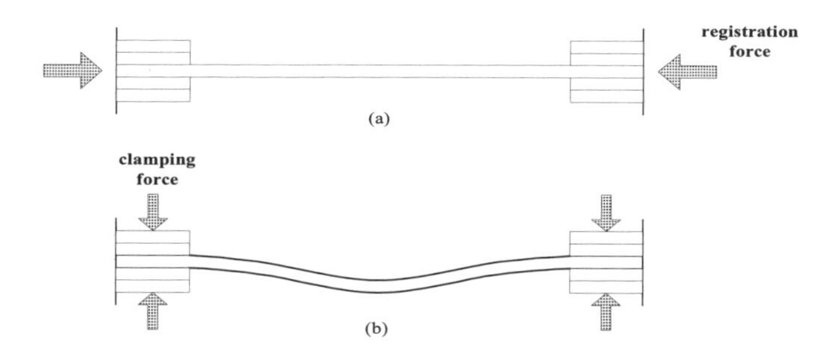

Fig. 14.11. Schematic of fin buckling due to excess registration force

Fig. 14.12. SEM micrographs of a microchannel heat exchanger with nonuniform flow channels (a) before and (b) after fin buckling (Courtesy: Albany Research Center)

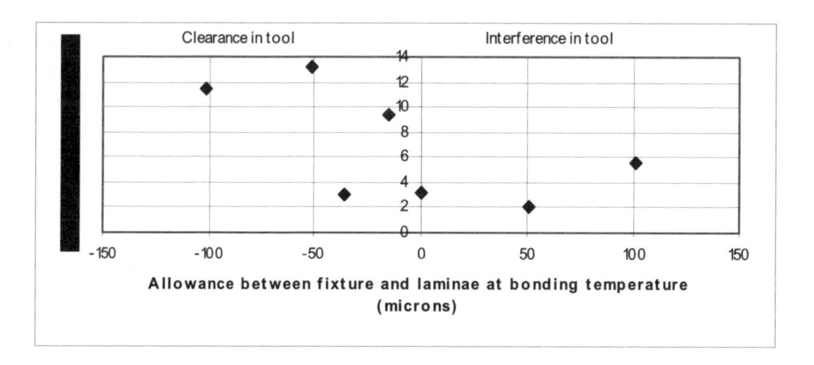

Fig. 14.13. Graph of misalignment versus allowance between fixture and laminae at bonding temperature

14.5.1 Analysis

Assumptions for the following derivation are reported. At some critical temperature called the registration temperature (T_r), the allowance between the fixture and the laminae will be zero. Under these conditions, W can be solved for in terms of W_s:

$$W = \frac{W_s \cdot (1 + \alpha_g \Delta T_r)}{(1 + \alpha_{ss} \Delta T_r)} \tag{14.5}$$

Where, ΔT_r is the difference between the registration temperature and room temperature. By definition, ΔT_r is defined as:

$$\Delta T_r = \Delta T_{tot} - \Delta T_n \tag{14.6}$$

Where, ΔT_{tot} is the total change in temperature from room temperature to the bonding temperature and ΔT_n is the temperature increase beyond the registration temperature at which interference between the laminae and the fixture occurs (i.e., the bonding temperature is above the registration temperature in order to produce a registration load). To determine if a fin will buckle, the mode of failure needs to be confirmed by calculating several loads. For the fixed-ended boundary condition shown in Fig. 14.11, the critical buckling load (P_c) can be calculated from:

$$P_c = \frac{4 \cdot \pi^2 \cdot E \cdot I}{W_s^2} \tag{14.7}$$

Where, E is Young's Modulus and I is the moment of inertia of the specimen. The second load to be considered is the yield load (P_y) defined as the load necessary to yield the fin in compression which is the product of the yield strength and the cross sectional area of the fin:

$$P_y = \sigma_p \cdot A_f \tag{14.8}$$

The final load to be considered is called the registration load (P_r) due to thermal expansion which occurs beyond the registration temperature. The registration load can be calculated from:

$$P_r = A_f \cdot E \cdot (\alpha_{ss} - \alpha_g) \cdot \Delta T_n \tag{14.9}$$

Beyond the registration temperature, the allowance in Eq.19.4 becomes an interference. To help relate the different load parameters, Fig. 14.14 shows the resultant interference limits at which fin buckling will begin as a function of the fin stiffness and thickness for a typical MECS device configuration (Wattanutchariya and Paul 2004). The shape of this plot shows that the load to plastically compress the material is larger than the critical buckling load up to a stiffness of about 2.0 x 10^7 kg/s^2 (i.e., thickness of 0.11 mm). Below this stiffness, the fin may buckle elastically if the registration force is less than the yield force. In this case, ΔT_n can be found by substituting the expression for P_y in Eq.19.8 for P_r in Eq.19.9 and solving for ΔT_n:

$$\Delta T_n = \frac{\sigma_p}{E \cdot (\alpha_{ss} - \alpha_g)} \tag{14.10}$$

However, above this stiffness, the fin may yield in compression without buckling if the registration load is larger than the yield load but below the critical load for fin buckling (Gere and Timoshenko 1997). In this case, ΔT_n can be found by substituting P_c in Eq.14.4 for P_r in Eq.14.6 and solving for ΔT_n:

$$\Delta T_n = \frac{P_c}{A_f \cdot E \cdot (\alpha_{ss} - \alpha_g)} \tag{14.11}$$

In either case, if the registration load exceeds both the critical load for fin buckling and the yield load, the fin will permanently buckle. Fig. 14.14 can be used to evaluate the tolerance limits of a bonding fixture to avoid fin buckling for different stiffness of fin layers. This relation also shows that as a fin layer becomes stiffer, buckling becomes less likely in the device because the interference limit in the stiffer fin is allowed to be greater. The magnitude of fin buckling as a function of interference has been derived.

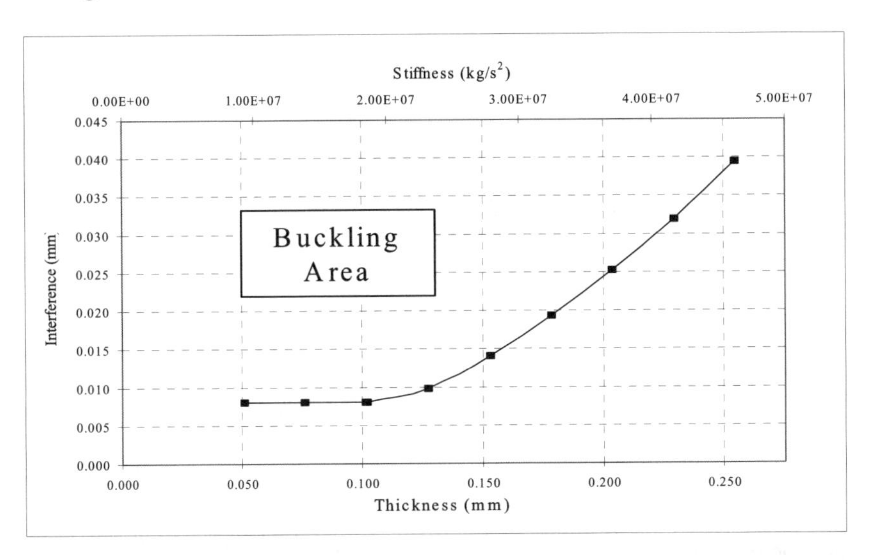

Fig. 14.14. Relationship between fin stiffness or thickness and the allowable fin-fixture interference at the critical point of buckling (Wattanutchariya and Paul 2004)

14.5.2 Results

Fig. 14.15 illustrates both the theoretical and experimental magnitude of maximum buckling deflection with respect to the fixture allowance at the diffusion temperature. Fin width and length were 8.8 and 20 mm respectively and fin thickness varied at 50.8, 76.2 and 101.6 μm. As can be seen from the graph, buckling begins for all three fin thickness at 10 μm of interference. The standard deviation of the channel-to-channel misalignment in interference was about 3.56 μm. This suggests that a fixture with a tolerance of ± 5 μm could on average yield a registration better than the tolerance of the fixture. There was no statistical evidence that the magnitude of warpage depends on the thickness of the shim under the scope of this experiment. In this case, the load at the proportional limit was found to be greater than the critical point of buckling; therefore, all specimens in this experiment were expected to have the same buckling behavior. The significance of this fact is that as the thickness of laminae continues to decrease, the fixture tolerances will not need to continue to decrease below this lower limit. This could be significant as the drive to decrease device dimensions continues.

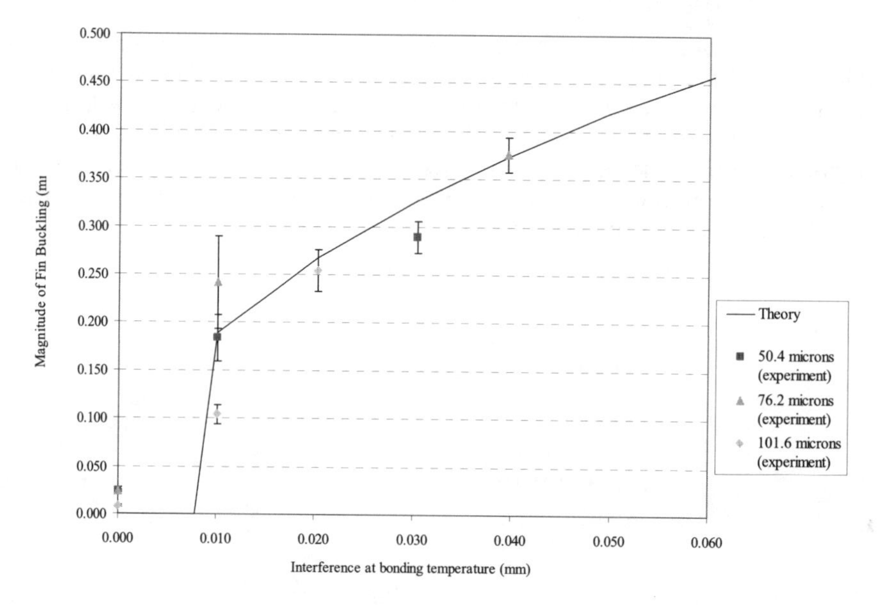

Fig. 14.15. Magnitude of fin buckling versus the interference between fixture and laminae at bonding temperature with 95% confidence interval of standard deviation

As an attempt to increase the tolerance limits of TEER fixtures, an investigation was performed to incorporate compliant mechanisms within TEER fixtures in an effort to increase the allowable tolerance of TEER fixtures (Wattanutchariya and Paul 2004). It was found that for 100 μm laminae, fixture tolerances can be increased from 10 μm to over 100 μm with the use of compliant fixture pins showing that it is possible to engineer TEER fixtures in a way to bring the fixture tolerances within conventional machining tolerances. These tolerance limits make the fabrication of compliant TEER tools more economical. This compliant TEER technique maintained sub-five µm layer-to-layer registration of buckle-free laminae.

14.6 Effects of Registration and Bonding on Microchannel Array Performance

Several previous investigations have found that the registration step is critical in microlamination. Wangwatcharakul (2002) found that laminae misregistration on the order of 20 µm can cause malfunction of out-of-plane microvalves on the order of 100 µm in diameter. Ashley reported that layer-to-layer bearing gaps on the order of 10 µm were necessary in order to operate the microscale turbine generator developed at MIT (Lohner et al. 1999). In each of these applications, precision alignment of laminae was necessary in order to produce a functional device. However, the effect of misregistration on microchannel array heat exchanger performance has been found to be negligible (Wattanutchariya 2002) though Paul, et al. (2000) found that poor registration in intermetallic microchannel devices resulted in an amplification in the fin warpage found within the device.

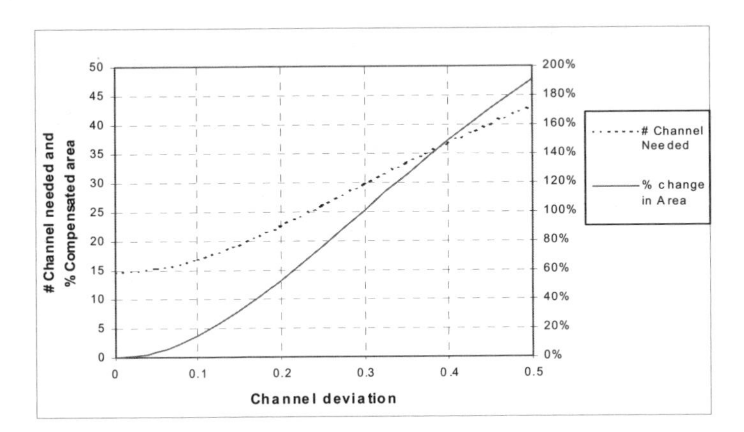

Fig. 14.16. The relation of channel deviation and the number of channel needed as well as percent increase in area to compensate the loss in heat transfer performance

As suggested, the major technological concern in bonding MECS devices is to eliminate fin warpage. Because of the difference in the cross-section of flow channels due to fin warpage, the working fluid will normally flow at different rates between channels. More fluid tends to flow in the larger cross-sectional channels resulting in flow maldistribution. This flow maldistribution results in a longer diffusion path for heat transfer to or from the wall, resulting in a less effective heat transfer coefficient for the total surface area as well as the overall heat transfer. Wattanutcharlya performed an analysis of the effect of flow maldistribution on microchannel heat exchanger performance. Fig. 14.16 illustrates the results of this analysis indicating the percent change in area and change in the total number of microchannels needed to compensate for an average percent channel deviation between adjacent channels within the heat exchanger. Here, percent channel deviation indicates the percent change in channel cross-section as a result of warpage. Fig. 14.16 shows that for a 20% channel deviation, the percent increase in the total number of channels, and, consequently, the heat transfer surface, would need to be 50% to compensate for the loss in heat exchanger performance caused by this deviation. This suggests that channel deviations on the order of 20% of the channel dimension can result in an increase in size of almost 50%.

14.7 Geometrical Constraints in Microchannel Arrays

In the case of heat transfer, the function of a typical microchannel heat exchanger is to transmit the heat from one fluid stream into a second fluid stream. Several different microchannel configurations can be used to accomplish this task. The most efficient heat transfer methods tend to interleave the two fluids in an alternating succession of cross-flow or counter-flow channels. In particular, counter-flow channels are known to provide better temperature distributions along the length of heat exchangers when compared to co-flow heat exchangers (Kakac and Liu 1997). Most MECS applications make use of counterflow microchannel arrays such as in Fig. 14.17. Whereas in the case of co-flow arrays relatively high aspect ratios have been successfully fabricated even in hard-to-produce materials, the aspect ratio is generally constrained in two-fluid microscale heat exchangers because of the more complex geometry.

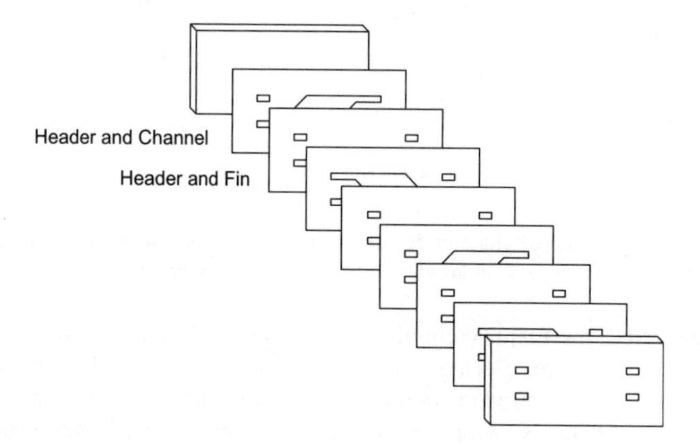

Fig. 14.17. An exploded view illustrating a microlamination procedure

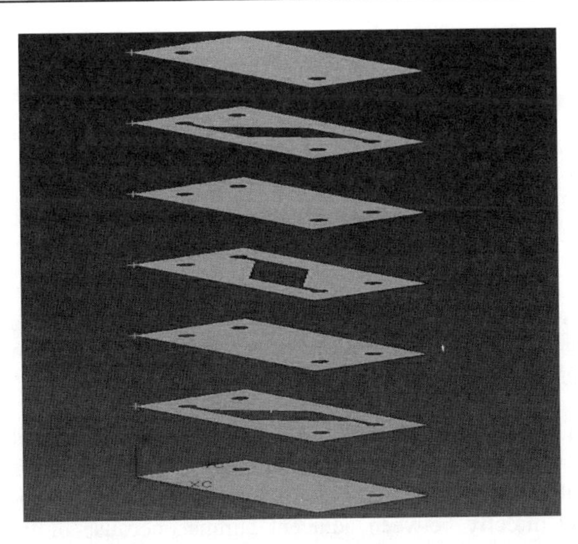

Fig. 14.18. An exploded view of the two-fluid counterflow microchannel array investigated in this study (Model courtesy of P. Kwon at Michigan State University)

The prototypical counter-flow array is produced by alternating layers of microchannels, which allow the two fluids to flow in opposite directions separated by fins. A concept of the through-cut lamina design for a microlaminated counter-flow microchannel array is shown in Fig. 14.18. It is assumed that once the laminae are stacked they are diffusion bonded according to Fig. 14.18. The plan view of the resulting monolith comprised of microchannels and fins is shown in Fig. 14.19. In order to diffusion bond the laminae, a uniform bonding pressure is applied on the stack at an elevated temperature. The pressure is transmitted uniformly through the device except at the necked of the microchannels where each individual microchannel interfaces with fluid headers (the gray regions where the cross-section AA is in Fig. 14.19. The lack of transmitted bonding pressure in these regions can cause the laminae to remain unbonded. In many cases, unbonded regions are observed in the cross-section of counterflow microchannel arrays at points A & B shown in Fig. 14.20, resulting in leakage within the device and mixing of the two fluids.

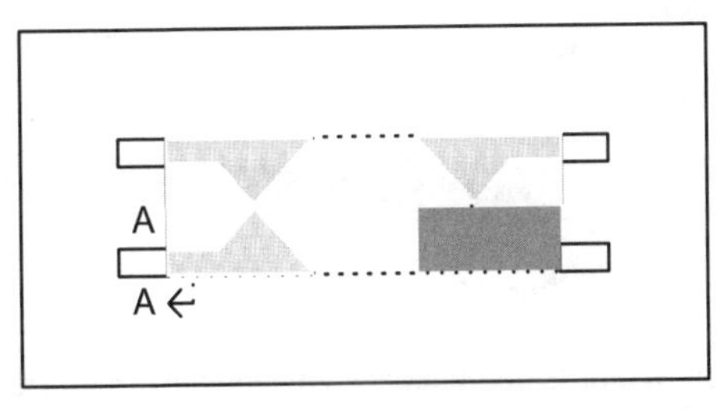

Fig. 14.19. Top view of a counterflow microchannel array comprised of microchannel and fin laminae. The-regions of the device are highlighted in gray

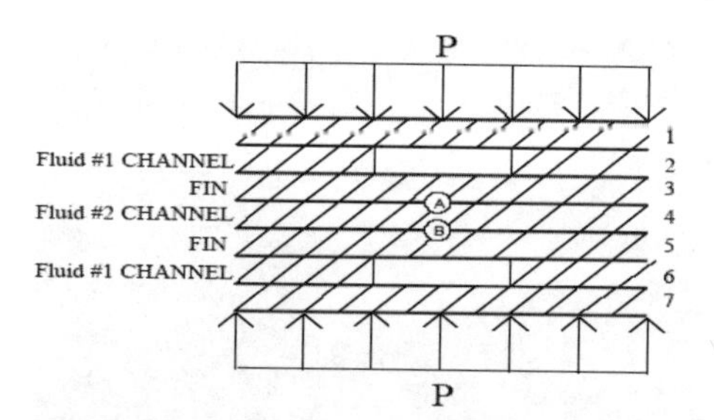

Fig. 14.20. Cross-section of a microchannel neck at cross-section AA in Fig. 14.19. As the width of the channels increase, the stress at points A and B eventually become zero and can even tensile yielding unbounded regions which leak

Fig. 14.21a shows a region within a counterflow microchannel device where pressure was directly transmitted through all of the laminae. Joints between laminae exhibit excellent bonding. Fig. 14.21b shows a region where pressure was not transmitted directly between adjacent laminae because of the counter-flow design. Fig. 14.21b exhibits unbonded regions. From the experience, the width of the necked has to be constrained to a dimension where bonding will occur. The limit on the width of the channel at which bonding takes place depends on the thickness of adjacent fin lamina. In other words, as the thickness of the fin laminae above and below the neck of the microchannels increases, the neck can be made wider. The ratio of the channel span in the necked region to the fin lamina thickness is defined as the fin aspect ratio. The fins in two-fluid microchannel heat exchangers have a certain fin aspect ratio limit beyond which poor bonding will occur within the device resulting in leakage and mixing of the two fluids in the device. Using finite element analysis, it has been found that for a given set of bonding conditions, the key bonding characteristic in all regions of the device must be that only compressive out of plane stresses exist. Maximum fin aspect ratios in the necked regions of counter-flow devices have been implemented up to 10:1 with the allowable fin aspect ratio generally decreasing with increasing fin thickness. This finding is counter-intuitive but has been verified by FEA.

Fig. 14.21. Cross section of a diffusion bounded counter-flow microchannel heat exchanger showing section through which the bonding pressure (a) was transmitted; and (b) was not transmitted. Cross section b resulted in leakage between the two fluids (Micrographs courtesy of Albany Research Center)

14.8 Economics of Microlamination

A key impediment to the proliferation of MECS technology will be its economical production. It is expected that along with the diversity of MECS products that will be developed, there will be a similar diversity with regards to how they are made. Fig. 14.22 shows the unit costs associated with four different microlamination platforms based on photochemical machining (PCM) or blanking/punching (BLK) combined with diffusion bonding (DB) or surface mount technology (SMT). Clearly, the unit cost of MECS devices is better when using blanking/punching and surface mount technology. These improvements are mainly due to decreased cycle times. However, surface mount technology provides the added benefit of reducing the number of laminae needed since the channel laminae are printed using solder paste.

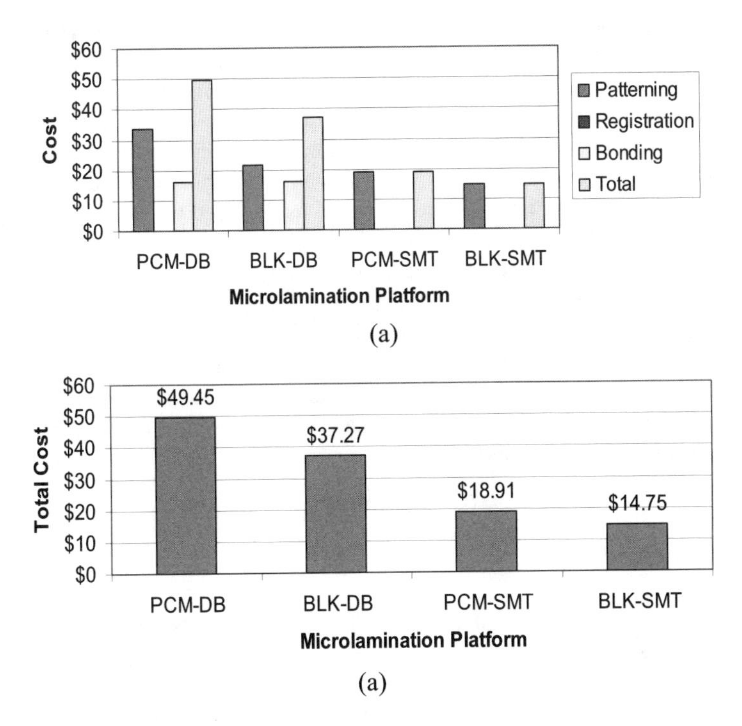

Fig. 14.22. Unit cost for producing a 50 mm x 50 mm x 100 laminae device at a production rate of 100,000 units/year using four different microlamination platforms: (a) when separated into patterning, registration and bonding costs; (b) estimated unit cost for four different platforms

There are two primary implications of this cost analysis. First, it appears that the cost drivers in microlamination processes are likely to be on the patterning side. This is primarily due to the fact that bonding processes are typically performed only once per device while patterning processes are required for each individual lamina. Second, a major strategy for reducing costs is therefore to reduce the number of laminae in a particular design (Porter et al. 2002). Consequently, when one thinks of designing MECS devices for cost reduction, the general principle would be to make the device substrates large and few in number. For device configurations that are small, it would make sense to pattern and bond devices in large substrate configurations and die cut individual devices after bonding to distribute patterning and bonding costs across multiple devices. In this sense, the economics of microlamination are similar to those in the IC industry following the trend toward increasingly larger substrate sizes. In a little more than ten years of MECS device fabrication, already device sizes in excess of 70 cm have been proposed compared with IC fabrication where it took more than 40 years to get over 30 cm. The continued development of high production volume processes for fabricating larger and larger device sizes will become increasingly more important.

As suggested, one promising avenue for addressing the economical production of metal MECS devices is the application of SMT (Sharma and Paul 2003). SMT is the practice and method of attaching leaded and non-leaded electrical components to the surface of conductive patterns in the electronic assembly industry. In addition to being an efficient, economical platform for creating solder joints, SMT also provides a platform for integrating electronics into MECS devices. This factor may become more critical as the need to integrate sensors and actuators within MECS devices grows. The bonding process in SMT requires a low temperature of about 300°C and occurs at atmospheric pressure. The reflow process takes 2-3 minutes, which is negligible when compared to techniques like diffusion bonding which takes hours to bond laminae. The printing and reflow processes can be easily automated as well. Finally, the low fabrication temperatures and pressures offered by surface mount technology will help to minimise warpage and residual stress in materials leading to better process control and more sophisticated geometries.

14.9 References

Ameel TA, Papautsky I, Warrington RO, Wegeng RS, Drost MK (2000) Miniaturisation technologies for advanced energy conversion and transfer systems, J. of Propulsion and Power 16:577-582

Ameel TA, Warrington RO, Wegeng RS, Drost MK (1997) Miniaturisation technologies applied to energy systems, Energy Conversion and Management 38:969-982

Ashley S (1997) Turbines on a dime, Mechanical Engineering 119:78-81

Becker EW, Ehrfeld W, Hagmann P, Maner A, Munchmeyer D (1986) Fabrication of microstructures with high aspect ratios and great structural heights by synchrotron radiation, galvanoforming, and plastic molding (LIGA process), Microelectronic Engineering 4:35-56

Beltrami I, Joseph C, Clavel R, Bacher J, Bottinelli S (2004) Micro- and nanoelectric-discharge machining, J. Mat Proc Tech. 49:263-265

Benson RS, Ponton JW (1993) Process miniaturisation: A route to total environmental acceptability?, Trans. IchemE. 71A:160-168

Brooks KP, PM Martin, MK Drost, Call CJ (1999) Mesoscale combustor/evaporator development. Proc. of the ASME, vol. 9, pp 1-50

Brunger WH, Kohlmann KT (1992), E-beam induced fabrication of microstructures. Proc. IEEE microelectromechanical systems, Germany, pp 168-170

Datta M, Romankiw LT (1989) Application of chemical and electrochemical micromachining in the electronics industry, J. Electrochemical Society 136:285-292

Datta M, Harris D (1997) Electrochemical micromachining: An environmentally friendly, high speed processing technology, Electrochimica Acta. 42:3007-3013

Daymo EA, Wiel DP, Fitzgerald SP, Wang Y, Rozmiarek RT, LaMont MJ, Tonkovich ALY (2000) Microchannel fuel porcessing for man portable power. AIChE Spring National Meeting, Atlanta, pp 1-5

Drost MK, Friedrich M (1997) Miniature heat pumps for portable and distributed space conditioning applications, Proc. of IECEC 2:1271-1274

Drost MK, Friedrich M, Martin C, J Martin, Hanna B (1999) Mesoscopic heat-actuated heat pump developmen. Proc. of the ASME vol. 39, New York

Drost MK, Wegeng RS, Martin PM, Brooks KP, Martin JL, Call C (2000) Microheater AIChE National Meeting. Technical Report, AIChE, Atlanta, pp 1-6

Ehrfeld W, Bley P, Gotz F, Hagmann P, Maner A, Mohr A, Moser HO, Munchmeyer D, Schelb W, Schmidt D, Becker EW (1987) Fabrication of microstructures using the LIGA process. Proc. IEEE Micro Robots and Teleoperators Workshop, p 160

Friedrich C, Kikkeri B (1995) Rapid fabrication of molds by mechanical micromilling: process development. Proc. SPIE, vol. 2640, pp 161-171

Gabriel M, Paul BK, Wilson RD, Alman DE (2001) Characterisation of metallic foil joints using diffusion bonding and diffusion soldering in microtechnology-based energy and chemical systems, Transactions of NAMRC XXIX, Gainesville

Gere JM, Timoshenko SP (1997) Mechanics of materials. Boston: PWS-KENT Pub. Co.

Holmes AS, Saidam SM, Lawes RA (1997) Low cost LIGA processes. IEE Microengineering Technologies and How to Exploit Them, IEE digest pp 54-59

Ihlemann J, Schmidt H, Wolff-Rottke B (1993) Excimer laser micromachining, Adv. Materials for Opt. Electr. 2:87-92

Kakac S, Liu H (1997) Heat exchangers-selection, rating, and thermal design. CRC Press, Boca Raton

Kanlayasiri K, Paul BK (2004) A nickel aluminide microchannel array heat exchanger for high-temperature applications, J. Mfg Processes 6:17-25

Kawano K, Minakami K, Iwasaki H, Ishizuka M (1998) Microchannel heat exchanger for cooling electrical equipment. Proc. of the ASME (Heat Transfer Division), vol. 361-3, pp 173-180

Koeneman PB, Busch-Vishniac IJ, Wood KL (1997) Feasibility of micro power supplies for MEMS, J. MicroElectoMechanical Sys. 6:355-362

Krause V, Treusch HG, Loosen P, Kimpel T, Biesenbach J, Kosters A, Robert F, Oestreicher H, Marchiano M, DeOdorico B (1994) Microchannel coolers for high power laser diodes in copper technology. Proc. of the SPIE, vol. 2148

Little WA, (1990) Microminiature refrigerators for Joule-Thomson cooling of electronic chips and devices. Advances in Cryogenic Engineering 35:1325-1333

Lohner KA, Chen KS, Ayon AA, Spearing SM (1999) Microfabricated silicon carbide microengine structures. Proc. of Material Resource Society Symposium, vol. 546, pp. 85-90

Martin PM, Matson DW, Bennett WD, Stewart DC, Lin Y (1999) Laser micromachined and laminated microfluidic components for miniaturised thermal, chemical and biological systems. Proc. of SPIE, vol. 3680, pp 826-833

Martin PM, Matson DW, Bennett WD (1999b) Microfabrication methods for microchannel reactors and separations systems, Chemical Engineering Communications 173:245-254

Martin PM, Matson DW,Bennett WD (1999) Microfabrication Methods for Microchannel Reactors and Separations Systems , Chemical Engineering Communications, 173, 245-254

Matson DW, Martin PM, Bennett WD, Stewart DC, Johnston JW (1997) Laser micromachined microchannel solvent separator. Proc. of Micromachining and Microfabrication Process Technology III, Austin

Matson DW, Martin PM, Stewart DC, Tonkovich ALY, White M, Zilka JL, Roberts GL (1999) Fabrication of microchannel chemical reactors using a metal lamination process. 3rd Int. Conf. on microreaction technology, Frankfurt, pp 1-11

Paul BK, Klimkiewicz M (1996) Application of an environmental scanning electron microscope to micromechanical fabrication, Scanning 18:490-496

Paul, B.K., R.B. Peterson, and W. Wattanutchariya (1999) The effect of shape variation in microlamination on the performance of high-aspect-ratio, metal microchannel arrays, Proc. of the 3rd Int. Conf. on Microreaction Tech., Frankfurt, pp 53-61.

Paul B, Peterson R (1999) Microlamination for microtechnology-based energy, chemical and biological systems. Proc. of the ASME, vol. 39, New York

Paul BK, Dewey T, Alman D, Wilson RD (2000) Intermetallic Microlamination for high-temperature reactors. 4th Int. Conf. nn Microreaction Tech., Atlanta, pp 236-243

Paul BK, Aramphongphun C, Chaplen F, Upson R (2003) An evaluation of packaging architectures for tissue-based microsystems, Transactions of NAMRC XXXI, Hamilton, Canada

Paul BK, Thomas J (2003) Thermally-enhanced edge registration (TEER) for aligning metallic microlaminated devices, J. Mfg Processes 5:185-193

Peng XF, Peterson GP, Wang BX (1994) Frictional flow characteristics of water flowing through rectangular microchannels, Experimental Heat Transfer 7:249-264

Peterson RB (1999) Numerical modeling of conduction effects in microscale counter flow heat exchangers, Microscale Thermophysical Engineering 3:17-30

Peterson RB (1998) Size limits for regenerative heat engines, Microscale Thermophysical Engineering 2:121-131

Peterson RB (2001) Small Package, Mechaninal Engineering

Pfahler J, Harley J, Bau H, (1990) Liquid transport on micron and submicron channels, sensors and Actuators A 23:431-434

Porter JD, Paul BK, Ryuh B (2002) Cost drivers in microlamination based on a high-volume production system design. ASME International Mechanical Engineering Congress and Exposition, New Orleans, Paper no. IMECE2002-32896

Rahman MM, Gui F (1993) Experimental measurements of fluid flow and heat transfer in microchannel cooling passages in a chip substrate, Advances in Electronic Packaging 4:685-692

Sharma N, Porter JD, Paul BK (2003) Understanding cost drivers in microlamination approaches to microsystem development. IIE IERC, Portland, Oregon

Thomas J, Paul BK (2002) Thermally-enhanced edge registration (TEER) for aligning metallic microlaminated devices, Transactions of NAMRC XXX:663-670

Tonkovich AY, Call CJ, Zilka JL (1998) Catalytic partial oxidation of methane in a microchannel chemical reactor, process miniaturisation. 2nd Int. Conf. on Microreaction Technology, American Institute of Chemical Engineers, New York

Toyoda N, Yamada I (2004) Optical thin film formation by oxygen cluster ion beam assisted depositions, Appl. Surf. Science 226:231-236

Tseng A, (2004) Recent developments in micromilling using focused ion beam technology, J. Micromech Microengr 14:15-34
Weigl BH, Yager P (1999) Microfluidic diffusion-based separation and detection, Science 283:346-347
Walsh ST, Boylan R, Warrington R, Elders J (1996) The emerging infrastructure for MEMS based sensor innovation. Proc. Sensors Expo, Anaheim, p 431
Warren et al. (1999) Mesoscale machines and electronics: There's plenty of room in the Middle. Proc. of the ASME, vol. 39, New York
Wangwatcharakul W (2002) Development of a passive micro-ball valve. MS thesis, Oregon State University
Wattanutchariya W, Kannlayasiri K, Paul BK (2003) Effect of machining methods on flow friction behavior in microchannels. Proc. of the 12th IE-Network Conf., Thailand, pp. 125-130
Wattanutchariya W, Paul BK (2004) Bonding fixture tolerances for high-volume metal microlamination based on fin buckling and laminae misalignment behavior, J. Intl Soc Prec. Engr. Nanotechology 28:117-128
Wattanutchariya W, Paul BK (2004) The effect of fixture compliance on thermally-enhanced edge registration in microlamination, accepted in J. Mfg SE
Wegeng RS, Call CJ, Drost MK (1996) Chemical System miniaturisation, AIChE, New Orleans
White FM (1994) Fluid mechanics. 3th edn, McGraw Hill, Singapore
Wu P, Little WA (1983) Measurement of friction factors for the flow of gases in very fine channels used for microminiature Joule-Thomson refrigerators, Cryogenics 273-277
Zemel JN, Harley JC, Pfahler JN, Urbanek W, Bau HH (1994) Fluid transport in microchannels. Proc. of the Symposium on Microstructures and Microfabricated Systems, vol. 94, pp 210-219

15 Sculptured Thin Films

J. A. Polo Jr.

Department of Physics and Technology, Edinboro University of Pennsylvania, Edinboro

15.1 Introduction

Thin films began to be produced in the mid 1800s following the development of vacuum systems and sources of electricity (Mattox 2003). The first recorded thin films, created by physical vapor deposition (PVD), were made by Grove (Grove 1852) in 1852 when he sputtered metallic films by applying a voltage between a wire and a silver substrate at a reduced pressure. Discovery of internal structure in films is attributed to Kundt (Kundt 1886) who noted the rotation of polarisation of light in transmission through thin films of iron, cobalt and nickel in 1884. Direct observation of physical structure, however, had to await the development of the transmission electron microscope (TEM), and was not made until 1966 by Niewenhuizen and Haanstra (Niewenhuizen and Haanstra 1966).

Thin films made under suitable conditions were observed to be composed of columnar structures on the nanometer scale. Electron micrographs, which they painstakingly made by producing transparent replicas for the TEM, are now made much more easily and routinely with the scanning electron microscope (SEM). Within the last 10 years or so the technology has developed to control the morphology of the columnar structure to produce sculptured thin films (STFs) (Lakhtakia and Messier 2004). Rotating and tilting the substrate while depositing the film can engineer the columnar structures engineered to a wide variety of nanowire shapes. Thus sculptured thin films have entered the growing number of technologies in the exploding arena of nanotechnology (Lakhtakia and Messier 2004).

STFs have a number of properties, both optical and mechanical, which make them suitable for exploitation. In fact, investigation into commercial applications of STF technology has already begun (Suzuki et al. 1999). Many other applications have been proposed and, with the intensity of research currently being carried out in this nascent technology, many more applications are sure to come.

In some ways STFs are complementary to another class of self-assembling systems, liquid crystals (LCs). They share many properties. Both can have either nematic or chiral structure, and they have similar optical properties. While STFs and LCs are both self-assembling systems, STF technology offers self-assembly with control. Although LCs can be manipulated to some degree, the versatility of STFs is unmatched. STF nanowires can be fabricated to virtually any preconceived shape, simply by programming the rotation of a substrate during deposition of the film. In addition, STFs can be made of a wide variety of materials and offer both mechanical and thermal stability not available in LCs.

On the other hand, the optical properties of LCs are easily manipulated electronically, and this has resulted in their ubiquitous use in numerical displays and computer screens seen today. In order to make STFs electronically addressable, the combination of STFs and LCs has been proposed (Robbie et al. 1999). This is made possible because of another important attribute of STFs, high porosity, which allows LCs and other fluids to be infused into them. Many of the proposed non-optical applications of STFs take advantage of the high degree of porosity in STFs.

15.2 STF Growth

15.2.1 Experimental and Phenomenological

Directing a collimated beam of vapor at a substrate in vacuum produces a STF. A schematic of the growth system is shown in Fig. 15.1. The basic components of the system are: vapor source, substrate holder with means of rotation, and crystal monitor to monitor the rate of deposition. Of course, the components are all enclosed in a vacuum chamber. The vapor may be produced either by evaporation or sputtering. If the vapor beam is produced by evaporation, either resistive heating or heating by electron beam may be used.

In order for columns to form, the substrate must be held at a temperature T is less than $0.3T_m$, where T_m is the melting point of the material, thereby limiting adatom surface diffusion. An additional requirement is that the vapor pressure be kept sufficiently low during deposition. At low vapor pressure, the mean free path of molecules within the vapor will be larger than the source to substrate distance, leading to a well-defined beam direction. With these conditions met, at normal incidence, the film grows with the so-called match stick morphology. The film is composed of tightly packed columns with rounded ends. The conditions can be summarised by the structural zone model of thin films developed by Movchan and Demchishin (1969), extended for sputtering by Thornton (1977), and later refined by Messier et al. (1984). In the structural zone model, the locus of points in the plot of sputtering potential and temperature leading to column formation is referred to as the M zone.

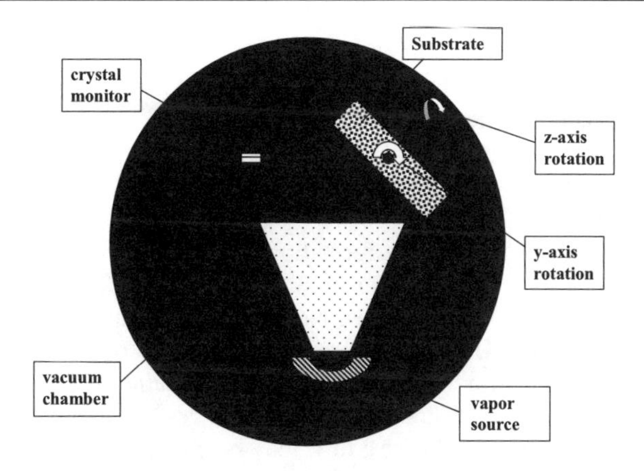

Fig. 15.1. Vacuum deposition system for fabricating STFs

With large oblique angles of incidence of the vapor beam, self-shadowing by spontaneously nucleated clusters on the substrate, combined with the low surface diffusion, lead to the formation of well separated columns. The nanocolumns grow at an angle from the surface, producing a columnar thin film (CTF). Column diameter ranges from 10 nm to 100 nm. The geometry is shown in Fig. 15.2.

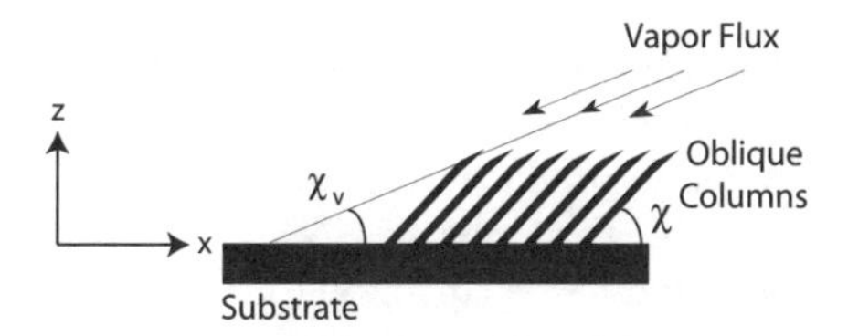

Fig. 15.2. Geometry for the formation of a columnar thin film

Two characteristics of the columns can be cited in general: they have flattened non-circular cross-sections, and the angle at which they are inclined to the surface of the substrate, χ, is greater than the vapor deposition angle, χ_v. It should be cautioned that many investigators measure the tilt of the column relative to the normal to the substrate rather than the plane of the substrate.

The tilt angle, χ, is often related to χ_v through the so-called tangent rule. It is written as

$$\tan \chi = m \tan \chi_v \tag{15.1}$$

Where, $m \geq 1$ is a constant. This form of the relationship between χ and χ_v was originally proposed by Nieuwenhuizen and Haanstra (1966) with m=2. Since that time, however, it has been necessary to modify the value of m to more accurately describe CTFs made of various materials under differing conditions

(Hodgkinson and Wu 1998, Hodgkinson et al. 2001). In any case, the relationship is not accurate at extremely low angles, as experiments show (Messier et al. 1997) that, contrary to Eq. 15.1, χ approaches a value near 20° as χ_v approaches 0°. An alternative description of the χ-χ_v relation, derived by geometrical arguments, has been proposed by Tait et al. (1993), but that too is not universal.

With proper conditions for column formation established, the substrate may be rotated about various axes during deposition to change the direction of growth, producing curved columns or nanowires. Nanowires of virtually any curvilinear form desired can be shaped by programming the rotation of the substrate as a function of the amount of film deposited, and hence the name *sculptured* thin film (Lakhtakia et al. 1996). The substrate may be rotated gradually and continuously, producing smoothly curving nanowires; or it may be changed abruptly, producing a kink in an otherwise smoothly curved or straight section of nanowire. The change in χ resulting from an abrupt change in substrate orientation has been shown to occur over a film thickness as small as ~3 nm (Niewenhuizen and Haanstra 1966). Since 3 nm is much less than optical wavelengths, the change in χ may be considered instantaneous for optical applications.

STFs can be classified by nanowire shape into two canonical classes: sculptured nematic thin film (SNTF), and thin film helicoidal bianisotropic medium (TFHBM) (Lakhtakia 1997). When the axis of rotation of the substrate lies in the plane of the substrate, the nanowires form two dimensional shapes and a SNTF results.

When the axis of rotation is perpendicular to the substrate, helicoidal structures are formed resulting in a TFHBM or chiral STF. Of course, the two types of rotation may be combined to produce even more complex shapes. Various shapes have been produced to date. Examples of SNTF shapes that have been produced are: zigzag, "C" shaped, and "S" shaped. STFs of the TFHBM type which have been made are: helical, square helical, slanted helical, and superhelical. Fig. 15.3 shows electron micrographs of a few STFs with various shapes. Fig. 15.3(a-c) are examples of SNTFs; Fig. 3.3(d) is a chiral STF; while Fig. 15.3(e) is an example of a STF composed of two different sections, with the bottom one chiral and the top one a chevron shaped SNTF. Growth rates and film thicknesses vary from one experimental set up to another. In some recent experiments Horn et al. (2004), using a rastered electron beam evaporation technique, have achieved growth rates as high as 0.4 μm/min and uniform column diameters for large thickness (> 3 μm) films. Furthermore, they were able to produce highly uniform STFs over a large area (75 mm diameter). Previous to this work, areas were more typically on the order of 1 cm^2. In addition to the basic STF growth technique just described, various refinements and modifications have been developed. Producing well-defined columns oriented normal to the surface is difficult with the basic growth method. Deposition at normal incidence to the substrate yields columns, which tend to expand in diameter, as they grow taller except under narrowly controlled conditions (Messier and Lakhtakia 1999). Furthermore, the columns are rather densely packed. Two solutions have been developed, both employing deposition at oblique angles.

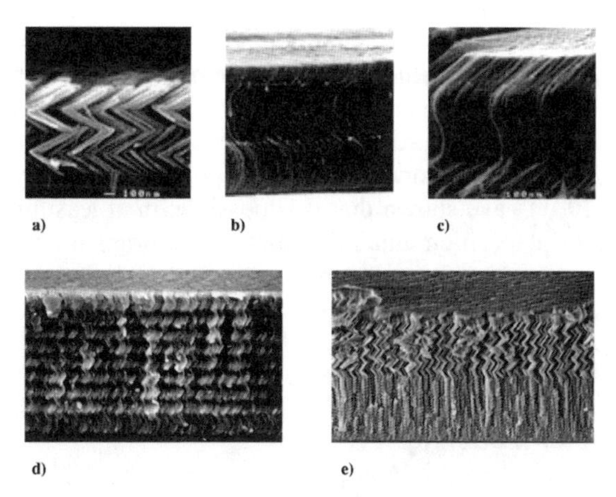

Fig. 15.3. Electron micrographs of STFs: (a) Chevron SNTF (b) "S" SNTF (c) Slant "S" SNTF (d) Chiral STF (e) Mixed chiral and chevron STF. Micrographs courtesy of Russell Messier, and Mark Thorn, Pennsylvania State University

In one case (Robbie et al. 1999), the substrate is rotated about the z–axis, as in the production of a chiral STF, but at a much faster rate. At high rotation rate, a chiral STF is not produced; rather, the coils of the helix merge to produce a vertical column with constant diameter. In the other method, serial bideposition (Robbie et al. 1998), deposition at a fixed χ_v is used, but the film is grown as a series of short depositions with a rotation of 180° about the z –axis between each deposition. The tilt, which would normally result from oblique deposition, is averaged out by deposition from both sides of the normal. The spacing of nanowires is related to χ. Thus, in addition to creating well-defined, constant diameter columns normal to the substrate, these methods also offer a degree of control over the density of the film. Serial bideposition also offers additional control over the nanowire morphology. Nanowire cross-sections become more flattened as χ_v is decreased, with the long axis of the cross-section perpendicular to the plane of incidence of the vapor flux. With bideposition, the cross-section of the nanowire can be flattened by depositing at low χ_v, yet a vertical column can still be produced. The anisotropy of the film in the x-y plane is thus controllable. This can be particularly important for optical applications, especially when high optical activity (Hodgkinson et al. 2000) is desired. Serial bideposition, however, is not limited to producing columns normal to the surface. The method can be altered to produce chiral STFs, slanted CTFs, and virtually any other shape by varying the series of rotations about the z-axis with angles other than 180°. Simultaneous bideposition, using two vapor sources, has been discussed extensively by Kranenburg and Lodder (1994). Their work is comprised of both experimental studies as well as computer simulations. They discuss using two sources with identical material to control morphology, as well as two sources with different materials to control local chemical composition. It may be advantageous in some instances to cover the STF with a final capping layer of material. This has been studied by Robbie and Brett (1997), with emphasis on avoiding stress fractures and preventing the void region of the film from being filled with material. They describe a method in which the angle of incidence relative to the film normal is decreased exponentially with time in order accomplish both objectives. The crystal structure of the material composing the columns is another property which can be important in some applications, catalysis for instance. Suzuki et al. (2001) have shown that it is possible, in at least one case (TiO_2), to anneal STFs after deposition and still maintain the original STF structure. X-ray diffraction revealed that the material within the STF, originally showing no diffraction peaks, was converted by the annealing process to the single anatase phase of TiO_2.

15.2.2 Computer Modeling

The modeling of STF growth is currently in its infancy with much of the work on CTFs. As yet, the ability to enhance the design of STFs by simulation has been limited (Lakhtakia and Messier 2004; Kaminska et al. 2004). Once a solid framework for handling CTFs in detail is developed, the application of the models to obtain a thorough understanding of STF growth should proceed swiftly. Several different approaches have been taken to understand the growth of CTFs; most seem to fall into two broad classes. The first type uses a simplified model to calculate the growth of the CTF (Meakin et al. 1986; Tait et al. 1993; Vick et al. 1999; Smy et al. 2000; Suzuki and Taga 2001; Kaminska et al. 2004). Typically, the model assumes: a nanoscale sized particle, often represented as squares, discs, or cubes restricted to discrete grid positions; a distribution of angles describing the trajectory of particles in the beam; some sort of sticking coefficient between the particle and film; and parameters characterizing limited diffusion or hopping on the film. Various calculational schemes are used to simulate the CTF, among them are: Monte Carlo, molecular dynamics, or a combination of methods. The advantages of such models are that: they are able to simulate the growth of a thin film in a reasonable amount of computation time, and they actually produce CTF and even STF structure. Some simulations are restricted to 2-D while others simulate a 3-D film. However, the models are a crude representation of physical reality and do not give specific information which would be helpful in engineering particular nanowire morphologies for particular materials. The models have been used to calculate χ as a function of χ_v with moderate success (Meakin et al. 1986; Tait et al. 1993). The porosity of the film has also been characterised using these methods. The other type of model works at the opposite extreme by simulating the growth of the thin film on the atomistic scale with atom by atom addition to the film (Kranenburg and Lodder 1994; Dong et al. 1996). Several simulations have been done using a Leonard-Jones potential and molecular dynamics calculation to simulate the growth. This type of calculation is extremely, computationally intensive. To make the calculations tractable, the simulations are often done in only two dimensions. Furthermore, due to long calculation times, the thickness of the film which can be grown in this type of calculation is very limited.

Although crude channels and column like structures begin to appear in these simulations, the size of the film simulated is too small to show the development of a true CTF. The calculations have been used to estimate χ for various situations as well as the porosity of the film. Though this type of calculation may be more physically realistic, the models simulate a generic film with no real insight into engineering particular characteristics in a real film. Furthermore, the physical realities of fabricating a film are often not accounted for, such as degree of beam collimation, and residual gas pressure. Given the impractical requirements of performing an atomistic type simulation of a real CTF over the vast range of length and time scales involved in growing a STF, it is almost certain that an understanding of STF growth will be obtained using a model which is a hybrid of atomistic and macroscopic models. This approach has been used with some success in other areas (Huang et al. 1998; Rahman et al. 2004; Shenoy 2004).

15. 3 Optical Properties

The optical properties of STFs offer a particularly rich area for study. In addition to the influence of the innumerable variations of nanowire morphology on the optical properties of the film, the large range of materials out of which STFs may be constructed, with their intrinsic electromagnetic properties, compound the number of possibilities. Dielectric materials have been given the greatest attention. Even within this restricted class of materials, there are many variations. For example, dielectric materials may be: highly absorptive or nearly non-absorptive, weakly dispersive or highly dispersive, linear or non–linear, chiral or non–chiral, gyrotropic or non-gyrotropic. STFs offer many opportunities and challenges in both theoretical and experimental areas.

15.3.1 Theory

The theoretical treatment of the electromagnetic properties of STFs can be divided essentially into two main tasks (Lakhtakia and Messier 2004): a description of the constitutive relations of the STF, and the solution of Maxwell's equations to describe the propagation of the electromagnetic wave. Once the description of the propagation of the wave has been obtained, various optical properties of the film such as reflectance, transmission, and optical rotation may be extracted easily.

15.3.1.1 Constitutive Properties

The width of the nanowires and voids in STFs are on the order of 50 nm, about an order of magnitude smaller than optical wavelengths. As described in many elementary texts (Jackson 1975), a composite medium, with a uniform composition and features much smaller than a wavelength, may be treated as a homogeneous material with average constitutive parameters. In fact, Maxwell's equations in materials represent such an averaging over atomic dimensions. On a larger scale, a thin layer of STF at a constant depth into the film may be considered a uniform composite of nanowire and void, since, at a given depth, all nanowires have the same morphology and orientation. Along the axis of inhomogeneity, however, the dimensions describing the STF features are on the order of an optical wavelength. Thus, in thin layers the STF may be treated as a homogeneous material; but the variation of parameters from one layer to the next necessitates treating the entire STF as a homogeneous material with constitutive parameters which vary continuously along the axis of inhomogeneity as the nature of the layer changes.

We will first look at finding average constitutive parameters of a single layer. As some applications of STFs involve infiltrating the void region with a fluid, the void region potentially has material in it. Assuming linear materials, the most general constitutive relations for the material in both naonowires and voids may be written in the form (Weiglhofer 2003)

$$\underline{D}(\underline{r}) = \varepsilon_0 \left[\underline{\underline{\varepsilon}}(\omega) \bullet \underline{E}(\underline{r},\omega) + \underline{\underline{\alpha}}(\omega) \bullet \underline{H}(\underline{r},\omega) \right] \tag{15.2}$$

$$\underline{B}(\underline{r}) = \mu_0 \left[\underline{\underline{\beta}}(\omega) \bullet \underline{E}(\underline{r},\omega) + \underline{\underline{\mu}}(\omega) \bullet \underline{H}(\underline{r},\omega) \right] \tag{15.3}$$

Where, the dyadics $\underline{\underline{\varepsilon}}(\omega)$ and $\underline{\underline{\mu}}(\omega)$ are the relative permittivity and permeability; and $\underline{\underline{\alpha}}(\omega)$ and $\underline{\underline{\beta}}(\omega)$ are the dyadics describing the magnetoelectric properties of the material. The permittivity and permeability of free space are $\varepsilon_0 = 8.854 \times 10^{-12}\ F \cdot m^{-1}$ and $\mu_0 = 4\pi \times 10^{-7}\ H \cdot m^{-1}$, respectively. The problem, then, is to determine a single set of four 3x3 complex valued dyadics ($\underline{\underline{\varepsilon}}$, $\underline{\underline{\alpha}}$, $\underline{\underline{\beta}}$, and $\underline{\underline{\mu}}$) describing the equivalent homogeneous material to replace the two sets of dyadics representing nanowire and void. The four dyadics are often combined for convenience to form one compact 6x6 matrix as,

$$\underline{\underline{C}} = \begin{bmatrix} \varepsilon_0 \underline{\underline{\varepsilon}} & \varepsilon_0 \underline{\underline{\alpha}} \\ \mu_0 \underline{\underline{\beta}} & \mu_0 \underline{\underline{\mu}} \end{bmatrix} \tag{15.4}$$

In this formalism, the fields are then combined to form a column 6–vector,

$$\underline{F} = \begin{bmatrix} \underline{E} \\ \underline{H} \end{bmatrix} \tag{15.5}$$

so that the right sides of Eq. 15.2 and Eq. 15.3, for instance, are combined to be compactly written as $\underline{\underline{C}} \bullet \underline{F}$.

Determining average constitutive parameters for composite materials, a process known as homogenisation, is a very active area of research. For all but the simplest cases, great effort must be expended to determine the constitutive parameters for an equivalent homogeneous material, referred to as a homogenised composite material (HCM). Various schemes have been devised to determine the HCM constitutive properties, each with its own limitations (Lakhtakia 1996; Michel 2000; Mackay 2003).

One of the earlier and simplest formalisms, Maxwell Garnett (1904), is limited to situations where one of the materials in the composite takes the form of a small particulate dispersed in a second material, the host, which must make up most of the volume of the composite. As can be seen from the electron micrographs, this is often not the situation for STFs, where nanowire and void can occupy comparable volumes.

Two different modifications of the basic Maxwell Garnett formalism have been developed to solve the problem of homogenizing composites with dense inclusions. One approach is to build up a dense concentration of particles in a number, N, of dilute particle additions or increments and is thus known as the Incremental Maxwell Garnett (IMG) formalism (Lakhtakia 1998). After each addition of particles, the composite is re–homogenised so that the next addition of particles is made into a homogeneous medium. Hence, each homogenisation satisfies the Maxwell Garnett requirement of a system of *dilute* particles in a host material.

The other modification to the Maxwell Garnett formalism is obtained from the IMG in the limit $N \rightarrow \infty$. In this case, a differential equation for the constitutive parameters as a function of concentration of particulate is obtained, replacing the iterative procedure of the IMG formalism. This formalism has been named the Differential Maxwell Garnet (DMG) formalism (Michel et al. 2001). Another approach currently being used to locally homogenise STFs (Sherwin and Lakhtakia 2001; Sherwin et al. 2002), which is not derived from the Maxwell Garnett method, is the Bruggeman formalism (Bruggeman 1935). Unlike the Maxwell Garnett formalism, this method places no restriction on the concentration of inclusions. In fact, this method is carried out by considering both materials of the composite as inclusions in the resultant HCM being sought. The constitutive parameters of the HCM are then adjusted so as to produce zero polarisation in the HCM–inclusion mixture. The HCM thus obtained represents the desired HCM characterizing the composite.

Finally, a more complete formalism that has received some attention is the Strong Property Fluctuation Theory (SPFT) (Tsang and Kong 1981; Zhuck 1994). The method is known for being quite difficult to carry out numerically, but is exact in the unapproximated form. It takes account of spatial correlations between components in the composite. Second order SPFT has been developed for linear bianiostropic HCMs (Mackay et al. 2000, 2001, 2001, 2002). It has also been extended to the weakly non-linear regime (Lakhtakia 2001, Lakhtakia and Mackay 2002). Whichever formalism is chosen, the polarisablity of a single inclusion particle must be calculated as a first step. This can be done relatively simply for ellipsoidally shaped inclusions. Lakhtakia et al. (1999, 2001) have developed a nominal model to describe a STF layer in which both nanowire and void are modeled as ellipsoids. With long, narrow ellipsoids, the aciculate nature of the nanowire can be captured.

It should be pointed out that the development of homogenisation procedures is not at the stage where absolute constitutive properties of a STF can be calculated from first principles. Nonetheless, there are several ways in which calculation of HCMs is useful. Calculation of trends, as various STF parameters are changed, may be useful in the design of STFs for specific purposes. Also, although local homogenisation may not give absolute numbers for the constitutive relations, it does give the form that the constitutive dyadics should take for various types of materials. Some proposed uses of STFs involve infiltrating the porous film with different types of fluids (Lakhtakia 1998; Ertekin and Lakhtakia 1999; Lakhtakia et al. 2001). Here, again, homogenisation is very useful in pointing to the correct form of the constitutive dyadics.

Rather than relying on homogenisation, most calculations of optical properties of STFs have used one of two approaches to obtaining constitutive parameters. One approach is to guess reasonable parameters. This is useful when gross behavior is desired. Furthermore, STFs made of the same materials, in the same way, at different labs will have different detail properties. Thus, STFs for a specific purpose will have to be fine tuned, and absolute numbers are of limited use. The other approach is to use principal indices of refraction extracted from experimental data on CTFs (Hodgkinson and Wu 1998; Hodgkinson et al. 2001; Chiadini and Lakhtakia 2004; Polo and Lakhtakia 2004). This has the added advantage of accounting for higher order influences on constitutive parameters, such as the change in nanowire cross-sectional morphology and material density as a function of χ.

So far, only the constitutive parameters of a single electrically thin layer of a STF have been discussed. With the shape and orientation of the nanowire changing over a larger length scale (comparable to optical wavelengths) due to sculpting, the variation of the constitutive properties of the HCM as a function of depth into the film must be described. STFs made of various types of materials have been modeled. For purposes of discussion, a dielectric material will be assumed here since: only one dyadic need be considered, expressions are reasonably compact, and other materials are handled similarly.

The constitutive parameters for the film can be written as,

$$\underline{D}(\underline{r}) = \varepsilon_0 \underline{\underline{S}}_z(z) \cdot \underline{\underline{S}}_y(\chi) \cdot \underline{\underline{\varepsilon}}_{ref}^{o} \cdot \underline{\underline{S}}_y^{-1} \cdot \underline{\underline{S}}_z^{-1}(z) \cdot \underline{E}(\underline{r}) = \underline{\underline{\varepsilon}}(z) \cdot \underline{E}(\underline{r}), \tag{15.6}$$

$$\underline{B}(\underline{r}) = \mu_0 \underline{H}(\underline{r}) \tag{15.7}$$

As the material is non-magnetic, Eq. 15.7 takes a simple form with a single scalar, μ_o, defining the isotropic magnetic constitutive relation. Eq. 15.6, on the other hand, is much more complex with several factors required to describe the permittivity *dyadic*.

In this expression, $\underline{\underline{\varepsilon}}^{o}_{ref}$, the *reference* permittivity dyadic, represents the permittivity dyadic for a layer of STF with the nanowires aligned along a reference direction, the $x-$ axis. In principle, this is the quantity which can be determined by homogenisation. As discussed above, other methods are often used. In any case, for the present situation, it takes the form,

$$\underline{\underline{\varepsilon}}^{o}_{ref} = \varepsilon_a \, \underline{u}_z \underline{u}_z + \varepsilon_b \, \underline{u}_x \underline{u}_x + \varepsilon_c \, \underline{u}_y \underline{u}_y \tag{15.8}$$

Where, $\underline{u}_x$, $\underline{u}_y$ and $\underline{u}_z$ are the Cartesian unit vectors. This is the same form of permittivity dyadic used to describe a bulk homogeneous bianisotropic material. This reflects the fact that the principal axes of the nanowire are, in general, all different due to the flattened cross-section of the nanowire. The permittivity along the length of the nanowire is represented by ε_b. The other two components (ε_a, ε_c) represent the permittivities along the principal axes through the cross-section. In canonical models of STFs, the permittivities along the principal axes are assumed to be independent of χ; whereas, in models relying on empirical determinations of the constitutive relations the permittivities are expressed as a function of χ. The rotation matrix, given below,

$$\underline{\underline{S}}_y(\chi) = (\underline{u}_x \underline{u}_x + \underline{u}_z \underline{u}_z)\cos\chi + (\underline{u}_z \underline{u}_x - \underline{u}_x \underline{u}_z)\sin\chi + \underline{u}_y \underline{u}_y \tag{15.9}$$

rotates $\underline{\underline{\varepsilon}}^{o}_{ref}$ by an angle χ to account for the tilt of the nanowire in the STF. For CTFs and chiral STFs, χ is constant, while in SNTFs $\chi=\chi(z)$. The second rotation matrix rotates the already inclined dyadic about the z axis. For a chiral STF, which has a constant rate of rotation as a function of z, it is written as,

$$\underline{\underline{S}}_z(z) = (\underline{u}_x \underline{u}_x + \underline{u}_y \underline{u}_y)\cos(\pi z/\Omega) + (\underline{u}_y \underline{u}_x - \underline{u}_x \underline{u}_y)\sin(\pi z/\Omega) + \underline{u}_z \underline{u}_z \tag{5.10}$$

in order to create the helical structure. For SNTFs, there is no rotation about the z axis and $\underline{\underline{S}}_z(z)$ may be considered to be $\underline{\underline{I}}$, the identity matrix.

15.3.1.2 Wave Propagation

With the constitutive relations obtained by one of the methods described above, the propagation of the wave in the STF can be characterised by solving Maxwell's curl equations. The STF will be considered to be an infinite slab occupying the space from z=0 to z=L, with the axis of inhomogeneity along the z direction. As the spectral response of the STF is most often desired, we will assume the fields to have a harmonic time dependence of ε^{-iwt}, where w is the angular frequency. In time harmonic form, Maxwell's curl equations are,

$$\left.\begin{aligned}\nabla\times\underline{E}(\underline{r}) &= i\omega\mu_0\underline{H}(\underline{r})\\ \nabla\times\underline{H}(\underline{r}) &= -i\omega\underline{\underline{\varepsilon}}(z)\bullet\underline{E}(\underline{r})\end{aligned}\right\} \quad (15.11)$$

With a plane wave incident on the film from z<0, the fields may be expressed in terms of their Fourier amplitudes as

$$\underline{E}(\underline{r}) = [e_x(z,\kappa,\psi)\underline{u}_x + e_y(z,\kappa,\psi)\underline{u}_y + e_z(z,\kappa,\psi)\underline{u}_z]e^{[i\kappa(x\cos\psi+y\sin\psi)]} \quad (15.12)$$

$$\underline{H}(\underline{r}) = [h_x(z,\kappa,\psi)\underline{u}_x + h_y(z,\kappa,\psi)\underline{u}_y + h_z(z,\kappa,\psi)\underline{u}_z]e^{[i\kappa(x\cos\psi+y\sin\psi)]} \quad (5.13)$$

Where, $k=k_0$ is the transverse wavenumber, with k_0 the free space wavenumber and θ the angle of incidence. The plane of incidence is described by the angle ψ, which is measured from the x direction. Substituting Eq. 15.12 and Eq. 15.13 into the Maxwell curl relations results in two sets of equations. The first is a set of two algebraic equations, which describe the z components of the fields in terms of the x and y components. The second is a set of four coupled ordinary differential equations involving the x and y amplitudes of the two fields. The four differential equations may be written as one compact matrix ordinary differential equation (Berreman 1972; Lakhtakia 1997) in the form,

$$\frac{d}{dz}\left[\underline{f}(z)\right] = i\left[\underline{\underline{P}}(z)\right]\cdot\left[\underline{f}(z)\right] \quad (15.14)$$

Where, the field amplitudes have been assembled into the column 4-vector

$$\left[\underline{f}(z)\right] = \begin{bmatrix} e_x(z) \\ e_y(z) \\ h_x(z) \\ h_y(z) \end{bmatrix} \quad (15.15)$$

and $\left[\underline{\underline{P}}(z)\right]$, referred to as the kernel, is a complex 4x4 matrix which depends on the form of the constitutive relations describing a particular STF (Lakhtakia 1997; Lakhtakia and Messier 2004).

In determining optical properties such as reflectance or transmission, it is convenient to recast the formulation in terms of a 4x4 matrix, called the matrizant, such that

$$\left[\underline{f}(z)\right] = \left[\underline{\underline{M}}(z)\right]\left[\underline{f}(0)\right] \quad (15.16)$$

Where, $\left[\underline{f}(0)\right]$ is the field 4–vector at the surface. Substituting the right hand side of Eq. 15.16 into the Eq. 15.14 yields an equivalent differential equation (Lakhtakia and Messier 2004),

$$\frac{d}{dz}\left[\underline{\underline{M}}(z)\right]=i\left[\underline{\underline{P}}(z)\right]\cdot\left[\underline{\underline{M}}(z)\right] \tag{15.17}$$

in terms of the matrizant. In this way, the boundary conditions may be stated generally once and for all as,

$$\left[\underline{\underline{M}}(0)\right]=\left[\underline{\underline{I}}\right] \tag{15.18}$$

and the differential equation may be solved solely in terms of the STF properties. The heart of the propagation problem, then, is the solution of Eq. 15.17. In general, $\left[\underline{\underline{P}}(z)\right]$ is a function of z, and no known closed form solutions of Eq. 15.17 exist. However, there are two special cases for which $\left[\underline{\underline{P}}(z)\right]$ is a constant. In these cases, the solution can be written formally as,

$$\left[\underline{\underline{M}}(z)\right]=exp\left(i\left[\underline{\underline{P}}\right]z\right),\quad (when\left[\underline{\underline{P}}\right]=constant) \tag{15.19}$$

One case is that of the CTF, which has a constant [P] because the constitutive relations are constant as a function of z. The other situation is for chiral STFs, but only occurs for normal incidence or, in other words, when k=0. In this case the constitutive relations are not constant due to the helical rotation. However, the problem can be worked in terms of transformed fields under the so-called Oseen transformation (Oseen 1933; Lakhtakia and Weiglhofer 1997) given by

$$\left.\begin{aligned}\underline{E}'(\underline{r})&=\underline{\underline{S}}_z^{-1}(z)\cdot\underline{E}(\underline{r})\\ \underline{H}'(\underline{r})&=\underline{\underline{S}}_z^{-1}(z)\cdot\underline{H}(\underline{r})\end{aligned}\right\} \tag{15.20}$$

Where, $\underline{\underline{S}}_z^{-1}(z)$ is the rotation matrix for chiral STFs described earlier. When written in terms of the transformed fields, the differential equation has a constant kernel for normal incidence. Once the transformed fields are determined, the inverse transformation can be used to obtain the physical fields. The problem is often worked in terms of the transformed fields at oblique angles of incidence, as well. Even though a closed form solution does not exist, the algebra is simplified significantly by using the transformation.

In the more common situation, when the kernel is not constant, several schemes have been used to obtain a solution to Eq. 15.17. One is an exact solution for chiral STFs written as an infinite polynomial series (Lakhtakia and Weiglhofer 1996; Polo and Lakhtakia 2004). In practice, the series must be terminated after a finite number of terms. The convergence of the series may be monitored as each term is added and the summation terminated when the desired accuracy has been obtained. The ability to monitor the convergence as the calculation proceeds may be an advantage over other numerical techniques for which the accuracy of the result is discerned by repeating the entire calculation under different approximation parameters (Polo and Lakhtakia 2004). The method is particularly advantageous if many calculations are to be performed for different situations.

For situations in which [P(z)] is periodic, a perturbational approach has been applied (Weiglhofer and Lakhtakia 1996, 1997) when the STF is weakly anisotropic. For stronger anisotropy, the coupled-wave method has been used (McCall and Lakhtakia 2000, 2001; Wang et al. 2002). A numerical method offering great versatility is the piecewise homogeneity approximation (Abdulhalim et al. 1986; Venugopal and Lakhtakia 2000). In this approximation, the STF is divided into a series of thin slices parallel to the surface. Each slice is treated as if it had a constant permittivity dyadic, with the value of the dyadic taken to be that of the STF at the center of the slice. With the approximation of constant permittivity dyadic, the matrizant for each single slice can be calculated using Eq. 15.19. The matrizant for the entire STF is then obtained as the sequential product of the matrizants of the slices. The accuracy of the method depends on the thickness of the slice used for the calculation, with thinner slices yielding higher accuracy but taking more computation time. Once the matrizant has been determined, the optical properties of the STF can be obtained by solving a boundary value problem with incident, reflected, and transmitted waves ((Lakhtakia and Messier 2004).

15.3.2 Characteristic Behavior

The optical characteristics of STFs have been both calculated and measured (McPhun et al. 1998; Wu et al. 2000). By far, the most studied effect is the circular Bragg phenomenon (CBP) of chiral STFs. Most work has been on linear dielectrics. Theoretical calculations have been carried out for non-linear dielectrics (Lakhtakia 1997; Venugopal and Lakhtakia 1998) and gyrotropic materials (Pickett and Lakhtakia 2002; Pickett et al. 2004). The effects of absorption and dispersion in dielectric STFs has also be studied (Venugopal and Lakhtakia 2000).

15.3.2.1 Chiral STFs

As with other periodic systems, chiral STFs exhibit strong reflection in narrow bands of wavelengths. The reflection bands, or Bragg peaks, are centered at a series of wavelengths forming a harmonic series. The range of wavelengths over which the reflection occurs is referred to as the Bragg regime or alternatively the Bragg zone. Normally, all of the peaks are very weak, except the one at the longest wavelength, 1st order Bragg peak, and the one at half that wavelength, 2nd order Bragg peak. These phenomena have been systematically studied for a range of parameters describing both the incident plane wave and the chiral STF by Venugopal and Lakhtakia (1999-2000). Several generalisations can be cited. Reflection of light in the 1st Bragg regime is like reflection from a normal mirror. The state of *linear* polarisation is maintained upon reflection; while the state of *circular* polarisation is reversed upon reflection. Unlike a normal mirror, however, the reflection only occurs in a restricted band of frequencies. Additionally, the reflection in the 1st Bragg regime is strongest at large angles of incidence and goes to zero as the angle of incidence approaches zero. As reflection in the 1st Bragg regime is similar to normal reflection, it has attracted little attention. Reflection in the 2nd Bragg regime is quite different. It is characterised by strong reflection of circularly polarised light matching the handedness of the material, a phenomenon known as the circular Bragg phenomenon (CBP). Light of the opposite circular polarisation is largely transmitted. Furthermore, unlike reflection from a normal mirror in which the handedness of the light is reversed upon reflection, the reflected light in the CBP maintains the same circular polarisation as the incident light. Because of these unusual properties, the CBP has received considerable attention. Fig. 15.4 shows an example of the calculated optical remittance spectrums (reflectances and transmittances) , in the vicinity of the 2nd order Bragg peak, for a typical chiral STF. In this example, the wave is at normal incidence and the chiral STF is right handed. The remittances are designated by: R_{RR}, R_{RL}, R_{LR}, R_{LL} and T_{RR}, T_{RL}, T_{LR}, T_{LL}. Here, R_{RL}, for instance, represents the fraction of the incident power density that is reflected as right circularly polarised (RCP) light when the incident wave is left circularly polarised (LCP). Both R_{RL} and R_{LR} are referred to as cross-polarised reflectances, since they describe a change in polarisation state upon reflection. On the other hand, R_{RR} and R_{LL} are referred to as co-polarised reflectances, since the polarisation state is preserved on reflection. The same terminology is used to describe the transmittances. In Fig. 15.4(a) the Bragg regime can be clearly seen in the spectrum of R_{RR} between λ_0=775 nm and λ_0=815 nm, where the reflectance is very high. Small but significant cross-polarised reflectances and transmittances can also be seen within the Bragg regime. The co-polarised reflectance R_{LL} however, remains very small within the Bragg regime, demonstrating the preferential reflection of RCP light from a right-handed chiral STF. The CBP is also evident in the transmission spectrum, shown in Fig. 15.4(b), with a dip in T_{RR} over the wavelengths corresponding to the peak in the R_{RR} spectrum. T_{LL} remains high in the Bragg regime indicating that LCP light is mainly transmitted. Outside of the Bragg regime, all of the remittances exhibit oscillations. These correspond to the Fabry-Perot oscillations (Hecht 2002) seen in normal homogeneous material and are a result of interference between the waves reflected from the two surfaces of the material. They can be related to material thickness and average index of refraction. Investigations of CBP characteristics as a function of various parameters yield several trends. At normal incidence, four parameters are useful: Ω, χ, ε_c and $\tilde{\varepsilon}_d$, where

$$\tilde{\varepsilon}_d \equiv \varepsilon_a \varepsilon_b / \left(\varepsilon_a \cos^2 \chi + \varepsilon_b \sin^2 \chi \right) \tag{15.21}$$

With these quantities, the central wavelength λ_0^{Br} of the Bragg peak and the full-width-at-half-maximum of the peak $\Delta\lambda_0^{Br}$ can be estimated for weakly absorbing chiral STFs as

$$\lambda_0^{Br} \approx \Omega\left(|\,\varepsilon_c\,|^{1/2} + |\,\tilde{\varepsilon}_d\,|^{1/2}\right) \tag{15.22}$$

$$\Delta\lambda_0^{Br} \approx 2\Omega\,\left|\,|\,\varepsilon_c\,|^{1/2} - |\,\tilde{\varepsilon}_d\,|^{1/2}\right| \tag{15.23}$$

It should be noted that when $|\,\varepsilon_c\,|=|\,\tilde{\varepsilon}_d\,|$, referred to as the pseudo-isotropic point, the width of the Bragg regime is zero and the CBP disappears. As the pseudo-isotropic point may have practical applications, it has been characterised by calculation for some selected materials using experimentally derived constitutive parameters (Lakhtakia 2002). The height of the Bragg peak is dependant on the number of helical cycles in the film and the degree of local birefringence. Greater peak heights are obtained for larger local birefringence. The peak height saturates as the thickness of the film increases, with the thickness of the film required to obtain a particular peak height dependant on the details of the parameters describing the particular chiral STF. With light incident at oblique angles, the Bragg peak shifts to shorter wavelengths, becomes narrower, and is reduced in height, as the angle of incidence is increased.

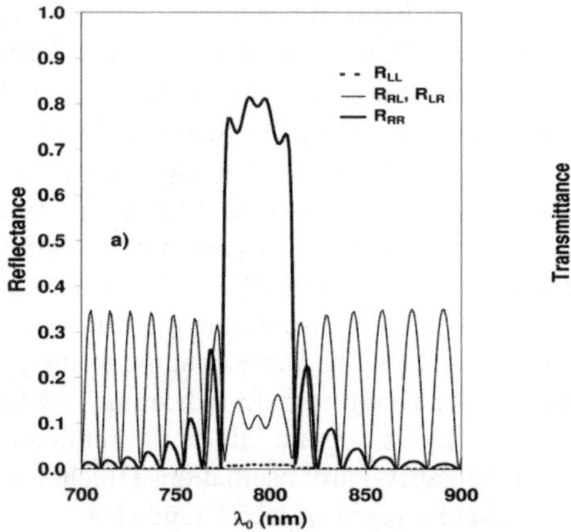

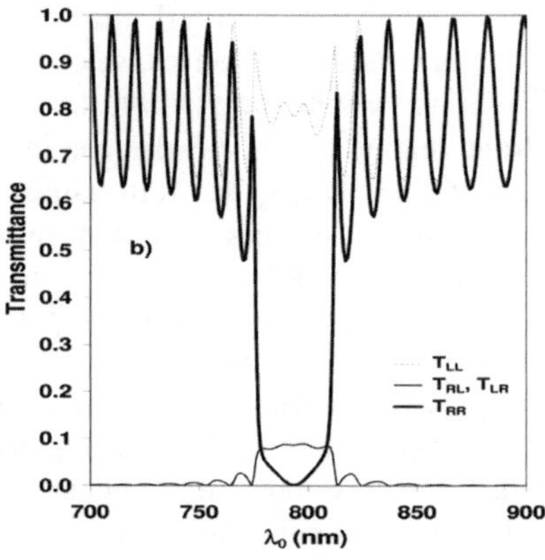

Fig. 15.4. Optical remittances of a right handed chiral STF with Ω=200 nm, L=60Ω, χ=63.6°, ε_a=1.9025, ε_b=2.0199, (a) reflectances (b) transmittances (Adapted from Polo and Lakhtakia 2004)

15.3.2.2 SNTFs

SNTFs have received somewhat less attention than chiral STFs. The local constitutive relations for a chiral STF may not be known exactly. However, because χ is constant, the local constitutive relations will remain constant. Thus the chiral STF may be modeled, at least in form, relatively easily by selecting reasonable constants. In a SNTF, however, χ is a function of z and, as a result, the local constitutive relations are expected to vary as a function of z. As mentioned earlier, this can be attributed to two causes: change in material density, and nanowire cross-sectional morphology. The dependences of the local constitutive relations on χ are not functions that one would reasonably expect to guess. However, limited experimental data obtained by Hodgkinson et al. (1998) on CTFs for some materials have become available from which the dependence of constitutive relations on χ have been inferred (Hodgkinson and Wu 1998; Hodgkinson et al. 2001; Chiadini and Lakhtakia 2004). Recently, remittances have been calculated for SNTFs resulting from deposition during which the vapor deposition angle, χ_v, is varied sinusoidally as a function of film thickness (Polo and Lakhtakia in press). When linearly polarised light is incident on the film in the x-z plane (plane parallel to planes in which the SNTF nanowires lie) the spectrums show multiple Bragg regimes. Unlike the case for chiral STFs, Bragg reflections of many orders have strong peaks. Even films with a small number of structural cycles are highly reflective, particularly for large modulation amplitudes about an average value of 90°. The spectral positions of the Bragg regimes differ for s polarisation (electric field perpendicular to the x-z plane) and p polarisation (electric field in the x-z plane). The height and width of the Bragg regimes grows with the amplitude of the oscillation of χ_v that is used to produce the SNTF. Like chiral STFs, the Bragg regimes shift to shorter wavelengths with incident angle.

15.3.2.3 Mechanical Properties

The stiffness of a chiral STF made of SiO has been measured and shown to behave as a highly linear spring (Brett et al. 1999). There is at least one theoretical investigation, which models the elastostatic response of chiral STFs (Lakhtakia 2002). Several works on acoustic wave propagation also exist (Lakhtkia 1995, 1996, 2000, 2003; Oldano et al. 2003). There are parallels between optical and acoustic phenomena, and Bragg phenomena similar to those observed in optics are also predicted for vibrational waves.

15.4 Applications

Many applications have been proposed for STFs. Although none has seemed to reach the stage of commercial application as yet, it would appear that it is only a matter of time. In some cases, STFs offer an advantage over current technologies, while other applications offer new possibilities.

15.4.1 Optical

STFs offer many possibilities as passive optical devices. The ability to engineer the birefringent properties of STFs and fabricate very compact devices with multiple layers allow for an extremely wide range of design possibilities not available with traditional crystal optics. The design of such devices, mainly with layers of CTFs is extensively explored in a book by Hodkinson and Wu (1997). Chiral STFs can be used to create highly reflective circular polarisation filters only a few helical cycles thick. Hodgkinson et al. (2002) have proposed a method to enhance the performance of these circular polarisation filters with the addition of anti-reflection coatings. Their calculations have demonstrated that the cross-polarised reflectance of chiral STFs within the Bragg regime can be almost entirely eliminated with proper design of the anti-reflection coatings, producing an extremely pure circularly polarised reflected beam. In applications requiring mirrors, which reflect both polarisations, two solutions have been found. Sequential layers of chiral STFs with opposite handedness have been shown to reflect both RCP and LCP light equally well independent of the order of the two cascaded layers (Lakhtakia and Venugopal 1998). Calculations on chiral STFs with χ modulated as a function of depth (Polo and Lakhtakia 2004) have shown that, with sufficient modulation amplitude, the film acts as a conventional mirror over a range of wavelengths. Furthermore, the width of the reflection regime can be controlled by changing the amplitude of modulation of χ.

Because of finite bandwidths and sharp roll-offs, the Bragg regimes of both chiral STFs and SNTF may be used, for circularly polarised and linearly polarised light respectively, as bandpass, bandstop and notch filters. Chiadini and Lakhtakia (2004) have theoretically explored creating circular polarisation filters with even wider bandwidths than available in a simple chiral STF by cascading several chiral STFs with a range of half-pitch values, Ω. In slanted chiral STFs, which because of their structure have a surface grating, circular polarisation beam splitters may be possible (Wang and Lakhtakia 2002). Rugate filters have been produced using both vertical columns with undulating density (Kaminska et al. 2004) as well as SNTFs with a modulated χ (McPhun et al. 1998). For extremely narrow bandwidth filters, the introduction of defects of various types into the periodic structure of STFs has been investigated theoretically (Hodgkinson et al. 2000; Polo and Lakhtakia in press) and experimentally (Hodgkinson et al. 2000) for both SNTFs and chiral STFs. The defect produces a feature in the Bragg regime known as a spectral hole. Over a very narrow range of wavelengths, $\lambda_0<0.1$ nm, the reflective Bragg effect is destroyed leading to a very sharp dip to nearly zero reflectance in the spectrum. Equivalently, a very sharp spike in the transmission spectrum arises with essentially 100% transmission.

The use of STFs as active optical elements has also been suggested and investigated. One possibility is to use the STF as a chemical sensor (Lakhtakia 1998; Hodgkinson et al. 1999). Due to high porosity, STFs are easily infiltrated with fluids which may alter the overall constitutive relations of the STF. Changes in constitutive relations then affect various optical properties of the STF, which can be monitored. An extremely sensitive sensor makes use of the shift in the spectral hole of a spectral hole filter. The effect has been demonstrated experimentally (Lakhtakia et al. 2001) with the redshift of a spectral hole in a STF exposed to moisture. The possibility of using a Šolc filer made with a STF as a sensor of gas concentration has been investigated theoretically (Ertekin and Lakhtakia 1999). The effect of fluid infiltration, as well as quasi-static electric and magnetic fields, on STFs have been proposed as means of creating dynamically tunable laser mirrors (Lakhtakia and Venugopal 1998). Mechanical control of chiral STFs using the piezoelectric effect has also been investigated in this regard (Wang et al. 2002, 2003). Kopp et al. (2003) have looked at the use of chiral STFs as the lasing medium of the laser. The states at the edge of the Bragg regime as well as the defect state created in a spectral hole filter were shown to have desirable properties. Infiltrating a STF with a liquid crystal may make a display device with the best properties of both worlds. The STF offers mechanical and thermal stability with engineered conformation which may be imposed on the infiltrated LC. The conformation in the STF is controllable over larger thickness than normally obtainable with LCs. LCs on the other hand offer electrical addressability. Both the imposition of order (Robbie et al. 1999) on the LC as well as electrical switching (Sit et al. 2000) have been demonstrated experimentally for chiral STFs infiltrated with a LC. The infiltration of a LC into a STF may also be useful in increasing the optical activity of a chiral STF and reducing degrading, environmental effects on the optical properties of the STF (Kennedy et al. 2001). Infiltrates may affect optical properties in another way. Lakhtakia and Horn (2003) have calculated the effects of columnar thinning in chiral STFs. They show that post deposition thinning of the STF by etching or bioreduction can change the optical characteristics of the film. It also allow sensing certain chemicals which react with the film.

15.4.2 Chemical

The high porosity and large surface area make STFs likely candidates for use in some chemical applications. Photocatalysis using STFs of TiO_2, a highly studied photocatalytic material, has been investigated experimentally for both gaseous and liquid phase reactants (Weinberger and Garber 1995; Suzuki et al. 2001). In the liquid phase work, χ is found to be the major factor influencing reaction rates, with the particular shape of the STF nanowire playing a very minor role. Another study investigated the enhancement of surface area in microchannels by coating them with both CTFs and chiral STFs (Harris et al. 2000).

15.4.3 Electronics

With increasing miniaturisation of electronics, there is a need for low permittivity materials. STFs may fulfill this need (Venugopal et al. 2000).

15.4.4 Biological

A number of biological applications may be possible (Lakhtakia 2002; Lakhtakia and Messier 2004). Many of the proposals exploit the porosity of the STF. Among them is the use of the porous structure of the STF as a nanosieve for virus entrapment. The porosity of STFs may also provide a useful surface treatment to promote the adherence of biological tissue. The use of STFs as microreactors for biochemicals has also been suggested. One study (Steltz and Lakhtakia 2003) looks at the generation of second harmonic radiation in a chiral STF as a means of sensing bio-luminescent reactions.

15.5 Concluding Remarks

Understanding of STFs has grown dramatically since the concept was first proposed in the mid 1990s, with vigorous activity still continuing in STF research. In addition to unsolved problems, new developments continue to bring additional opportunities. For instance, the recent combination of STF and electron beam lithography techniques by Horn et al. (2004) to produce STFs with transverse architecture with consistent characteristics over large areas opens up a whole new line of possible applications. Fig. 15.5 shows examples of simple lines of chiral STF produced by this technique. More complex 2-D patterns such as checkerboards have also been produced. In their most recent work (Horn et al. 2004), they have been able to control the locations of individual nanowires on the substrate by creating a regular array of circular posts 60 nm in diameter which serve as nucleation sites.

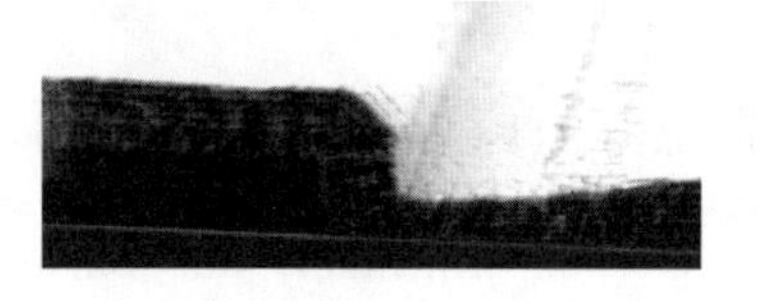

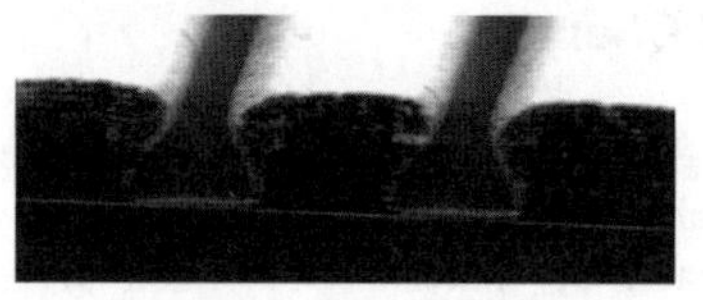

Fig. 15.5. STFs with transverse architecture (Courtesy: Mark Horn, Pennsylvania State University)

Some areas of existing research are surely to see more activity. To date, most work on the optical properties of STFs has focused on frequency domain results. As society reaches forever faster rates of data transmission, pressure to pursue an understanding, in the time domain, of short pulse propagation in STFs will undoubtedly increase. Some work in this area has already begun (Geddes and Lakhtakia 2001; 2003). The ultrasonic use of STFs has been explored theoretically, but, due to the unavailability of STFs with sufficiently large scale features, experimental work is non-existent. With continual improvements and new techniques in STF fabrication, this is also likely to change.

Other areas for potential development (biological, electronic, and chemical applications for examples) are interdisciplinary in nature. With time, the right parties will come together to carry out more work in these exciting areas. Whatever the focus – basic understanding, new applications, or commercial implementation of existing ideas – new and unexpected developments in sculptured thin films are sure to continue at least into the near future.

15.6 References

Abdulhalim I, Weil R, Benguigui L (1986) Dispersion and attenuation of the eigenwaves for light propagation in helicoidal liquid crystals, Liq. Cryst. 1:155–167

Berreman DW (1972) Optics in stratified and anisotropic media: 4x4-matrix formulation, J. Opt. Soc. Am. 62:502

Brett MJ, Seto MW, Sit JC, Harris KD, Vick D, Robbie K (1999) Glancing angle deposition: recent research results. In: Lakhtakia A and Messier RF (eds) Engineered nanostructural films and materials. Proc. SPIE 3790, pp114–118

Bruggeman DAB (1935) Berechnung verschiedener physikalischer Konstanten von heterogenen Substanzen. I. Dielektrizitatskonstanten und Leitfahigkeiten der Mischkorper aus isotropen Substanzen, Ann. Phys. Leipzig 24:636-679

Chiadini F, Lakhtakia A (2004) Gaussian model for refractive indexes of columnar thin films and Bragg multilayers, Optics Commun. 231:257-261

Chiadini F, Lakhtakia A (2004) Design of wideband circular-polarisation filters made of chiral sculptured thin films, Microwave Opt. Technol. Lett. 42:135–138

Dong L, Smith RW, Srolovitz DJ (1996) A two-dimensional molecular dynamics simulation of thin film growth by oblique deposition, J. Appl. Phys. 80:5682–5690

Ertekin E, Lakhtakia A (1999) Sculptured thin film Šolc filters for optical sensing of gas concentration, Eur. Phys. J. Appl. Phys. 5:45–50

Geddes JB, Lakhtakia A (2001) Reflection and transmission of optical narrow-extent pulses by axially excited chiral sculptured thin films, Eur. Phys. J. Appl. Phys. 13:3-14; corrections, 16:247

Geddes JB, Lakhtakia A (2003) Effects of carrier phase on reflection of optical narrow-extent pulses from axially excited chiral sculptured thin films, Optics Commun. 225:141-150

Grove WR (1852) On the electrochemical polarity of gases, Phil. Trans. Royal Soc. B 142:87

Harris KD, Brett MJ, Smy TJ, Backhouse C (2000) Microchannel surface area enhancement using porous thin films, J. Electrochem. Soc. 147:2002-2006

Hecht E (2002) Optics. 4th edn, Addison Wesley, San Francisco

Hodgkinson IJ, Wu Qh (1997) Birefringent thin films and polarizing elements. World Scientific, Singapore

Hodgkinson IJ, Wu Qh (1998) Empirical equations for the principal refractive indices and column angle of obliquely deposited films of tantalum oxide, titanium oxide, and zirconium oxide, Appl. Opt. 37:2653-2659

Hodgkinson IJ, Wu Qh, Collet S (2001) Dispersion equations for vacuum-deposited tilted-columnar biaxial media, Appl. Opt. 40:452–457

Hodgkinson IJ, Wu Qh, Thorn KE, Lakhtakia A, McCall MW (2000) Spacerless circular-polarisation spectral-hole filters using chiral sculptured thin films: theory and experiment, Optics Commun. 184:57-66

Hodgkinson IJ, Wu Qh, McGrath KM (1999) Moisture adsorption effects in biaxial and chiral optical thin film coatings. In: Lakhtakia A and Messier RF (eds) Engineered nanostructural films and materials. Proc. SPIE, vol. 3790, pp 184–194

Hodgkinson I, Wu Qh, Knight B, Lakhtakia A, Robbie K (2000) Vacuum deposition of chiral sculptured thin films with high optical activity, Appl. Opt. 39:642–649

Hodgkinson IJ, Wu Qh, Arnold M, McCall MW, Lakhtakia A (2002) Chiral mirror and optical resonator designs for circularly polarised light: suppression of cross-polarised reflectances and transmittances, Optics Commun. 210:201-211

Horn MW, Pickett MD, Messier R, Lakhtakia A (2004) Selective growth of sculptured nanowires on microlithographic lattices, J. Vac. Sci. Technol. B 22:3426–3430

Horn MW, Pickett MD, Messier R, Lakhtakia A (2004) Blending of nanoscale and microscale in uniform large-area sculptured thin-film architectures, Nanotechnology 15:303-310

Huang H, Gilmer GH, Rubia TD de la (1998) An atomistic simulator for thin film deposition in three dimensions, J. Appl. Phys. 84:3636–3649

Jackson JD (1975) Classical Electrodynamics. 2nd edn, Wiley, New York

Kaminska K, Suzuki M, Kimura K, Taga Y, Robbie K (2004) Simulating structure and optical response of vacuum evaporated porous rugate filters, J. Appl. Phys. 95:3055–3062.

Kennedy SR, Sit C, Broer DJ, Brett MJ (2001) Optical activity of chiral thin film and liquid crystal hybrids, Liquid Crystals 28:1799–1803

Kopp VI, Zhang Z, Genack AZ (2003) Lasing in chiral photonic structures, Progress in Quantum Electronics 27:369–416

Kranenburg K van, Lodder C (1994) Tailoring growth and local composition by oblique-incidence deposition: a review and new experimental data, Mater. Sci. Eng. R 11:295-354.

Kundt A (1886) Ueber Doppelbrechung des Lichtes in Metallschicten, welche durch Zerstauben einer Kathode hergestellt sind, Ann. Phys. Chem. Lpz. 27:pp59–62

Lakhtakia A (1995) Wave propagation in a piezoelectric, continuously twisted, structurally chiral medium along the axis of spirality, Appl. Acoust. 44:25–37; corrections, 44:385

Lakhtakia A, Weiglhofer WS (1996) Simple and exact analytic solution for oblique propagaion in a cholesteric liquid crystal, Microwave Opt. Technol. Lett. 12:245–248

Lakhtakia A (Ed) (1996) Selected Papers on Linear Optical Composite Materials. SPIE Optical Engineering Press, Bellingham

Lakhtakia A (1996) Exact analytic solution for oblique propagation in a piezoelectric, continuously twisted, structurally chiral medium, Appl. Acoust. 49:225–236

Lakhtakia A, Messier R, Robbie K, Brett MJ (1996) Sculptured thin films (STFs) for optical, chemical and biological applications, Innovat. Mater. Res. 1:165–176

Lakhtakia A (1997) On second harmonic generation in sculptured nematic thin films (SNTFs), Optik 105:115-120

Lakhtakia A (1997) Director-based theory for the optics of sculptured thin films, Optik 107:57-61

Lakhtakia A (1998) On determining gas concentrations using thin-film helicoidal bianisotropic medium bilayers, Sens. Actuat. B: Chem. 52:243-250

Lakhtakia A, Weiglhofer WS (1997) Further results on light propagation in helicoidal bianisotropic mediums: oblique propagation. Proc. R. Soc. London A 453:93-105; errata: 454:275

Lakhtakia A, Venugopal VC (1998) Dielectric thin-film helicoidal bianisotropic medium bilayers as tunable polarisation independent laser mirrors and notch filters, Microwave Opt. Technol. Lett. 17:135–140

Lakhtakia A (1998) Incremental Maxwell Garnett formalism for homogenizing particulate composite media, Microwave Opt. Technol. Lett. 17:276-279

Lakhtakia A, Venugopal VC (1999) On Bragg reflection by helicoidal bianisotropic mediums, Int. J. Electron. Commun. 53:287–290

Lakhtakia A, Sunal PD, Venugopal VC, Erekin E (1999) Homogenisation and optical response properties of sculptured thin films. In: Lakhtakia A and Messier RF (eds) Engineered nanostructural films and materials, Proc. SPIE, vol. 3790, pp77-83

Lakhtakia A (2000) Shear axial modes in a PCTSCM. Part VI: simpler transmission spectral holes, Sensors and Actuators A 87:78–80

Lakhtakia A (2001) Application of strong permittivity fluctuation theory for isotropic, cubically nonlinear, composite mediums, Optics Commun. 192:145–151

Lakhtakia A, McCall MW, Sherwin JA, Wu Qh, Hodgkinson IJ (2001) Sculptured-thin-film spectral holes for optical sensing of fluids, Optics Commun. 194:33–46

Lakhtakia A (2002) Pseudo-isotropic and maximum-bandwidth points for axially excited chiral sculptured thin films, Microwave Opt. Technol. Lett. 34:367–371

Lakhtakia A (2002) Microscopic model for elastostatic and elastodynamic excitation of chiral sculpured thin films, J. Compos. Mater. 36:1277–1297

Lakhtakia A (2002) Sculptured thin films: accomplishments and emerging uses, Mat. Sci. and Eng. C 19:427–434

Lakhtakia A, Horn MW (2003) Bragg-regime engineering by columnar thinning of chiral sculptured thin films, Optik 114:556–560

Lakhtakia A (2003) Shear axial modes in a PCTSCM. Part VII: Reflection spectral holes, Sens. Actuators A 104:188–190

Lakhtakia A, Messier R (2004) Sculptured thin films: Nanoengineered morphology and optics, SPIE Press, Bellingham

Lakhtakia A, Messier R (2004) Sculptured thin films, nanometer structures theory, modeling, and simulation. Lakhtakia A (ed) SPIE Press, Bellingham

Mackay TG, Lakhtakia A, Weiglhofer WS (2001) Strong-property-fluctuation theory for homogenisation of bianisotropic composites: formulation, Phys. Rev. E 62:6052–6064; erratum, 63:049901

Mackay TG, Lakhtakia A, Weiglhofer WS (2001) Ellipsoidal topology, orientation diversity and correlation length in bianisotropic composite mediums, Arch Elecktron Ubertrag 55:243–251

Mackay TG, Lakhtakia A, Weiglhofer WS (2001) Homogenisation of similarly oriented, metallic ellipsoidal inclusions using the bilocal-approximated strong-property-fluctuation theory in electromagnetic homogenisation, Optics Commun. 197:89–95

Mackay TG, Lakhtakia A, Weiglhofer WS (2002) Homogenisation of isotropic, cubically nonlinear, composite mediums by the strong-permittivity-fluctuation theory: third-order considerations, Optics Commun. 204:219–228

Mackay TG (2003) Homogenisation of linear and nonlinear complex composite materials. In: Wieglhofer WS, Lakhtakia A (eds) Introduction to complex mediums for optics and electromagnetics, SPIE Press, Bellingham

Mattox DM (2003) The foundations of vacuum coating technology. Noyes Publications, William Andrews Publishing, Norwich

Maxwell Garnett JC (1904) Colours in metal glasses and in metallic films, Phil. Trans. Royal Soc. London A 203:385–420

McCall MW, Lakhtakia A (2000) Development and assessment of coupled wave theory of axial propagation in thin-film helicoidal bianisotropic media. Part 1: reflectances and transmittances, J. Mod. Opt. 47:973-991; corrections, 50:2807

McCall MW, Lakhtakia A (2001) Development and assessment of coupled wave theory of axial propagation in thin-film helicoidal bianisotropic media. Part 2: dichroisms, ellipticity transformation and optical rotation, J. Mod. Opt. 48:143-158

McPhun AJ, Wu Qh, Hodgkinson IJ (1998) Birefringent rugate filters, Electronics Letters 34:360–361

Meakin P, Ramanlal P, Sander LM, Ball RC (1986) Ballistic deposition on surfaces, Phys. Rev. A 34:5091–5103

Messier R, Giri AP, Roy RA (1984) Revised structure zone model for thin film physical structure, J. Vac. Sci. Technol. A 2:500-503

Messier R, Gehrke T, Frankel C, Venugopal VC, Otaño W, Lakhtakia A (1997) Engineered sculptured nematic thin films, J. Vac. Sci. Technol. A 15:2148-2152

Messier R Lakhtakia A (1999) Sculptured thin films–II. Experiments and applications, Mat. Res. Innovat. 2:217–222

Michel B (2000) Recent developments in the homogenisation of linear bianisotropic composite materials. In: Sing ON, Lakhtakia A (eds) Electromagnetic fields in unconventional materials and structures, Wiley, New York, pp 39–82

Michel B, Lakhtakia A, Weiglhofer WS, Mackay TG (2001) Incremental and differential Maxwell Garnett formalisms for bianisotropic composites, Compos. Sci. Tech. 61:13–18

Movchan BA, Demchishin AV (1969) Study of the structure and properties of thick vacuum condensates of nickel, titanium, tungsten, aluminum oxide and zirconium dioxide, Phys. Met. Metallogr. 28:83–90

Niewenhuizen JM, Haanstra HB (1966) Microfractography of thin films, Philips Tech. Rev. 27:87-91

Oldano C, Reyes JA, Ponti S (2003) Twist defects in helical sonic structures, Phys. Rev. E 67:56624

Oseen CW (1933) The theory of liquid crystals, J. Chem. Soc. Faraday Trans. II 29:883

Pickett MD, Lakhtakia A (2002) On gyrotropic chiral sculptured thin films for magneto-optics, Optik 113:367–371

Pickett MD, Lakhtakia A, Polo JA (2004) Spectral responses of gyrotropic chiral sculptured thin films to obliquely incident plane waves, Optik 115:393–398

Polo JA, Lakhtakia A (2004) Comparison of two methods for oblique propagation in helicoidal bianisotropic mediums, Optics Commun. 230:369–386

Polo JA, Lakhtakia A (2004) Tilt-modulated chiral sculptured thin films: An alternative to quarter-wave stacks, Optics Commun. 242:13–21

Polo JA, Lakhtakia A (in press) Sculptured nematic thin films with periodically modulated tilt angle as rugate filters

Robbie K, Brett MJ (1997) Sculptured thin films and glancing angle deposition: Growth mechanics and applications, J. Vac. Sci. Technol. A 15:1460–1465

Robbie K, Sit JC, Brett MJ (1998) Advanced techniques for glancing angle deposition, J. Vac. Sci. Technol. B 16:1115–1122

Robbie K, Broer DJ, Brett MJ (1999) Chiral nematic order in liquid crystals imposed by an engineered inorganic nanostruture, Nature 399:764–766

Robbie K, Shafai C, Brett MJ (1999) Thin films with nanometer scale pillar microstructure, J. Mater. Res. 14:3158–3163

Rahman TS, Gosh C, Trushin O, Kara A, Karim A (2004) Atomistic studies of thin film growth nanomodeling. In: Lakhtakia A and Maksimenko SA (eds), Proc. SPIE, vol. 5509, pp 1–14

Sherwin JA, Lakhtakia A (2001) Nominal model for the structure-property relations of chiral dielectric sculptured thin films, Math. Comput. Model. 34:1499-1514; corrections, 35:1355–1363

Sherwin JA, Lakhtakia A, Hodgkinson IJ (2002) On calibration of a nominal structure–property relationship model for chiral sculptured thin films by axial transmittance measurements, Optics Commun. 209:369–375

Shenoy VB (2004) Mechanics at small scales, Nanomodeling. In: Lakhtakia A, Maksimenko SA (eds), Proc. SPIE, vol. 5509, pp46–60

Sit JC, Broer DJ, Brett MJ (2000) Liquid crystal alignment and switching in porous chiral thin films, Adv. Mater. 12:371–373

Smy T, Vick D, Brett MJ, Dew SK, Wu AT, Sit JC, Harris KD (2000) Three-dimensional simulation of film microstructure produced by glancing angle deposition, J. Vac. Sci. Technol. A 18:2507–2512

Steltz EE, Lakhtakia A (2003) Theory of second-harmonic-generated radiation from chiral scultured thin films for bio-sensing, Optics Commun. 216:139–150

Suzuki M, Taga Y (2001) Numerical study of the effective surface area of obliquely deposited thin films, J. Appl. Phys. 90:5599–5605; erratum, 91:2556

Suzuki M, Ito T, Taga Y (1999) Recent progress of obliquely deposited thin films for industrial applications. In: Lakhtakia A, Messier RF (eds) Engineered Nanostructural Films and Materials, Proc. SPIE, vol. 3790, pp 94–105

Suzuki M, Ito T, Taga Y (2001) Photoctalysis of sculptured thin films of TiO_2 Appl. Phys. Lett. 78, pp 3968–3970

Tait RN, Smy T, Brett J (1993) Modelling and characterisation of columnar growth in evaporated films, Thin Solid Films 226:196–201

Thornton JA (1977) High rate thick film growth, Annu. Rev. Mater. Sci. 7, pp239-260

Tsang L, Kong JA (1981) Scattering of electromagnetic waves from random media with strong permittivity fluctuations, Radio Sci. 16:303–320

Venugopal VC, Lakhtakia A (1998) Second harmonic emission from an axially excited slab of a dielectric thin-film helicoidal bianisotropic medium, Proc. R. Soc. Lond. A 454:1535-1571; corrections, 455:4383

Venugopal VC, Lakhtakia A (2000) Electromagnetic plane-wave response characteristics of non-axially excited slabs of dielectric thin-film helicoidal bianisotropic mediums, Proc. R. Soc. Lond. A 456:125–161

Venugopal VC, Lakhtakia A, Messier R, Kucera JP (2000) Low-permittivity materials using sculptured thin film technology, J. Vac. Sci. Technol. B 18:32–36

Venugopal VC, Lakhtakia A (2000) On absorption by non-axially excited slabs of dielectric thin-film helicoidal bianisotropic mediums, Eur. Phys. J. Appl. Phys. 10:173-184

Vick D, Friedrich LJ, Dew SK, Brett MJ, Robbie K, Seto M, Smy T (1999) Self-shadowing and surface diffusion effects in obliquely deposited thin films, Thin Solid Films 339:88–94

Wang F, Lakhtakia A, Messier R (2002) Towards piezoelectrically tunable chiral sculptured thin film lasers, Sens. Actuators A 102:31–35

Wang F, Lakhtakia A, Messier R (2002) Coupling of Rayeigh-Wood anomalies and the circular Bragg phenomenon in slanted chiral sculptured thin films, Eur. Phys. J. Appl. Phys. 20:91–104; errata, 24:91

Wang F, Lakhtakia A, Messier R (2003) On piezoelectric control of the optical response of sculptured thin films, J. Modern Optics 50:239–249

Weiglhofer WS, Lakhtakia A (1996) Oblique propagation in a cholesteric liquid crystal: 4x4 matrix perturbational solution, Optik 102:111-114

Weiglhofer WS (2003) Constitutive characterisation of simple and complex mediums. In: Weiglhofer WS, Lakhtakia A (eds) Introduction to complex mediums for optics and electromagnetics, pp 27–61

Weinberger BR, Garber RB (1995) Titanium dioxide photoctalysts produced by reactive magnetron sputtering, Appl. Phys. Lett. 66:2409–2411

Wu Qh, Hodgkinson IJ, Lakhtakia A (2000) Circular polarisation filters made of chiral sculptured thin films: experimental and simulation results, Opt. Eng. 39:1863–1868

Zhuck NP (1994) Strong–fluctuation theory for a mean electromagnetic field in a statistically homogeneous random medium with arbitrary anisotropy of electrical and statistical properties, Phys. Rev. B 50:15636-15645

16 e-Beam Nanolithography Integrated with Nanoassembly: Precision Chemical Engineering

P. M. Mendes and J. A. Preece

University of Birmingham, Edgbaston, Birmingham, United Kingdom

16.1 Introduction

The minimum feature size of electronic devices used in microprocessors has shrunk to the nanoscale (<100 nm) in 2004 (http://www.intel.com/research/silicon/nanotechnology.htm). These electronic devices still work on multielectron processes. However, interest in single electron (quantum) devices has been of academic interest for a number of years (Daniel 2004), and is increasingly becoming of interest to industry as conventional microelectronic devices shrink toward the quantum regime. In addition, apart from the drive to create smaller structures on native SiO_2 for use in nanoelectronics, a significant challenge in current nanoscience and nanotechnology is to spatially self-organise self-assembled nanoscale components (such as metal nanoparticles, carbon nanotubes, proteins, cells, organic molecules, polymers, etc.) onto surfaces to fabricate functional nanostructured systems for electronic, optoelectronic, biological, or sensing applications (Mendes et al. 2003; Ball 2000; Dagani 2000). Precise control over the relative position and orientation of the nanocomponents is frequently required in such systems to obtain useful properties. Moreover, the integration and the stability of interfaces to these nanostructures from the micron-length and macroscopic scales are key to the success of future applications. Nanostructuring surfaces by top-down nanolithography techniques, and subsequently building into the third dimension utilising bottom-up self-organisation of self-assembled nanoentities, is an attractive approach for creating such a bridge between macroscopic systems and the nanoscale dimensions that many modern technologies demand (Dagani 2000; Whitesides 2002). The integration of both top-down and bottom-up approaches has been termed precision chemical engineering (Mendes 2004), and is based on the concept that surfaces are written with precision engineering (i.e. in a spatially controlled fashion to direct the surface chemistry), coupled with the engineering that is involved in fabricating three-dimensional architectures from the surface using molecular and condensed phase entities. Besides requiring lithographic techniques with nanometre resolution, ultra-high resolution patterning also needs materials that allow a controlled modification of their physical and/or chemical properties.

However, it is becoming increasingly clear that the photolithographic process for creating structures on the native oxide of silicon is rapidly reaching a physical limit in terms of the lateral resolution that is achievable. This limitation is a result of (i) the diffraction limit, (ii) limitations of the optics, and (iii) the materials used to transfer patterns to the surface (Silverman 1997). Thus, several methodologies have been developed in recent years that may be the successor to conventional photolithography. In this review we will discuss electron beam (e-beam) lithography (Roberts 1987), and in particular the development of the organic based resist materials that have been used, starting with polymeric materials, through molecular materials and finally self-assembled monolayers (SAMs).

16.2 Electron-Beam Radiation

The use of polymeric materials as electron beam resists is a very mature field of research and even technology, and it is not intended to give an in depth review of this area in this chapter. However, we will review the limitations of polymeric materials, and how the field has developed such that organic molecular materials have started to be investigated and subsequently organic self-assembled monolayers (SAMs). The SAM system is the one that will be reviewed in detail in this Chapter.

16.2.1 Polymeric Materials

Polymeric materials that are sensitive to electrons, such as poly(methyl methacrylates) (Rai-Choudhury 1997), copolymers of glycidyl methacrylate (Rai-Choudhury 1997) and ethyl methacrylate, and polysulphones (Roberts 1987) have been used as electron beam resists. The first two are referred to as positive tone resists in which the e-beam fragments the polymer backbone leading to an enhanced solubility of the fragmented area relative to the non-fragmented. The polysulphones are referred to as negative tone resists in which the polymer chains are cross-linked by the e-beam, resulting in a lower solubility of the non-irradiated area, relative to the irradiated area. The differential solubilities allows the SiO_2 surfaces coated with the polymers to have the e-beam written pattern, to be transferred to the underlying SiO_2, after washing off the more soluble component, and plasma or chemically etching away the exposed SiO_2. Although PMMA has led to structures of less than 10 nm being created on SiO_2 surfaces, there are limitations to its use. These limitations include the low etch durability of PMMA (i.e. the etchants that etch the SiO_2 also etch the PMMA at a similar rate) (Fujita 1996). This limitation can be overcome by the use of high etch durability polymeric composite resists such as SAL-601 (Dobisz 1998). However, this polymer composite has poor lateral resolution leading to structures of ~60 nm (Dobisz 1997). In addition polymeric materials by virtue of their large size, relative to molecular materials, will limit their own ultimate resolution to the size of the polymers that are used (Tada 1996). Thus, attention has recently turned to using low molecular organic materials (~1 kDa) as e-beam resists. Properties such as cross-linking or selective cleavage upon e-beam irradiation can be designed, in order to improve the negative/positive tone resist properties and etch durability, as well as improve the final resolution of the structures by virtue of molecular materials having a much lower molecular weight and size.

16.2.2 Molecular Materials

Low molecular weight organic resists initially were not suitable for e-beam lithography because thin films made of them on SiO_2 were polycrystalline (Yoshiiwa 1996). Recently, organic materials have been described that form amorphous films and are sensitive to electrons. The third allotrope of carbon, C60, has been shown to be a negative tone e-beam resist, leading to 20 nm structures (Tada 1997). The draw back of this material was its low solubility and relatively poor sensitivity toward electrons compared to the polymeric materials. However, these limitations have been overcome to some extent by the chemical modification of C60 (Fig. 16.1(a)), which has improved the solubility, inhibited the tendency to form polycrystalline materials (Robinson 1998) and achieve sensitivities toward electrons which are almost comparable to PMMA.

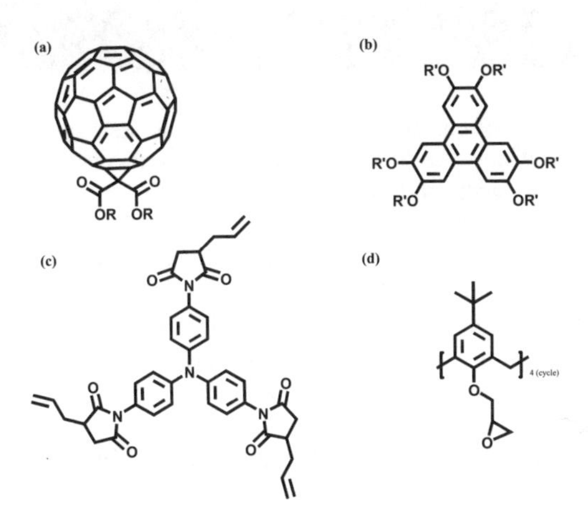

Fig. 16.1. Chemical structures of (a) C60 derivatives (b) triphehylene derivatives (Ref. Robinson 1999,2000), (c) a 4,4',4''-tris(allylysuccinimido)-triphenylamine (from Ref. (Yoshiiwa 1996), and (d) a calixarene derivative (from Ref. Sailor 2004)

Another approach to overcoming the polycrystallinity problem has been the use of liquid crystalline materials based on triphenylenes (Fig. 16.1(b)), which also have large extended π electron framework and high carbon content like the C60 derivatives, which are thought to be important in terms of the sensitivity and etch durability (Robinson 1999). These liquid crystalline materials were able to surpass the comparative resolution of PMMA, but the sensitivity was not comparable (Robinson 2000). Fig. 16.2 illustrates the e-beam patterned structures that can be created with these triphenylene derivatives.

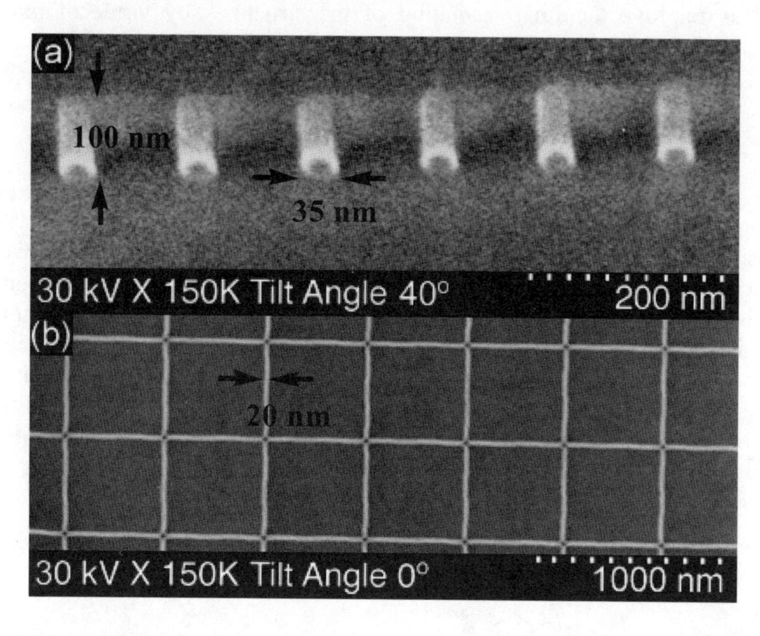

Fig. 16.2. Scanning electron micrographs of e-beam written structures to thin films of hexapentyloxytriphenylene on SiO_2 surfaces after washing to leave the crosslinked triphenylene structures (adapted from Ref. Robinson 1999)

Other classes of molecular materials that have received interest as e-beam resist materials are calixarenes (Schock 1997), 1,3,5-tris[4-(4-toluenesulfonyloxy) phenyl)-benzene (Yoshiiwa 1996) and 4,4',4''-tris(allylysuccinimido)-triphenylamine (Yoshiiwa 1996) (Fig. 16.1(c)). The calixarenes appear to be making some running in terms of sensitivity and resolution of features especially when used as a chemically amplified resist (Sailor 2004) (Fig. 16.1(d)). The limitations of these molecular materials are that a thin film of several 10 to 100s of molecules thick is formed by drop casting or spin coating of a solution onto the SiO_2 surface, followed by solvent evaporation. Thus, resolution of the structures formed is limited by the thickness of the film, as a result of electron scattering effects in the multilayer film (Rai-Choudhury 1997). The electron scattering results in the electron exposure to propagate radially away from the focus of the original e-beam incident source, and scales with the thickness of the resist film. In addition, a number of processing steps are required pre- and post- e-beam exposure (Chen 1993). The use of self-assembled monolayers, i.e. one molecule thick, as resists overcomes the back-scattered electron problem, as well as making processing potentially simpler.

16.3 Self-Assembled Monolayers

Self-assembled monolayers (SAMs) are attractive ultrathin organic films for patterning high resolution features on a number of technologically relevant surface substrates. SAMs are molecular assemblies that are formed spontaneously by the adsorption of a surfactant with a specific affinity of its headgroup to a substrate (Ulman 1991,1996). The attraction of these systems as ultrathin resists is driven primarily because SAMs eliminate depth of focus, transparency issues and can be prepared with a discrete number of well-defined chemical functional groups to permit further nanoscale materials attachment. Patterning SAMs can be achieved using a variety of lithographic techniques, which include (i) stamping or molding methods, i.e. soft lithography (Xie 1998), (ii) scanning probe microscopy (SPM)-based techniques such as 'scanning probe electrochemistry' (Sugimura 2002), 'dip-pen nanolithography' (Piner 1999), 'nanografting' (Xu 1997, Liu 2002) and 'nanoshaving' (Liu 2000), which utilise the mechanical interaction of an SPM tip with a surface as well as (iii) radiative sources including ultraviolet (UV)/visible light, X-ray and electron-beams (e-beam).

E-beam irradiation can either induce removal, crosslinking, damage, or modification of an internal and/or terminal functional group of SAMs (Process A-Scheme 1). Such systems can then be used in a second stage as general templates to direct the assembly of nanoscale components, to either the unirradiated (Processes B-Scheme 1) or the irradiated (Processes C-Scheme 1) regions, to create three-dimensional nanostructured surfaces.

Until recently, most work on e-beam irradiation of SAMs was focused on simple aliphatic and aromatic SAMs, and lateral dimensions down to 5 nm have been fabricated (Fig. 16.3) (Lercel 1996). E-beam irradiation of aliphatic SAMs (alkanethiols and alkylsilanes) has been shown to cause the loss of the orientational and conformational order (Zharnikov 1999, 2000), partial dehydrogenation, formation of C=C bonds, appearance of oxygenated functional groups, and partial desorption of the film fragments. Regarding aromatic SAMs (biphenylthiol on Au and hydroxybiphenyl on Si), electron irradiation induced primarily C-H bond scissions in the aromatic unit, with subsequently crosslinking between the neighbouring aromatic moieties. Structures with a lateral size below 20 nm have been fabricated. The increased etch resistance of the electron-irradiated aromatic SAMs make them a negative-tone electron resist for lithographic applications, whereas aliphatic SAMs, where electron-induced damage dominates over crosslinking, can be used as positive-tone e-beam resists.

Beyond the principal differences between aliphatic and aromatic SAMs, the molecular orientation and packing density also affect the extent of the irradiation-induced modification (Frey 2002). Damage introduced into the film has also been found to be preferentially localised near the monolayer-vacuum interface, with the susceptibility to damage at this interface significantly increasing as the chain length increases (Olsen 1998). At the same time, the extent of the irradiation-induced desorption was found to be significantly enhanced by the introduction of sulphur-derived functionalities into the alkyl chain, as long as they were placed close to the SAM-vacuum interface (Heister 2004).

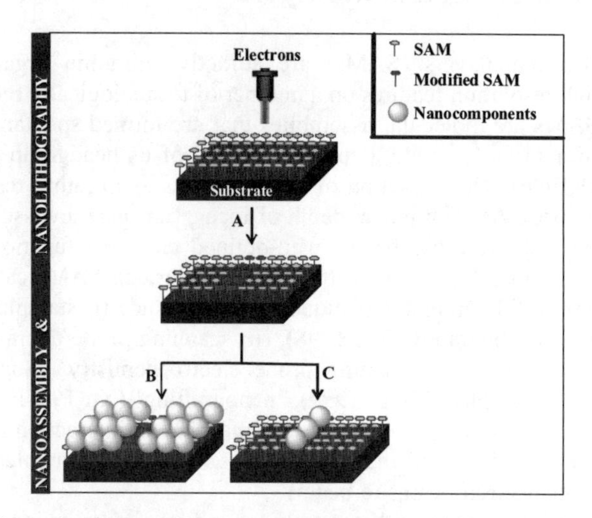

Scheme 1. Schematic diagram illustrating precision chemical engineering by e-beam irradiation of SAMs (Process A-nanolithography), followed by selective self-organisation of self-assembled nanocomponents onto templated SAMs, *i.e.* nanoassembly. Nanoassembly can occur on the unirradiated (Process B) or irradiated areas (Process C)

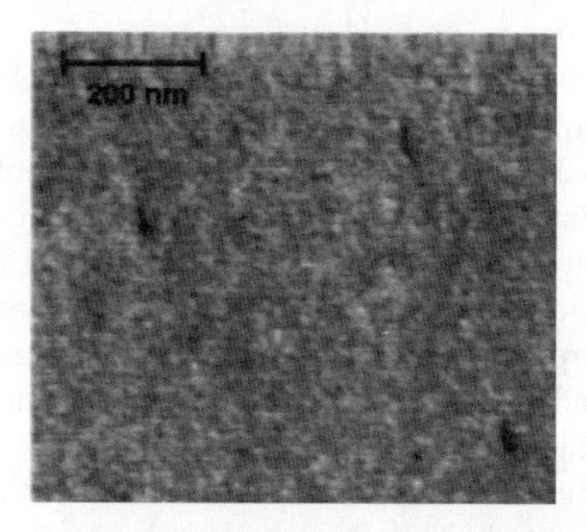

Fig. 16.3. AFM image of an array of ~ 5 nm dots generated in octadecylsiloxane SAM by a focused e-beam

Trimethylsilyl SAMs on Si/SiO_2 can be removed by e-beam irradiation revealing the native SiO_2, which can be used for adsorption of a second silane. This approach allowed the formation of a NH_2-terminated SAM without an intermediate development step by adsorption of aminopropyltriethoxysilane to the exposed SiO_2 surface (Krupke 2002, Liu 1999). Carbon nanotubes were shown to align well with the NH_2 functionalised patterns (10-50 nm) on the surface. Similar templates have also been useful for attachment of a variety of materials relevant to biotechnology including 20 nm aldehyde-, and 40 nm carboxylic acid-functionalised polystyrene particles and biotin on silicon and gold substrates (Harnett 2001).

More interestingly, chemoselective processes promoted by e-beam irradiation have also been demonstrated. Grunze and co-workers (Eck 2000, Geyer 2001) demonstrated that nitro groups are selectively reduced to amine groups by irradiation of 4-nitro-1,1-biphenyl-4-thiol SAMs on gold substrates with low and high energy electrons, while the aromatic biphenyl layer is dehydrogenated and crosslinked. It has been suggested that the source of hydrogen atoms required for reduction of the nitro groups are generated by the electron-induced dissociation of the C-H bonds in the biphenyl moieties. Chemical patterns were generated by irradiation through a mask defining lines with a width down to 70 nm. Direct-write e-beam lithography achieved 20 nm resolution lines, which were suggested to be further reduced by using a highly focused e-beam (Golzhauser 2001). The templates were used for the surface immobilisation of fluorinated carboxylic acid anhydrides, rhodamine dyes and polystyrene brushes on the regions terminated by the NH_2 groups.

A similar molecular transformation on nitro-terminated aromatic monolayers on Si/SiO_2 substrate caused by low (Jung 2003, La 2002) and high energy electron beams has also been reported. Templates of reactive amino sites within the nitro-terminated monolayers were further employed to guide the assembly of 16 nm gold nanoparticles (Fig.16.4), to immobilise biotin and subsequently to bind Cy3-Tagged Streptavidin, and to initiate the ring-opening polymerisation of aziridine to form a hyperbranched polymer (Jung 2003). Cleavage of an internal disulfide group in a SAM of phenyl (3-trimethoxysilylpropyl) disulfide by high energy electrons has also been described (Wang 2003).

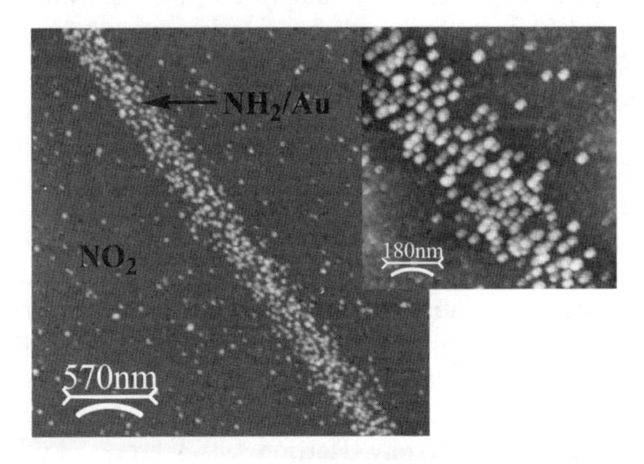

Fig. 16.4. AFM images of citrate-stabilised gold nanoparticles attached preferentially to the amino sites within the nitro-terminated monolayer

The resulting SH functionalised nanopatterns (35 nm) were further developed by reaction with N-(1-pyrene)maleimide. Brandow and co-workers (Dressick 2001) described an alternative e-beam patterning approach based on noncovalent entrapment of amine ligands in solvent-imprinted nanocavities on aromatic siloxane SAMs, and subsequent selective displacement of these ligands using high energy electrons. Amine ligands on the unirradiated regions were used to bind a catalyst capable of initiating the selective deposition of an electroless nickel film. Non-metallised features of 40 nm were generated in the nickel film.

Similar immobilisation systems have also been demonstrated by damaging/partially removing a functionalised monolayer, rendering that area of the monolayer inert for further attachment (Harnett 2000, Maeng 2003). NH_2 patterned monolayers (80 nm) were used as templates for the site-selective immobilisation of palladium colloids, 20 nm aldehyde modified polystyrene spheres and 40 nm NeutrAvidin protein-coated polystyrene spheres onto the unirradiated areas of the monolayers (Harnett 2000). As discussed above, biomolecules and polymerisation of aziridine can be selectively immobilised on templates of reactive amino sites generated by e-beam irradiation on nitro-terminated monolayers (Jung 2003). Although demonstrated with micro scale resolution, a reverse system for immobilisation of these nanocomponents was shown to be more efficient. This system consisted of transforming an internal imine group of a benzaldimine monolayer into a nonhydrolyzable secondary amine group by low energy electron beams. Hydrolysis of the imine groups at the unirradiated regions yielded reactive primary amine functionalities for further modification (Jung 2003). This work clearly suggests that the efficiency of the chemoselective process promoted by the e-beam irradiation is crucial for the success of nanopatterning and subsequently building into the third dimension.

E-beam irradiation of a SAM exhibiting an initiator for atom transfer radical polymerisation has also been shown to create microscale structures of polymer brushes (Maeng 2003). For instance, the C-Br atom in a benzyl bromide moiety was selectively and homolytically cleaved by low energy e-beam through a mask, and the pattern was vertically amplified and transformed into patterned polystyrene and poly(methyl methacrylate) brushes at the unirradiated regions (Maeng 2003). E-beam irradiation has also been demonstrated to promote the fragmentation of furoxan on a furoxan imine monolayer to release nitrogen oxide and the concomitant generation of a carbon-carbon triple bond containing product on the surface (Kim 2003).

It is clear from the discussion above that e-beam lithography is capable of producing high-resolution patterns on SAMs with high reliability and precise control over the location of the feature. Moreover, the chemoselective reactions induced by e-beam irradiation widen the window of resist choice (Table 16.1). In contrast to an optical lithography system, the e-beam is not diffraction-limited, but it exhibits a low throughput. Nonetheless, since e-beam technology is commonly used in semiconductor manufacturing, several new parallel techniques/tools such as projection e-beam lithography (Harriott 1997), arrays of e-beam microcolumns (Chang 1996) and multibeam sources (Baum 1999, Pei 1999) have been developed.

Table 16.1. SAM systems exploited by e-beam irradiation

NANOLITHOGRAPHY				& NANOASSEMBLY	
Substrate	*SAM*	*Modified SAM*	*Resolution*	*Nanocomponents*	
				Irradiated	*Unirradiated*
Au Si/SiO_2 Ti/TiO_2	-S-$(CH_2)_nCH_3$ $(-O)_3Si-(CH_2)_n-CH_3$ 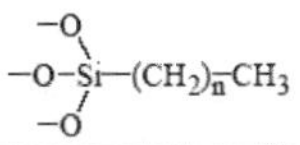	Disordered; De-hydrogenation; Desorption;	5 nm	SAM→Bio-molecules; Carbon nanotubes; Polymer particles[a]	-
Au Si	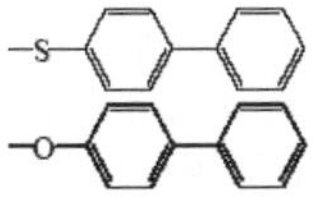 -S-(biphenyl) -O-(biphenyl)	Crosslinked	20 nm	-	-
Au Si/SiO_2	-S-R-NO_2 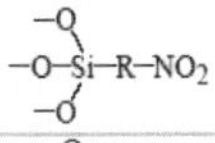 $(-O)_3Si-R-NO_2$	-S-R-NH_2 $(-O)_3Si-R-NH_2$ 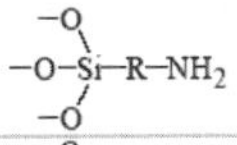	20 nm	Polymers; Organic moieties; Metal nanoparticles; Biomolecules; Dyes	-
Si/SiO_2	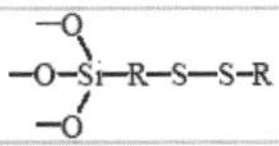 $(-O)_3Si-R-S-S-R$	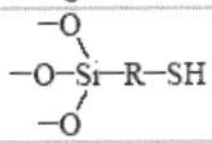 $(-O)_3Si-R-SH$	35 nm	Organic moieties	-
Si/SiO_2	Entrap of 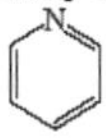 (pyridine)	Displacement	40 nm	-	Metals
Si/SiO_2	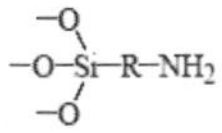 $(-O)_3Si-R-NH_2$	Damage	80 nm	-	Metal and polymer nanoparticles
Si/SiO_2	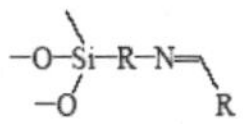 $-O-Si-R-N=CH-R$	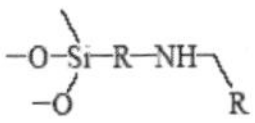 $-O-Si-R-NH-CH_2-R$	μ scale	-	Biomolecules; Polymers
Si/SiO_2	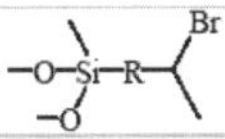 $-O-Si-R-C(CH_3)_2Br$	Damage	μ scale	-	Polymers
Si/SiO_2	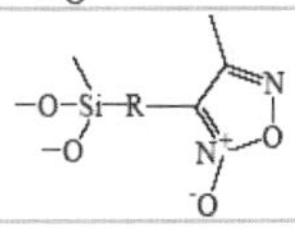 $-O-Si-R-$(furoxan ring: N, N^+-O^-, O)	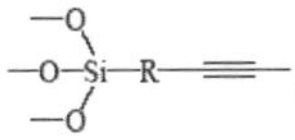 $(-O)_3Si-R-C\equiv C-$	-	-	-

[a]These systems involve immobilisation of materials after formation of a second SAM on the patterned surface

16.4 Summary and Outlook

The most interest for the future is the possibility of creating functional nanostructures by combining bottom-up self-organisation of self-assembled nanoentities, which occur in solution, with top-down lithographic approaches such as e-beam lithography. The integration of the top-down and bottom-up methodologies, that has been termed precision chemical engineering, and is representing a new paradigm for creating nanostructured surfaces, and is being actively pursued by a number of groups within the scientific and engineering communities. Significant progress in understanding the chemical and physical modifications of SAMs by e-beam irradiation has been achieved over the past decade. First demonstrations of SAM-based nanolithography exist using this technique, but their development into practical commercial methods for low cost, high-volume processing still requires great efforts. Several factors have to be considered as a collective in order to achieve such a major task. Future work should attempt to establish the ultimate resolution limits for patterning of SAMs using e-beam exposure sources. Sufficiently low exposure doses for nanopatterning will be also necessary to ensure adequate throughput during manufacture. Additionally, irradiation must yield well-defined surface products in sufficient quantities such that useful amounts of the desired nanoscale components can be subsequently and selectively attached onto the surface. A minimal number of post-irradiation high selective chemical attachment steps (ideally, a single step) using inexpensive, simple reagents will also be beneficial to minimise process cost.

16.5 References

Ball P (2000) Chemistry meets computing, Nature 406:118-20

Baum A, Arcuni P, Aebi V, Presley S, Elder M (1999) Prototype negative electron affinity-based multibeam electron gun for lithography and microscopy, J Vac Sci Technol B 17:2819-2822

Chang THP, Thomson MGR, Kratschmer E, Kim HS, Yu ML, Lee KY, Rishton SA, Hussey BW, Zolgharnain S (1996) Electron-beam microcolumns for lithography and related applications, J Vac Sci Technol B 14:3774-3780

Chen W, Ahmed H (1993) Fabrication of 5–7 nm wide etched lines in silicon using 100 keV electron-beam lithography and polymethylmethacrylate resist, Appl Phys Lett 62:1499-501

Dagani R (2000) Building from the bottom up, C&EN, October 16:27-32

Daniel M-C, Astruc D (2004) Gold nanoparticles: Assembly, supramolecular chemistry, quantum-size-related properties, and applications toward biology, catalysis, and nanotechnology Chem. Rev. 10:293-346

Dobisz EA, Fedynshyn TN, Ma D, Shirley LM, Bass R (1998) Electron-beam nanolithography, acid diffusion, and chemical kinetics in SAL-601, J Vac Sci Technol B 16:3773-3778

Dobisz EA, Marrian CRK (1997) Control in sub-100 nm lithography in SAL-601, Vac Sci Technol B 15:2327-2331

Dressick WJ, Chen M-S, Brandow SL, Rhee KW, Shirey LM, Perkins FK (2001) Imaging layers for 50 kV electron beam lithography: Selective displacement of noncovalently bound amine ligands from a siloxane host film, Appl Phys Lett. 78:676-678

Dressick WJ, Nealey PF, Brandow SL (2001) Fabrication of patterned surface reactivity templates using physisorption of reactive species in solvent-imprinted nanocavities. Proc. of the SPIE, vol. 4343, pp 294-330

Eck W, Stadler V, Geyer W, Zharnikov M, Gölzhäuser A, Grunze M (2000) Generation of surface amino groups on aromatic self-assembled monolayers by low energy electron beams-A first step towards chemical lithography, Adv Mater 12:805-808

Frey S, Rong H-T, Heister K, Yang Y-J, Buck M, Zharnikov (2002) Response of biphenyl-substituted alkanethiol self-assembled monolayers to electron irradiation: Damage suppression and odd-even effects, Langmuir 18:3142-3150

Fujita J, Oshnishi Y, Ochiai Y, Matsui S (1996) Ultrahigh resolution of calixarene negative resist in electron beam lithography, Appl Phys Lett. 68:1297-1299

Geyer W, Stadler V, Eck W, Zharnikov M, Gölzhäuser A, Grunze M (1996) Electron-induced crosslinking of aromatic self-assembled monolayers: Negative resists for nanolithography, Appl Phys Lett. 75:2401-2403

Geyer W, Stadler V, Eck W, Zharnikov M, Gölzhäuser A, Grunze M (2001) Electron induced chemical nanolithography with self-assembled monolayers, J Vac Sci Technol B 19:2732-2735

Gölzhäuser A, Eck W, Geyer W, Stadler V, Weimann T, Hinze P, Grunze M (2001) Chemical nanolithography with electron beams, Adv Mater 13:806-809

Harnett CK, Satyalakshmi KM, Craighead G (2000) Low-energy electron-beam patterning of amine-functionalised self-assembled monolayers, Appl Phys Lett. 76:2466-2468

Harnett CK, Satyalakshmi KM, Craighead HG (2001) Bioactive templates fabricated by low-energy electron beam lithography of self-assembled monolayers, Langmuir 17:178-182

Harriott LR (1997) Scattering with angular limitation projection electron beam lithography for suboptical lithography, J Vac Sci Technol B 15:2130-2135

Heister K, Frey S, Ulman A, Grunze M, Zharnikov M (2004) Irradiation sensitivity of self-assembled monolayers with an introduced 'weak link', Langmuir 20:1222-1227

http://www.intel.com/labs/features/si08042.htm?iid=TechTrends+featured_65nanometer&.

http://www.intel.com/research/silicon/nanotechnology.htm

Hutt DA, Leggett GJ (1999) Static secondary ion mass spectrometry studies of self-assembled monolayers: Electron beam degradation of alkanethiols on gold, J Mater Chem 9:923-928

Jung YJ, La Y-H, Kim HJ, Kang T-H, Ihm K, Kim K-J, Kim B, Park JW (2003) Pattern formation through selective chemical transformation of imine group of self-assembled monolayer by low-energy electron beam, Langmuir 19:4512-4518

Kim CO, Jung JW, Kim M, Kang T-H, Ihm K, Kim K-J, Kim B, Park JW, Nam H-W, Hwang K-J (2003) Low energy electron beam irradiation promoted selective cleavage of surface furoxan, Langmuir 19:4504-4508

Krupke R, Malik S, Weber HB, Hampe O, Kappes MM, Löhneysen Hv (2002) Patterning and visualizing self-assembled monolayers with low-energy electrons, Nano Lett. 2:1161-1164

Küller A, Eck W, Stadler V, Geyer W, Gölzhäuser A (2003) Nanostructuring of silicon by electron-beam lithography of self-assembled hydroxybiphenyl monolayers, Appl Phys Lett. 82:3776-3778

La Y-H, Kim HJ, Maeng IS, Jung YJ, Park JW (2002) Differential reactivity of nitro-substituted monolayers to electron beam and x-ray irradiation, Langmuir, 18:301-303

Lercel MJ, Craighead HG, Parikh AN, Seshadri K, Allara DL (1996) Sub-10 nm lithography with self-assembled monolayers, Appl Phys Lett. 68:1504-1506

Lercel MJ, Redinbo GF, Pardo FD, Rooks M, Tiberio RC, Simpson P, Sheen CW, Parikh AN, Allara DL (1994) Electron beam lithography with monolayers of alkylthiols and alkylsiloxanes, J Vac Sci Technol B 12:3663-3667

Lercel MJ, Rooks M, Tiberio RC, Craighead HG, Sheen CW, Parikh AN, Allara DL (1995) Pattern transfer of electron beam modified self-assembled monolayers for high-resolution lithography, J Vac Sci Technol B 13:1139-1143

Lercel MJ, Tiberio RC, Chapman PF, Craighead HG, Sheen CW, Parikh AN, Allara DL (1993) Self-assembled monolayer electron-beam resists on GaAs and SiO_2, J Vac Sci Technol B 11:2823-2828

Lercel MJ, Whelan CS, Craighead HG, Seshadri K, Allara DL (1996) High-resolution silicon patterning with self-assembled monolayer resists, J Vac Sci Tech. B 14:4085-4090

Liu G-Y, Xu S, Qian Y (2000) Nanofabrication of self-assembled monolayers using scanning probe lithography, Acc Chem Res 33:457-466

Liu J, Casavant MJ, Cox M, Walters DA, Boul P, Lu W, Rimberg AJ, Smith KA, Colbert DT, Smalley RE (1999) Controlled deposition of individual single-walled carbon nanotubes on chemically functionalised templates, Chem Phys Lett. 303:125-129

Liu J-F, Cruchon-Dupeyrat S, Garno JC, Frommer J, Liu G-Y (2002) Three-dimensional nanostructure construction via nanografting: positive and negative pattern transfer, Nano Lett, 2:937-940

Maeng IS, Park JW (2003) Patterning on self-assembled monolayers by low-energy electron-beam irradiation and its vertical amplification with atom transfer radical polymerisation, Langmuir 19:4519-4522

Mendes PM, Chen Y, Palmer RE, Nikitin K, Fitzmaurice D, Preece JA (2003) Nanostructures from nanoparticles, J Phys: Condens Matter, 15:S3047-63

Mendes PM, Jacke S, Kritchley K, Plaza J, Chen Y, Nikitin K, Palmer RE, Preece JA, Evans SD, Fitzmaurice D (2004) Gold nanoparticle patterning of silicon wafers using chemical e-beam lithography, Langmuir 20:3766-3768

Mendes PM, Preece JA (2004) Precision chemical engineering: Integrating nanolithography and nanoassembly, Curr Opin Colloid 9:236–248

Müller HU, Zharnikov M, Völkel B, Schertel A, Harder P, Grunze M (1998) Low-energy electron-induced damage in hexadecanethiolate monolayers, J Phys Chem B 102:7949-7959

Olsen C, Rowntree PA (1998) Bond-selective dissociation of alkanethiol based self-assembled monolayers adsorbed on gold substrates, using low-energy electron beams, J Chem Phys 108:3750-3764

Pei Z, McCarthy J, Berglund CN, Chang TPH, Mankos M, Lee KY, Yu ML (1999) Thin-film gated photocathodes for electron-beam lithography, J Vac Sci Technol B 17:2814-2818

Piner RD, Zhu J, Xu F, Hong S, Mirkin CA (1999) Dip-pen nanolithography, Science 283:661-663

Rai-Choudhury P (1997) (ed) Handbook of microlithography: Micromachining and icrofabrication. IEE London, vol. 1

Rieke PC, Baer DR, Fryxell GE, Engelhard MH, Porter MS (1993) Beam damage of self-assembled monolayers, J Vac Sci Technol A 11:2292-2297

Roberts ED (1987) The chemistry of the semiconductor industry. In: Moss SJ (ed), Ledwith, Pub. Blackie, London, p 198

Roberts ED (1987) The chemistry of the semiconductor Industry. Blackie and Sons Ltd., London

Robinson APG, Palmer RE, Tada T, Kanayama T, Allen MT, Preece JA, Harris KDM (1999) 10 nm scale electron beam lithography using a triphenylene derivative as a negative/positive tone resist, J Phy D: Appl Phys 32:L75-78

Robinson APG, Palmer RE, Tada T, Kanayama T, Allen MT, Preece JA, Harris KDM (2000) Polysubstituted derivatives of triphenylene as high resolution electron beam resists for nanolithography, J Vac Sci Technol B 18:2730-276

Robinson APG, Palmer RE, Tada T, Kanayama T, Allen MT, Preece JA, Harris KDM (2000) A triphenylene derivative as a novel negative/positive tone resist of 10 nanometer resolution, Microelectronic Engineering 53:425-428

Robinson APG, Palmer RE, Tada T, Kanayama T, Preece JA (1998) A Fullerene derivative as an electron beam resist for nanolithography, Appl Phys Lett. 72:1302-1304

Robinson APG, Palmer RE, Tada T, Kanayama T, Shelley EJ, Philp D, Preece JA (1999) Exposure mechanism of fullerene derivative electron beam resists, Chem Phys Lett. 312:469-474

Robinson APG, Palmer RE, Tada T, Kanayama T, Shelley EJ, Preece JA (1999) Fullerene derivatives as novel resist materials for fabrication of MEMS devices by electron beam lithography. Mat Res Soc Symp Proc., vol.546, pp 219-24

Robinson APG, Hunt MRC, Palmer RE, Tada T, Kanayama T, Preece JA, Philp D, Jonas U, Diederich F (1998) Electron beam induced fragmentation of fullerene derivatives, Chem Phys Lett. 289:586-590

Sailer H, Ruderisch A, Kern D.P., Schurig V (2004) A chemically amplified calix[4]arene-based electron-beam resist, Microelectronic Engineering 73-74:228-232

Schmelmer U, Jordan R, Geyer W, Eck W, Gölzhäuser A, Grunze M, Ulman A (2003) Surface-initiated polymerisation on self-assembled monolayers: Amplification of patterns on the micrometer and nanometer scale. Angew Chem Int ed, vol. 42, pp 559-563

Schock K-D, Prins FE, Strahle S, Kern DP (1997) Resist processes for low-energy electron-beam lithography, J Vac Sci Technol B 15:2323-2326

Seshadri K, Froyd K, Parikh AN, Allara DL, Lercel MJ (1996) Craighead, Electron-beam-induced damage in self-assembled monolayers, J Phys Chem 100:15900-15909

Silverman JP (1997) X-ray lithography: Status, challenges, and outlook for 0.13 μm, J Vac Sci Technol B 15:2117-2124

Sugimura H, Hanji T, Hayashi K, Takai O (2002) Surface modification of an organosilane self-assembled monolayer on silicon substrates using atomic force microscopy: Scanning probe electrochemistry toward nanolithography, Ultramicroscopy 91:221-26

Tada T, Kanayama T (1996) Nanolithography using fullerene films as an electron beam resist, Jpn J Appl Phys 35:L63-65

Tada T, Kanayama T (1997) C60-Incorporated nanocomposite resist system, J Photopolym Sci Technol 10:651-656

Tada T, Uekusu K, Kananyama T, Nakayama T, Chapman R, Cheung WY, Eden L, Hussain I, Jennings M, Perkins J, Philips M, Preece JA, Shelley EJ (2001) Improved sensitivity of multi-adduct derivatives of fullerene, J Photopolym Sci Techn 14:543-546

Tada T, Uekusu K, Kanayama T, Nakayama T, Chapman R, Cheung WY, Eden L, Hussain I, Jennings M, Perkins J, Philips M, Preece JA, Shelley EJ (2002) Multi-adduct derivatives of C60 for electron beam nano-resists, Microelectronic Engineering 61:737-743

Ulman A (1996) Formation and structure of self-assembled monolayers, Chem Rev. 95:1533-1554

Ulman A (1991) An Introduction to Ultrathin Organic Films. Academic Press Ltd., UK

Wang X, Hu W, Ramasubramaniam R, Bernstein GH, Snider G and Lieberman M (2003) Formation, characterisation, and sub-50-nm patterning of organosilane monolayers with embedded disulfide bonds: An engineered self-assembled monolayer resist for electron-beam lithography, Langmuir 19:9748-9758

Weimann T, Geyer W, Hinze P, Volker S, Eck W, Gölzhäuser A (2001) Nanoscale patterning of self-assembled monolayers by e-beam lithography, Microelectron Eng. 57-58:903-907

Whitesides GM, Grzybowski B (2002) Self-Assembly at all scales, Science 295:2418-21

Xia Y, Whitesides GM (1998) Soft lithography. Angew Chem Int Ed, vol. 37, pp 550-575

Xu S, Liu G-y (1997) Nanometer-scale fabrication by simultaneous nanoshaving and molecular self-assembly, Langmuir 13:127-129

Yoshiiwa M, Kageyama H, Shirota Y, Wakaya F, Gamo K, Takai M (1996) Novel class of low molecular-weight organic resists for nanometer lithography, Appl Phys Lett. 69:2605-2607

Zharnikov M, Frey S, Heister K, Grunze M (2000) Modification of alkanethiolate monolayers by low energy electron irradiation: Dependence on the substrate material and on the length and isotopic composition of the alkyl chains, Langmuir 16:2697-2705

Zharnikov M, Geyer W, Gölzhäuser A, Frey S, Grunze M (1999) Modification of alkanethiolate monolayers on Au-substrate by low energy electron irradiation: Alkyl chains and the S/Au interface, Phys Chem Chem Phys. 1:3163-3171

Zharnikov M, Grunze M (2002) Modification of thiol-derived self-assembling monolayers by electron and x-ray irradiation: Scientific and lithographic aspects, J Vac Sci Technol B 20:1793-1807 and references there in

17 Nanolithography in the Evanescent Near Field

M. M. Alkaisi and R. J. Blaikie

MacDiarmid Institute of Advanced Materials and Nanotechnology, University of Canterbury, New Zealand

17.1 Introduction

The ability of nanotechnology to manipulate systems down to the scale of individual atoms and molecules offers exciting potential for novel devices and materials. It is expected that significant progress in science and technology will be achieved. As a result, considerable interest in developing new nanofabrication techniques for manufacturing nanoscale structures has emerged worldwide. Techniques such as nano-imprint lithography, microcontact printing, immersion lithography and electron projection lithography offer alternatives to conventional electron beam nanolithography or recently developed EUV lithography. Many of these alternatives offer advantages of low cost and high throughput for mass production of nanostructured materials and devices.

The resolution of conventional projection optical lithography is limited by diffraction, making it extremely challenging to fabricate sub-100 nm structures even using deep UV light sources and advanced wavefront engineering. However, by working in the optical near field the conventional diffraction limit can be overcome and nanoscale patterning can be achieved. Whilst this idea is not new it is only relatively recently that detailed studies have been performed. The first demonstrations of sub-wavelength resolution near field nanolithography used serial scanning techniques which is not suitable for high throughput lithography process, however in this chapter we will be presenting the implications of using evanescently decaying components in the near field of a photomask as a parallel lithography tool for the fabrication of nanoscale structures.

Both experimental and simulation results for evanescent near field optical lithography (ENFOL) will be discussed. The ENFOL technique allows for sub-diffraction-limited resolution to be achieved with optical lithography by keeping a shadow mask in intimate contact with an ultra-thin photoresist layer during exposure – this ensures that the photoresist receives exposure from the evanescently decaying diffracted orders in the near field of the mask. Features as small as 70 nm on a 140 nm period have been patterned using broadband UV illumination (365-600 nm), and subsequently transferred these patterns into silicon using reactive ion etching (RIE) (Blaikie et al. 1999; Alkaisi 2001). This resolution is below the conventional diffraction limit for projection optical lithography and illustrates the promise for extending optical lithography into the sub-100 nm realm. Simulation studies have shown that resolution down to 20 nm should be possible with this technique (McNab and Blaikie 2000), an enticing prospect for the nanolithographer.

The rest of this chapter will be structured as follows. We will first briefly review the historical development of near field contact lithography, outline the principles of the ENFOL technique and give details of the techniques required for the fabrication of conformable near field masks. Experimental results demonstrating the resolution that can be achieved with ENFOL exposure into ultra-thin photoresists are presented. Details of the subtractive and additive pattern transfer processes that have been developed are presented. Simulation results are also shown. This indicates that the resolution limits for ENFOL have yet to be reached experimentally. Finally, advanced near field lithography techniques such as evanescent interferometeric lithography, planar lens and surface plasmon lithography will be discussed.

17.2 Historical Development

An essential requirement for near field photolithography techniques such as ENFOL is to maintain a conformable photomask in intimate contact with the photoresist during exposure. The first work using conformable photolithography masks was reported by Smith in 1969 for fabricating surface acoustic wave (SAW) devices. Mechanical pressure was used to maintain contact and interdigital electrodes with 2.5 μm periods were replicated. In later work Smith demonstrated 400 nm linewidths on an 800 nm pitch using the same technique but this time using vacuum pressure to attain intimate contact between the mask and substrate. A distinct advantage of this system is the insensitivity of the linewidth to exposure time when operating in true intimate contact. In that paper Smith hints at the prospect of achieving higher resolutions – "Although no effort has been made to explore the ultimate limitation of the conformable mask photolithography, our observations lead us to speculate that narrower linewidths than reported here are probably possible, but may require the exercise of finer control over photoresist thickness and exposure parameters". These words turned out to be quite accurate, although were not realised until nearly a decade later, when White et al. demonstrated patterning of 150 nm features with a 157 nm laser source (White et al. 1984). A further important advantage of the conformable mask that is cited is the absence of damage or wear through repeated use (Smith et al. 2000), an improvement on the conventional photomasks which suffer degradation with use.

Goodberlet has used a modification of the conventional conformable mask, where the absorber is embedded into the mask substrate. His embedded amplitude mask (EAM) consists of a substrate material made from fused silica, with embedded Cr absorber patterns. There are a number of advantages of the having the absorber embedded: a flat mask would be expected to reduce local deformations around the absorber areas, the narrow regions between the absorber also improve waveguiding with their higher refractive index, and the planarity of the mask may protect the absorber and minimise particle contamination. The EAM mask has been used to reproduce 100 nm line and space structures, using a source with a wavelength of 220 nm. Pattern placement errors have been evaluated by Goodberlet by measuring the in-plane distortions of a double exposure with the mask misaligned slightly between the first and second exposure. An average displacement of 58 nm was observed over an area of 2 cm^2. Attempts at multi-level alignment are also in progress. More recently Goodberlet (2002) has demonstrated patterning features down to 45 nm using this technique.

In another technique, light-coupling masks (LCM) for conformable contact photolithography have been developed by a group at IBM's Research Laboratory in Zurich (Schmid et al. 1998). The masks are made of an elastomeric material, similar to that developed for microcontact printing (Rogers et al. 1997), which allows for conformal contact. The mask consists of a surface relief pattern and relies on this pattern to induce local optical modes; this waveguiding action amplifies the intensity in the protruding mask regions, which results in contrast in the underlying resist (Paulus et al. 2001). Another contrast mechanism occurs when the wavelength is much smaller that the features on the mask; if the depth of the surface relief results in a phase change close to π the destructive interference that results at the mask edge results in a strongly unexposed region. The same effect is observed in near field phase shifting masks (PSMs). Results for LCM exposure demonstrate that 240 nm period is possible with a low surface relief pattern (70 nm deep) using 248 nm wavelength exposure. In this case the low surface relief and small size of the patterns means that the second exposure mechanism is not significant.

Fig. 17.1 illustrates the four possible mask configurations for near field lithography. A simulation study comparing the performance of these different techniques has been performed (Paulus etal.2000); whilst the contrast, depth of focus and minimum feature size are slightly improved for the embedded absorber and recessed absorber compared to the protruding absorber, the simplicity of mask fabrication makes the latter technique the best choice in many cases.

An alternative to using a conformable mask with a rigid substrate is to use a rigid mask with a conformable substrate. Ono et al. have replicated 500 nm period structures with such an arrangement. They formed substrates from 1 μm silicon diaphragms and performed a contact exposure with standard chrome on glass mask using vacuum pressure and a mercury arc lamp source. However, the use of conformable substrates limits the range of devices that can be fabricated. Near field schemes using conformable masks are much more prominent, and other experimental and simulation studies have appeared in the literature where resolution beyond the diffraction limits have been demonstrated (Haefliger and Stemmer 2004; Luo and Ishihara 2004).

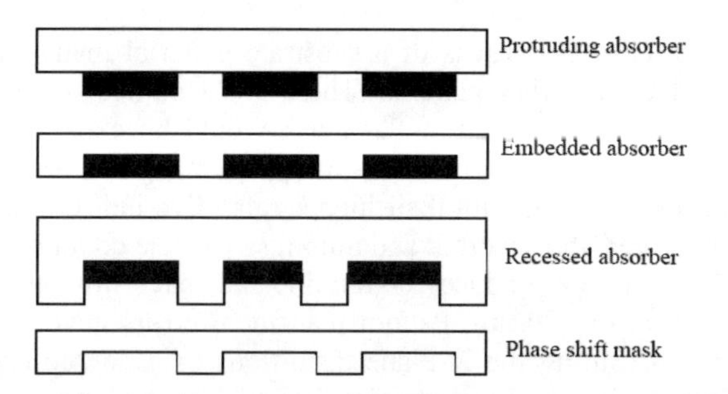

Fig. 17.1. Illustration of the four possible configurations of near field masks, where dark features represent the absorber: (a) is the protruding absorber (b) embedded absorber (Goodberlet 2000), (c) recessed absorber (Schmid et al. 1998; Paulus et al. 2001), and (d) chromeless phase shift mask (Levenson 1993; Alkaisi et al. 1998)

17.3 Principles of ENFOL

The principle of the ENFOL technique is illustrated in Fig. 17.2. A conformable membrane mask is held in intimate contact with a photoresist-coated substrate, such that the mask-substrate gap g (including the photoresist thickness) is much less than the wavelength of the exposing light source. Under these conditions the optical field in the thin resist contains high spatial frequency evanescent components together with propagating diffracted components. The presence of these high frequency components makes resolution beyond the diffraction limit of projection lithography possible.

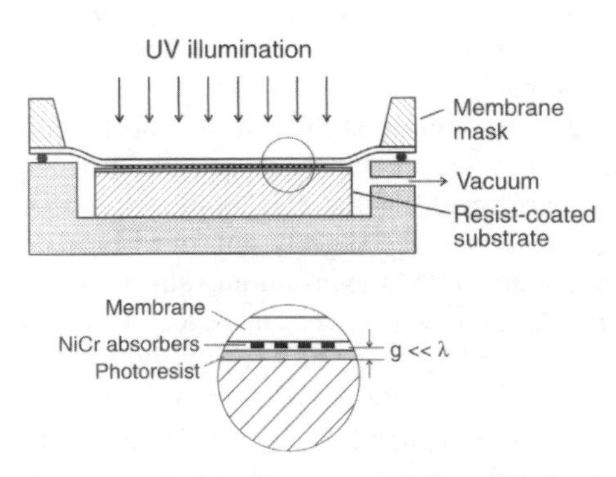

Fig. 17.2. Schematic diagram illustrating the evanescent near field optical lithography (ENFOL) process. Photolithography is performed through a conformable mask held in intimate contact with a resist layer that is much thinner than the illuminating wavelength

As can be seen from Fig. 17.2, the two main requirements for ENFOL are intimate mask-substrate contact over a large area and ultra-thin (sub-100 nm) resists. Using a conformable mask held in contact with the resist by external pressure or the use of a vacuum can achieve the former, however the exact requirements for the resist thickness are not easy to determine. Trade-offs exist between the required resolution and the ease of performing pattern transfer with the resist after exposure and development. The process latitude for the exposure can also be severely degraded if the resist becomes too thin.

17.4 Mask Requirements and Fabrication

In this section we will describe the fabrication of a typical conformable mask suitable for near field exposure. Conformable amplitude masks have been fabricated on either 2 μm thick low stress silicon nitride Si_xN_y films grown on silicon substrates using low-pressure chemical vapour deposition (LPCVD), or on thicker (100–150 μm) glass or fused silica masks; fabrication of the silicon nitride masks will be presented in detail here. Membranes of silicon nitride were formed by wet etching of the backside of the silicon substrate in 50% by weight potassium hydroxide (KOH) solution. Typically, a membrane 5×5 mm^2 square is formed in the centre of a larger sample, leaving an outer silicon supporting ring. The membrane mask-making process is illustrated in Fig .17.3.

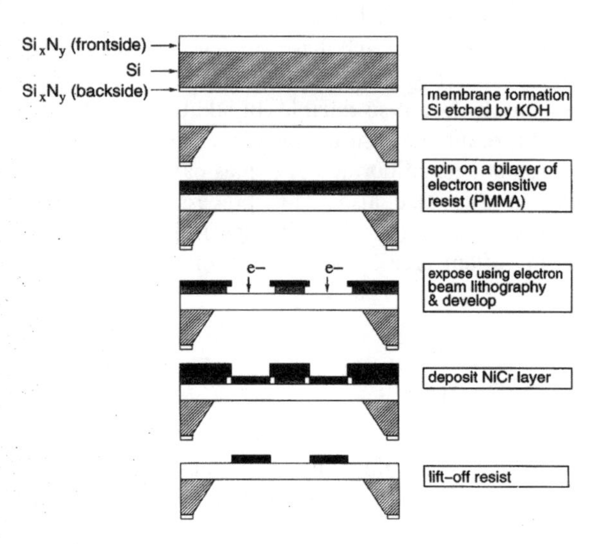

Fig. 17.3. Mask fabrication process, showing steps taken for forming the membrane and patterning the absorber features

Pattering of the mask patterns onto the Si_xN_y was performed using electron-beam lithography (EBL) onto membranes coated with a bilayer of polymethyl-methacrylate (PMMA). The bilayer resist was used to obtain an undercut profile desirable for the lift off metallisation process that was used to form the mask patterns. Arrays of 30 nm thick NiCr metal gratings of varying duty cycle were used to define the opaque regions on the mask. The linewidths in the gratings vary from 2 μm down to 60 nm. A typical amplitude mask with 130 nm gratings, isolated lines, and dense lines of NiCr patterns on a silicon nitride membrane is shown in Fig. 17.4.

Fig. 17.4. Scanning electron microscope image of a typical membrane mask patterned with 25-nm thick NiCr gratings. The pattern on the left is isolated 130 nm lines on 390 nm period, the middle pattern contains 260 nm wide lines on 390 nm period, and the right-hand side pattern is a grating with 260 nm period (130 nm lines and spaces)

Other types of mask material, including glass or fused silica substrates have also been successfully employed, and these are generally more robust than the thin nitride membrane masks. However, obtaining conformal contact over large areas with these thicker masks in not so reliable. In addition, a subtractive process has been developed for forming the absorber patterns, using reactive-ion etching (RIE) of tungsten in a sulphur hexafluoride (SF_6) plasma; EBL in PMMA is used for pattern definition in this case also. This process reduces the occurrence of unwanted protrusions (so-called 'lift-off tags') that can sometimes prevent intimate contact in certain areas, leading to poor definition of sub-wavelength patterns.

17.5 Pattern Definition

ENFOL exposures were performed using a conventional mask aligner with mercury arc lamp having broadband illumination with wavelengths between 313 and 600 nm.

17.5.1 Exposure Conditions

A commercial *g*-line photoresist (Shipley S-1805) can be used, but has to be thinned to provide sub-wavelength thickness when spun onto a silicon substrate. A vacuum chuck ensures that the substrate and membrane mask are in intimate contact. Optimum results for reproducing 50 nm features were obtained using photoresist diluted 1:4 in PGMEA (60 nm thickness after spin coating) and an exposure time of 70 s. It should be noted that no strong feature-size dependent exposure times have been observed in these experiments. This has been demonstrated by the ability to pattern micron-scale and nanometre-scale features at the same time and from the same mask

17.5.2 Resist Requirements

Photoresist thickness is a major factor that determines the resolution limits and quality of patterns defined by ENFOL. The optimum resist thickness of 60 nm is required to ensure that the exposure is performed in the evanescent near field of the mask. For thinner resist layers, pattern transfer using RIE becomes difficult and for thicker resist layers the resolution degrades due to diffraction. Fig. 17.5(a) shows a partially resolved pattern exposed for 70 s into a 125 nm thick resist layer. The equivalent exposure into a 60 nm resist layer is shown in Fig. 17.5(b). Similar resolution degradation is observed if the mask and substrate are not held in conformable contact. These results illustrate the need to have significant exposure from the evanescent field close to the aperture plane of the mask in order to attain sub-wavelength features.

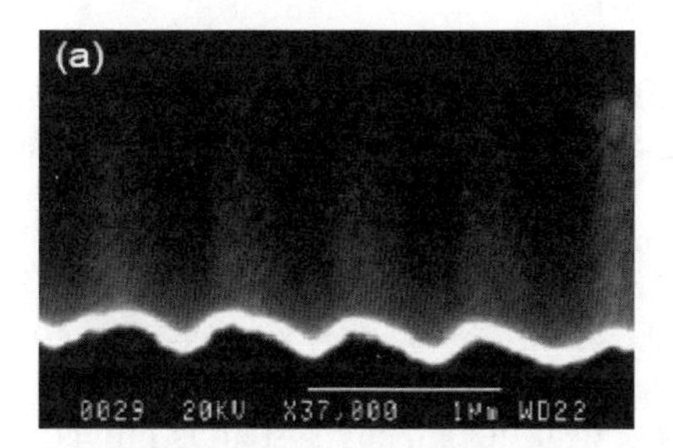

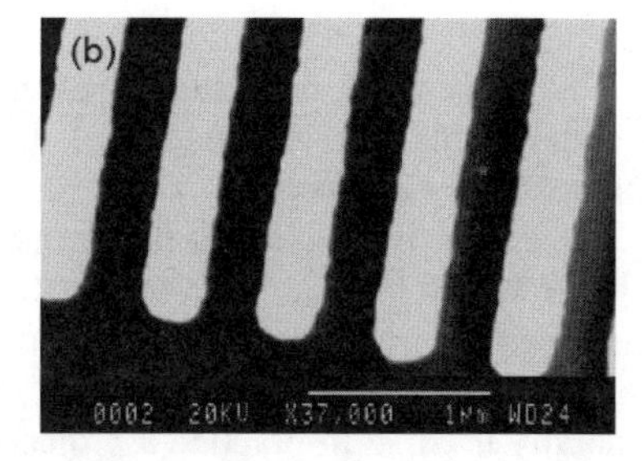

Fig. 17.5.(a) Partially resolved pattern formed by ENFOL exposure into 125 nm thick resist, (b) Pattern clearly resolved, obtained by exposing resist with the optimum thickness of 60 nm. Subtractive pattern transfer has been used to transfer the pattern in (a) and (b) into the silicon substrate to a depth of 100 nm

17.5.3 Overcoming the Diffraction Limit

Scanning electron microscope (SEM) images of high-resolution ENFOL-defined patterns are shown in Fig. 17.6. The patterns have been dry etched approximately 100 nm deep into silicon to facilitate imaging. Fig. 17.6(a) shows a 280 nm period grating formed by exposure through a mask with 70 nm isolated lines. After exposure, development and dry etching, the line width has been reduced to less than 50 nm, and there is some line edge roughness (LER) evident. It should be noted that the width of these lines is less than $\lambda_{min}/7$, where λ_{min} is the minimum exposing wavelength (365 nm). The period is approximately 0.77 λ_{min}.

Fig. 17.6(b) shows a 280 nm period grating formed by exposure through a mask with 70 nm apertures, the negative of the mask pattern used for Fig. 17.6(a). These 70 nm apertures are less than $\lambda_{min}/5$ wide, however, the exposure time for this pattern is the same as for patterns with large features. This indicates that there is no significant attenuation of the incident light through these sub-wavelength apertures. High transmission through sub wavelength sized circular aperture arrays has been previously reported (Ebbesen et al. 1998), and was attributed to plasmon resonances. In the ENFOL case, we have rectangular aperture that are long with respect to the wavelength, and high transmission of one polarisation will always be possible. Fig. 17.6(c) shows a 140 nm period grating formed in resist after exposure and development. This was exposed though a mask with 70 nm lines and 70 nm apertures. The lines in this pattern are discontinuous in places due to LER, however continuous lines are obtained for all patterns with periods greater than 200 nm. The grating period for the pattern shown in Fig. 17.6(c) is well below the diffraction limit for an equivalent projection lithography system. For normal incidence coherent illumination at wavelength λ the minimum resolvable period p_{min} in a projection optical lithography system is (Rai-Choudhury 1997),

$$p_{\min} = \frac{\lambda}{NA} \tag{17.1}$$

Where *NA* is the numerical aperture of the system. The maximum ***possible*** *NA* is equal to the refractive index of the imaging medium, *n*, which is 1.6 for the photoresist we are using.

Therefore, the best result that projection lithography into this photoresist could achieve would be $p_{min} = 228$ nm for the shortest available exposure wavelength in our source ($\lambda = 365$ nm). The structure shown in Fig. 17.6(c) has a period 40% smaller than this diffraction limit. Index-matched projection lenses are not generally used so in practice a projection optical lithography system is limited to $NA < 1$. If, together with this, we consider the peak exposing wavelength transmitted through the masks ($\lambda = 436$ nm) then $p_{min} = 436$ nm and the structure shown in Fig. 17.6(c) has a period which is approximately 1/3 of the diffraction limit for the equivalent projection system.

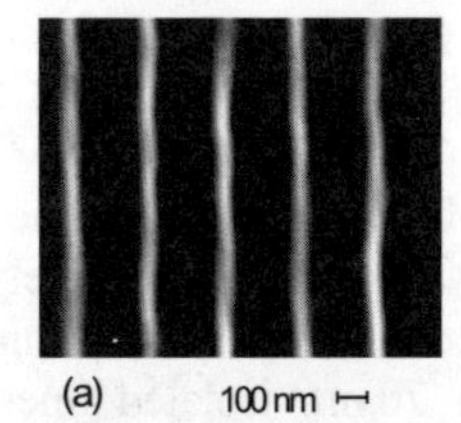

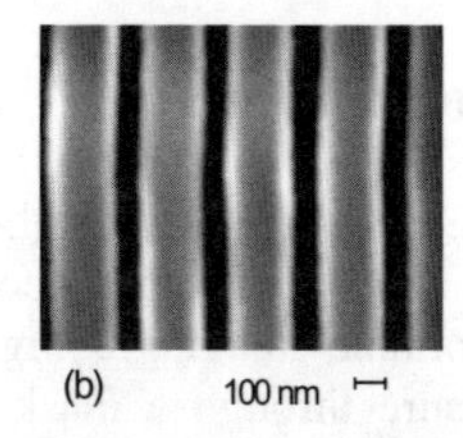

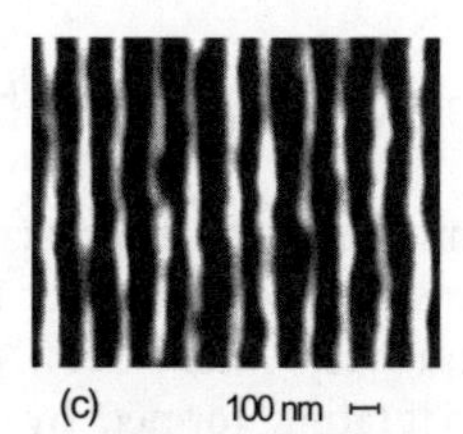

Fig. 17.6. Scanning electron microscope images of patterns produced using ENFOL and subsequent reactive ion etching: (a) 280 nm period grating exposed through a mask with 70 nm lines, (b) 280 nm period grating exposed through a mask with 70 nm apertures, and (c) 140 nm period grating exposed through a mask with 70 nm lines and 70 nm apertures

ENFOL's ability to pattern nanometre-scale and micron-scale features at the same time using the same exposure conditions is illustrated in Fig. 17.7. This figure shows a SEM micrograph for a grating structure of 2 μm wide lines with 1 μm spaces exposed and processed simultaneously with gratings that have periods down to 140 nm. These patterns show no pinhole defects and are uniform over the $40 \times 40\ \mu m^2$ area of each grating on the substrate, and are uniform between fields distributed over the $5 \times 5\ mm^2$ patterned area.

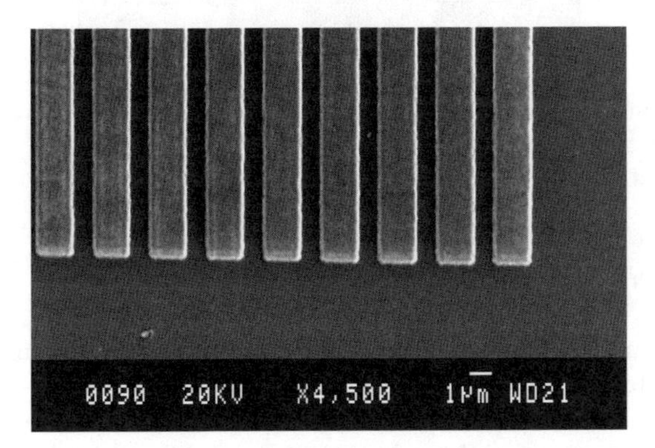

Fig. 17.7. Scanning electron microscope image of large feature patterns with 2 μm lines and 1 μm spaces produced using ENFOL. Patterns with feature sizes down to 70 nm have also been exposed and produced simultaneously from the same mask

17.6. Pattern Transfer

Transferring the photoresist patterns onto the substrate is an essential step in device fabrication. The challenge with ENFOL is performing pattern transfer using the ultra-thin resist layers that are necessary to achieve sub-diffraction limited resolution. Both subtractive (etching) and additive (lift off) pattern transfer techniques compatible with the ultra-thin resists, will be described in this section.

17.6.1 Subtractive Pattern Transfer

Reactive ion etching (RIE) has been used for subtractive pattern transfer into silicon substrates, where the resist acts as an etch mask. Etching is particularly challenging with ultra-thin photoresist, since the gases used to etch the silicon also etch the resist. Using cryogenic temperatures suppresses the resist etching and a successful dry etching process has been developed as detailed in Fig. 17.8. A short SF_6 plasma etch was carried out at 173 K to transfer the resist pattern 120 nm deep into the silicon substrate. An anisotropic etch profile was achieved with an etch rate of 10 nm/s under the following conditions: SF_6 gas flow rate of 100 sccm, power density of 0.45 W/cm^2, etch pressure of 55 mTorr, and dc bias of –153 V.

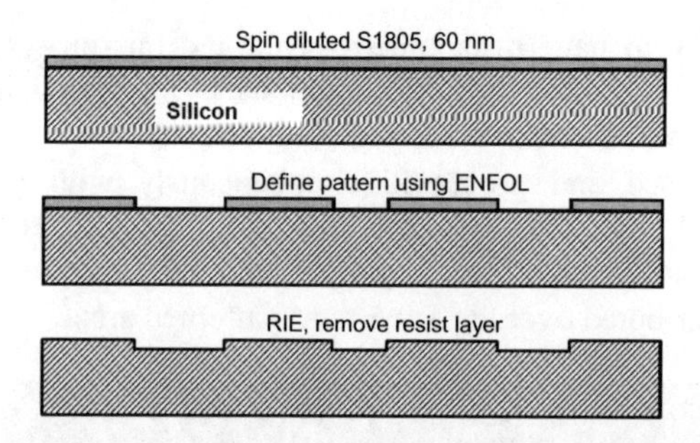

Fig. 17.8. Schematic diagram illustrating processing steps for subtractive pattern transfer using RIE. The pattern is defined first with the 60 nm thin resist and subjected to SF_6 plasma etching at 173°K

17.6.2 Additive Pattern Transfer

Additive pattern transfer involves the addition of metal films through a resist mask onto a substrate; the photoresist layer is then removed or lifted–off. Successful additive pattern transfer is not possible with ultra-thin resist. To achieve lift-off the resist should have a slightly undercut profile, and the resist thickness should be significantly greater than the metal being deposited. A trilayer scheme has been developed which improves the resist profile for lift off processes. The trilayer scheme consists of a top imaging layer (the ultra-thin photoresist), a barrier layer that acts as a hard etch mask (SiO_x) and an anti-reflection coating (ARC) bottom layer. Fig. 17.9 illustrates the trilayer additive pattern transfer technique that has been developed for ENFOL. In this process a 130 nm thick film of a commercial ARC (XLT, Brewer Science) is first spun on the silicon substrates. This is then baked on a hot plate for 20 sec at 95°C followed by a one-minute bake at 170°C. Following this a silicon oxide SiO_x barrier layer is thermally evaporated to a thickness of 16 nm, and the final ultra-thin photoresist imaging layer is then spun to give a uniform thickness of 45 nm. The sample is now ready for ENFOL exposure and development. Fig. 17.10(a) shows an atomic force microscope (AFM) image of a 270-nm period grating exposed into the imaging photoresist layer of a trilayer sample. As before, a clear sub-diffraction-limited pattern is evident, although there is a significant degree of LER. This AFM image of the photoresist prior to any pattern transfer process indicates that the LER is related to the resist not to any subsequent RIE. After exposure and development of this imaging layer, the substrate is subjected to two RIE steps, to dry etch the SiO_x and the ARC respectively. The RIE recipes are given in Table 17.1. Fig. 17.10(b) shows a cross-sectional view of a 270 nm period grating after these RIE steps have been performed, and the transfer of the pattern into the full trilayer thickness is evident. The resist now has sufficient surface relief depth to allow lift-off of relatively thick films, however the required undercut sidewall profile of the etched trenches cannot be resolved with the AFM. This can only be determined from the success of the resultant lift-off. Any number of different materials could be deposited in the lift-off stage, and for this work a 30 nm NiCr film was then deposited by thermal evaporation and lifted off in MF320 developer. Fig. 17.10(c) shows an AFM image of the final 270 nm period NiCr gratings obtained with this trilayer pattern transfer technique. No degradation in pattern quality has resulted from the pattern transfer and lift-off processes, so this should be capable of being used for finer-period structures if the LER issues in the ENFOL exposure can be resolved.

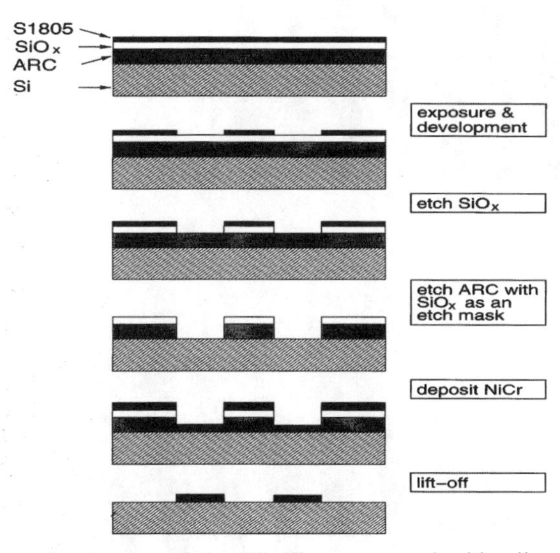

Fig. 17.9. Schematic diagram of the lift-off process used with trilayer additive pattern transfer scheme. The anti-reflection coating layer is 130 nm thick, SiO_x layer is 16 nm thick and the NiCr is 30 nm thick

Table 17.1. RIE recipes for SiO_x and ARC pattern transfer processes

Gas	Etch rate (nm/min)	Pressure (mT)	Flow rate (sccm)	Power density (W/cm^2)	DC self bias (V)	Temp (°C)
CHF_3	20	3	15	0.45	506	22
O_2	90	6	20	0.45	489	22

17.7 Simulations

The resolution achievements of scanning near field optical microscopy of better than λ/40 (Betzig 1991) suggest the potential for equivalent resolutions in near field lithography. An understanding of the near field mechanisms responsible for evanescent exposure are required to guide future ENFOL studies and to explore its potential as a nanolithography technique. Simulations have been undertaken to study the near field region behind conducting gratings.

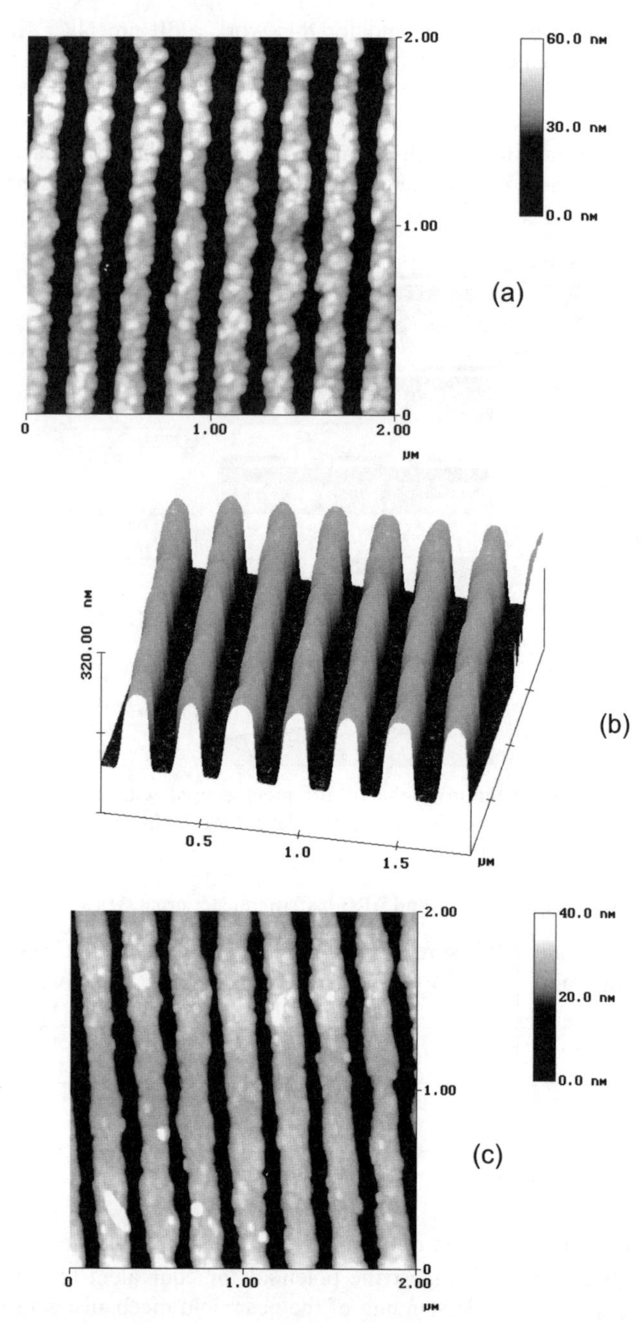

Fig. 17.10. AFM images of a 270 nm period grating: (a) in photoresist following exposure and development (note it is only 45 nm thick), (b) in ARC and SiO_x illustrating the improved profile following dry etching, (c) in 30 nm of NiCr following lift-off

17.7.1 Simulation Methods and Models

The two-dimensional multiple multipole program (MMP) (Hafner 1990) has been used to investigate diffraction in the evanescent near field of metallic gratings, suspended in various dielectric media. MMP is a semi-analytic technique, in that the electromagnetic fields are approximated by a set of basis functions that are solutions to Maxwell's equations. The program numerically sets the coefficients to these basis functions to minimise the errors for the boundary conditions at matching points distributed along the boundary. Once the coefficients have been determined a full vector solution can be found at any point. Finite-difference time domain (FDTD) techniques have also been used to study ENFOL and related techniques, and these are in agreement with the MMP results. Simulations are performed for chrome transmission gratings of pitch p and thickness t suspended in a medium with index of refraction n. Fig. 17.11 illustrates a typical geometry that has been simulated and defines the Cartesian coordinate system and illumination polarisation. In general only the TM polarisation is considered as for gratings of this scale very little light TE light is transmitted due to the polarising nature of the grating (McNab and Blaikie 2000). A unit cell consisting of one period of the grating is computed while specialised periodic boundaries take into account the effects of surrounding conductors to obtain a solution for effectively an infinitely long grating. Note that, an evanescent exposure is defined here as an exposure where all the transmitted diffracted orders are evanescent. The zeroth propagating order is also generally present but does not provide contrast for photolithography as it does not contain any modulation in the x direction; it simply appears as an intensity offset. For incident illumination normal to the grating an evanescent exposure occurs when $p < \lambda/n$, where λ/n is the effective wavelength.

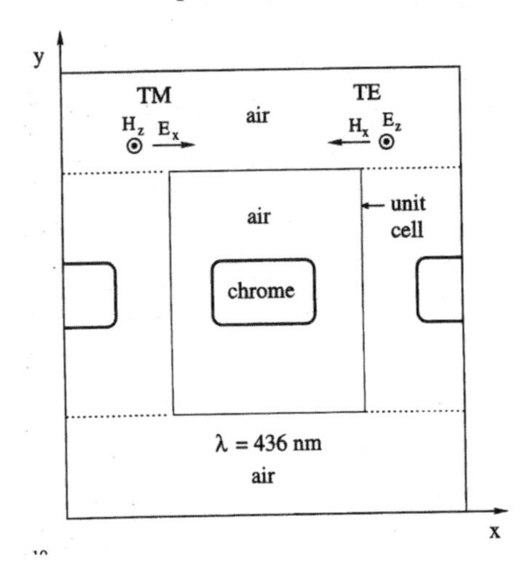

Fig. 17.11. Simulation model illustrating suspended chrome gratings mask to simulate exposure into a substrate held in intimate contact with the mask. The TE and TM polarisation orientations are indicated

17.7.2 Intensity Distribution

Fig. 17.12 shows the 2D normalised intensity $|E|^2/|E_{in}|^2$, where E_{in} is the incident electric field, for an evanescent exposure of a 140 nm pitch grating with Cr conductors (ε = -13.24 + *i*14.62), 40nm thick suspended in free space ($n = 1$).

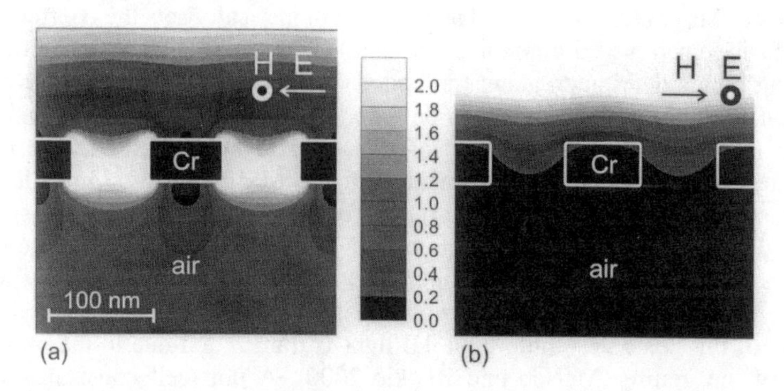

Fig.17.12. Normalised intensity distribution for a 140 nm pitch, 40 nm thick Cr grating in free space. The grating is illuminated from above by 436 nm TM polarised light (electric field E and magnetic field H directions are indicated). Contour plots of the normalised electric field intensity are shown $|E|^2/|E_{in}|^2$, where E_{in} is the incident electric field. The scale varies linearly from 0 (black) to 2.0 (white) in 10 linear steps

The grating is illuminated at normal incidence by TM polarised light with a wavelength λ = 436 nm. Line-plots of the intensity are presented in Fig. 17.13 for the same exposure conditions as in Fig. 17.13 extracted at the exit aperture of the grating, and then 10, 20 and 50 nm below the grating. These figures illustrate two of the characteristics of an ENFOL exposure – the decaying amplitude of the intensity as we move further below the grating, and strong edge enhancements that are evident close to the exit aperture of the grating, such as can be seen in Fig. 17.12(a).

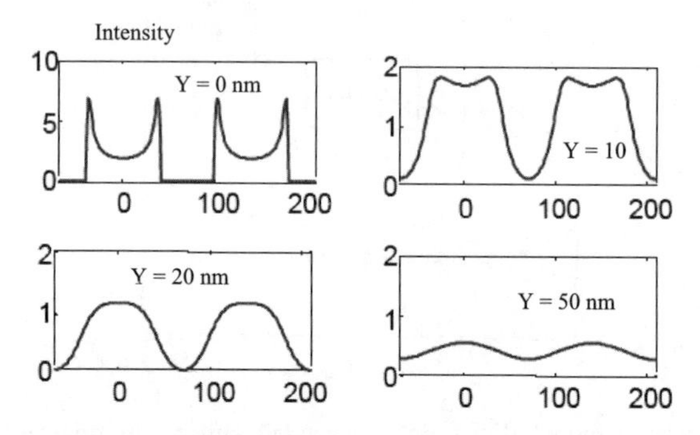

Fig. 17.13. Normalised intensity line plots for the simulation in Fig. 17.11. The line plots are taken at y = 0 (the exit plane of the grating), then at 10 nm, 20 nm, and 50 nm below the exit plane of the grating mask respectively

17.7.3 Depth of Field (DOF)

The depth of field (DOF) is a critical parameter for a lithography technique that relies on evanescent field components to expose photoresist. It defines the depth at which "sufficient" contrast is available for an exposure; but this is not a fixed value, it varies with different resist chemistries and different exposure conditions. We define the contrast k in the image at a distance y below the mask to be,

$$k(y) = \frac{I_{\max}(y) - I_{\min}(y)}{I_{\max}(y) + I_{\min}(y)} \tag{17.2}$$

W here I_{max} and I_{min} are the maximum and minimum intensities in the x-z pl-ane a distance y below the mask. We will define the depth of field, DOF_k, as the depth below the grating at which the contrast k falls below a specified value.

Simulations of a conducting Cr grating suspended in free space have been performed and Fig. 17.14 plots DOF_k versus the grating pitch for various k factors. These results show a linear relationship between grating pitch and depth of field, which is due to the near field nature of the exposure. In addition, the depth of field is greater for smaller k factors, as might be expected. This implies that the use of high contrast resists (which allow imaging at low k factors) will ease the ultra-thin-resist constraint imposed by the ENFOL technique. However, even for a contrast factor as low as $k = 0.3$, resist thicknesses below 100 nm are required to achieve resolution below 300 nm. The simulations also show that high contrast images are present in the near field regions of gratings with pitches as small as 20 nm, although in these cases the depth of field is less than 10 nm. This indicates the prospect for resolution of 20 nm pitch gratings in 3 nm of resist for $k = 0.5$. We believe that such resolution is experimentally achievable using new generation resist chemistries such as surface layer imaging resists (Herndon et al 1999) or self-assembled monolayer resists (Friebel et al 2000).

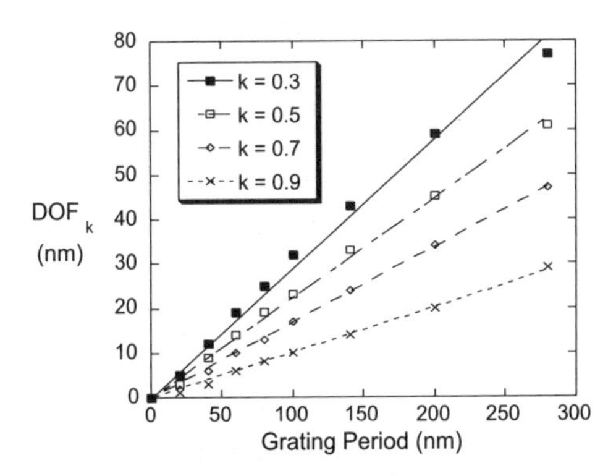

Fig. 17.14. Depth of Field (DOF_k) versus pitch p for simulated gratings plotted for k values of 0.3, 0.5, 0.7, and 0.9

The linear nature of the relationship between DOF_k and grating pitch can be qualitatively understood for an evanescent exposure by considering y^I_m, the depth at which each diffracted component decays to 1/e of its initial intensity (Loewen and Popov 1997),

$$y_m^I = \frac{1}{4\pi\sqrt{\frac{m^2}{p^2} - \frac{n^2}{\lambda^2}}} \tag{17.3}$$

Where, m is the diffracted order. From Eq. 17.3 it is evident that the higher the diffracted order, the faster its intensity decays. This has an impact on the exposure at the depth defined by DOF_k. For high values of k there exists a significant contribution from higher order ($m > 1$) evanescent components while at low k values there is generally only a significant contribution from the ±1 orders providing contrast. For gratings with deep sub-wavelength pitches $p \ll \lambda/n$, the $m = \pm 1$ components will dominate, and the linear relation $y^I_m \approx p/4\pi$ results from Eq. 17 3. Note that this is independent of λ and n, so we expect the resolution of ENFOL to be independent of these parameters for deep sub-wavelength gratings. This opens the prospect of choosing an exposure wavelength for optimal resist performance, by-passing the usually difficult task of designing resist performance for an optimal exposure wavelength. The choice of resist thickness is crucial for a successful exposure, firstly to obtain sufficient contrast and secondly to ensure adequate process latitude. It is clear from Fig. 17.14 that the resist thickness should be chosen according to the smallest feature pitch to be patterned. To improve process latitude, operation at a high k is preferable which requires reducing the resist thickness further for a given resist system.

Operating at low k values also increases the likelihood of exposure variation. As we move further away from the grating where the higher order evanescent components have disappeared, the intensity is dominated by the zeroth diffracted order, with transmission coefficient $T_0 = |E_0|^2/|E_{in}|^2$ where E_0 is the zeroth order electric field component. This is modulated by the contribution from the evanescent ±1 diffracted orders as shown in Fig. 17.12(d) for example. The exposure conditions are then heavily dependent on the magnitude of the T_0 component, which can fluctuates with factors such as changes in the duty cycle of the grating mask (% of conductor width to the grating pitch), as well as being sensitive to grating resonances. Variations in T_0 would make it difficult to expose gratings of different duty cycles in a single exposure. These effects are quantified in Fig. 17.15 for a 140 nm period grating suspended in air, in which $DOF_{0.5}$ and the maximum intensity at this depth are plotted. Increasing the duty cycle improves the DOF at the expense of exposure intensity. The improvement with larger duty cycles is due to the decreasing magnitude of the zeroth diffracted order relative to the other diffracted orders, which provide the contrast for the exposure. However, while an improvement in contrast is obtained, a longer exposure is required to compensate for a reduction in the intensity.

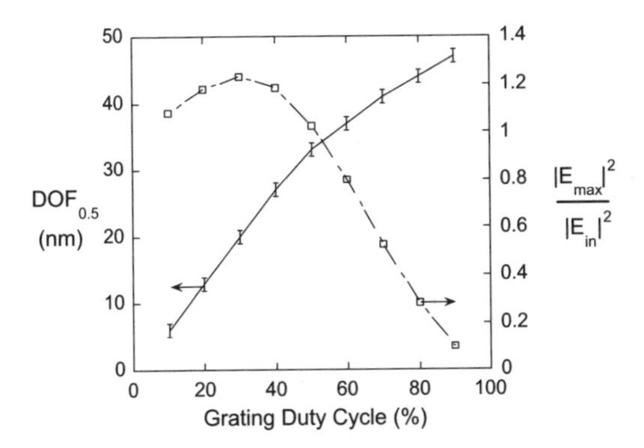

Fig. 17.15. Depth of Field $DOF_{0.5}$ for a 140 nm pitch grating (left axis) and the maximum intensity at this depth (right axis) versus duty cycle. The grating is suspended in free space and illuminated by 436 nm wavelength TM polarised light

17.7.4 Exposure Variations due to Edge Enhancements

Close to the exit aperture of the grating, high intensity enhancements occur at the grating conductor edges, as can be seen in Fig. 17.13(a) for example. These enhancements are due to the sharp discontinuity at the conductor corners, which encourages charge concentration in a manner analogous to a static electromagnetic problem. The static analogy becomes more reasonable as the grating period becomes much smaller than the wavelength. These enhancements are only evident with TM illumination as it is in the TM case that the sharp discontinuity in the x-component of electric field exists. The surface charge distributions that give rise to these field enhancements are known as surface plasmons – they give rise to electromagnetic fields with their maximum field at the surface and a characteristically exponential decay away from the surface (Raether 1998).

To investigate the effects of these edge enhancements simulations were carried out for a 140 nm pitch grating, in conditions identical to that of Fig. 17.11, except that the radius of curvature of the conductor corners was varied from 1 to 10 nm in steps of 1 nm. An increase in amplitude of the edge enhancements is evident as the radius is decreased, consistent with the static analogy. This is shown in Fig. 17.16, where we see the peak intensity at the exit aperture for a grating with a 1 nm radius is 25 times the incident intensity and 6 times greater than for a 10 nm radius. The relationship between the radius and peak intensity follows a power-law decay, as may be expected from a simple electrostatic analogy.

The edge enhancements are relatively short-range. In 20 nm the peaks become less intense in amplitude than at the centre of the aperture, and their decay is independent of the radii of curvature. In the far field an increase in T_0 is evident at the expense of other diffracted orders as the radii are increased, as is shown in Fig. 17.16. This acts to reduce the contrast and, over the range of radii simulated, the zeroth transmitted order increases by 4%. This suggests that for reliable lithography, maintaining as small a radius of curvature as possible on the mask edges is desirable. Line broadening also results from increasing the radii due to light leakage around the conductor corners and the creation of electric field components in the x direction over a broader region. Variability in the conductor profile during the production of the mask will therefore produce unpredictable duty cycles in the exposed resist.

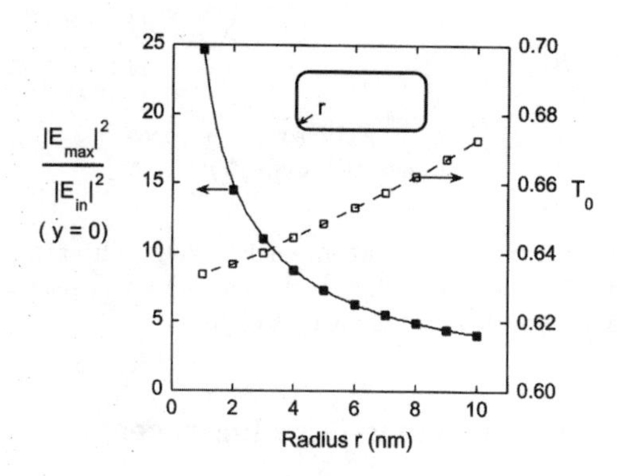

Fig. 17.16. Peak intensity at the grating exit aperture (left axis) and zeroth order intensity coefficient T_0 (right axis) versus the grating conductor radii of curvature r. The simulation model is the same as that in Fig. 17.11

ENFOL is also seen as a strong contender for printing onto curved surfaces due to the conformability of the mask. When printing onto curved surfaces, the light cannot be assumed to be collimated, and now a range of incident angles must be considered. Changes in the angle of incidence can result in placement biases in the lithography, but if the curvature of the substrate is known these can be accounted for in the design of the mask. The placement biases however will be susceptible to the conductor profile, particularly at angles varying greatly from normal incidence. Larger conductor radii can exaggerate the exposure broadening, making the mask profile more critical. To achieve acceptable tolerances for the lithography, the largest incident angle of light onto the surface to be printed needs to be determined which in turn specifies the control required on the conductor profile, for a particular photoresist thickness.

17.8 Nanolithography using Surface Plasmons

Surface plasmons have attracted renewed interest recently particularly because they can give rise to some interesting electromagnetic phenomena (www.surfaceplasmonoptics.org). Ebbesen *et al.* demonstrated high transmission through an array of sub-wavelength apertures due to coupling between surface plasmons on either side of the metallic film. This transmission exceeded the Bethe limit for single apertures (Bethe 1944) by several orders of magnitude. A large amount of work has since been carried out, and sophisticated sub-wavelength structures have been devised to enhance the throughput from *single* sub-wavelength apertures and to provide a highly collimated far-field radiation pattern from such apertures.

In addition, surface plasmons have been identified as having a very important role to play in designing new materials that exhibit a *negative index of refraction* (Shelby 2001). It has been predicted that a lens made from such material will have resolution beyond the conventional diffraction limit – the so-called 'Perfect Lens' (Pendry 2000). Such a material requires both negative permeability and negative permittivity, however a near-field demonstration of perfect lensing should be possible in a material with only one of these negative, such as a metal illuminated close to its plasma frequency. Recent theoretical work also suggests that metal nanowire composites could be used to realise negative refraction at optical wavelengths.

Surface plasmons play an important role in ENFOL, and by manipulating their generation and propagation some interesting new near-field lithography modes can be developed and explored (Blaikie 2001; Luo 2004). Some of these are reviewed here, although it should be noted that this field is very much in its infancy at this point in time.

17.8.1 Evanescent Interferometric Lithography (EIL)

The principle of evanescent interferometric lithography (EIL) is based on the nature of interference between evanescent diffracted orders, particularly close to a resonance of the grating that is illuminated with UV light (Blaikie 2001).

Fig. 17.17 compares the intensity distribution for two identical TM-illuminated chrome gratings. Both have a period of 270 nm and are imbedded in a medium with a refractive index $n = 1.6$. For this grating period and refractive index the cutoff for first order diffraction is at a wavelength of 432 nm, and this is the illumination wavelength chosen for Fig.17.16(a). In this case the ± 1 diffracted orders, which propagate at close to a grazing angle, interfere to form a period-halved intensity distribution beneath the mask. Zeroth order diffraction is suppressed (the so-called Wood's anomaly) which results in a high contrast in the interference pattern. This near field interference mimics that obtained in conventional interferometic lithography, and the period halving effect is useful for giving a 2× reduction of the grating.

In Fig. 17.17(b) the illumination wavelength has been increased to 438 nm, for which the ± 1 diffracted orders are now evanescent. No far field diffraction would be obtained for this grating, however a strong, high-contrast interference pattern is retained in the near field. The intensity and contrast in this interference pattern decay exponentially away from the mask; in this case the depth of field is more than 300 nm, which would be adequate for exposure of conventional resist layers. The most interesting point to note about the EIL exposure is the intensity enhancement that can be obtained. The peak near field intensity in Fig. 17.17(b) is more than five times the incident intensity, which would result in a significant lowering of exposure time compared with conventional interferometric lithography.

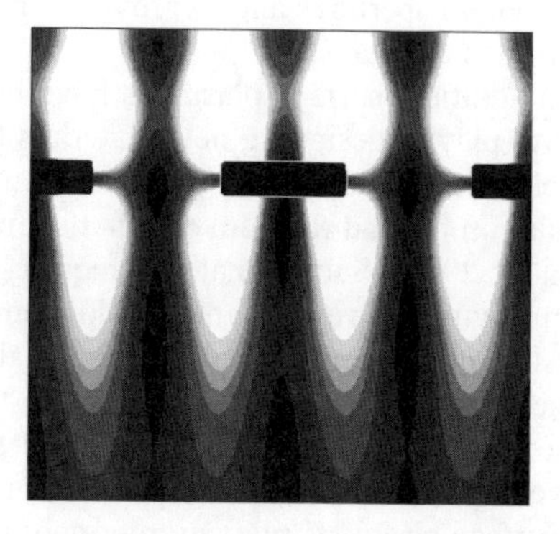

Fig. 17.17. Simulations of TM exposure through 280-nm period Cr gratings embedded in a dielectric material with refractive index n = 1.6. Exposure wavelengths are (a) 432 nm and (b) 438 nm

17.8.2 Planar Lens Lithography (PLL)

Following Pendry's prediction that a planar silver layer excited near its plasma frequency could produce a sub-diffraction-limited resolution image in the near field simulations and experiments have been performed to test this for near field lithography (Blaikie 2002; Melville 2004). The system that has been studied is shown in Fig. 17.18. Within a free-space background medium a metal grating of period p and thickness t_1 is illuminated from above with transverse magnetic (TM) polarised light of wavelength λ. At a distance f_1 beneath the exit plane of the grating there is a planar layer (PL) of thickness t_2. In a simplistic negative refraction picture the presence of this layer is predicted to cause an image of the mask pattern to be formed a distance $f_2 \approx f_1$ below its bottom surface (see ray diagram). Silver (Ag) is chosen for the metallic material as its complex permittivity, $\varepsilon = \varepsilon' + j\varepsilon''$, has a negative real part and small complex part for blue/UV wavelengths used for photolithography. For example, at $\lambda = 341$ nm its complex permittivity is $\varepsilon = -1 + 0.4$, (Johnson and Christy 1972) which is closely index-matched to the surrounding free-space medium. This is the illumination wavelength chosen for this work – tuning the illumination wavelength so that the layer is index-matched to some other dielectric medium will produce similar results.

Simulation results comparing the near field intensity distribution with and without a silver NFPL are shown in Fig. 17.19. Exposures are for a mask with period $p = 140$ nm, which is sub-diffraction limited. Fig. 17.19(a) shows the case with $f_1 = 20$ nm and $t_2 = 40$ nm. Two sub-diffraction-limited images of the mask are observed, the first in the center of the PL, and the second at a distance of $f_2 = 23$ nm below it. This second image could be used for nanolithography by placing a photoresist in this plane. This second image is superimposed on a background intensity that decays with distance below the PL, making the image difficult to discern. This arises because the mask period is sub-diffraction-limited, and the image is constructed from evanescent fields beneath the NFPL. Nonetheless, the peak intensity in the second image is 74% of the incident intensity, so exposure times for this new technique should be similar to those for conventional contact lithography with the same source and resist. The intensity distribution for the non-PL case shown in Fig. 17.19(b) shows no similar lensing effects. There is a region directly beneath the mask in which a shadow image forms, but the intensity and visibility of this image decays smoothly with distance. This simple near field intensity distribution has been used to obtain sub-diffraction-limited resolution, but there is a requirement of intimate mask-resist contact for higher resolution.

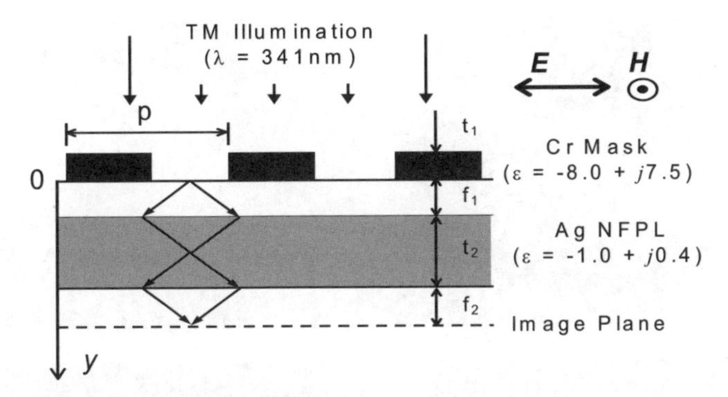

Fig. 17.18. Geometry & illumination condition for the near field planar lensing simulations

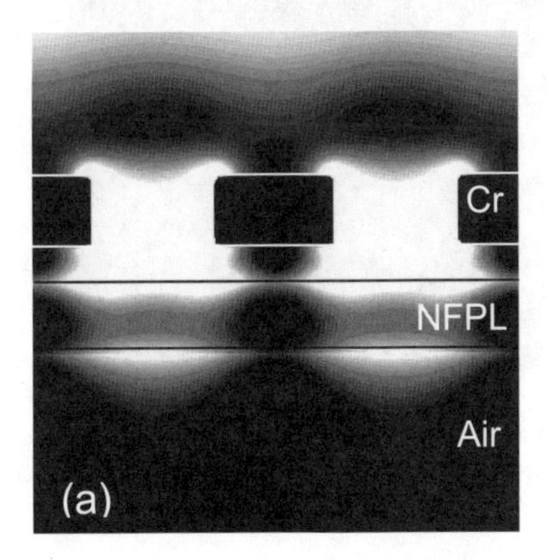

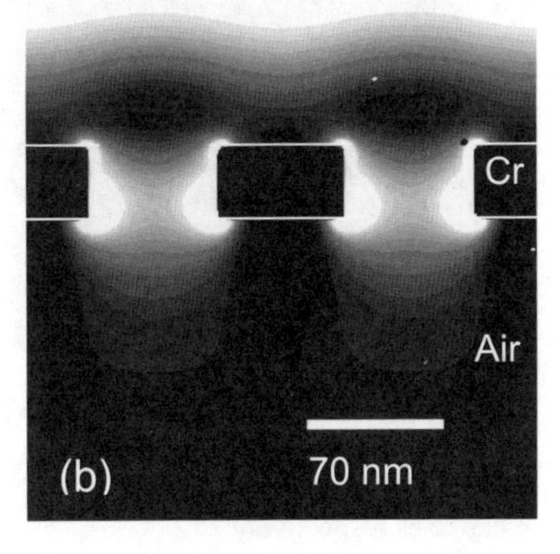

Fig. 17.19. Near field intensity profiles for a 140 nm period grating illuminated at 341 nm, (a) with a near field planar lens (NFPL) and (b) without a NFPL. The normalised intensity is plotted from 0 (black) to 2 (white) in linear steps of 0.1

Experimental verification of PLL has been achieved (Melville 2004), although not yet with sub-diffraction-limited resolution. Fig. 17.20 shows comparative AFM images of 1 μm period gratings exposed with and without the PL layer. The exposure time was 120 s and the development time was 8 s in both cases. These AFM images clearly show that the silver PL is effective in forming a near field imaging of the the mask, whereas in the proximity exposure all resolution for the 1 μm-period grating is lost. Whilst there is some degree of overexposure for the proximity exposure of Fig. 17.20(a), faithful reproduction of the mask object is not achieved even for shorter exposures (Melville 2004). This is the first experimental demonstration of near field imaging through a silver layer, and efforts are now underway to determine the resolution of the technique. By reducing the thickness of the silver and spacers layers improved resolution can be obtained, and sub-wavelength resolution has been achieved by using a 50 nm thick silver layer, as shown in Fig. 17.21. The fidelity of the micron-scale feature is much improved in this case, and resolution down to a period of 250 nm is evident.

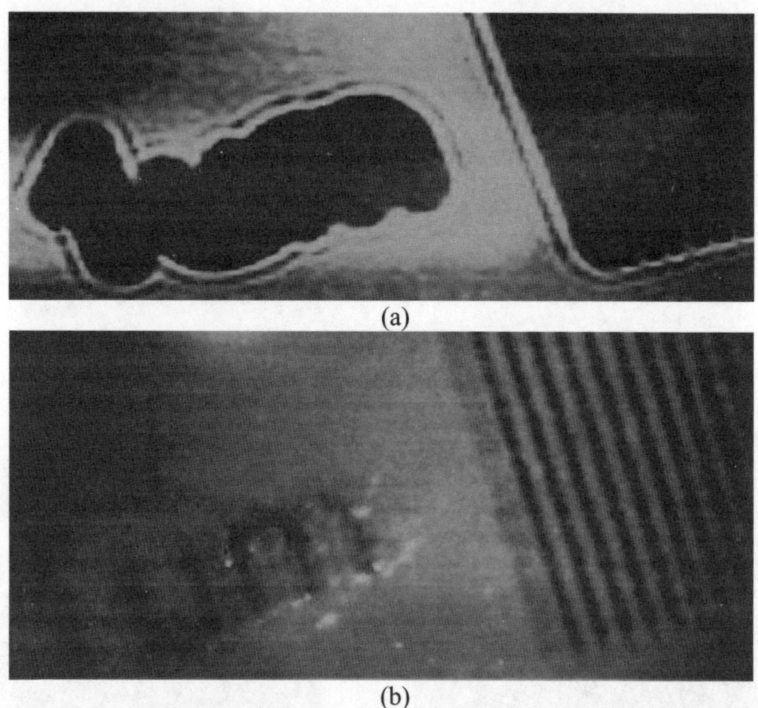

Fig. 17.20. Atomic force microscope (AFM) images of 1-micron period gratings exposed into 50 nm thick photoresist: (a) exposure performed with a 120 nm thick PMMA spacer; (b) exposure performed through a PMMA/silver/PMMA stack with 60nm/120nm/60nm thicknesses respectively

Simulations for the exposure geometries of Fig. 17.20 have been performed, and the results are shown in Fig. 17.22. In this case the FDTD method was used to perform these simulations. The simulation results for the proximity exposure are shown in Fig. 17.22(a), and it is clear that the pattern in the resist layer is of low contrast. There are some near field interference features apparent in this simulation, which may have been expected to give rise to discernable patterns in the developed resist. However the simulation is only for a single wavelength, and for the broadband source used for the exposure many such interference pattern will superimpose to produce a featureless final exposure pattern. For the PL exposure shown in Fig. 17.22(b) there is a much clearer image present in the resist layer, as observed in the experimental results. In this case the broadband nature of the exposure does not change the simulation results significantly, as the self-filtering nature of the silver PL only allows significant transmission at this wavelength.

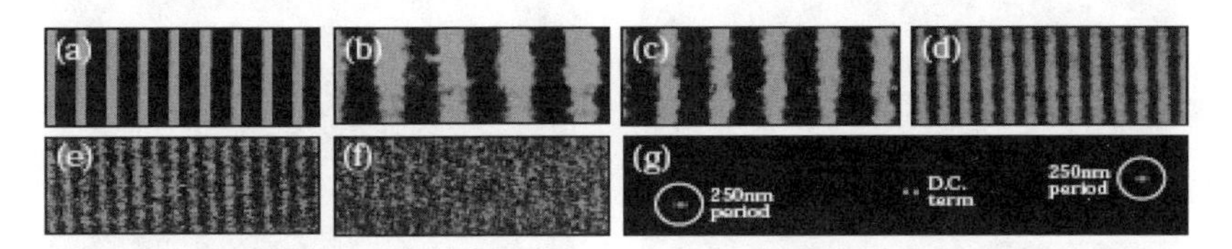

Fig. 17.21. AFM scans show resist profiles exposured through a 25 nm|50 nm|10 nm – PMMA|Ag|SiO_2 mask stack. Feature fidelity at (a) 1 μm, (b) 500 nm, (c) 420 nm, (d) 350 nm, (e) 290 nm, and (f) 250 nm periods are shown. Part (g) shows the two-dimensional Fourier transform of (f), confirming the presence of the 250 nm period features

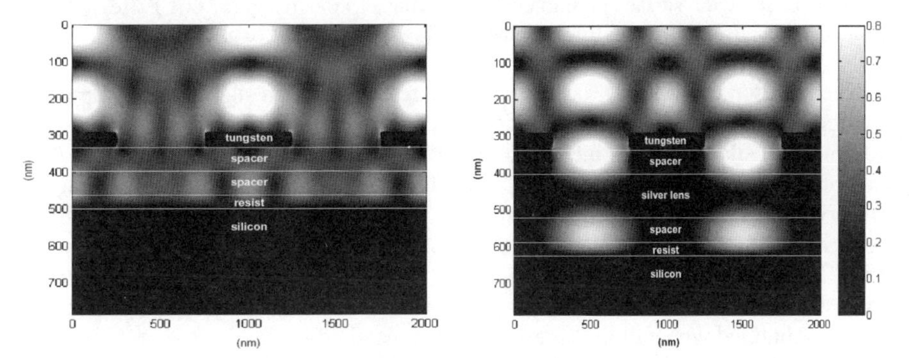

Fig. 17.22. Finite difference time domain (FDTD) simulations for the exposures shown in Fig. 17.20. The illumination wavelength is 341 nm in both cases and the spacer material is PMMA

17.8.3 Surface Plasmon Enhanced Contact Lithography (SPECL)

Another interesting effect was observed during the study of PLL. It was noted that the quality of the image *above* the planar lens could be enhanced by surface plasmons on the underlying metallic film, and this effect has been termed Surface Plasmon Enhanced Contact Lithography (SPECL). This is illustrated in Fig. 17.23, which compares the near field intensity profiles for two 140 nm period Cr gratings illuminated at 341 nm, embedded in a medium with refractive index of 1.6.

Fig. 17.23(a) shows the usual ENFOL case, where the material beneath the mask has a uniform refractive index (or is an index-matched stack). In this case the image has poor contrast at the exit plane of the mask, has a limited depth of focus, and that at depths greater than 80 nm beneath the mask an image reversal occurs. These effects reduce the process latitude for faithful patterning in this case.

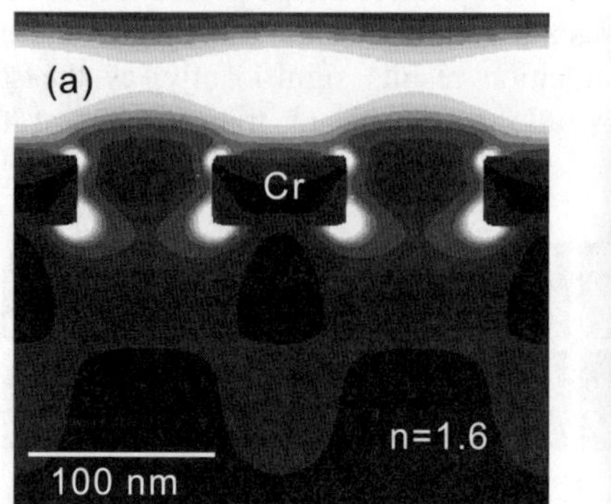

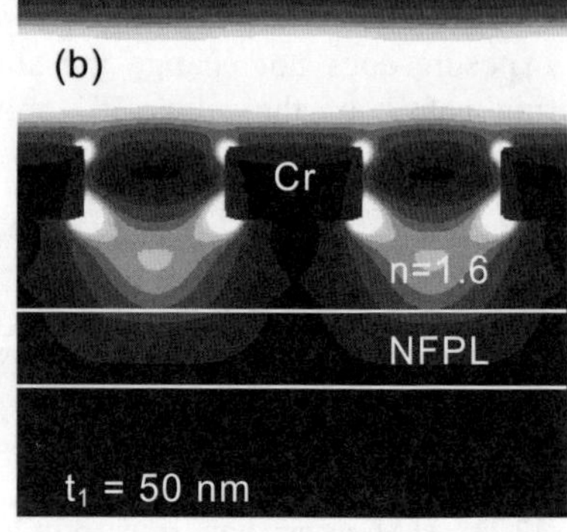

Fig. 17.23. Near field intensity profiles for a 140 nm pitch grating embedded in a medium with refractive index of 1.6 and illuminated at 341 nm: (a) with the entire underlying material index matched with n = 1.6; and (b) with a 40 nm thick silver layer 50nm beneath the mask to show the surface plasmon enhanced contact lithography (SPECL) effect. The normalised intensity is plotted from 0 (black) to 2 (white) in linear steps of 0.2

The addition of a SPECL layer within the resist stack can significantly improve the depth of field and process latitude, as is shown in Fig. 17.23(b). In this case a 40 nm thick silver layer is included 50 nm beneath the mask. The contrast at the exit plane of the mask is improved, and high contrast without image reversal is preserved throughout the entire resist layer. This improvement is attributed to the generation of surface plasmons on the underlying silver layer, which illuminate from beneath in addition to the incident illumination from above. This goes against conventional wisdom that dictates that the best resolution in a multilayer resist is obtained when all the layers are index matched.

The improvement in process latitude is shown further in Fig. 17.24, which plots expected pattern linewidth versus depth for the two intensity distributions shown in Fig. 17.23. These are extracted from iso-intensity contours at $I = 0.4I_0$, $0.5I_0$ and $0.6I_0$, a ±20% variation about the middle value. In the case of the index-matched substrate (Fig. 17.24(a)) the linewidth for the $I = 0.5I_0$ exposure increases by more than a factor of three between depths of 15 nm and 35 nm beneath the mask. This compares to the SPECL situation in which the linewidth increase is less than 20%. Variations in exposure intensity will also be less critical for SPECL. A 20% increase in exposure intensity will increase the linewidth 30 nm beneath the mask by more than 40% for the index matched substrate exposure, whereas the equivalent increase for SPECL is less than 20%. This is encouraging for the development of a stable and repeatable process.

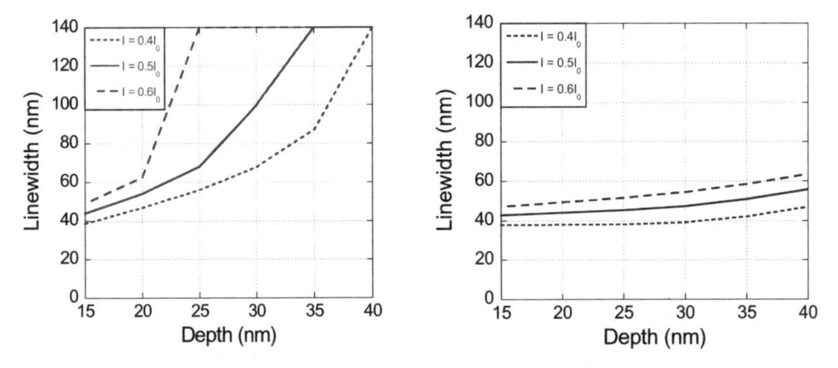

Fig. 17.24. Linewidth as a function of depth beneath the exit plane of the mask for the two intensity distributions shown in Fig. 17.23(a) index matched substrate; (b) a 40 nm thick silver layer 50 nm beneath the mask. Linewidths are shown for the iso-intensity contours $I = 0.4I_0$ (dotted), $0.5I_0$ (solid) and $0.6I_0$ (dashed)

17.9 Conclusions

In the optical near field region the resolution limits for imaging are well below those for the far field. We have performed experimental and simulation studies of the implications of this for optical lithography, and in our ENFOL technique resolution below on fifth of the wavelength can routinely be achieved. Simulations indicate that there is scope to improve this resolution by at least another factor of two, which would allow sub-50 nm patterning without the need to use expensive deep-ultraviolet light sources; all of the experimental work presented here has been achieved with standard, low-cost mercury lamps.

The presence of surface plasmons on the metallo-dielectric interfaces in ENFOL-like exposures also opens a number of interesting possibilities. Firstly, the high electric fields associated with near-field plasmonic light confinement reverse the usual constrain of low intensity for near field imaging, showing that exposure times need not be sacrificed for resolution in the near field. Secondly, interference between evanescent orders can be used for interferometric lithography (EIL), resulting in exposed features smaller than the mask features; so near-field lithography is not solely limited to 1× reproduction. Thirdly, the tuned plasmonic properties of a silver film exposed near the i-line of mercury can be used to 'project' a near field image (PLL); so the constraint of hard contact can potentially be relaxed. Finally, plasmonic effects in an underlying layer of a resist stack can also be used to improve the depth of focus. These techniques are still in their infancy, and further work is needed to explore their true potential.

17.10 References

Alkaisi MM, Blaikie RJ, McNab SJ, Cheung R, Cumming DRS (1999) Sub-diffraction-limited pattering using Evanescent near field optical lithography, Appl. Phys. Lett. 75:5360

Alkaisi MM, Blaikie RJ, McNab SJ (2000) 70 nm features on 140 nm period using evanescent near field optical lithography, Microelectronic Eng. 53:237

Alkaisi MM, Blaikie RJ, McNab SJ (2001) Nanolithography in the evanescent near field, Adv. Mater. 13:877

Alkaisi MM, Blaikie RJ, McNab SJ (1998) Nanolithography using wet etched silicon nitride phase masks, J. Vac. Sci. Technol. B 16:3929

Austin MD, Ge H, Wu W, Li M, Yu Z, Wasserman D, Lyon SA, Chou SY (2004) Fabrication of 5 nm linewidth and 14 nm pitch features by nanoimprint lithography, Appl. Phys. Lett. 84:5299

Baida FI, Labeke D (2003) Three-dimensional structures for enhanced transmission through a metallic film: Annular aperture arrays, Phys. Rev. B 67:155314

Bender M, Otto M, Hadam B, Vratzov B, Spangenberg B, Kurz H (2000) Fabrication of Nanostructures using a UV-based imprint technique, Microelectronic Eng. 53:233

Bethe HA (1944) Theory of diffraction by small holes, Phys. Rev. 66:163

Betzig E, Trautman JK, Harris TD, Weiner JS, Kostelak RL (1991) Breaking the diffraction barrier - optical microscopy on a nanometric scale, Science 251:1468

Blaikie RJ, Alkaisi MM, McNab SJ, Cumming DRS, Cheung R, Hasko DG (1999) Nanolithography using optical contact exposure in the evanescent near field, Microelectronic Eng. 46:85

Blaikie, RJ, McNab SJ (2001) Evanescent interferometric lithography, Appl. Opt. 40:1692

Blaikie RJ, McNab SJ (2002) Simulation study of ‘perfect lenses’ for near-field optical nanolithography, Microelecron. Eng. 61:97

Bouchiat V, Esteve D (1996) Lift-off lithography using an atomic force microscope, Appl. Phys. Lett. 69:398

Chou SY, Krauss PR, Renstrom PJ (1995) Imprint of sub-25nm vias and trenches in polymers, Appl. Phys. Lett., 67:3114

Chou SY, Krauss PR, Zhang W, Gou L, Zhuang L (1997) Sub-10nm imprint lithography and applications, J. Vac. Sci. Technol. B 5:2897

Davy S, Spajer M (1996) Near field optics: Snapshot of the field emitted by a nanosource using a photosensitive polymer, Appl. Phys. Lett. 69:3306

Ebbesen TW, Lezec HJ, Ghaemi HF, Thio T, Wolff PA (1998) Extraordinary optical transmission through sub-wavelength hole arrays, Nature 391:667

Fischer UC, Zingsheim HP (1981) Sub-microscopic pattern replication with visible-light, J. Vac. Sci. Technol. 19:881

Friebel S, Aizenberg J, Abad S, Wiltzius P (2000) Ultraviolet lithography of self-assembled monolayers for submicron patterned deposition, Appl. Phys. Lett. 77:2406

Gallatin GM, Houle FA, Cobb JL (2003) Statistical limitations of printing 50 and 80 nm contact holes by EUV lithography, J. Vac. Sci. Technol. B 21:3172

Goodberlet JG (2000) Patterning 100nm features using deep-ultraviolet contact photolithography, Appl. Phys Lett. 76:667

Goodberlet JG, Bryan LD (2000) Deep ultraviolet contact photolithography, Microelectronic Eng. 53:95

Goodberlet JG, Kavak H (2002) Patterning sub-50 nm features with near-field embedded-amplitude masks, Appl. Phys.Lett. 81:1315

HaefligerD A Stemmer, (2004) Ultramicroscopy, vol. 100, p 457

Hafner C (1990) The generalised multiple multipole technique for computational electromagnetic. Artech, Boston, Mass
Herndon MK, Collins RT, Hollingsworth RE, Larson PR, Johnson MB (1999) Near-field scanning optical nanolithography using amorphous silicon photoresists, Appl. Phys. Lett. 74:141
Hibbins AP, Sambles JR, Lawrence CR (2002) Gratingless enhanced microwave transmission through a subwavelength aperture in a thick metal plate, Appl. Phys. Lett. 81:4661
http://www.surfaceplasmonoptics.org
Johnson PB, Christy RW (1972) Optical constants of the noble metals, Phys. Rev. B 6:4370
Khang DY, Lee HH (2000) Room temperature imprint lithography by solvent vapour treatment, Appl. Phys. Lett., 76, 870
Li M, Chen L, Chou S (2001) Direct three-dimensional patterning using nanoimprint lithography, Appl. Phys. Lett. 78:3322
Li M, Chen L, Zhang W, Chou SY (2003) Pattern transfer fidelity of nanoimprint lithography on six-inch wafers, Nanotechnology 14:33
Levenson M (1993) Wavefront engineering for photolithography, Phys. Today 7:28
Lezec HJ, Degiron A, Devaux E, Linke RA, Martin-Moreno L, Garcia-Vidal FJ, Ebbesen TW (2002) Beaming light from a subwavelength aperture, Science 297:820
Loewen E, Popov E (1997) Diffraction gratings and applications. Marcel Dekker
Luo XG, Ishihara T (2004) Sub-100-nm photolithography based on plasmon resonance, Jpn. J. Appl. Phys. part1 6B:4017
Martin-Moreno L, Garcia FJ, Lezec HJ, Degiron A, Ebbesen TW (2003) Theory of highly directional emission from a single subwavelength aperture surrounded by surface corrugations, Phys. Rev. Lett. 90:167401
McNab SJ, Blaikie RJ, Alkaisi MM (2000) Analytical study of gratings patterned by evanescent near field optical lithography, J. Vac. Sci. Technol. B 18:2900
McNab SJ, Blaikie RJ (2000) Contrast in the evanescent near field of lambda/20 period gratings for photolithography, Appl. Opt. 39:20
Melville DOC, Blaikie RJ, Wolf CR (2004) Submicron imaging with a planar silver lens, Appl. Phys. Lett. 84:4403
Melville DOS, Blaikie RJ (2004) Near-field optical lithography using a planar silver lens, J. Vac. Sci. Technol. B 22:3470
Ono T, Esashi M (1998) Subwavelength pattern transfer by near field photolithography, Jpn. J. Appl. Phys. 37:6745
Paulus M, Schmid H, Michel B, Martin OJF (2001) Contrast mechanisms in high-resolution contact lithography: A comparative study, Microelectronic Eng. 57:109
Pendry JB (2000) Negative refraction makes a perfect lens, Phys. Rev. Lett. 85:3966
RaiChoudhury P (1997) (ed) Microlithography, micromachining, and microfabrication. SPIE Press, Washington, vol. 1, pp 31-33
Raether H (1998) Surface plasmons on smooth and rough surfaces and on gratings, vol. 111, Springer Tracts in Modern Physics. Springer-Verlag
Rogers JA, Paul KE, Whitesides GM (1997) Imaging the irradiance distribution in the optical near field, Appl. Phys. Lett. 71:3773
Rogers JA, Paul KE, Jackman RJ, Whitesides GM (1997) Using an elastomeric phase mask for sub-100nm photolithography in the optical near field, Appl. Phys. Lett. 70:2658
Schmid H, Biebuyck H, Michel B, Martin OJF, Piller NB (1998) Light coupling masks: an alternative lensless, approach to high resolution optical contact lithography, J. Vac. Sci. Technol B 16:3422

Shelby RA, Smith DR, Schultz S (2001) Experimental verification of a negative index of refraction, Science 292:77

Smith HI (1969) Methods for fabricating high frequency surface acoustic transducers, Rev. Sci. Instrum. B 40:729

Smith HI (1974) Fabrication techniques for surface acoustic wave and thin film optical devices. Proc. of the IEEE, vol. 62, p 1361

Smith HI, Efremow N, Kelly PL (1974) Photolithographic contact printing of 4000Å linewidth patterns, J. Electrochemical Soc. 121:1503

Thio T, Lezec HJ, Ebbesen TW, Pellerin KM, Lewen GD, Nahata A, Linke RA (2002) Giant optical transmission of sub-wavelength apertures: physics and applications, Nanotechnology 13:429

White JC, Craighead HG, Howard RE, Jackel RE, Behringer LD, Epworth RW, Henderson DHI, Sweeney JE (1984) Sub-micron, vacuum ultraviolet contact lithography with an F_2 excimer laser, Appl. Phys. Lett. 44:22

Yamashita H, Amemiya I, Takeuchi K, Masaoka H, Takahashi K, Ikeda A, Kuroki Y, Yamabe M (2003) Complementary exposure of 70 nm SoC devices in electron projection lithography, J. Vac. Sci.Technol. B 21:2645

18 Nanotechnology for Fuel Cell Applications

D. S. Mainardi[1] and N.P. Mahalik[2]

[1]Institute for Micromanufacturing, Louisiana Tech University
[2]Department of Mechatronics, Gwangju Institute of Science and Technology, Republic of South Korea

18.1 Current State of the Knowledge and Needs

A fuel cell is an electrochemical system that generates electricity by converting chemical energy into electrical energy. In general form, a fuel cell consists of an anode and a cathode separated by an electrolyte, which acts as an ion carrier. An oxidant is fed to the cathode to supply oxygen, while a fuel is fed to the anode to supply hydrogen. When hydrogen is used as fuel, it is electro-oxidised in the anode producing electrons and protons according to Eq.18.1.

$$2H_2 \rightarrow 4H^+ + 4e^- \qquad (18.1)$$

Electrons are transported through an external circuit to the cathode while protons go through the electrolyte. Finally, electrons and protons are recombined together with O_2 in the cathode to give H_2O (Eq.18.2).

$$O_2 + 4e^- + 4H^+ \rightarrow 2H_2O \qquad (18.2)$$

One of the best known and currently most promising fuel cells is the Proton-Exchange Membrane (PEM) fuel cell (Fig. 18.1), which has been proposed as an automotive power source due to its relatively light weight, low operating temperature (60°-100°C), high power density, relatively quick start-up, and rapid response to varying loads (Mehta et al. 2003). Also PEM fuel cells are considered as alternatives to replace batteries in portable electronic equipment such as laptop computers (Jung et al. 1998), cellular phones (Hockaday et al. 1999), and hand-held devices. Significant investment in the fuel cell technology is taking place resulting in considerable progress regarding their reliability, increased power density, improved manufacturing technologies, and materials cost reduction. Nevertheless, to meet the fuel cell's full application potential, there is still significant work to be done, since further cost reduction is required.

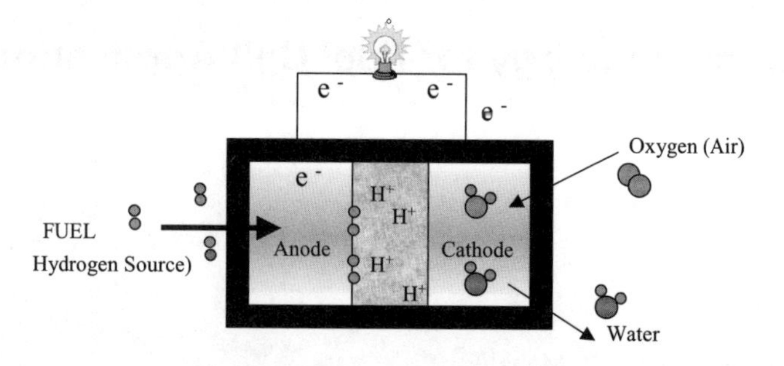

Fig. 18.1. Proton-Exchange Membrane (PEM) fuel cell

For a fuel cell to be effective, strong acidic or alkaline solutions and high temperatures and pressures are needed. Most fuel cells use platinum as catalyst (Adzic 1998; Markovic et al. 2002), which is expensive, limited in availability, and easily poisoned by carbon monoxide (CO), a by-product of many hydrogen production reactions in the fuel cell anodic chamber (Markovic et al. 2002). In PEM fuel cells, the type of fuel used dictates the appropriate type of anode catalyst needed. In this context, tolerance to CO is an important issue. It has been shown that the PEM fuel cell performance drops significantly with a CO concentration of only several parts per million, due to the strong chemisorption force of CO onto the anode catalyst (Mehta et al. 2003). For anode catalyst materials, platinum-based binary and tertiary platinum/ruthenium-based alloys seem to offer the best performance when CO poisoning is of concern (Toda et al. 1999; Neergat et al. 2001; Markovic et al. 2002).

Current fuel cell cathode catalysts are also made of platinum and platinum-based alloys, which exhibit higher catalytic activity for the oxygen electroreduction reaction (ORR) in pure acid electrolytes than pure platinum (Toda et al. 1999; Neergat et al. 2001). The enhanced electrocatalysis can be explained by an electronic factor, i.e., the change of the d-band vacancy in Pt upon alloying and/or by geometric effects such as the Pt coordination number and Pt-Pt distance (Toda et al. 1999). Both effects should enhance the reaction rate for oxygen adsorption and breaking of the O-O bond during the reduction reaction (Toda et al. 1999). Although, the ORR is enhanced by platinum-based catalysts, still the oxygen kinetics is significantly slow. The exchange current density of oxygen electroreduction on platinum is only about 1.0 nA/cm^2 in acidic electrolyte, which is six orders of magnitude lower than that of hydrogen oxidation (Markovic et al. 2002). This sluggish oxygen reduction reaction on the Pt catalyst causes a large overpotential at the cathode (Adzic 1998), requiring a much larger amount of Pt in the cathode than in the anode to achieve the kinetics requirements. In addition, when platinum-based catalysts are considered, further platinum reduction is needed due to its near-term availability and high cost. Hence, there is a need to enhance the oxygen electroreduction kinetics for which new or improved cathode catalysts must be designed. Numerous studies have been aimed toward reducing the amount of platinum required in current fuel cells where platinum-based nanoparticles are considered due to their unique properties.

18.2 Nanoparticles in Heterogeneous Catalysis

Nano-sized materials offer a large variety of potential applications, especially in the field of catalysis due to their characteristic high-surface areas and distinctive properties. Chemical reactions are known to take place on the surface of catalysts, which become more active towards a specific reaction as their surface areas increase. Hence, catalysts at the nanoscale (nanocatalysts) efficiently provide the way to make many chemical processes. Understanding their electronic and dynamic properties has become one of the most active areas in modern physical chemistry, and tremendous advances have been achieved due to the permanent interaction between theory and experiments for the past several decades.

In the field of heterogeneous catalysis metal nanoparticles have achieved good selectivity towards specific reactions due to their unusual chemical and physical properties (Aiken-III et al. 1999). The observed catalytic enhancement in metal alloys is determined by several variables such as their overall composition, preparation of the catalyst, and external conditions (Campbell 1990; Toshima et al. 1998; Aiken-III et al. 1999; Mainardi et al. 2001). Changes in these macroscopic variables and processes are responsible for changes at the micro and nanoscopic levels, which in turn define the reactivity behavior of the material. Proper control of the surface chemistry, geometry, and composition at the nano- and macroscales are needed when designing heterogeneous metal catalysts (Bazin et al. 2000). This effect is most efficiently accomplished by alloying two or three metals, a process that causes variations in shape, structure, and surface atomic distribution in the nanoclusters depending on their size, shape, and overall composition (Mainardi et al. 2001).

In many cases, the active components of dispersed metal catalysts are small nanoparticles with well-defined crystallographic surfaces, and therefore nanoparticle properties, rather than the bulk properties are responsible for the observed catalytic characteristics. In an attempt to reduce the amount of platinum used in fuel cells, small atomic clusters of platinum deposited on carbon supports were proposed (Bockris et al. 1993; Huang et al. 2003). For such supported nanocatalysts, metal-metal and metal-adsorbate interactions were observed to be key factors for controlling their activity and reactivity (Combe et al. 2000; Huang et al. 2002). Those interactions are responsible for the specific crystallographic surfaces that are formed on the overall nanoparticle surface where catalytic reactions take place. Hence, a major component in the formulation of a molecular basis to surface-chemical reactivity is the study of single crystal surfaces of catalytically active nanoparticles.

The interaction of oxygen with transition metal surfaces has been a topic of high interest in surface science due to its technological applications. In particular, the importance of platinum as a catalyst in many oxidation and reduction reactions has lead to numerous studies on the O_2/Pt (111) system (Puglia et al. 1995; Eichler et al. 1997; Adzic et al. 1998; Nolan et al. 1999; Mainardi et al. 2003). A significant amount of studies has been oriented towards the understanding of adsorption and reactions of molecular oxygen on Pt (111) surfaces (Mainardi et al.

2003) and clusters (Li et al. 2001). The Pt (111) surface consist of a two-dimensional hexagonal lattice and contains six adsorption sites per unit cell, three independent bridge, two hollow sites, and one top site (Fig. 18.2).

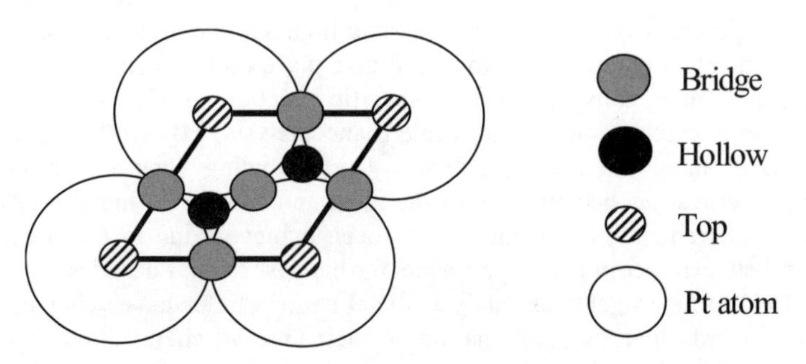

Fig. 18.2. Pt (111) unit cell containing 1-top, 3-bridge and 2-hollow adsorption sites

It is well established that oxygen adsorption on platinum occurs through different adsorption states, which have been thoroughly investigated using various spectroscopic techniques. Four adsorption states have been found for O_2/Pt (111) at various substrate temperatures using X-ray photoemission spectroscopy (XPS), ultraviolet photoemission (UPS), near edge X-ray absorption fine structure (NEXAFS), auto ionisation, and Auger electron spectroscopy (Puglia et al. 1995), and are summarised in Fig. 18.3.

Oxygen molecules were found to physisorb on the surface at about 25°K, and two molecular chemisorbed states were identified as precursors for the thermally activated dissociation process (Puglia et al. 1995). The first of these molecular states is weakly chemisorbed at 90°K and it is adsorbed in a bridge site with a saturation coverage of 0.23 (molecules per Pt surface atom). The second state, chemisorbed at 135°K is more strongly bonded to the Pt substrate and seems to be adsorbed in hollow or hollow-bridge sites at fractional coverages of 0.15 (Puglia et al. 1995). Moreover, adsorbed atomic oxygen has been detected on platinum threefold adsorption sites with saturation coverage of 0.25 prevailing at temperatures above 150°K.

Other authors studied oxygen adsorption and desorption on the Pt (111) and on stepped Pt(S)-12(111)x(111) surfaces over the 100 to 1300°K temperature range using thermal desorption (TD), Auger electron spectroscopy, XPS, and low energy electron diffraction (LEED) (Gland 1980). These authors have reported little or no dissociation of adsorbed molecular oxygen occurring at 100°K, and adsorption of oxygen above 170°K proceeding rapidly to a (2x2) structure (Gland 1980). Studying the kinetics and energetics of oxygen adsorption on Pt (111), the maximum O_2 desorption rate was found at ~140°K using TPD, isothermal desorption, Auger spectroscopy, LEED and isotopic measurements (Winkler et al. 1988). Similar findings regarding the existence of adsorbed oxygen states of oxygen was reported on the basis of EELS, UPS, and TD spectroscopy techniques over the temperature range 100°K to 1400°K (Gland et al. 1980).

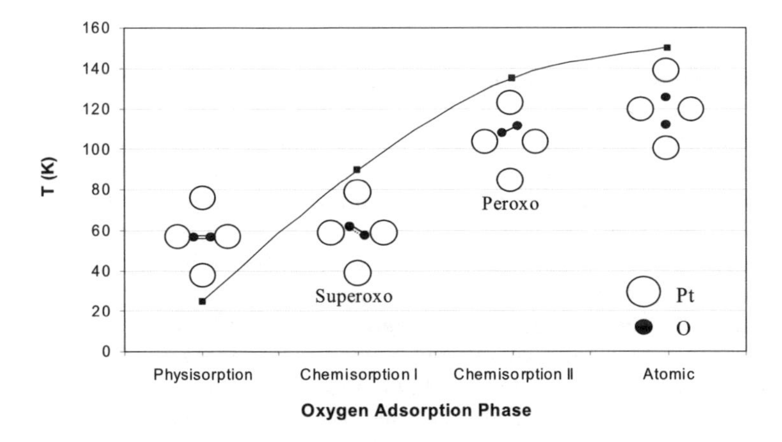

Fig. 18.3. O_2 adsorption states found on Pt (111) surfaces at various substrate temperatures

Microscopic details of the oxygen adsorption process on platinum surfaces have been obtained through *ab initio* studies from which two distinct, but energetically almost degenerate chemisorbed precursors were identified on Pt (111) (Fig. 18.3): a superoxo-like paramagnetic precursor formed at the bridge site (Eichler et al. 1997), with the molecule parallel to the surface, and a second peroxo-like non-magnetic precursor formed in the threefold hollow, with the atom slightly canted in a top-hollow-bridge geometry (Eichler et al. 1997). Cluster models provide the advantage that the local chemistry can be analysed in geometries that most closely resemble active sites. Studies on $Pt_{21}O_2$ indicated that the O_2 lying-down orientations are favored over the upright orientations, and chemisorption on the bridge site with the O-O axis along the bridge is the most stable configuration (Zhou et al. 1992). Density Functional Theory calculations on Pt clusters also showed the existence of peroxo and superoxo states and provided insights into the effect of an applied electric field on the geometries and energies of the chemisorption states (Li et al. 2001).

Experiments and *ab initio* calculations thus provide microscopic details about possible states of oxygen adsorption and dissociation. However, in order to understand the oxygen adsorption and reduction reactions in fuel cell cathode catalysts, studies on the time evolution of the rates of oxygen adsorption and desorption on platinum surfaces, and their dependence on pressure, temperature, and applied voltage are needed.

18.3 Oxygen Electroreduction Reaction on Carbon-Supported Platinum Catalysts

Over the past years, the oxygen reduction reaction (ORR) on platinum has been reviewed many times; however, there is not consensus yet about which electroreduction mechanism should prevail under given process conditions (Adzic 1998; Anderson 2002; Markovic et al. 2002). Numerous past and current studies have been published towards improving the understanding of this important electro-catalytic system (Adzic 1998; Anderson et al. 2000; Calvo et al. in press).

Due to the existence of kinetic, mass transport, and design limitations, fuel cells do not operate at equilibrium (Jarvi et al. 1998). The cathode reduction reactions are initiated at potentials lower than their equilibrium potential, and the oxidation reactions at the anode are initiated at potentials higher than their equilibrium potential. The shift in potentials at the cathode and anode are named cathodic and anodic overpotentials respectively. Because of these shifts, there is a decrease of the potential difference across the fuel cell. Therefore its electrical efficiency, defined as the actual potential difference divided by the equilibrium potential difference, is diminished (Jarvi et al. 1998). The standard reversible potential for O_2 reduction is 1.23 V on the hydrogen scale (Bard et al. 2001). However, due to the aforementioned effects the fuel cell cathode has a working potential of around 0.8 V (Anderson 2002), so that its overpotential is around 0.4 V (Adzic 1998). Several reasons may be responsible for this overpotential, such as the presence of adsorbed molecules that block the approach of O_2 molecules to the active surface sites where they can undergo reduction (Stamenkovic et al. 2001; Anderson 2002).

Due to their effect on the kinetics of fuel cell reactions, the study of the adsorption of electrolyte components on metal electrodes has become one of the major topics in surface electrochemistry. Nafion, a polymer electrolyte patented by Dupont, is frequently used as the electrolyte in fuel cell and electrolysis applications due to its excellent chemical and mechanical stability and high ionic conductivity (Delime et al. 1998). Typically, a thin layer of polymer electrolyte is applied to the catalyst surface by brushing a solution of polymer onto the electrode surface. Reduction of oxygen thus occurs at the interface electrode/electrolyte. Thus, the blockage of catalyst sites by adsorbed electrolyte components is important in explaining the substantial differences in the performance of platinum as catalyst for O_2 reduction in various electrolyte solutions.

Much interest has been focused on chemisorbed anions, such as bisulfate and halide, in order to elucidate the potential dependence of surface coverage by anions that are present in electrolyte solution (Markovic et al. 1999a; Stamenkovic et al. 2001). Experimental results indicate that bisulfate anions suppress the surface activity for O_2 reduction; however it does not affect the reaction pathway for the ORR on Pt (111), since hydrogen peroxide is not detected on the electrode in the potential range where bisulfate anions may be present on the catalyst surface. In contrast to the effect of bisulfate anion, peroxide formation is observed on Pt (111) modified in the presence of adsorbed bromide anion (Markovic et al. 1999a). Hence, the catalytic activity of the Pt (111) surface for the ORR is dramatically affected by adsorbed anions (Koper 1998; Markovic et al. 1999a; Stamenkovic et al. 2001), that can suppress both the adsorption of molecular oxygen and the availability of consecutive pairs of platinum sites involved in the breaking of the O-O bond. Moreover, the presence of adsorbed non-reactive species may even affect the reaction pathway for the ORR on Pt (111) contributing to both sluggish oxygen electroreduction, and poor catalyst utilisation. Inhibition of the oxygen reduction mechanism by oxides of Pt has been reported, and specific reduction paths such as the formation of hydrogen peroxide are usually associated with impurity adsorption. Markovic et al. observed a decrease in surface activity for O_2 reduction on Pt (111) in weak perchloric acid solution, concluding that OH(ads) formed on Pt (111) from the oxidation of water (Eq.18.3) is the cause for such phenomenon (Markovic et al. 1999b, 2002).

$$H_2O(ads) \rightarrow OH(ads) + e^- + H^+(aq) \tag{18.3}$$

These authors found that this reduction in surface activity occurs after the electrode potential is increased from 0.0 V to 0.4 V (vs. SHE) in the underpotential deposited (upd) hydrogen potential region, through the 0.4 V to 0.6 V double layer region, and finally, along the oxidative branch of the reversible butterfly region (Markovic et al. 1999b; Markovic et al. 1999c). Spectroscopy techniques also show evidence that OH(ads) is the product of the water oxidation reaction (Iwasita et al. 1996). Evidently, the mechanism of oxygen electroreduction is complex, and the lack of understanding of the factors that control the overpotential makes it difficult the intents of proposing alternative more efficient catalyst materials. In order to study the complete ORR process taking place at the cathode of the PEM fuel cell, information on the catalyst surface structure as well as the rates of the reactions and preferred adsorption sites of the species involved in such mechanism is needed. When electrocatalyst particles are in the nano-size regime, a relatively large fraction of metal atoms is exposed on a surface, sometimes causing unexpected metal-metal and metal-adsorbate interactions. These interactions clearly affect the equilibrium and the kinetics of growth and stabilisation of the nanoparticles and their effect on reactivity (Combe et al. 2000; Huang et al. 2002).

The effect of adsorbed electrolyte components on the surface of the catalysts, and metal-support interactions also play a significant role in the performance of the nanocatalyst (Horch et al. 1999). Regardless the tremendous efforts made in the past two decades in searching for highly active catalysts, current PEM fuel cell catalysts, which consist of platinum or its alloy nanoparticles supported on *carbon black*, still suffer from the following three major drawbacks, such as:

- Low resistance to carbon monoxide poisoning leading to rapid anode catalyst deactivation
- Sluggish O_2 electroreduction leading to high overpotential at the cathode
- Poor catalyst utilisation leading to high fuel cell costs

Hence, there is need for the development of highly active catalysts able to offer high resistance to carbon monoxide, increase the oxygen electroreduction kinetics, and overall lead to significant fuel cell cost reduction. Novel nanostructured supports able to enhance the performance of nanocatalysts for the oxygen electroreduction reaction may provide potential competitive catalysts for fuel cell applications.

18.4 Carbon Nanotubes as Catalyst Supports

Carbon nanotubes (CNTs) have extraordinary electrical, thermal, mechanical, and chemical properties (Dresselhaus et al. 2001). With the extremely high specific surface areas that are comparable to carbon black (CB), they may become promising PEM fuel cell catalyst supports (Li et al. 2003; Wang et al. 2004) substantially improving the Pt catalyst activity via metal-support interactions (Arico et al. 2003). Recent studies have shown that Pt nanoparticles supported on multi-walled CNTs (Pt/MWCNT) have much higher catalytic activity for oxygen electroreduction than Pt nanoparticles supported on CB (Pt/CB) (Li et al. 2003).

When nanoparticles are deposited on supports, their physical and chemical properties strongly depend not only on their particle size, but also their surface structure and the characteristics of the metal/substrate interface. Nanoparticle-support interactions can be fully understood by studying the morphology and structures of the supported nanoparticles. Applied quantum chemical methods (Koch et al. 2001), based on solutions to the Schrödinger equation, can provide the quantum-chemical basis for such studies. Additional information about the dynamics of nanoparticle-support interactions, surface structure, dynamics and thermodynamics of supported nanoparticles can be obtained using Molecular Dynamics (MD) methods (Allen et al. 1987). These methods do not explicitly include the presence of electrons; instead, effective force fields (Table 18.1) are used to represent the electronic effects. The physical and thermodynamic properties are obtained using the laws of classical statistical mechanics, which provide rigorous mathematical expressions that correlate macroscopic properties to the distribution and motion of atoms in a N-body system.

To simulate supported Pt nanoparticle morphologies and structures, preliminary studies were conducted for a Pt nanoparticle containing 74 atoms (average size ~2.5 nm) supported on a carbon slab (CB), and on a multi-walled carbon nanotube (MWCNT) using MD methods. The MWCNT consists of three concentric carbon nanotubes having diameters of 2, 4, and 6 nm respectively, and the CB consists of two parallel carbon layers in a volume of 8.3 x 8.3 x 1.4 nm, both corresponding to a total of 9,527 and 4,500 carbon atoms respectively. Simulations are carried out at constant temperature (300K) using a Berendsen thermostat. The relaxation time is 0.4 ps, small enough for the thermodynamic equilibrium to be reached, where the simulated system approaches the canonical ensemble (NVT) at constant number of atoms N, volume V and temperature T, without periodic boundary conditions (Allen et al. 1987). The equations of motion are integrated using the Verlet leapfrog method with a time step of 0.001 ps. The simulation length is 600 ps with 400 ps equilibration steps and 200 ps for the production stage. In order to reduce the computational load, a static substrate was considered in these simulations by fixing the positions of the C atoms.

The many-body Sutton-Chen (Sutton et al. 1990) and the classical 12-6 Lennard-Jones (LJ) potentials (Cramer 2002) were used to describe metal-metal, and metal-substrate and carbon-carbon interactions respectively. Although developed for bulk metals, the SC model has been applied to optimise the structure of transition metal clusters with good agreement with experimental observations (Doye et al. 1998). The SC potential (Eq.18.4) describes reliably static and dynamic properties of transition and noble metals, such as bulk module and elastic constants (Sutton et al. 1990), as well as surface energies, stress tensor components, and surface relaxation of fcc metals (Todd et al. 1993), and has been successfully applied for the description of nanoclusters (Doye et al. 1998; Huang et al. 2002; Huang et al. 2003). The potential energy (U_t) in the SC model is (Sutton et al. 1990) given by Eq.18.4.

$$U_t = \varepsilon\left[\frac{1}{2}\sum_{i\neq j}\sum\left(\frac{a}{r_{ij}}\right)^n - c\sum_i \rho_i^{1/2}\right] \quad (18.4)$$

With the first term on the right hand side representing the pair-wise repulsive potential and the second term the metallic binding energy with the local electron density given by (Eq.18.5), where r_{ij} is the distance between atoms *i* and *j,* c is a dimensionless parameter, ε is an energy parameter, and a is a length parameter.

$$\rho_i = \sum_{i\neq j}\left(\frac{a}{r_{ij}}\right)^m \quad (18.5)$$

The mixture parameters of the SC potential can be obtained from the Lorentz-Berthelot (Allen et al. 1987) mixing rules from those of the pure components, using the geometric and arithmetic means for the energy and length parameters.

Table 18.1. Force field parameters. Pt-Pt SC and C-C LJ parameters are taken from (Sutton et al. 1990) and (Bhetanabotla et al. 1990), respectively.

A	B	Field	ε (eV)	σ(Å)	a	n	m	c
Pt	Pt	SC	0.019833		3.92	10.0	8.0	34.408
C	C	LJ	0.002413	3.400				
Pt	C	LJ	0.040922	2.936				

Although accurate descriptions of metal-carbon substrate interactions are not currently available, the classical 12-6 Lennard-Jones (LJ) potential has been successfully used to describe the interactions of metallic clusters including platinum (Liem et al. 1995) and silver (Rafii-Tabar et al. 1997) with a graphite substrate. Hence, in these calculations the use of the LJ potential is well justified for the slab support. Even when the MWCNT support has a curved graphitic surface structure, its size (length and outer diameter) is large compared to the nanoparticle size thus minimizing curvature effects on the Pt nanoparticle morphology. Hence, using the LJ potential to describe metal-substrate interactions may be a good first approximation. Preliminary results at 300°K suggest that the Pt nanoparticle stable geometrical shape is faceted with well defined edges and angles for Pt/CB (Fig. 18.4(a)), while it is more spherical-like for Pt/MWCNT (Fig. 18.4(b)). These differences are attributed to a competition between metal-metal and metal support interactions in both cases (Fig. 18.4(a-b)).

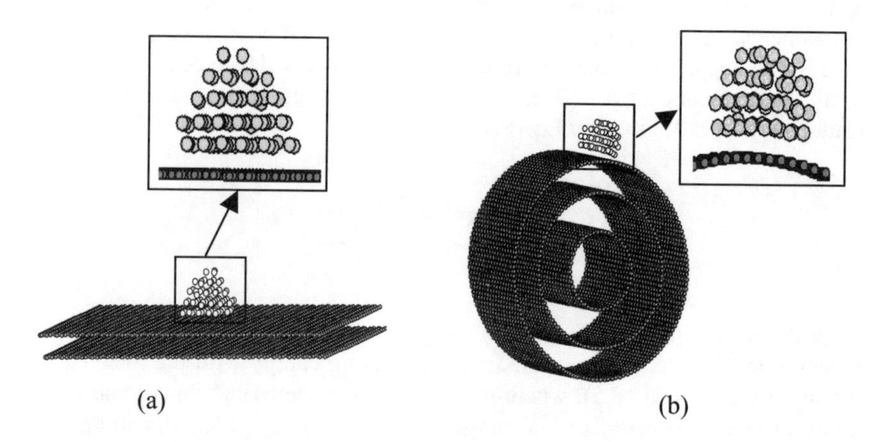

Fig. 18.4. Molecular Dynamics results at 300K corresponding to a 74-Pt atom nanoparticle supported on (a) carbon slab, and (b) MWCNT supports

It is interesting to note that these preferred arrangements of Pt atoms form 5-layer and 4-layer stacks on the CB and MWCNT respectively in the direction perpendicular to the supports (later referred as the z direction). The preferred layered arrangements can be quantified by computing the Pt density profiles, $\rho(z)$ (number of Pt atoms per unit length), for the nanoparticle on different supports as function of the distance to the substrate in the z direction (Fig. 18.5 and Fig. 18.6). These plots help in understanding how Pt atoms distribute inside the nanoparticle permitting the direct visualisation of its atomic structure.

Fig. 18.7 and Fig. 18.8 show the integrated Pt atomic population in the z direction starting from the layer next to the support to the complete nanoparticle (all layers). In addition, these figures also show the detailed layer-by-layer structure corresponding to the Pt layers parallel to the supports, starting from the layer next to it up to the topmost layer (L1), as well as the top view of the complete Pt nanoparticle.

The various Miller planes found on the overall nanoparticle surface suggest the type of crystallographic faces that are more exposed for catalysis applications (Fig. 18.9 and Fig. 18.10). Perfectly flat (111) and (100) fcc Miller planes are observed on the Pt/CB overall surface (Fig. 18.9). However, uneven (due to curvature effects) (111) and "mixed" (100)+(111) fcc Miller planes are found on the Pt/MWCNT surface (Fig. 18.10).

Different surface planes often display different catalytic activities. For instance, Pt (111) is known to be much more active than Pt (100) in the oxidation of CO and methanol, while Pt (100) is found to be more active for oxygen reduction. Platinum nanoparticles show (111) and (100) crystallographic flat or uneven planes on their overall surfaces irrespective of the nature of the carbon support.

"Mixture" (111) and (100) surfaces are found on the Pt nanoparticle supported on the MWCNT. However, it is still not known what catalytic effect this surface would have on the oxygen reduction reaction.

The results presented here aim to show clear differences on nanoparticles morphologies and structures depending upon the characteristics of the support. Nanoparticle surface characteristics, including exposed crystallographic phases, may play a key role on the catalytic activity and can be specially controlled and tailored by selecting a specific support.

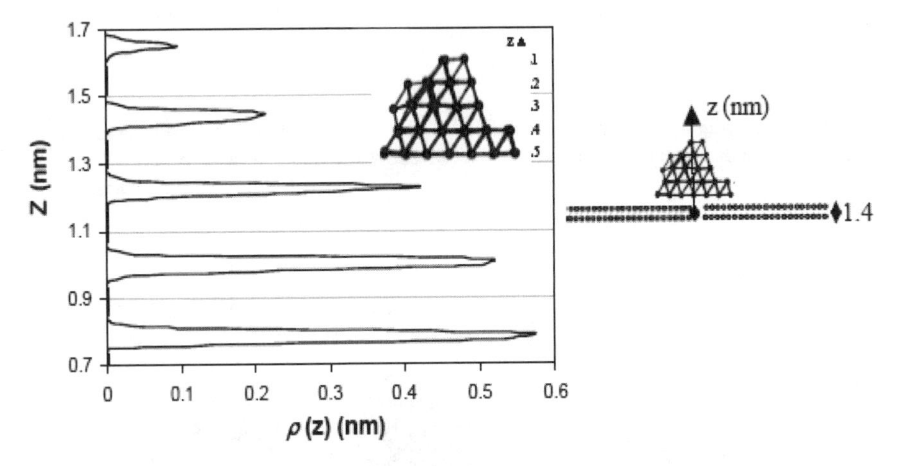

Fig. 18.5. Density profiles for a 74-Pt nanoparticle at 300K as a function of the distance to the substrate, measured in the z-direction, perpendicular to the carbon slab substrate (z = 0 at the center of CB)

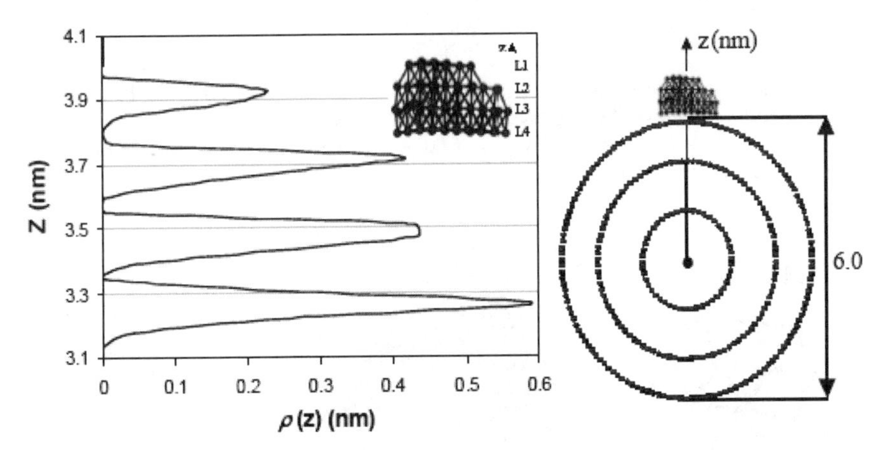

Fig. 18.6. Density profiles for a 74-Pt nanoparticle at 300K as a function of the distance to the substrate, measured in the z-direction, perpendicular to the multi-walled carbon nanotube substrate (z = 0 at the center of MWCNT)

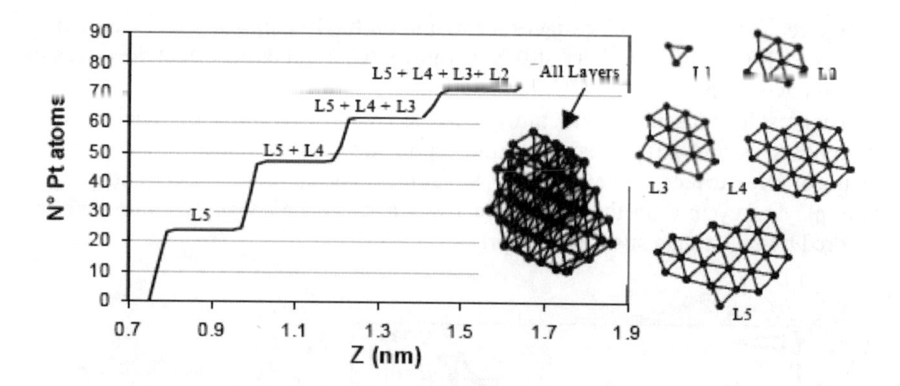

Fig. 18.7. Integrated Pt atomic population in the z direction for a 74-Pt nanoparticle supported on a carbon slab at 300 K. Snapshots show the detailed layer-by-layer in the z-direction (L1= topmost layer), as well as the top view of the complete nanoparticle (All layers)

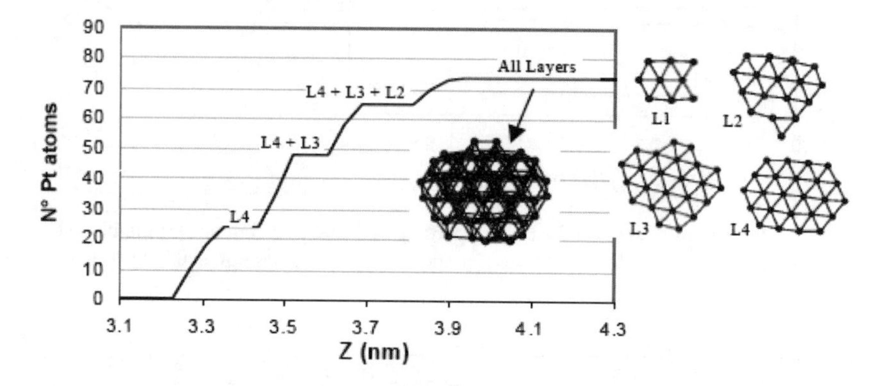

Fig. 18.8. Integrated Pt atomic population in the z direction for a 74-Pt nanoparticle supported on a multi-walled carbon nanotube at 300 K. Snapshots show the detailed layer-by-layer in the z-direction (L1= topmost layer), as well as the top view of the complete nanoparticle (All layers)

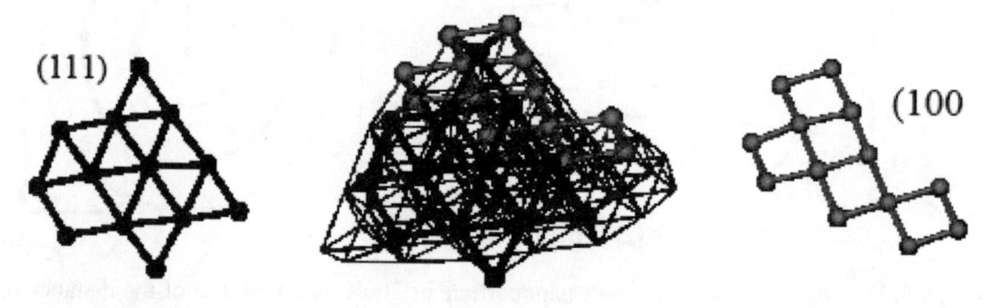

Fig. 18.9. Representative Miller planes found on the overall 74-Pt nanoparticle supported on a carbon slab

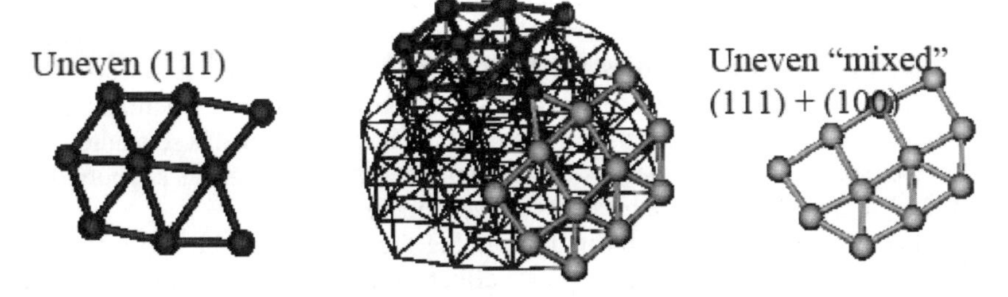

Fig. 18.10. Representative Miller planes found on the overall 74-Pt nanoparticle supported on a multi-walled carbon nanotube

18.5 Concluding Remarks

Fuel cells have been known for a long time and still are under development. Several challenges to widespread implementation of fuel cell technology are needed, although novel inexpensive and long-lasting electrocatalyst materials are major factors in design and development. In all fuel cell technologies, independent of the fuel, operating temperature, and the type of material used as the cathode electrocatalyst, the overpotential for the reduction of oxygen at operating currents is significantly high due to the slow oxygen electrochemical kinetics. As a result of this sluggish kinetics, cell voltage is decreased and therefore fuel cells loss in efficiency. Current fuel cell cathode catalysts are carbon-supported Pt or Pt-based alloys nanoparticles. While noble metal cost is less critical for applications in which small power sources are needed, platinum costs and supplies constraint for large-scale applications. The high overpotential for oxygen reduction, however, is a long-standing problem and, so far, research on the fundamental processes of oxygen reduction and catalysis has not yielded a breakthrough. Thus, new alternatives for finding cheaper materials with lower overpotential are needed.

Recent advances in the application of nanostructured carbon-based materials have suggested the possibility of using carbon nanotubes as novel electrocatalyst supports. Studies have shown that Pt nanoparticles supported on carbon nanotubes display remarkably higher electrocatalytic activity toward the reduction of oxygen than Pt nanoparticles supported on carbon black, which would contribute to substantial cost reduction in PEM fuel cells. The finite size of nanoscale materials such as carbon nanotubes positively influences the thermodynamics and kinetics of oxygen reduction due to their length scale and specific properties. Overall, their unique characteristics encourage the use of nanosize catalyst materials instead of their bulk counterparts to enhance the oxygen electroreduction performance. To gain a better understanding of the catalyst surface and mechanistic aspects of oxygen reduction by MWCNT (Multi-wall Carbon Nanotube) supported Pt nanoparticles, the effect of nanoparticle size and external factors, such as temperature, needs to be investigated. Thus, detailed studies of different cluster morphologies are needed to understand the preferred atomic arrangements (shape) as function of external conditions and characteristics of the MWCNT support. Additionally, in order to understand the connection of microscopic information with the time evolution of the rates of oxygen reduction reactions, their dependence on pressure and temperature needs to be elucidated. These studies will help elucidate the effect of electrode potential when Pt nanocatalysts supported on MWCNT are used as cathode catalysts in fuel cell devices.

Researchers need to focus on understanding the unique surface and interfaces of nanocomposites and how they affect the energetics, kinetics, and thermodynamics of oxygen kinetics. For this purpose, a combined experimental-theoretical approach would be appropriate to provide a new platform for the design of more selective and impurity-tolerant catalysts for the fuel cell technology.

18.6 References

Adzic RR (1998) Recent advances in the kinetics of oxygen reduction. In: Lipkowski J and Ross PN (ed) Electrocatalysis, Wiley-VCH, pp 19–242

Adzic RR, Wang JX (1998) Configuration and site of O_2 adsorption on the Pt (111) Electrode Surface, J. Phys. Chem. B 102:8988-8993

Aiken-III JD, Finke RG (1999) A review of modern transition-metal nanoclusters: Their synthesis, characterisation, and applications in catalysis, J. Mol. Cat. A: Chemical 145:1-44

Allen MP, Tildesley DJ (1987) Computer simulation of liquids. Oxford University Press

Anderson AB (2002) O_2 Reduction and co-oxidation at the Pt-electrolyte interface: The role of H_2O and OH adsorption bond strengths, Electrochim. Acta 47:3759-3763

Anderson AB, Albu TV (2000) Catalytic effect of Pt on O_2 reduction, an Ab initio model including electrode potential dependence, J. Electrochem. Soc. 147:4229–4238

Arico AS, Antonucci PL, Antonucci V (2003) Metal-support interaction in low-temperature fuel cell electrocatalysts. In: Wieckowski A, Savinova ER, Vayenas CG (eds) Catalysis and Electrocatalysis at Nanoparticle Surfaces, New York, Marcel Dekker

Bard AJ, Faulkner LR (2001) Electrochemical methods: Fundamentals and applications. New York

Bazin D, Mottet C, Treglia G (2000) New opportunities to understand heterogeneous catalysis processes on nanoscale bimetallic particles through synchrotron radiation and theoretical studies. Appl. Catal. A, vol. 200, pp 47-54

Bhetanabotla VR, Steele WA (1990) Computer simulation study of melting in dense oxygen layers on graphite, Phys. Rev. B 41:9480-9488

Bockris JOM, Khan SUM (1993) Surface electrochemistry: A molecular level approach. New York, Plenum Press

Calvo S, Mainardi DS, Jansen APJ, Lukkien JJ, Balbuena PB (in press) Test of a mechanism for O_2 electroreduction on Pt (111) via dynamic Monte Carlo simulations. In: Jungst RG, Weidner JW, Liaw BW and Nechev K (eds) Power Sources Modeling, Pennington, NJ

Campbell CT (1990) Bimetallic surface-chemistry, Annu. Rev. Phys. Chem. 41:775-837

Combe N, Jensen P, Pimpinelli A (2000) Changing shapes in the nanoworld. Phys. Rev. Lett. 85:110–113

Cramer CJ (2002) Essentials of computational chemistry: Theories and models. Great Britain, John Wiley & Sons Ltd

Delime FJ, Leger JM, Lamy C (1998) Optimisation of platinum dispersion in Pt-PEM Electrodes: Application to the electro-oxidation of ethanol, J. Appl. Electrochem. 28:27

Doye JPK, Wales DJ (1998) Global minima for transition metal clusters described by sutton-chen potentials, New J. Chem. 12:733-744

Dresselhaus MS, Dresselhaus G Avouris P (Eds.) (2001) Carbon nanotubes: Synthesis, structure, properties and applications. Springer-Verlag, New York

Eichler A, Hafner J (1997) Molecular precursors in the dissociative adsorption of O_2 on Pt (111), Phys. Rev. Lett. 79:4481–4488

Gland JL (1980) Molecular and atomic adsorption of O_2 on the Pt (111) and Pt (S)-12 (111) X (111) surfaces, Surf. Sci. 93: 487-514

Gland JL, Sexton BL, Fisher GB (1980) Oxygen interactions with the Pt (111) surface, Surf. Sci. 95:587-602

Hockaday R, Navas C (1999) Micro-fuel cells for portable electronics, Fuel cells 10:9-12

Horch S, Lorensen HT, Helveg S, Laegsgaard E, Stensgaard I, Jacobsen KW, Norskov JK, Besenbecher F (1999) Enhancement of surface self-diffusion of platinum atoms by adsorbed hydrogen, Nature 398:134-136

Huang SP, Balbuena PB (2002) Platinum nanoclusters on graphite substrates: A molecular dynamics study, Mol. Phys. 100:2165-2174

Huang SP, Mainardi DS, Balbuena PB (2003) Structure and dynamics of graphite-supported bimetallic nanoclusters, Surf. Sci. 545:163-179

Iwasita T, Xia X (1996) Adsorption of water at Pt (111) electrode in $HclO_4$ solutions: The potential of zero charge, J. Electroanal. Chem. 411:95-102

Jarvi TD, Stuve EM (1998) Fundamental aspects of vacuum and electrocatalytic reactions of methanol and formic acid on platinum surfaces. In: Lipkowski J and Ross PN (eds) Electrocatalysis, Wiley-VCH, Inc. pp 75-153

Jung DH, Lee CH, Kim CS, Shin DR (1998) Performance of a direct methanol polymer electrolyte fuel cell, J. Power Sources 71:169-173

Koch W, Holthausen MC (2001) A chemist's guide to density functional theory. Wiley-CVH

Koper M, (1998) A lattice-gas model for halide adsorption on single-crystal electrodes, J. Electroanal. Chem. 450:189-201

Li T, Balbuena PB (2001) Computational studies of the interactions of O_2 with Pt clusters, J. Phys. Chem. B 105:9943-9952

Li W, Liang C, Zhou W, Qiu J, Zhou Z, Han H, Wei Z, Sun G, Xin Q (2003) Preparation and characterisation of multi-walled carbon nanotube-supported Pt for cathode catalysts of direct methanol fuel cells, J. Phys. Chem. B 107:6292-6299

Liem SY, Chan KY (1995) Simulation study of platinum adsorption on graphite using the sutton-chen potential, Surf. Sci. 328:119-128

Mainardi DS, Balbuena PB (2001) Surface segregation in bimetallic nanoclusters: geometric and thermodynamic effects, Int, J. Quant. Chem. 85:580-591

Mainardi DS, Calvo S, Balbuena SB, Jansen APJ, Lukkien JJ (2003) Dynamic Monte Carlo simulations of O_2 adsorption and reaction on Pt (111), Chem. Phys. Lett. 382:553-560

Markovic NM, Gasteiger HA, Grgur N, Ross PN (1999a) Oxygen reduction reaction on Pt(111): Effects of bromide, J. Electroanal. Chem. 467:157-163

Markovic NM, Ross PN (1999b) Electrocatalysis at well-defined surfaces: Kinetics of oxygen reduction and hydrogen oxidation/evolution on Pt (hkl) electrodes. In: Wieckowski A (ed) Interfacial Electrochemistry: Theory, experiment and applications, New York, Marcel Dekker, pp 821-841

Markovic NM, Ross PN (2002) Surface science studies of model fuel cells electrocatalysts. Surf. Sci. Rep. 286: 1 - 113

Markovic NM, Schmidt TJ, Grgur BN, Gasteiger HA, Behm RJ, Ross PN (1999c) Effect of temperature on surface processes at the Pt (111)-liquid interface: Hydrogen adsorption, oxide formation, and CO oxidation, J. Phys. Chem. B 103:8568-8577

Mehta V, Cooper JS (2003) Review and analysis of PEM fuel cell design and Manufacturing, J. Power Sources 114:32-53

Neergat N, Shukla AK, Gandhi KS (2001) Platinum-based alloys as oxygen-reduction catalysts for solid-polymer-electrolyte direct methanol fuel cells, J. Appl. Electrochem. 31(4):373-378

Nolan PD, Lutz BR, Tanaka PL, Davis JE, Mullins CB (1999) Molecularly chemisorbed intermediates to oxygen adsorption on Pt (111): A molecular beam and electron energy-loss spectroscopy study, J. Chem. Phys. 111:3696-3704

Puglia C, Nilsson A, Hernnas B, Karis O, Bennich P, Martensson N (1995) Physisorbed, chemisorbed and dissociated O_2 on Pt (111) studied by different core level spectroscopy methods, Surf. Sci. 342:119

Rafii-Tabar H, Kamiyama H, Cross M (1997) Molecular dynamics simulation of adsorption of Ag particles on a graphite substrate, Surf. Sci. 385:187-19

Stamenkovic V, Markovic NM, Ross PN (2001) Structure-relationships in electrocatalysis: oxygen reduction and hydrogen oxidation reactions on Pt (111) and Pt (100) in solutions containing chloride ions, J. Electroanal. Chem. 500:44-51

Sutton AP, Chen C (1990) Long-range finnis-sinclair potentials. Phil. Mag. Lett. 61:139-146

Toda T, Igarashi H, Uchida H, Watanabe M (1999) Enhancement of the electroreduction of oxygen on Pt alloys with Fe, Ni, and Co, J. Electrochem. Soc. 146:3750-3756

Todd BD, Lynden-Bell RM (1993) Surface and bulk properties of metals modeled with sutton-chen potentials, Surf. Sci. 281:191-206

Toshima N, Yonezawa T (1998) Bimetallic nanoparticles-novel materials for chemical and physical applications, New. J. Chem.:1179-1201

Wang C, Waje CM, Wang X, Tang JM, Haddon RC, Yan Y (2004) Proton exchange membrane fuel cells with carbon nanotube based electrodes. Nano Lett. 4:345-348

Winkler A, Guo AX, Siddiqui HR, Hagans PL, Yates JT (1988) Kinetics and energetics of oxygen adsorption on Pt (111) and Pt (112): A comparison of flat and stepped surfaces, Surf. Sci. 201:419

Zhou R, Cao P (1992) Molecular cluster analysis of O_2 adsorption and dissociation on Pt (111), Phys. Lett. A 169:167

19 Derivatisation of Carbon Nanotubes with Amines: A Solvent-free Technique

V. A. Basiuk[1] and E. V. Basiuk[2]

[1]Instituto de Ciencias Nucleares, UNAM, Mexico
[2]Centro de Ciencias Aplicadas y Desarrollo Tecnológico, UNAM, Mexico

19.1 Introduction

A very low solubility of carbon nanotubes (CNTs) hampers many of their potential applications. Especially this is true for multi-walled carbon nanotubes (MWNTs). One of the most efficient ways to increase CNT solubility and dispersibility is their covalent derivatisation (Bahr and Tour, 2002; Basiuk and Basiuk, 2004; Hirsch, 2002; Sun et al. 2002). All known types of chemical reactions employed for this purpose up to now can be divided into two main groups: (a) defect-group derivatisation; and (b) covalent sidewall derivatisation. The first group of reactions involve oxygenated functionalities (mainly carboxylic groups) formed on CNT tips, as well as on their sidewalls to some degree, as a result of oxidative treatment with strong mineral acids (Bahr and Tour 2002; Basiuk and Basiuk 2004; Chen et al. 1998; Hirsch 2002; Jia et al. 1999; Liu et al. 1998; Sun et al. 2002). Among these, the most extensively explored is the formation of amide derivatives between carboxylic groups on oxidised CNT tips and long-chain amines (Ausman et al. 2000; Hamon et al. 1999; Wong et al. 1998a,b). Traditionally, the reaction is performed through chemical activation of the carboxylic groups with thionyl chloride or carbodiimides in an organic solvent medium. For example, in the case of thionyl chloride and commonly used octadecylamine, the reaction can be schematically represented as follows:

$$\mathrm{CNT{-}COOH} + \mathrm{SOCl_2} \rightarrow \mathrm{CNT{-}COCl} \xrightarrow{\mathrm{H_2N(CH_2)_{17}CH_3}} \mathrm{CNT{-}CONH(CH_2)_{17}CH_3}$$

While this reaction by itself is efficient and fast, such inevitable auxiliary operations as CNT filtering and centrifuging are not. And in addition, the use of organic solvents produces harmful wastes and cannot be particularly welcomed. From common considerations, for the functionalisation of CNTs one can use the same chemical approaches that have been developed for other poorly soluble inorganic materials, for example silica materials. A decade ago we performed systematic studies on the use of the gas-phase chemical derivatisation for the synthesis of chemically modified silicas, mainly for liquid chromatographic applications (Basiuk and Chuiko 1990, 1993; Basiuk and Khil'chevskaya 1991; Basyuk 1991; Basyuk and Chuiko 1990). Among the derivatising reagents tested were polyazamacrocyclic ligands (Basyuk 1991; Basyuk and Chuiko 1990), pyrimidine bases (Basiuk and Chuiko 1993; Basyuk 1991), and solid carboxylic acids (Basiuk and Chuiko, 1990; Basiuk and Khil'chevskaya, 1991). Most of those compounds are poorly volatile under ambient temperature and pressure. Nevertheless, decreasing the pressure to a moderate vacuum and, on the other hand, increasing the temperature to >150ºC provided efficient formation of the chemically bonded surface derivatives. In particular, the reaction between silica-bonded aminoalkyl groups and vaporous carboxylic acids to form surface amides proceeded smoothly at 150–180ºC *without* chemical activation of the carboxylic groups, it was relatively fast (0.5–1 h), and provided high yields of the amide derivatives (>50% based on the starting surface concentration of aminoalkyl groups). Excess derivatising reagent was spontaneously removed from the reaction zone. In addition, there was no need to use an (organic) solvent medium: this feature is attractive not only from an ecological point of view, but also in that it helps to avoid undesirable particle aggregation of the material derivatised.

Bearing the above advantages of the gas-phase derivatisation in mind, we systematically worked on the development of a similar procedure for chemical derivatisation of single-walled CNTs (SWNTs) and MWNTs with amines.

19.2 Experimental Design

Experimental setup for the gas-phase derivatisation procedure can be variable, but should necessary contain the following basic elements: (1) a pump capable of producing vacuum of 10^{-1}-10^{-2} Torr; (2) a vessel serving as reactor; (3) a heating element such as heating mantle.

In most of our experiments we employed the custom-made Pyrex glassware shown in Fig. 19.1 (Basiuk et al. 2002a). In a typical experiment, 100 mg of CNTs were placed into the bigger reactor 11. To remove volatile contaminants (adsorbed from the environment) from the nanotube material, the reactor was pumped out to a vacuum of ca. 10^{-2} Torr (valves 2 and 5 open; 1, 3 and 4 closed), and its bottom was heated for 0.5 h at 100-120ºC by means of heating mantle 13. Then the reactor was cooled, open and ca. 50 mg of amine was dropped to the bottom containing CNTs. After pumping the reactor out to ca. 1 Torr at room temperature, its valve 2 was closed, and the bottom was heated at 150-170ºC for 1-2 h. During this procedure, amine evaporated, reacted with CNTs, and its excess condensed a few centimeters above the heating mantle. The high derivatisation temperature not only facilitates the reaction, but also helps to minimise the amount of amine physically adsorbed on the nanotube material. After finishing the procedure, valve 2 was open again for 15-60 min to remove volatiles. Then the heating mantle was removed, the reactor cooled and disconnected from the manifold. Before taking the derivatised nanotubes out, the upper reactor part with condensed excess amine on the wall was wiped with cotton wool wet with ethanol. For milligram-scale derivatisation procedures, smaller reactor 12 can be used in a similar way.

19.3 Direct Amidation of Carboxylic Functionalities on Oxidised SWNT Tips

As it was mentioned above, the most extensively explored approach to the synthesis of amide derivatives between carboxylic groups on oxidised CNT tips and long-chain amines relies upon preliminary chemical activation of the carboxylic groups with thionyl chloride or carbodiimides in an organic solvent medium, followed by the addition of amine reagent (Ausman et al. 2000; Bahr and Tour, 2002; Basiuk and Basiuk 2004; Hamon et al. 1999; Hirsch 2002; Sun et al. 2002; Wong et al. 1998a,b). In our approach (Basiuk et al. 2002a,b), we attempted to substitute the *chemical* activation with *thermal* activation of terminal carboxylic groups, in other words, to synthesise amide derivatives directly according to the following general scheme:

$$\text{SWNT—COOH} + \text{HNR}^1\text{R}^2 \rightarrow \text{SWNT—CO—NR}^1\text{R}^2 + \text{H}_2\text{O}$$

where HNR^1R^2 is an aliphatic amine (nonylamine, dipentylamine, dodecylamine, octadecylamine, 4-phenyl butylamine, etc.).

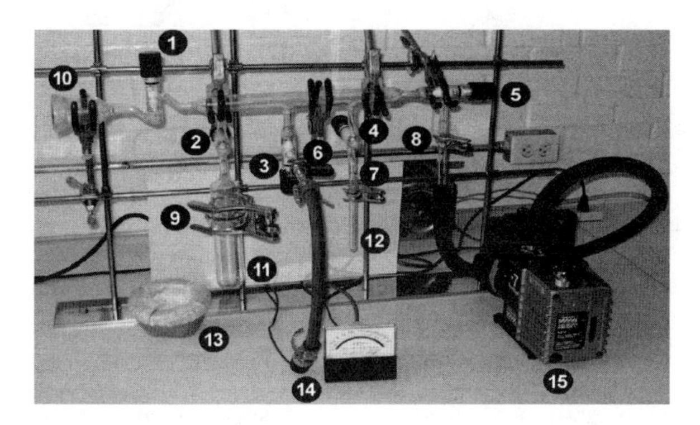

Fig. 19.1. Vacuum manifold designed for the gas-phase derivatisation of CNTs with amines. (1-5) Teflon valves; (6-8) 10-mm I.D. O-ring joints; (9,10) 41.4-mm I.D. O-ring joints; (11) gram-scale reactor; (12) milligram-scale reactor; (13) heating mantle; (14) vacuum gauge; and (15) vacuum pump

Attempting spectral characterisation of the amide-derivatised SWNTs, we analysed applicability of infrared (IR) spectroscopy. In particular, we compared IR spectra of the gas phase-synthesised nonylamine derivative to those of oxidised

SWNT samples impregnated with nonylamine under room temperature and different nonylamine:SWNTs ratios. After comparing spectral band intensities, we concluded that IR spectra of oxidised SWNTs treated with amines under both liquid-phase and solvent-free conditions cannot correspond to amide derivatives on SWNT tips, due to a very low concentration of the terminal groups relative to the whole sample mass, which implies a negligible contribution to the IR spectra (Basiuk et al. 2002a). The bands detectable in the case of long-chain amines correspond to amine molecules physisorbed due to strong hydrophobic interactions of their hydrocarbon chains with SWNT walls. According to molecular-mechanics modeling (MM+ force field implemented in HyperChem Version 5.1 package), energetically preferable adsorption sites are the channels inside SWNTs (Basiuk 2003; Basiuk et al. 2002a).

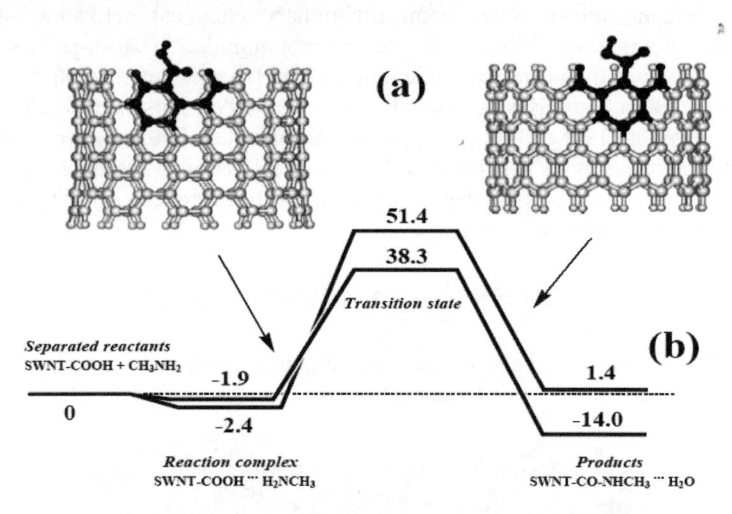

Fig. 19.2. (a) Monocarboxy-substituted fragments of (10,10) armchair (left) and (16,0) zigzag (right) SWNTs used in two-level ONIOM calculations of the model gas-phase reaction with methylamine (Basiuk 2003). The highlighted (dark) atoms and the ones constituting methylamine molecule were treated at the B3LYP/6-31G(d) level of theory; the remaining SWNT atoms were treated with UFF molecular mechanics. (b) Calculated potential energy surfaces for the two SWNT models, including reaction complexes, transition states, and products (water molecules hydrogen-bound to the amide derivatives); energies in kcal mol^{-1}

Temperature-programmed desorption mass spectrometry provided additional information on the chemical state of the amines (Basiuk et al. 2002a). Heating the amine-treated SWNTs at >200°C caused cleavage of alkenes from the amine residues: nonene and pentene formed in the case of nonylamine and dipentylamine, respectively. For a short-chain amine (dipentylamine), only one chemical form was detected, whereas two forms (amide and physisorbed amine) were distinguished for the SWNTs treated with nonylamine. The content of physisorbed nonylamine was estimated to be about one order of magnitude higher than the amide content.

An important aspect of CNT reactivity is the existence of different forms of the nanotubes: zigzag, armchair, and chiral forms. All of these species are apparently present in any single CNTs sample grown by most methods. For the particular case of direct amidation reaction, one can expect that the reactivity of carboxylic groups toward amines might be different depending on whether the nanotubes have an armchair or zigzag structure. Studying the specificity of such a sort can be considered as a possible pathway to selective derivatisation of different forms of CNTs, and to their further separation: for example, due to different solubility of derivatised and nonderivatised nanotubes. However, this goal seems to be very complicated from the experimental point of view and requires further sophistication of the methods of handling and observing individual CNTs.

We tried to give some insight by using the two-level ONIOM (combined quantum mechanics/molecular mechanics) calculations (Basiuk, 2003, 2004; Basiuk et al. 2001, 2002). As model reaction systems, we considered monocarboxylated short fragments of zigzag and armchair SWNTs interacting with methylamine as the simplest aliphatic amine. Two of such models, armchair (10,10) and zigzag (16,0), are shown in Fig. 19.2 along with the corresponding potential energy surfaces calculated at the B3LYP/6-31G(d):UFF level of theory (energies relative to the level of separated reactants; the ONIOM method implemented in Gaussian 98W package; Basiuk, 2003).

To begin with, the energies of the transition states differ significantly: 38.3 and 51.4 kcal mol^{-1} for the armchair and zigzag SWNTs, respectively. Energy for the armchair reaction complex (-1.9 kcal mol^{-1}) is by 0.5 kcal mol^{-1} higher than the one for its zigzag counterpart (-2.4 kcal mol^{-1}). Finally, the armchair product energy is -14.0 kcal mol^{-1} (highly exothermic reaction) whereas the energy of zigzag product is 1.4 kcal mol^{-1} (endothermic reaction). This result shows that the direct formation of amides on armchair SWNT tips is much more advantageous thermodynamically, as compared to the direct amide formation on zigzag SWNT tips. The same trend was found with other SWNT models and theoretical levels used for the ONIOM calculations (Basiuk, 2004; Basiuk et al. 2001, 2002).

Although it is too premature to discuss practical implications of the data obtained (and besides that, as any theoretical results, they have to be taken with a certain precaution), one can envision the use of this or other chemical reactions (e.g., esterification with alcohols; Basiuk 2002) for selective derivatisation of different forms of CNTs (armchair, zigzag, or chiral nanotubes). In the case of direct amidation reaction with long-chain amines, the derivatised nanotubes would acquire a higher solubility than underivatised CNTs, which could open a route to their facile separation.

19.4 Direct Amine Addition to Closed Caps and Wall Defects of Pristine MWNTs

The amidation reactions considered in the previous section have to do neither with graphene sheet chemistry, nor with fullerene-cap chemistry, but instead with the chemistry of carboxylic acids. On the other hand, all known covalent sidewall derivatisation reactions employ the chemistry of graphene sheet (Bahr and Tour 2002; Dyke and Tour 2003; Hirsch 2002; Hu et al. 2003, Peng et al. 2003; Stevens et al. 2003), although fullerene caps of pristine CNTs are reactive sites as well.

It seems quite surprising that despite a wide use of the term 'tubular fullerenes' for CNTs, no method for their derivatisation has been developed until recently, which would be based precisely on the chemistry of closed fullerene caps, and not graphene sheet and carboxylic acid chemistry. It becomes even more surprising when we realise that within the rich fullerene chemistry there is a very appropriate reaction, employing the same amine reagents. The reaction was discovered more than a decade ago (Hirsch et al. 1991; Wudl et al. 1992).

Both primary and secondary amines (which are all neutral nucleophiles) add onto C_{60} at room temperature by reacting with fullerene dissolved in liquid amines or in their solutions in dimethylformamide, dimethylsulfoxide, chlorobenzene, etc. (Goh et al. 2000; Hirsch et al. 1991; Nigam et al. 1995; Schick et al. 1995; Wudl et al. 1992). The reaction stoichiometry varies significantly depending on the size of amine molecule. For smaller amine molecules such as 2-methylaziridine, the average amine:C_{60} ratio can reach as much as 10:1 (Nigam et al., 1995). Recently we studied a solvent-free reaction of silica-supported C_{60} with vaporous nonylamine at 150°C, which produces a mixture of addition products as well (Basiuk et al. 2003).

Quantum chemical DFT calculations at B3LYP/STO-3G level of theory showed that the addition reaction preferably takes place across the 6,6 bonds of C_{60} pyracylene units, and not across the 5,6 bonds. According to C:N ratio found from elemental analysis, the average number of nonylamine molecules attached to C_{60} is 3. At the same time, field-desorption mass spectrometric study detected a series of molecular and fragment ions due to the adducts with up to six nonylamine moieties chemically bound to C_{60}. We attempted extending this aspect of fullerene chemistry to CNTs and explored the possibility to perform a similar direct amination reaction on closed caps of CVD-produced MWNTs and to improve dispersibility/solubility of the latter in this way (Basiuk et al. 2004).

We used unbranched nonylamine, dodecylamine and octadecylamine (ODA) as organic amine reagents. They have increasingly long linear hydrocarbon radicals and are commonly used (especially ODA) in the tip amidation of CNTs in order to enhance their solubility (Bahr and Tour 2002; Basiuk and Basiuk 2004; Hirsch 2002). Although we usually employ the gas-phase procedure, the reaction can be equally performed by baking MWNTs/amine mixture in a sealed vial (a solvent-free process as well; Basiuk et al. 2002b); in that case, excess amine should afterwards be removed by washing or evacuating/heating.

Besides two low-intensity bands at about 750 and 900 cm^{-1} due to ODA, IR spectra of ODA-MWNTs did not show clearly pronounced changes as compared to the starting nanotube material. This is a result of commonly very poor quality of infrared spectra of CNTs, on one hand, and of a low concentration and consequently negligible spectral contribution of the organic moieties, on the other hand. We also performed thermogravimetric analysis for ODA-MWNTs, and found that the steepest weight loss of ca. 5% due to organics decomposition is observed in a temperature interval of 250-400°C. If to take into account the high aspect ratios of 10^3 typical for MWNTs, this weight loss cannot correspond to ODA molecules attached to the fullerene caps only. Therefore we concluded that a major ODA fraction is distributed over MWNT sidewalls. These ODA molecules might be either (1) simply physisorbed on the sidewals, or (2) chemically attached there, in the way similar to their attachment to the fullerene caps. The first explanation seems less likely, since physisorbed ODA molecules are expected to remove at lower temperatures than 250-400°C. The latter suggestion is supported with observation by Chattopadhyay et al. (2003) that TGA decomposition of ODA physically adsorbed on SWNTs is occured in a temperature interval of 150-300°C. The second explanation is hardly compatible with the chemistry of *ideal*, defectless graphene sheet. However, this might be possible if to admit the presence of such defects as, for example, five-membered rings typical for spherical fullerenes.

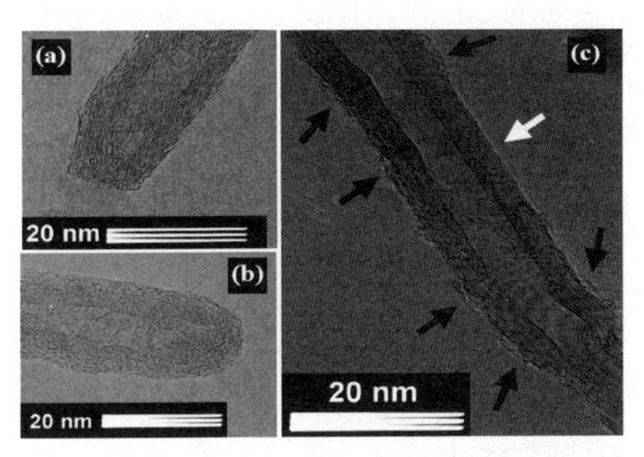

Fig. 19.3. HRTEM microphotographs showing closed caps of starting MWNTs (a), of ODA-MWNTs (b), and sidewalls of ODA-MWNTs (c). Black arrows point to an amorphous material originating from ODA molecules bound to the defect sites; white arrow points to the almost ideal sidewalls where no similar material can be observed. Reprinted with permission from Basiuk et al. (2004). Copyright 2004 American Chemical Society

Direct observations of ODA-MWNTs by means of high-resolution transmission electron microscopy (HRTEM) gave an insight. From Fig. 19.3(a) one can see the starting MWNTs to be composed of ca. 10 coaxial tubes. Their closed caps have an irregular shape, however somewhat fragmented graphene sheet fragments of the outer shell are relatively ordered. While similar closed caps were clearly distinguished in ODA-MWNTs, they turned to be covered with ca. 2-nm layer of some amorphous material (Fig. 19.3(b)). Its origin can be explained by electron beam burning of ODA molecules chemically bound to the defect sites (pentagons). In addition, HRTEM observation of the sidewalls of ODA-MWNTs revealed similar amorphous formations (Fig. 19.3(c)), but they were concentrated at the sites with well-pronounced curvature, where the number of defects must be relatively high. On the contrary, almost ideal sidewalls were found to be free of any additional material.

In order to suggest an explanation of the results obtained, we employed theoretical calculations by AM1 semi-empirical method (implemented in HyperChem Version 5.1 and Gaussian 98W packages). We tested energetic feasibility of ODA addition to different types of carbon atoms (Basiuk et al. 2004). Two SWNT structures were used as simplified nanotube models (Fig. 19.4), which are armchair (5,5) with both ends closed and zigzag (10,10) SWNT with one closed and one open end (dangling bonds were filled with hydrogen atoms). Formation energies calculated for different isomeric methylamine monoadducts are specified in Fig. 19.4. Addition positions are denoted as *n*N-*m*H, where 'N' and 'H' denote the sites of C_{SWNT}–$NHCH_3$ and C_{SWNT}—H bond formation, respectively; *n* and *m* are according to the numbering schemes in the same figure.

Both models have a five-membered ring right on the tip (including atoms 1, 2 and three remaining equivalent atoms; Fig. 19.4), but the distribution of other pentagons was very different. The closed cap of armchair (5,5) model was C_{60} fullerene hemisphere: it contained pyracylene units, and methylamine must preferentially add across their 6,6 bonds (connecting two pentagons), and not across the 5,6 bonds (Basiuk et al. 2003). Indeed, 1N-3H and 3N-1H additions turned to be most exothermic and equally possible (-22.9 kcal mol^{-1} relative to the level of reactants), with 4N-5H slightly less exothermic (-19.7 kcal mol^{-1}). Several positions were found where the reaction is still slightly exothermic, 1N-2H, 3N-4H, 4N-3H, 4N-6H, 6N-4H (all additions across the 5,6 bonds), as well as 6N-8H and 8N-6H (additions across the 6,6 bonds not belonging to pyracylene units). Exothermicity for the latter series can probably be associated to spherical curvature of the cap. However the spherical curvature disappeared beyond carbon atoms 6 and 7. The remaining combinations (8N-9H, 9N-8H, and so on) imply amine additions to graphene sheet of cylindrical curvature, and naturally turned to be highly endothermic (about 20-30 kcal mol^{-1}) and therefore unlikely.

The closed cap of zigzag (10,0) nanotube ended with five-membered ring as well (atoms 1, 2 and three remaining equivalent atoms; Fig. 19.4). However, contrary to the previous case, it did not contain pyracylene units at all. All pentagons were separated by two C—C bonds belonging to benzene rings. Nevertheless, several positions were found where the addition is highly exothermic (>25 kcal mol^{-1}); they are 1N-2H, 4N-5H, 4N-6H and 6N-4H. All of them belong to pentagons. There were other combinations concluding with 7N-6H with gradually increasing energy, which still were located within the region of spherical curvature. And starting with 7N-8H addition, the formation energies were positive of >10 kcal mol^{-1}. The theoretical results were interpreted in the following way.

- The presence of pyracylene units in the closed nanotube caps is not crucial for the amine addition to be possible.
- Site-specificity of the reaction does depend on the mutual position of five-membered rings. If the caps contain pyracylene units, the addition preferentially takes place on their 6,6 bonds; if they do not, the preferential reaction sites are C—C bonds of the pentagons.

- In any event, ideal CNT sidewalls composed of benzene rings only (derived from ideal graphene sheet), turn to be inert with respect to amines.

Since introducing pentagonal defects into the ideal sidewalls causes distortion of the cylindrical curvature (kinking), one can expect that these defects would be reactive towards amines as well. The high organic content of ODA-MWNTs found by TGA apparently results from the enhanced reactivity of the real sidewalls containing numerous pentagonal defects.

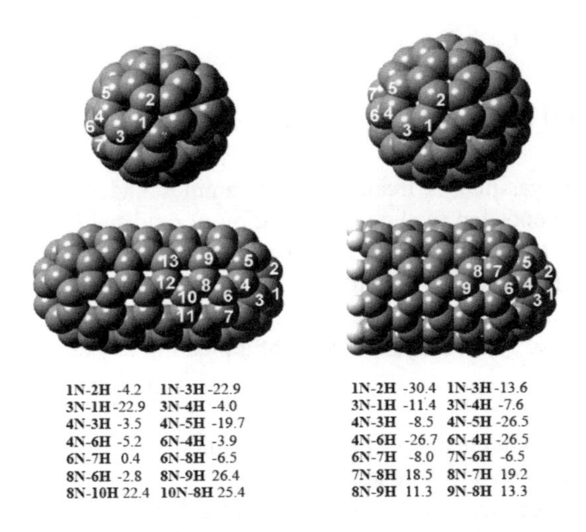

Fig. 19.4. Structures optimised with AM1 semi-empirical method and atom numbering scheme for armchair (5,5) (left) and zigzag (10,10) (right) SWNT models, used to study theoretically the reaction of methylamine monoaddition, along with AM1 formation energies (in kcal mol^{-1}) for isomeric monoadducts (Basiuk et al. 2004). Addition positions are denoted as *n*N-*m*H, where 'N' and 'H' denote the sites of C_{SWNT}–$NHCH_3$ and C_{SWNT}–H bond formation, respectively

Fig. 19.5. Dispersions of starting MWNTs (right) and ODA-MWNTs (left) in isopropanol 2 h after ultrasonication. ODA-MWNT sample did not show visible changes for more than one month

So far it remains unclear whether heptagonal defects, existing on CNT sidewalls along with five-membered rings, can undergo similar addition reactions. Solubility/dispersibility of the amine-derivatised MWNTs was tested by ultrasonicating them in isopropanol for 20 min, using starting MWNTs as reference. Fig. 19.5 illustrates the behaviour of ODA-MWNTs. While underivatised MWNTs precipitated almost immediately after ultrasonication, and the solvent became almost completely clear, ODA-MWNT solutions/dispersions acquired stable brown coloration and did not exhibit visible changes for more than one month.

19.5 Conclusions

The solvent-free (gas-phase) treatment with amines under temperatures of ca. 150°C is a simple one-step method of amidation of oxidised CNTs and amination of pristine carbon nanotubes. In the latter case, amines add on the closed caps and five-membered rings of the wall defects. Thus, this reaction is the most direct link between CNT and fullerene chemistry, contrary to all derivatisation methods designed up to now. The main advantages of amidation and amination procedures are as follows:

- No additional chemical activation is required, contrary to the common methods of carboxylic group amidation.
- The procedures are relatively fast, usually less than 2 h.
- Unreacted amine is spontaneously removed from the product under heating/pumping out.
- No (organic) solvent medium is used. This feature is especially attractive from an ecological point of view.
- A variety of amines can be employed, with the condition of their sufficient thermal stability and volatility under reduced pressure.

The method proposed can be useful not only for simple solubilisation/dispersion of CNTs, but also for the synthesis of chemical linkers for other compound immobilisation on the nanotubes, via employing bifunctional and polyfunctional amines.

19.6 References

Ausman KD, Piner R, Lourie O, Ruoff RC, Korobov M (2000) Organic solvent dispersions of single-walled carbon nanotubes: Toward solutions of pristine nanotubes, J Phys Chem B 104:8911

Bahr JL, Tour JM (2002) Covalent chemistry of single-wall carbon nanotubes, J Mater Chem 12:1952

Basiuk VA (2002) Reactivity of carboxylic groups on armchair and zigzag carbon nanotube tips: A theoretical study of esterification with methanol, Nano Lett. 2:835
Basiuk VA (2003) ONIOM studies of chemical reactions on carbon nanotube tips: Effects of the lower theoretical level and mutual orientation of the reactants, J Phys Chem B 107:8890
Basiuk VA (2004) Theoretical studies of amidation reaction at carbon nanotube tips by means of the ONIOM technique: Expanding the higher level, J Nanosci Nanotechnol 4:1095
Basiuk VA, Basiuk (Golovataya-Dzhymbeeva) EV (2004) Chemical derivatisation of carbon nanotube tips. In: Nalwa HS (ed) Encyclopedia of Nanoscience and Nanotechnology, American Scientific Publishers, vol. 1, p 761
Basiuk VA, Chuiko AA (1990) Gas-phase synthesis, properties and some applications of acylamide stationary phases for high-performance liquid chromatography, J Chromatogr 521:29
Basiuk VA, Chuiko AA (1993) Selectivity of bonded stationary phases containing uracil derivatives for liquid chromatography of nucleic acid components, J Chromatogr Sci. 31:120
Basiuk VA, Khil'chevskaya EG (1991) Gas-phase acylation of aminopropyl silica gel in the synthesis of some chemically bonded silica materials for analytical applications, Anal Chim Acta 255:197
Basiuk VA, Basiuk EV, Saniger-Blesa J-M (2001) Direct amidation of terminal carboxylic groups of armchair and zigzag single-walled carbon nanotubes: A theoretical study, Nano Lett. 1:657
Basiuk EV, Basiuk VA, Bañuelos J-G, Saniger-Blesa J-M, Pokrovskiy VA, Gromovoy TYu, Mischanchuk AV, Mischanchuk BG (2002a) Interaction of oxidised single-walled carbon nanotubes with vaporous aliphatic amines, J Phys Chem B 106:1588
Basiuk VA, Kobayashi K, Kaneko T, Negishi Y, Basiuk EV, Saniger-Blesa J-M (2002b) Irradiation of single-walled carbon nanotubes with high-energy protons, Nano Lett. 2:789
Basiuk (Golovataya-Dzhymbeeva) EV, Basiuk VA, Shabel'nikov VP, Golovatyi VG, Flores JO, Saniger JM (2003) Reaction of silica-supported fullerene C_{60} with nonylamine vapor, Carbon 41:2339
Basiuk EV, Monroy-Peláez M, Puente-Lee I, Basiuk VA (2004) Direct solvent-free amination of closed-cap carbon nanotubes: A link to fullerene chemistry, Nano Lett. 4:863
Basyuk VA (1991) Preparation and properties of functionalised sorbents based on bromobutyliminopropyl silica gel for high-performance liquid chromatography, J Anal Chem USSR - Engl Tr. 46:401
Basyuk VA, Chuiko AA (1990) Vapor-phase immobilisation of aza-macrocyclic ligands on silica surface, Doklady Chem - Engl Tr. 310:15
Chattopadhyay D, Galeska I, Papadimitrakopoulos F (2003) A route for bulk separation of semiconducting from metallic single-wall carbon nanotubes, J Am Chem Soc. 125:3370
Chen J, Hamon MA, Hu H, Chen Y, Rao AM, Eklund PC, Haddon RC (1998) Solution properties of single-walled carbon nanotubes, Science 282:95
Dyke CA, Tour JM (2003) Solvent-free functionalisation of carbon nanotubes, J Am Chem Soc. 125:1156

Goh SH, Lee SY, Lu ZH, Huan CHA (2000) C_{60}-containing polymer complexes: Complexation between multifunctional 1-(4-methyl)piperazinyl-fullerene or N-[(2-piperidyl)ethyl]aminofullerene and proton-donating polymers, Macromol Chem Phys. 201:1037

Hamon MA, Chen J, Hu H, Chen Y, Itkis ME, Rao AM, Eklund PC, Haddon RC (1999) Dissolution of single-walled carbon nanotubes, Adv Mater 11:834

Hirsch A (2002) Functionalisation of single-walled carbon nanotubes. Angew Chem Int Ed., vol. 41, p 1853

Hirsch A, Li Q, Wudl F (1991) Globe-trotting hydrogens on the surface of the fullerene compound $C_{60}H_6(N(CH_2CH_2)_2O)$. Angew Chem Int Ed Engl, vol. 30, p 1309

Hu H, Zhao B, Hamon MA, Kamaras K, Itkis ME, Haddon RC (2003) Sidewall functionalisation of single-walled carbon nanotubes by addition of dichlorocarbene, J Am Chem Soc. 125:14893

Jia Z, Wang Z, Liang J, Wei B, Wu D (1999) Production of short multi-walled carbon nanotubes, Carbon 37:903

Liu J, Rinzler AG, Dai H, Hafner JH, Bradley RK, Boul PJ, Lu A, Iverson T, Shelimov K, Huffman CB, Rodriguez-Macias F, Shon YS, Lee TR, Colbert DT, Smalley RE (1998) Fullerene pipes, Science 280:1253

Nigam A, Shekharam T, Bharadwaj T, Giovanola J, Narang S, Malhotra R (1995) Lattice-type polymers from an adduct of [60]fullerene and 2-methylaziridine, J Chem Soc Chem Comm. 15:1547

Peng H, Alemany LB, Margrave JL, Khabashesku VN (2003) Sidewall carboxylic acid functionalisation of single-walled carbon nanotubes, J Am Chem Soc. 125:15174

Schick G, Kampe KD, Hirsch A (1995) Reaction of [60]fullerene with morpholine and piperidine – Preferred 1,4-additions and fullerene dimer formation, J Chem Soc Chem Comm. 19:2023

Stevens JL, Huang AY, Peng H, Chiang IW, Khabashesku VN, Margrave JL (2003) Sidewallamino-functionalisation of single-walled carbon nanotubes through fluorination and subsequent reactions with terminal diamines, Nano Lett. 3:331

Sun Y-P, Fu K, Lin Y, Huang W (2002) Functionalised carbon nanotubes: Properties and applications, Acc Chem Res. 35:1096

Wong SS, Joselevich E, Woolley AT, Cheung CL, Lieber CM (1998a) Covalently functionalised nanotubes as nanometre-sized probes in chemistry and biology, Nature 394:52

Wong SS, Woolley AT, Joselevich E, Cheung CL, Lieber CM (1998b) Covalently-functionalised single-walled carbon nanotube probe tips for chemical force microscopy, J Am Chem Soc. 120:8557

Wudl F, Hirsch A, Khemani KC, Suzuki T, Allemand PM, Koch A, Eckert H, Srdanov G, Webb HM (1992) Survey of chemical reactivity of C-60, electrophile and dieno polarophile par excellence, Am Chem Soc Symp Ser. 481:161

20 Chemical Crosslinking in C_{60} Thin Films

E. V. Basiuk[1], E. A. Zauco[1] and V. A. Basiuk[2]

[1]Centro de Ciencias Aplicadas y Desarrollo Tecnológico, UNAM, Mexico
[2]Instituto de Ciencias Nucleares, UNAM, Mexico

20.1 Introduction

Various procedures have been developed for the modification of fullerene films (Karachevtsev et al. 2002; Käsmaier et al. 1996; Li et al. 2003; Rao et al. 1993; Rhee et al. 2003; Talyzin et al. 2002; Yogo et al. 2002; Zhao et al. 1994). During the last decade, there has been a great interest to polymerise C_{60} fullerene films on various substrates, photopolymerisation being one of the most common methods. This interest was stimulated with great potential technological possibilities of polymerised fullerene films. The photopolymerisation changes the solubility of fullerene films, and influences their physical and chemical characteristics (Käsmaier et al. 1996; Li and Pittman, 2003; Zhokhavets et al. 2003). On the other hand, photoirradiation induces not only the fullerene polymerisation, but also the formation of other carbonaceous phases (Käsmaier et al. 1996). Thus, finding an alternative way is highly desirable.

Among the whole rich chemistry of fullerenes, their interactions with amines gained a special attention (Adamov and Vojinovic-Miloradov, 1998; Bernstein and Foote, 1999; Dresselhaus et al. 1993; Goh et al. 2000; Hirsch et al. 1991; Lobach et al. 1995; Nigam et al. 1995; Qiao et al. 2000a,b; Schick et al. 1995; Wudl et al. 1992). A direct addition reaction of amines onto C_{60} molecules was discovered more than a decade ago (Hirsch et al. 1991; Wudl et al. 1992). Both primary and secondary amines (which are all neutral nucleophiles) add onto C_{60} at room temperature by reacting with fullerene dissolved in liquid amines or in their solutions in dimethylformamide, dimethylsulfoxide, chlorobenzene, etc. (Goh et al. 2000; Hirsch et al. 1991; Nigam et al. 1995; Schick et al. 1995; Wudl et al. 1992). The reaction stoichiometry varies significantly depending on the size of amine molecule. For smaller amine molecules such as 2-methylaziridine, the average amine:C_{60} ratio can reach as much as 10:1 (Nigam et al. 1995).

Recently we described a solvent-free reaction of silica-supported C_{60} with vaporous nonylamine at 150°C, which produces a mixture of addition products as well (Basiuk et al. 2003). According to C:N ratio found from elemental analysis, the average number of nonylamine molecules attached to C_{60} is 3. Nevertheless, field-desorption mass spectrometric study detected a series of molecular and fragment ions due to the adducts with up to six nonylamine moieties chemically bound to C_{60}.

In the present paper, we tested applicability of the gas-phase technique to the reaction of fullerene C_{60} thin film with 1,8-diaminooctane (or 1,8-octanediamine), a representative of long-chain aliphatic diamines, which can be expected to react with two or more neighbour fullerene molecules simultaneously and thus to act as a cross-linking agent (Fig. 20.1). Our main interest was in the preparation of relatively insoluble C_{60} films, which can be similar to the photopolymerised fullerene films in their stability, but at the same time the undesirable formation of other carbonaceous phases would be avoided.

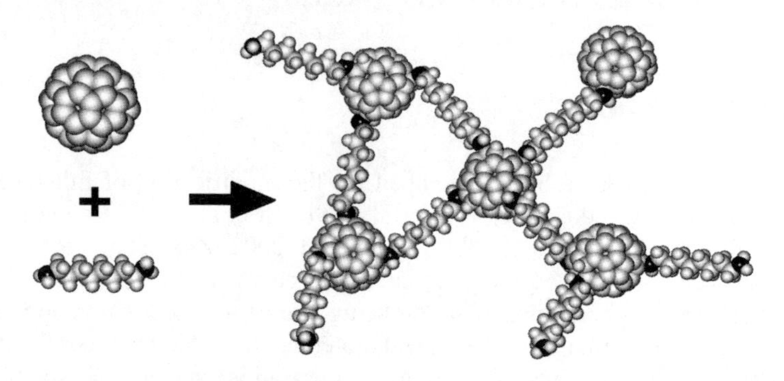

Fig. 20.1. Crosslinking of C_{60} fullerene molecules with aliphatic diamines (exemplified by 1,8-diaminooctane). Dark atoms are nitrogen atoms of amino groups

20.2 Experiment

Fullerene C_{60} powder (from MER Corp., 99.5% purity) and 1,8-diaminooctane (from Aldrich, 98% purity) were used.

20.2.1 Analytical Instruments

The silicon-supported samples were analysed by atomic force microscopy (AFM) in contact mode by means of an AutoProbe CP instrument from Park Scientific Instruments. Size scan varied between 1 and 5 µm. A cantilever with a 0.05 N/m force constant and a 22 kHz resonant frequency was used.

Thermogravimetric analyses (TGA) were performed on a DuPont Thermal Analyser 951, with a heating ramp of 10ºC min^{-1} until 1000ºC, under airflow of 100 mL min^{-1}. Infrared (IR) spectra were recorded by using a Nicolet 5SX FTIR spectrophotometer, in air at an atmospheric pressure of 560 Torr and at room temperature. Raman spectra were recorded by using a Thermo-Nicolet Almega Dispersive Raman instrument (λ=532 nm) under the same conditions.

UV-Visible spectra were recorded on an Analytical Instruments Model DT 1000CE spectrophotometer. X-Ray diffraction patterns were obtained on a Siemens D5000 instrument (0.010° step, 0.6 s step time, 2θ range 2-70°, Cu K_β radiation).

20.2.2 Deposition of Fullerene Films

Fullerene C_{60} was deposited onto (100) silicon wafers by sublimation, without heating the substrates. The deposition was performed in a vacuum chamber at a pressure of 6.2×10^{-6} Torr. Average film thickness varied between 100 and 200 nm.

20.2.3 Reaction with 1,8-Diaminooctane

In the experiments, we used the gas-phase derivatisation technique (Basiuk et al. 2002), employing the custom-made Pyrex glassware described in detail in the previous chapter of this volume (Fig. 19.1). The sample (fullerene film on a silicon substrate) was placed into the reactor 11. Then a few milligrams of 1,8-diaminooctane were added directly to the reactor bottom, avoiding direct contact with the sample. After pumping the reactor out to ca. 1 Torr at room temperature, its valve 2 was closed, and the bottom was heated at about 150°C for 3 h. During this procedure, 1,8-diaminooctane evaporated and reacted with C_{60}. The diamine excess condensed a few centimeters above the heating mantle. The high derivatisation temperature not only facilitates the reaction, but also helps to minimise the amount of diamine physically adsorbed on the sample. After finishing the treatment, the reactor valve 2 was open again for ca. 1 h to pump out unreacted diamine. Then the heating mantle was removed, the reactor cooled and disconnected from the manifold. Before unloading the derivatised sample, the upper reactor part with the excess diamine condensed was wiped with cotton wool wet with ethanol. Along with the chemical modification of C_{60} films, we have similarly performed the reaction of pristine fullerene powder with 1,8-diaminooctane, as a control experiment. It was essentially melting together the two reagents.

20.3 Results and Discussion

20.3.1 (1,8)-Diaminooctane-derivatised C_{60} Powder

Fullerene C_{60} pristine powder exhibits only one sharp and uniform weight loss between 500°C and 650°C (Fig. 20.2(b)). The derivatised C_{60} powder begins to lose the weight after 300°C (Fig. 20.2(a)). First the weight loss is relatively slow up to almost 500°C, then becomes sharper and straighter, and the material decomposes completely under further heating up to 650°C.

In the infrared absorption spectra of crystalline C_{60} powder reacted with 1,8-diaminooctane, we observed remarkable changes as compared to the spectrum of pristine C_{60} (Fig. 20.3(a)). The latter exhibits four characteristic bands at 526, 576, 1181 and 1427 cm^{-1} (Dresselhaus et al., 1993). The derivatisation product has a richer chemical structure, and in addition to the above four bands, it shows several intense broad peaks (Fig. 20.3(b)) near 715, 1096 (ν_{C-C}), 1643 (δ_{NH}), 2855 and 2922 cm^{-1} (ν_{CH}). The region of 500-800 cm^{-1} is a 'fingerprint' region for fullerene derivatisation, and new IR features are observed when the C_{60} skeleton is partially broken (Iwasa et al. 1994), for example in the formation of cross-linked fullerene molecules as reported by Sun and Reed (2000). In our case one can also see that instead of only one band at 1425 cm^{-1} (C=C skeleton vibration), there is another sharp peak at 1383 cm^{-1}. Peak position for this mode is known to be very sensitive to such modifications of C_{60} molecule as polymerisation (Sun and Reed, 2000).

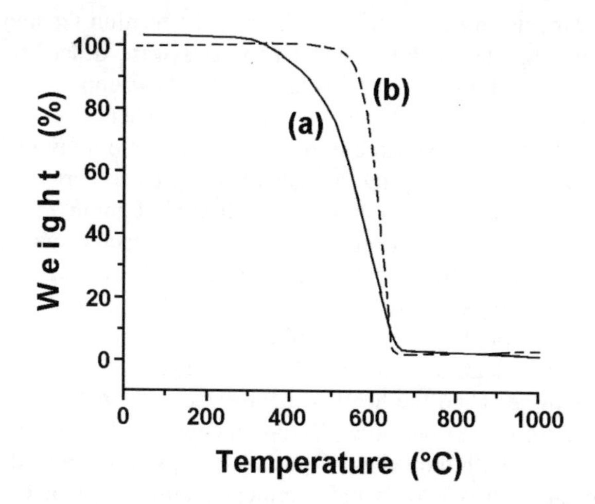

Fig. 20.2. TGA curves for C_{60} derivatised with 1,8-diaminooctane (a) in comparison with pristine C_{60} powder (b)

20.3.2 1,8-Diaminooctane-derivatised C_{60} Films

Pristine C_{60} and its films are soluble in toluene at room temperature, whereas the samples (both crystalline C_{60} and fullerene films on silicon) after reacting with 1,8-diaminooctane become poorly soluble.

AFM studies (Fig. 20.4) show clearly large differences in the film morphology before and after the diamine treatment. The pristine films are typically composed of well-shaped C_{60} clusters, of ca. 50 nm average diameter (Fig. 20.4(a)). The clusters become diffuse after the modification, and the surface morphology is smoothed (Fig. 20.4(b)). The changes observed in the film morphology can be explained by the formation of chemical linkages between neighbouring C_{60} molecules. Since the diamine linkers are relatively big and cannot fit the spaces between C_{60} molecules, the latter have to reaccommodate, making the film less dense and ordered. Changes of inter-C_{60} distances in the fullerene films and polymers usually manifest in X-ray diffractograms (Bennington et al. 2000; Cheng et al. 2001, 2003; Jayatissa et al. 2004; Katz et al. 2000). However in our case, we were not able to detect such manifestations, since the C_{60} films obtained by vapour deposition onto cold substrates turn to be X-ray amorphous. A similar observation was reported previously by Katz et al. (2000).

As it could be expected, the IR spectrum of pristine C_{60} film (Fig. 20.5(a)) exhibits four sharp F_{1u} bands at 527, 575, 1182 and 1428 cm^{-1}, which correspond to isolated C_{60} molecules with icosahedral symmetry and sp^2 hibridisation (Iwasa et al. 1994; Sun and Reed, 2000). It is known that compounds with a higher sp^2 content have relatively 'clear' IR spectra with sharper high-intensity bands (Iwasa et al. 1994; Sun and Reed, 2000). On the contrary, IR spectra of the compounds with a higher sp^3 content usually have lower-intensity bands. As it can be seen in Fig. 20.5(b), intensities for all four bands at 527, 575, 1182 and 1428 cm^{-1} strongly decrease after treating the C_{60} films with 1,8-diaminooctane.

The 'fingerprint' region for C_{60} derivatisation is between 500 and 800 cm^{-1}. Indeed, a noticeable band rearrangement can be observed around 470-570 cm^{-1}, which is indicative of strong distortion in the fullerene molecule.

Many spectral features can appear in the region of 1000-2000 cm^{-1} due to C=C vibrations of the carbon skeleton, and they all are sensitive to polymerisation of C_{60} molecules (Iwasa et al. 1994). What we can see in the spectra of C_{60} films chemically modified by 1,8-diaminooctane is, in particular, that the band around 960-965 cm^{-1} becomes more intense. A series of small bands appear between 1250 and 1750 cm^{-1}, which point to a lowering of icosahedral symmetry of C_{60} molecules with breaking its ideal structure during the formation of new covalent bonds. As it was mentioned above, another notable feature within this range is decrease in intensity and broadening of the sharpest peak due to C=C stretching vibrations at 1428 cm^{-1}. Comparison of these spectra with IR spectra for different polymerised C_{60} forms (Eklund et al. 1995) reveals some similarity with cross-linked structures reported by other groups, where the band at 1428 cm^{-1} does not disappear totally, only have a lower intensity, but the band at 1182 cm^{-1} cannot be seen anymore. Finally, despite of generally very noisy background in the higher-frequency region, one can notice that IR absorption at ca. 3000 cm^{-1} increases due to ν_{CH} vibrations in 1,8-diaminooctane moieties. Raman spectra of the pristine C_{60} films (Fig. 20.6(a)) exhibit clearly six of the ten primary intramolecular Raman-active modes at 272, 498, 714, 1430, 1467 and 1572 cm^{-1}, which are consistent with weak intermolecular interactions and icosahedral symmetry of an isolated C_{60} molecule. After treating the films with 1,8-diaminooctane we have found an evident intensity reduction for some bands, in particular for those at 250, 918, 1037, around 1300, 1504 and 1572 cm^{-1} (Fig. 20.6(b)). A similar behaviour of the band at 1469 cm^{-1}, reported by Karachevtsev et al. (2002), was attributed to lowering the C_{60} symmetry. The bands at 908, 981, 1018 and 1037 cm^{-1} become 'softened'. In the region of 1520-1660 cm^{-1} we observed smoothing of the complex band picture typical for pristine C_{60} films, with the formation of one well-shaped band at 1572, which is commonly known as the G-band of disordered sp^2 carbon atoms (Käsmaier et al. 1996).

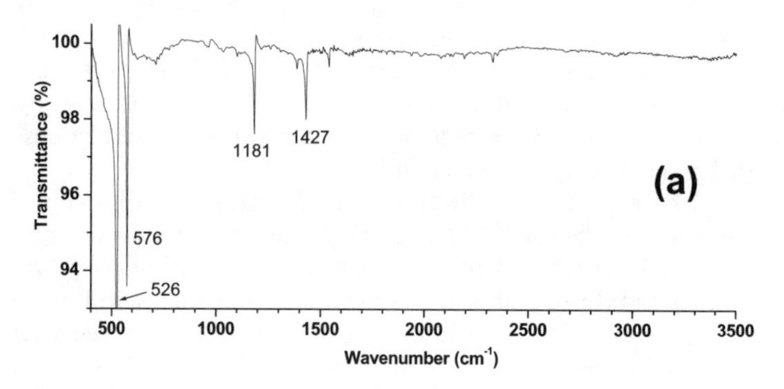

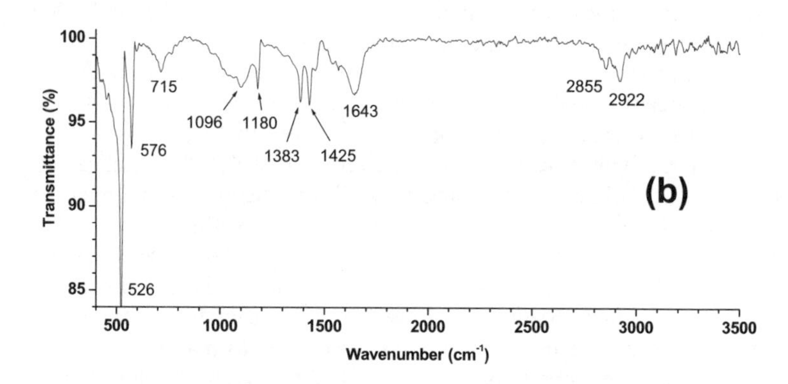

Fig. 20.3. IR spectrum of (a) pristine crystalline C_{60} powder and (b) C_{60} derivatised with 1,8-diaminooctane

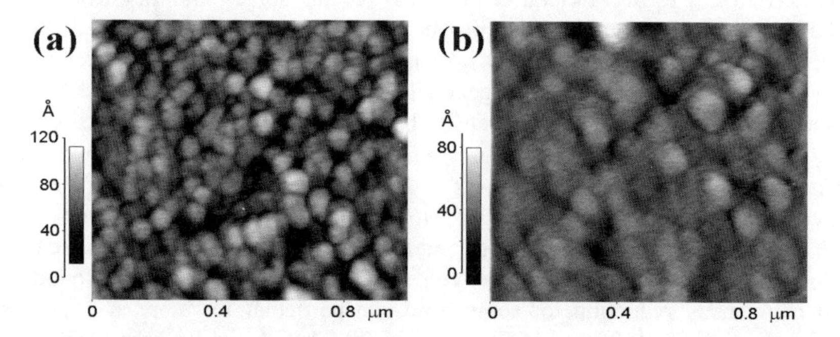

Fig. 20.4. AFM images of C_{60} films deposited onto Si substrates (a) before and (b) after derivatisation with 1,8-diaminooctane

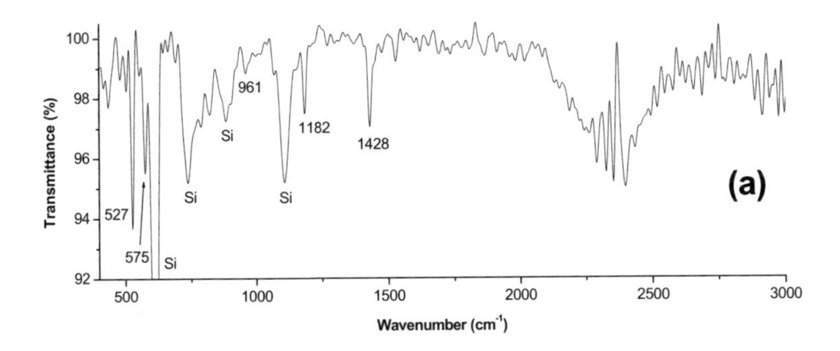

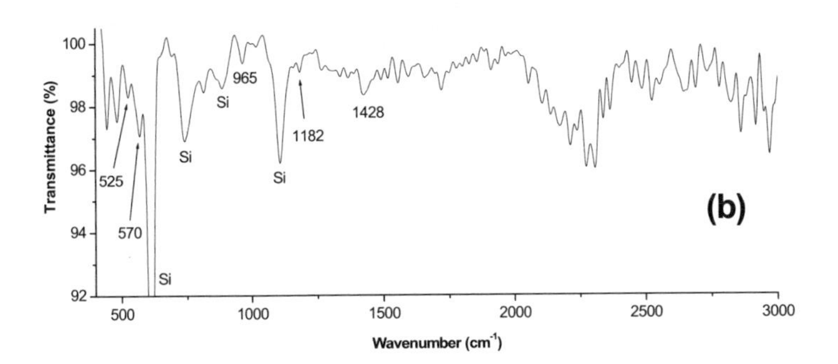

Fig. 20.5. IR spectra of C_{60} films deposited onto Si substrates (a) before and (b) after derivatisation with 1,8-diaminooctane. Bands due to Si substrate are marked with 'Si'

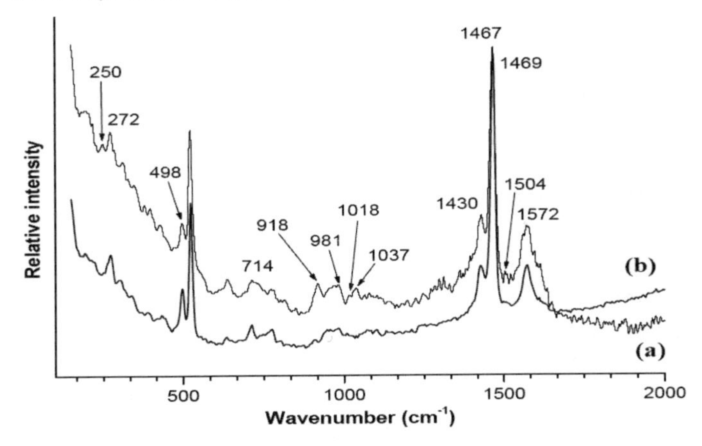

Fig. 20.6. Raman spectra of C_{60} films deposited onto Si substrates (a) before and (b) after derivatisation with 1,8-diaminooctane

The pentagonal-pinch mode (A_g) at 1467-1469 cm^{-1} is very sensitive to the modification by polymerisation. In our case we have not detected any strong frequency shift. The small band at 272 cm^{-1} broadened, that can be explained by the formation of intermolecular vibration modes: Raman spectral changes of this kind are usually observed in the region between 50 and 333 cm^{-1} for phototransformed fullerene films (Ferrari and Robertson, 2000).

It was not possible to measure UV-visible absorption spectra directly for fullerene samples deposited onto silicon wafers, since the latter are not transparent. That is why we substituted Si substrates for quartz glass, and performed the same vapour deposition and 1,8-diaminooctane derivatisation procedures. As it can be seen in Fig. 20.7, the pristine C_{60} film shows three absorption bands at 343, 447, 561 and 620 nm. The treatment with 1,8-diaminooctane results in an appreciable decrease of all of them. These changes can be explained by conversion of some sp^2 carbon atoms into sp^3 carbon atoms, which do not contribute anymore into the absorption at higher wavelengths.

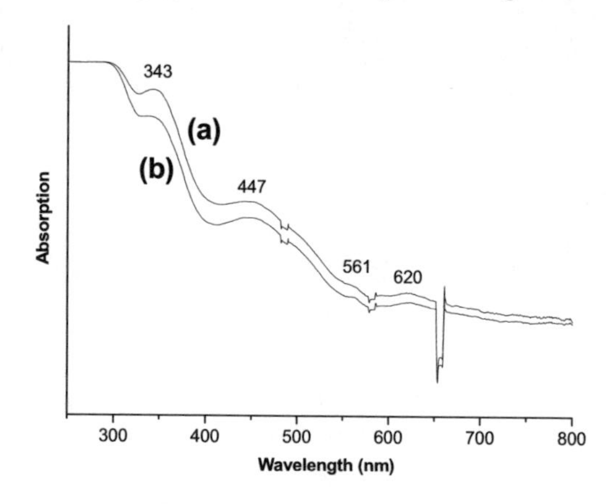

Fig. 20.7. UV-Visible spectra for quartz substrates with deposited C_{60} films (a) before and (b) after C_{60} derivatisation with 1,8-diaminooctane

20.4 Conclusions

The treatment of C_{60} fullerene films on silicon substrates with 1,8-octanediamine vapour at ca. 120°C reduces dramatically fullerene solubility in toluene, indicating the transformation of C_{60} into a different solid phase of polymeric nature. Compared to the spectra of pristine C_{60}, IR spectra of the transformed phase exhibit new lines along with a decrease or total disappearance of some IR-active modes typical for C_{60}, due to the lowering icosahedral symmetry of C_{60} molecule. Notable changes in their Raman spectra were observed as well. As found by AFM measurements, the diamine treatment changes morphology of the fullerene films.

20.5 References

Adamov J, Vojinovic-Miloradov M (1998) New materials on the basis of amino derivatives of fullerene C_{60}, Adv Mater Proc., Mater Sci Forum, vol. 282, p 101

Basiuk EV, Basiuk VA, Bañuelos JG, Saniger-Blesa JM, Pokrovskiy VA, Gromovoy TY, Mischanchuk AV, Mischanchuk BG (2002) Interaction of oxidised single-walled carbon nanotubes with vaporous aliphatic amines, J Phys Chem BL106:1588-1590

Basiuk (Golovataya-Dzhymbeeva) EV, Basiuk VA, Shabel'nikov VP, Golovatyi VG, Flores JO, Saniger JM (2003) Reaction of silica-supported fullerene C_{60} with nonylamine vapor, Carbon 41:2339

Bennington SM, Kitamura N, Cain MG, Lewis MH, Wood RA, Fukumi AK, Funakoshi K (2000) In situ diffraction measurement of the polymerisation of C60 at high temperatures and pressures, J Phys: Condens Matter 12:L451

Bernstein R, Foote CS (1999) Singlet oxygen involvement in the photochemical reaction of C_{60} and amines. Synthesis of an alkyne-containing fullerene, J Phys Chem A 103:7244

Chen R-S, Lin Y-J, Su Y-C, Chiu K-C (2001) Surface morphology of C_{60} polycrystalline films from physical vapor deposition, Thin Solid Films 396:103-105

Chen W-R, Tang S-J, Su Y-C, Lin Y-J, Chiu K-C (2003) Effects of substrate temperature on the growth of C_{60} polycrystalline films by physical vapor deposition, J Crystal Growth 247:401-406

Dresselhaus MS, Dresselhaus G, Eklund PC (1993) Fullerenes, J Mater Res, Vol. 8, p. 2054

Eklund PC, Rao AM, Zhou P, Wang Y, Holden JM (1995) Photochemical transformation of C_{60} and C_{70} films, Thin Solid Films 257:185-190

Ferrari AC, Robertson J (2000) Interpretation of Raman spectra of disordered and amorphous carbon, Phys Rev B 61:14095

Goh SH, Lee SY, Lu ZH, Huan CHA (2000) C_{60}-containing polymer complexes: Complexation between multifunctional 1-(4-methyl)piperazinyl-fullerene or N-[(2-piperidyl)ethyl]aminofullerene and proton-donating polymers, Macromol Chem Phys 201:1037

Hirsch A, Li Q, Wudl F (1991) Globe-trotting hydrogens on the surface of the fullerene compound $C_{60}H_6(N(CH_2CH_2)_2O)_6$. Angew Chem Int ed Engl, vol. 30, p 1309

Iwasa Y, Arima T, Fleming RM, Siegrist T, Zhou O, Haddon RC, Rothberg LJ, Lyons KB, Carter HL, Jr, Hebard AF, Tycko R, Dabbagh G, Krajewski JJ, Thomas GA, Yagi T (1994) New phases of C_{60} synthesised at high pressure, Science 264:1570

Jayatissa AH, Gupta T, Pandya AD (2004) Heating effect on C_{60} films during microfabrication: structure and electrical properties, Carbon 42:1143

Karachevtsev VA, Vork OM, Plochotnichenko AM, Peschanskii AV (2002) Structural and electron properties of molecular nanoestructures, AIP, p 17

Katz EA, Faiman D, Shtutina S, Isakina A (2000) Deposition and structural characterisation of high quality textured C_{60} thin films, Thin Solid Films 368:49-52

Käsmaier A, Lätsch S, Hiraoka H (1996) Irradiation of solid C_{60} films with pulsed UV-laser-light: Fabrication of a periodic submicron C_{60} structure and transformation of C_{60} into a different carbon phase, Appl Phys A 63 :305

Li GZ, Pittman CU (2003) Photoluminescence of fullerene-doped copolymers of methyl methacrylate during laser irradiation, J Mater Sci 38:3741

Li Y, Rhee JH, Singh D, Sharma SC (2003) Raman spectroscopy measurements of interface effects in C_{60}/copper-oxide/copper, Mat Res Soc Symp Proc, vol. 734, pp 9-31

Lobach AS, Goldshleger NF, Kaplunov MG, Kulikov AV (1995) Near-IR and ESR studies of the radical anions of C_{60} and C_{70} in the system fullerene: Primary amine, Chem Phys Lett. 243:22

NIgam A, Shekharam T, Bharadwaj T, Giovanola J, Narang S, Malhotra R (1995) Lattice-type polymers from an adduct of [60]fullerene and 2-methylaziridine, J Chem SocChem Comm 15:1547

Qiao JL, Gong QJ, Dong C, Jin WJ (2000a) Strong room temperature photoluminescence from the novel adducts of C_{60} with aliphatic amines, Chinese Chem Lett. 11:713

Qiao JL, Gong QJ, Du LM, Jin WJ (2000b) Spectroscopic study on the photoinduced reaction offullerene C_{60} with aliphatic amines and its dynamics — Strong short wavelength fluorescence from the adducts, Spectrochim Acta A 57:17-20

Rao AM, Zhou P, Wang K-A, Hager GT, Holden JM, Wang Y, Lee W-T, Bi X-X, Eklund PC, Cornett DS, Duncan MA, Amster IJ (1993) Photoinduced polymerisation of solid C_{60} films, Science 259:955

Rhee JH, Singh D, Li Y, Sharma SC (2003) Crystal structure of C_{60} following compression under 31.1 GPa in diamond anvil cell at room temperature, Solid State Comm 127:295

Schick G, Kampe KD, Hirsch A (1995) Reaction of [60]fullerene with morpholine and piperidine – Preferred 1,4-additions and fullerene dimer formation, J Chem Soc Chem Comm. 19:2023

Sun D, Reed CA (2000) Crystal engineering a linear polymer of C_{60} fullerene via supramolecular pre-organisation, Chem Comm, p 2391

Talyzin AV, Dubrovinsky LS, Le Bihan T, Jansson U (2002) Pressure-induced polymerisation of C_{60} at high temperatures: An in situ Raman study, Phys Rev B 65:245413

Wudl F, Hirsch A, Khemani KC, Suzuki T, Allemand PM, Koch A, Eckert H, Srdanov G, Webb HM (1992) Survey of chemical reactivity of C-60, electrophile and dieno polarophile par excellence, Am Chem Soc Symp Ser 481:161-164

Yogo A, Majima T, Itoh A (2002) Damage and polymerisation of C_{60} films irradiated by fast light and heavy ions, Nucl Instr Meth Phys Res B 193:299-302

Zhao YB, Poirier DM, Pechman RJ, Weaver JH (1994) Electron stimulated polymerisation of solid C_{60}, Appl Phys Lett. 64:577-575

Zhokhavets U, Goldhahn R, Gobsch G, Al-Ibrahim M, Roth H-K, Sensfuss S, Klemm E, Egbe DAM (2003) Anisotropic optical properties of conjugated polymer and polymer/fullerene films, Thin Solid Films 444:215-215

Index